Ferrates

Synthesis, Properties, and Applications in Water and Wastewater Treatment

ACS SYMPOSIUM SERIES **985**

Ferrates

Synthesis, Properties, and Applications in Water and Wastewater Treatment

Virender K. Sharma, Editor
Florida Institute of Technology

**Sponsored by the
ACS Division of Environmental Chemistry, Inc.**

American Chemical Society, Washington, DC

ISBN: 978-0-8412-6961-3

The paper used in this publication meets the minimum requirements of American National Standard for Information Sciences—Permanence of Paper for Printed Library Materials, ANSI Z39.48-1984.

Distributed by Oxford University Press

PRINTED IN THE UNITED STATES OF AMERICA

Foreword

The ACS Symposium Series was first published in 1974 to provide a mechanism for publishing symposia quickly in book form. The purpose of the series is to publish timely, comprehensive books developed from ACS sponsored symposia based on current scientific research. Occasionally, books are developed from symposia sponsored by other organizations when the topic is of keen interest to the chemistry audience.

Before agreeing to publish a book, the proposed table of contents is reviewed for appropriate and comprehensive coverage and for interest to the audience. Some papers may be excluded to better focus the book; others may be added to provide comprehensiveness. When appropriate, overview or introductory chapters are added. Drafts of chapters are peer-reviewed prior to final acceptance or rejection, and manuscripts are prepared in camera-ready format.

As a rule, only original research papers and original review papers are included in the volumes. Verbatim reproductions of previously published papers are not accepted.

ACS Books Department

Contents

Preface.....xi

Synthesis and Characterization

1. **Recent Advances in Fe(VI) Synthesis**.....2
   Stuart Licht and Xingwen Yu

2. **The Role of the Electrode and Electrolyte Composition in the Anode Dissolution Kinetics. Comparison of the Ferrate(VI) Synthesis in the Solutions of NaOH and KOH of Various Ratio**.....52
   Zuzana Mácová, Karel Bouzek, and Virender K. Sharma

3. **Electrochemical Ferrate(VI) Synthesis: A Molten Salt Approach**.....68
   M. Benová, J. Híveš, Karel Bouzek, and Virender K. Sharma

4. **Electrochemical Behavior of Fe(VI)–Fe(III) System in Concentrated NaOH Solution**.....81
   Cun Zhong Zhang, HongBo Deng, Tingting Zhao, Feng Wu, Wei Liu, Shengmin Cai, Kai Yang, and Virender K. Sharma

5. **Preparation of Potassium Ferrate by Wet Oxidation Method Using Waste Alkali: Purification and Reuse of Waste Alkali**.....94
   Jiang Chengchun, Liu Chen, and Wang Shichao

6. **New Processes for Alkali Ferrate Synthesis**.....102
   L. Ninane, N. Kanari, C. Criado, C. Jeannot, O. Evrard, and N. Neveux

7. **Higher Oxidation States of Iron in Solid State: Synthesis and Their Mössbauer Characterization**.....112
   Yurii D. Perfiliev and Virender K. Sharma

8. **Thermal Stability of Solid Ferrates(VI): A Review..............................124**
Libor Machala, Radek Zboril, Virender K. Sharma, Jan Filip, Oldrich Schneeweiss, János Madarász, Zoltán Homonnay, György Pokol, and Ria Yngard

9. **A Fluorescence Technique to Determine Low Concentrations of Ferrate(VI). Determination of Micromolar Fe(VI) Concentrations for Laboratory Investigations....................................145**
Nadine N. Noorhasan, Virender K. Sharma, and J. Clayton Baum

Properties

10. **Aqueous Ferrate(V) and Ferrate(IV) in Alkaline Medium: Generation and Reactivity...158**
Diane E. Cabelli and Virender K. Sharma

11. **Identification and Characterization of Aqueous Ferryl(IV) Ion.......167**
Oleg Pestovsky and Andreja Bakac

12. **Ferrate(VI) Oxidation of Nitrogenous Compounds............................177**
Michael D. Johnson, Brooks J. Hornstein, and Jacob Wischnewsky

13. **Kinetics and Product Identification of Oxidation by Ferrate(VI) of Water and Aqueous Nitrogen Containing Solutes..........................189**
James D. Carr

14. **Recent Advances in Fe(VI) Charge Storage and Super-Iron Batteries...197**
Stuart Licht and Xingwen Yu

15. **Transformation of Iron(VI) into Iron(III) in the Presence of Chelating Agents: A Frozen Solution Mössbauer Study: Mössbauer Investigation of the Reaction between Iron(VI) and Chelating Agents in Alkaline Medium..255**
Zoltán Homonnay, Nadine N. Noorhasan, Virender K. Sharma, Petra Á. Szilágyi, and Ernő Kuzmann

Applications

16. **Electrochemical Fe(VI) Water Purification and Remediation...........268**
Stuart Licht, Xingwen Yu, and Deyang Qu

17. **Evaluating the Coagulation Performance of Ferrate: A Preliminary Study**....292
Khoi Tran Tien, Nigel Graham, and Jia-Qian Jiang

18. **The Use of Ferrate(VI) Technology in Sludge Treatment**....306
Jia-Qian Jiang and Virender K. Sharma

19. **Evaluation of Ferrate(VI) as a Conditioner for Dewatering Wastewater Biosolids**....326
Hyunook Kim, Yuhun Kim, Virender K. Sharma, Laura L. McConnell, Alba Torrents, Clifford P. Rice, Patricia Millner, and Mark Ramirez

20. **Ferrate(VI) Oxidation of Recalcitrant Compounds: Removal of Biological Resistant Organic Molecules by Ferrate(VI)**....339
Virender K. Sharma, Nadine N. Noorhasan, Santosh K. Mishra, and Nasri Nesnas

21. **Heterogeneous Photocatalytic Reduction of Iron(VI): Effect of Ammonia and Formic Acid. Enhancement of Photocatalytic Oxidation of Ammonia and Formic Acid in Presence of Iron(VI)**....350
Virender K. Sharma and Benoit V. N. Chenay

22. **Degradation of Dibutyl Phthalate in Aqueous Solution by a Combined Ferrate and Photocatalytic Oxidation Process**....366
X. Z. Li, B. L. Yuan, and Nigel Graham

23. **Preparation and Properties of Encapsulated Potassium Ferrate for Oxidative Remediation of Trichloroethylene Contaminated Groundwater**....379
B. L. Yuan, M. R. Ye, and H. C. Lan

24. **Oxidation of Nonylphenol Using Ferrate**....390
Myongjin Yu, Guisu Park, and Hyunook Kim

25. **Preparation of Potassium Ferrate for the Degradation of Tetracycline**....405
Shih-fen Yang and Ruey-an Doong

26. **Removal of Estrogenic Compounds in Dairy Waste Lagoons by Ferrate(VI)**....421
Jarrett R. Remsberg, Clifford P. Rice, Hyunook Kim, Osman Arikan, and Chulhwan Moon

27. Use of Ferrate(VI) in Enhancing the Coagulation of Algae-Bearing Water: Effect and Mechanism Study....................................436
Wei Liu and Yong-Mei Liang

28. Combined Process of Ferrate Preoxidation and Biological Activated Carbon Filtration for Upgrading Water Quality...............448
Jun Ma, Chunjuan Li, Yingjie Zhang, and Ran Ju

29. Enhanced Removal of Cadmium and Lead from Water by Ferrate Preoxidation in the Process of Coagulation...........................458
Jun Ma, Wei Liu, Yingjie Zhang, and Chunjuan Li

30. Potential of Ferrate(VI) in Enhancing Urban Runoff Water Quality...467
Umid Man Joshi, Rajasekhar Balasubramanian,
and Virender K. Sharma

Indexes

Author Index...479

Subject Index..481

Preface

This book is derived from a symposium sponsored by the American Chemical Society (ACS) Divisions of Environmental Chemistry, Inc. and Inorganic Chemistry, Inc., *Ferrates: Synthesis, Properties, and Applications in Water and Wastewater Treatment*, which was organized for the 232nd ACS National Meeting in San Francisco, California, September 10–14, 2006. Papers were solicited with a call for papers, as well as with direct contact with researchers. The symposium was international in nature and 33 experts from the United States, Europe, and Asia gave oral presentations.

In recent years, the higher oxidation states of iron (Ferrates) are of interest because of their involvement in reactions of environmental, industrial, and biological importance. A number of high-quality papers addressing these reactions are being published in high-impact journals. New ferrate chemistry is still being developed and new analytical techniques are used to characterize the ferrate species. Applications of ferrate to treat common pollutants and emerging contaminants such as arsenic, estrogens, pharmaceuticals, and personal-care products are being explored. This book is timely because ferrate is emerging as a green chemistry chemical for organic synthesis and for treating toxins in water. A main objective of the symposium was to bring together experts with a background in fundamentals and applications of ferrate chemistry. Many contributors covered the multidisciplinary theme during the symposium.

Chapters for the book were solicited from the symposium's oral presentations as well as from additional researchers in the field to provide a balanced presentation. This book is organized into three sections: synthesis and characterization, properties, and applications in water and wastewater treatment. Chapter 1 provides a comprehensive review of recent advances in Fe(VI) synthesis. Chapters 2–4 are devoted to electrochemical synthesis of Fe(VI). Chapters 5 and 6 give wet- and dry-oxidation methods to synthesize Fe(VI). Chapter 7 gives an overview of oxidation states of iron with valences from +4 to +8 in solid states. Chapters 8 and 9 present a review of the thermal decomposition of

Fe(VI) salts and determination of low concentrations of Fe(VI), respectively. Chapters 10–13 are devoted to the generation and properties of aqueous high-oxidation states. Chapter 14 gives a review on recent advances in Fe(VI) charge storage and super-iron batteries. Chapter 15 presents a frozen solution Mössbauer spectroscopy technique to learn the aqueous chemistry of Fe(VI). The final section of the book is devoted to the application of Fe(VI) as an oxidant, coagulant, and disinfectant in the treatment of water, wastewater, and sludge. Chapters 16–19 present examples of treatment of water and biosolids. Chapter 20 gives the potential of ferrate to oxidize recalcitrant compounds. Chapters 21 and 22 demonstrate the combination of ferrate and a photocatalytic oxidation process to enhance the degradation of pollutants. Chapters 23–26 are on the removal of pharmaceuticals and estrogens in water by Fe(VI). Chapters 27–30 deal with the use of Fe(VI) to enhance the removal of algae, organics, and metals and thus to improve the water quality.

This book is the first comprehensive compilation of the chemistry and application of ferrates. The contents of the book should appeal to fundamental scientists and environmental scientists and engineers who are engaged in understanding the chemistry of high-valent iron and in applications of chemical oxidants, coagulants, and disinfectants to treat contaminants in water, wastewater, and industrial effluents.

Acknowledgments

I gratefully acknowledge the peer reviewers whose comments contributed significantly to improve the final versions of the chapters.

Virender K. Sharma
Chemistry Department
Florida Institute of Technology
150 West University Boulevard
Melbourne, FL 32901
vsharma@fit.edu

Ferrates

Synthesis, Properties, and Applications in Water and Wastewater Treatment

Synthesis and Characterization

Chapter 1

Recent Advances in Fe(VI) Synthesis

Stuart Licht and Xingwen Yu

Department of Chemistry, University of Massachusetts Boston, Boston, MA 02125

The synthesis and analysis of a range of Fe(VI) compounds are presented. Fe(VI) compounds have also been variously referred to as ferrates or super-iron compounds. Fe(VI) salts with detailed syntheses in this paper include the alkali Fe(VI) salts high purity Cs_2FeO_4, Rb_2FeO_4, and $K_xNa_{(2-x)}FeO_4$, low purity Li_2FeO_4, as well as high purity alkali earth Fe(VI) salts $BaFeO_4$, $SrFeO_4$, and also Ag_2FeO_4. Two conventional, as well as two improved Fe(VI) synthetic routes are presented. The conventional syntheses include solution phase oxidation (by hypochlorite) of Fe(III), and the synthesis of less soluble super-irons by dissolution of FeO_4^{2-}, and precipitation with alternate cations. The new routes include a solid synthesis route for Fe(VI) salts and the electrochemical synthesis (include *in-situ* & *ex-situ* synthesis) of Fe(VI) salts. Fe(VI) analytical methodologies summarized are FTIR, ICP, titrimetric, UV/VIS, XRD, Mössbauer and a range of electrochemical analyses. Fe(VI) compounds have been explored as energy storage cathode materials in both aqueous and non-aqueous phase in "super-iron" battery configurations, as well as novel oxidants for synthesis and water treatment purification. Preparation of reversible Fe(VI/III) thin film towards a rechargeable super-iron cathode is also presented. In addition, the preparation of unusual $KMnO_4$ and zirconia coatings on Fe(VI) salts, via organic solvent deposition, is summarized. These coatings can stabilize and activate Fe(VI) salts in contact with alkaline media.

The fascinating chemistry of hexavalent iron, Fe(VI) is not as established as that for ferrous, Fe(II), ferric, Fe(III) or zero valent (metallic) iron chemistry. As a strong oxidant, Fe(VI), is formed in aqueous solutions as FeO_4^{2-}, which has been investigated for several decades as a potentially less hazardous alternative to the chlorination purification of water (*1-3*). The field of Fe(VI) compounds for charge storage was introduced in 1999, and at that time the term super-iron was coined to refer to the class of materials which contain "super-oxidized" iron in the unusual hexavalent state *(4)*. The charge transfer chemistry of super-iron salts in both aqueous and non-aqueous media has been probed (*4-32*).

In conventional syntheses, high purity, stable K_2FeO_4 is prepared, from alkaline hypochlorite oxidation of Fe(III). Less soluble Fe(VI) salts are prepared by precipitation, upon addition of various salts to solutions containing dissolved FeO_4^{2-} (*5,6*). In addition to probing these syntheses, we have also introduced several routes to improved Fe(VI) salt synthesis, including solid-state syntheses and direct electrochemical synthesis of Fe(VI) salts. The conventional synthesis for $BaFeO_4$ utilizing solution phase reactants is generalized on the left side of Scheme 1. The scheme of the center illustrates an alternative solid synthesis, which uses only solid state reactants. The right side of the scheme illustrates the direct electrochemical synthesis of Fe(VI) salts.

The one-step direct electrochemical synthesis (*in-situ* electrochemical synthesis) of solid Fe(VI) salts has significant advantages in shorter synthesis time, simplicity, and reduced costs (no chemical oxidant is required). At potentials greater than 0.6 volt versus SHE (the standard hydrogen electrode) in alkaline media, an iron metal anode is directly oxidized to FeO_4^{2-}. When the electrolyte contains an Fe(VI) precipitating cation, the generated FeO_4^{2-} is rapidly isolated as a solid, and stabilized ferrate salt. As represented on the right side of Scheme 1, oxidation of an iron anode in a conductive, stabilizing alkaline electrolyte, containing the dissolved metal cation. In the illustrated case barium, yields by direct precipitation, the pure, stabilized $BaFeO_4$, (*7-9*).

In the center of the Scheme 1, the use of solid state reactants has several Fe(VI) synthetic advantages. Fe(VI) solution phase degradation to Fe(III) (*1*) is avoided, and fewer preparatory steps reduces requisite synthesis time, and can increase the yield of the Fe(VI) salt synthesis. For example, in the conventional synthesis of $BaFeO_4$, both K_2FeO_4 and $Ba(OH)_2$ are reacted in the aqueous phase, and $BaFeO_4$ is generated due to the higher alkaline insolubility of barium ferrate(VI) compared to that of potassium ferrate(VI). In the solid synthesis, ,the reactants such as K_2FeO_4 and barium oxide alone are stable, but fully react upon grinding together, forming a dough-like paste; KOH is removed, isolating the Fe(VI) salt. In the barium example, water, bound in the salt as the hydrate $BaO{\cdot}4H_2O$ is necessary and sufficient to drive the reaction, and forms an unusually pure (> 98%) and stable Fe(VI) salt (*10*).

This paper describes synthesis details of a variety of high purity Fe(VI) salts, Fe(VI/III) reversible thin films (*11, 12*) and Mn coated (*5*) and zirconia coated Fe(VI) salts (*13*). This includes electrochemical, and solid phase

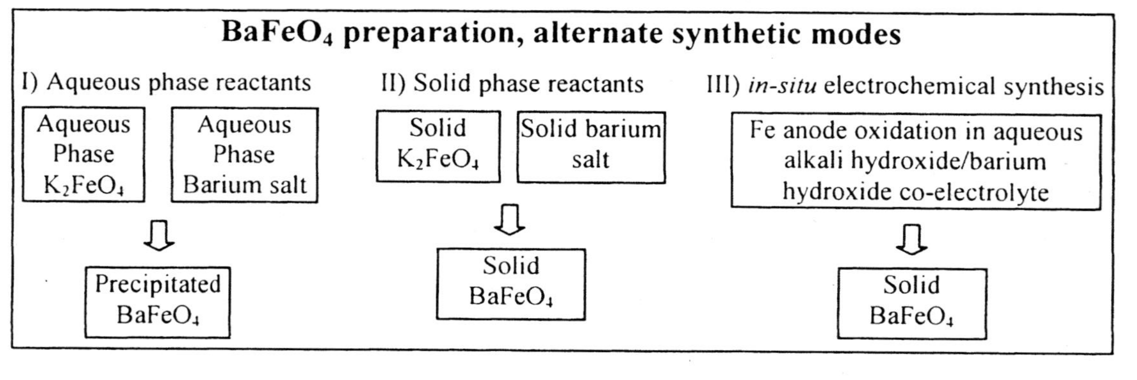

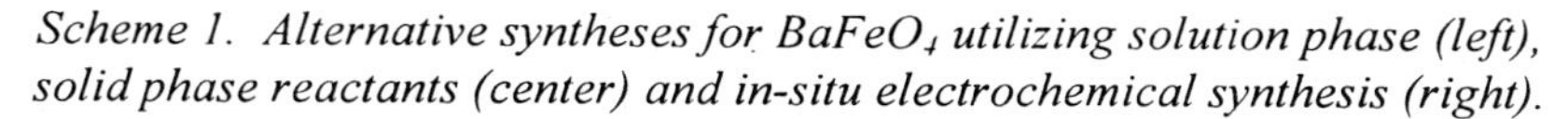

Scheme 1. Alternative syntheses for $BaFeO_4$ utilizing solution phase (left), solid phase reactants (center) and in-situ electrochemical synthesis (right).

syntheses, as well as the conventional synthesis of K_2FeO_4, and the conventional, precipitation from solution, syntheses of a variety of Fe(VI) oxides including: the alkali Fe(VI) salts high purity Cs_2FeO_4, Rb_2FeO_4, and $K_xNa_{(2-x)}FeO_4$ (*14*), and low purity Li_2FeO_4 (*15*), as well as the high purity alkali earth Fe(VI) salts $BaFeO_4$ (*5, 7-10*), $SrFeO_4$ (*16*), and also Ag_2FeO_4 (*17*). Topics presented in this paper include:

1. Alkaline hypochlorite synthesis of high purity, stable K_2FeO_4.
2. Solution precipitation synthesis of Cs_2FeO_4, Rb_2FeO_4, $K_xNa_{(2-x)}FeO_4$, and Li_2FeO_4.
3. Solution precipitation synthesis of $SrFeO_4$, and $BaFeO_4$.
4. Solution precipitation synthesis of Ag_2FeO_4.
5. Solid synthesis of $BaFeO_4$.
6. Direct electrochemical synthesis of $BaFeO_4$.
7. Preparation of Fe(III/VI) thin films.
8. Preparation of Mn or Zr coated Fe(VI) salts.
9. Fe(VI) Analysis.

1. Alkaline Hypochlorite Synthesis of High Purity, Stable K_2FeO_4

This section focuses on the chemical preparation of high purity potassium ferrate. Fe(VI), or ferrate(VI), compounds may be chemically synthesized to a high degree of purity, from a variety of ferric salts, in a variety of alkaline hydroxide media. Hypochlorite is a particularly effective oxidizing agent in the synthesis. However, the synthesis can be a challenge: for example, small variations in the filtration, purification, and drying processes, can lead to decomposition and diminish salt purity and stability. We have been scaling up the syntheses in a step-wise fashion to the half kilogram range. This paper summarizes syntheses yielding 80 to 100 grams of 96.5 to 99.5% pure K_2FeO_4, and the products of these syntheses are demonstrated to have a lifetime on the order of years.

One liter of KOH concentrated solution is prepared with Barnstead model D4742 deionized water from 0.620 kg of KOH pellets from Fruitarom, Haifa, Israel (Analytical reagent KOH with ~14% water, <2% K_2CO_3, <0.05% Na, < 0.03% NH_4OH, and 0.01% or less of other components). The solution is converted to potassium hypochlorite by reaction with chlorine. The Cl_2 is generated in-house within a 2 liter Woulff (spherical) flask (made by Schott of Duran glass) with fritted glass connections. The glass connections are attached to a 1 liter dropping flask with pressure equalizer inlet (with a burette controlled liquid inlet and another connection is to a gas-outlet). In the Woulff flask is 0.25 kg $KMnO_4$ (99% CP grade, Fruitarom), and from the dropping flask 1.13 liter of

37% HCl (AR grade, Carlo-Erba) is added drop-wise to the $KMnO_4$ to generate chlorine in accord with:

$$KMnO_4 + 8HCl \rightarrow MnCl_2 \cdot 4H_2O + KCl + 5/2Cl_2 \qquad (1.1)$$

Droplets, HCl and water are removed from the evolved Cl_2, through a series of 2 liter Dreschel (gas washing) flasks connected in series. The first and third are empty (to prevent backflow); the second contains water (to remove HCl), the fourth contains 95-98% H_2SO_4 (to remove water), and the fifth flask contains glass wool (to remove droplets). The evolved, cleaned Cl_2 flows into a reaction chamber (a sixth Dreschel flask containing the concentrated KOH solution, and surrounded by an external ice-salt bath) where it is stirred into concentrated KOH solution. Excess gas is trapped within a final flask containing waste hydroxide solution. Chlorination of the KOH solution generates hypochlorite, which is continued until the weight of the concentrated KOH solution has increased by 0.25 kg, over a period of approximately 90 minutes, in accord with:

$$2KOH + Cl_2 \rightarrow KClO + KCl + H_2O \qquad (1.2)$$

This hypochlorite solution is cooled to 10°C. Alkalinity of the solution is increased, and KCl removed, through the addition of 1.46kg KOH pellets, added slowly with stirring, to permit the solution temperature to rise to no more than 30°C. Stirring is continued for 15 minutes, and the solution is cooled to 20°C. The precipitated KCl is removed by filtration through a 230 mm diameter porcelain funnel using a glass microfibre filter (cut from Whattman 1820-915 GF/A paper).

A ferric salt is added to the hypochlorite solution, reacting to Fe(VI), as a deep purple FeO_4^{2-} solution. An external ice-salt bath surrounds the solution to prevent overheating. Specifically, to the alkaline potassium hypochlorite solution at 10°C, is added 0.315 kg ground $Fe(NO_3)_3 \cdot 9H_2O$ (98% ACS grade, ACROS). In alkaline solution, the ferric nitrate constitutes hydrated ferric oxides or hydroxides, summarized as:

$$Fe(NO_3)_3 \cdot 9H_2O + 3OH^- \rightarrow Fe(OH)_3 + 9H_2O + 3NO_3^- \qquad (1.3)$$

which is oxidized by hypochlorite to form the Fe(VI) anion, FeO_4^{2-}, in solution:

$$Fe(OH)_3 + 3/2ClO^- + 2OH^- \rightarrow FeO_4^{2-} + 3/2Cl^- + 5/2H_2O \qquad (1.4)$$

During the ferric addition, a surrounding ice-salt bath is applied to maintain the solution temperature below 35°C. Following this addition, the solution is stirred for 60 minutes, with the solution temperature controlled at 20°C. For potassium salts, the overall reaction is summarized by equations 1.3 and 1.4 as:

$$Fe(NO_3)_3 \cdot 9H_2O + 3/2KClO + 5KOH \rightarrow K_2FeO_4 + 3/2KCl + 3KNO_3 + 23/2H_2O \quad (1.5)$$

Following this, the KOH concentration of the resultant Fe(VI) solution is increased to precipitate K_2FeO_4. Specifically into this solution is stirred 1.25 liter of 0°C, 9.6 M KOH. After 5 minutes, the suspension is (simultaneously) filtered onto two 120 mm P-1 sintered Duran glass filters (Schott).

The two precipitates are dissolved in 1.6 liter of 2.57 M KOH, and quickly filtered, through a funnel with 2 layers of GF/A filter paper of 230 mm diameter, directly into 1.7 liter of 0°C 12 M KOH. The solution is stirred for 15 minutes at 3°C, and then the solution is filtered onto a 90 mm P-2 sintered Duran glass filter (Schott). The wet K_2FeO_4 is dissolved in 0.850 liter of 0°C 2.57 M KOH solution, and quickly filtered on 2 sheets of filter paper GF/A 150 mm diameter, in a filtering flask which contains 2.7 liters of a 12 M KOH solution.

From this point, two grades of K_2FeO_4 are produced. The first generates higher yield, 90 g K_2FeO_4, at a purity of 96-97%. The second generates 80 g of K_2FeO_4 at even higher purity 97-98.5%. In both procedures, the wet K_2FeO_4 is redissolved in 0.850 liter of 0°C 2.57 M KOH solution, and quickly filtered on 2 sheets of filter paper GF/A 150 mm diameter, into a filtering flask containing 2.7 liter 12 M KOH solution. The resulting suspension is stirred for 15 minutes at 0°C and is filtered through a P-2 sintered glass filter. This redissolution/filtering step is repeated in the second (highest purity) procedure. In either procedure, on the same filter, the precipitate is successively rinsed: 4× (four times with) 0.16 liter n-hexane; 2×0.08 liter isopropyl alcohol; 8×0.15 liter methanol, and finally 3×0.080 liter diethyl ether. The K_2FeO_4 is dried for 30-60 minutes under room temperature vacuum (at 2-3 mbar).

The percentage of iron is determined by ICP, and the percentage the original iron containing material which is converted to solid Fe(VI) salt was determined redundantly by UV/vis analysis and by the chromite method. These analytical methodologies are described at the end of this paper. K_2FeO_4, prepared as described above, is particularly robust, and the long-term stability (over 1 year) is presented in Figure 1.1. As shown in the figure inset, K_2FeO_4 appears to be stable whether sealed under dry N_2 or sealed in air, and is also stable under acetonitrile (and a variety of other organic electrolytes). K_2FeO_4, chemically synthesized to a purity of over 97-99%, tends to fall to ~96.5% purity, at which point no further fall is observed for the duration of the experiment (over 1 year).

2. Solution precipitation synthesis of Cs_2FeO_4, Rb_2FeO_4, $K_xNa_{(2-x)}FeO_4$, and Li_2FeO_4

The dried K_2FeO_4 product that has been synthesized as described as section 1 may be used for Cs_2FeO_4, $Rb(K)FeO_4$, $Na(K)FeO_4$ and Li_2FeO_4 synthesis directly or after storage. The Cs_2FeO_4 and $Rb(K)FeO_4$ salts were synthesized

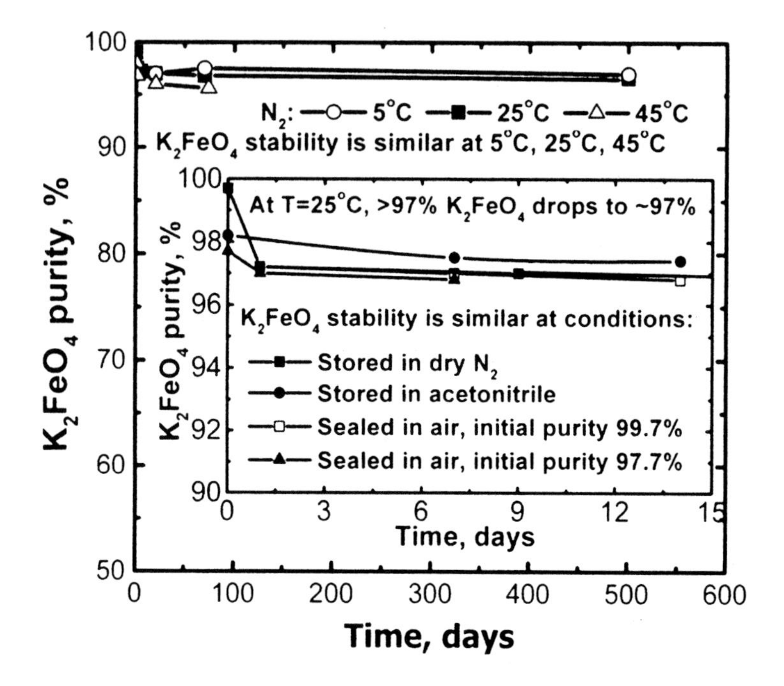

Figure 1.1. The long term stability of K_2FeO_4, measured after sealing in a variety of conditions. (Reference 5)

from potassium ferrate, by driving formation via their hydroxide reaction, in the respective cesium or rubidium hydroxide solution. In these media, effective Fe(VI) cesium or rubidium salts are observed in the precipitation reaction:

$$K_2FeO_{4\text{-aq}} + 2MOH_{aq} \rightarrow M_2FeO_{4\text{-solid}} + 2KOH_{\text{-aq}}; \qquad M{=}Cs \text{ or } Rb \qquad (2.1)$$

Or in the case of partial replacement:

$$K_2FeO_{4\text{-aq}} + xMOH_{aq} \rightarrow M_xK_{(2-x)}\, FeO_{4\text{-solid}} + xKOH_{\text{-aq}} \qquad (2.2)$$

Specifically, to a solution of 1.7 g K_2FeO_4 in 7ml 0.5% aqueous KOH, 4.4 g of $CsOH{\cdot}xH_2O$ (from Aldrich 19833-1) dissolved in 1.4 ml deionized water, at 0°C, was added, while maintaining 0°C using an ice bath. After 5 minutes of vigorous stirring and filtration on a sintered glass funnel No. 2, the precipitate was washed sequentially with the following organic solvents: n-hexane 4×5ml (4 washings, each with 5ml), iso-propanol 2×5ml, methanol 8×5ml, and finally 3×5ml diethylether. The resultant Cs_2FeO_4 was dried for 2 hours under room temperature vacuum (at 2-3 mbar) yielding 2.4 g Cs_2FeO_4.

ICP analysis provides evidence that the conversion from the potassium to the cesium Fe(VI), Cs_2FeO_4, salt has been complete. Specifically, that the material contains less than 0.6 percent equivalents of potassium relative to cesium, and that the Fe comprises the expected 0.27 (±0.01) weight fraction for the Cs_2FeO_4 compound. Chromite analysis, confirms the Fe(VI) valence state of the iron, and by this analysis that the material is 99.2% pure.

Rb(K)FeO_4 was synthesized in a similar manner, using a solution of 2.5 g K_2FeO_4 in 9 ml 0.5% KOH and 12 ml RbOH (50% RbOH aqueous solution, Aldrich 24369-8). The mixture was cooled to 0°C, and after 5 minutes of vigorous stirring was filtered. The precipitate was washed sequentially as in the preparation of the Cs_2FeO_4 salt. The resultant Rb(K)FeO_4 was dried for 2 hours under vacuum, yielding 2.4 g of 98.7% purity $Rb_{1.7}K_{0.3}FeO_4$ with the iron oxidation state determined by chromite analysis, and the relative Rb to K ratio determined by ICP.

Na(K)FeO_4 is synthesized from a partial conversion of K_2FeO_4 into Na_2FeO_4:

$$K_2FeO_{4\text{-aq}} + x\text{NaOH}_{\text{aq}} \rightarrow Na_xK_{(2\text{-}x)}FeO_{4\text{-solid}} + x\text{KOH}_{\text{-aq}} \tag{2.3}$$

A maximum Na:K ratio was obtained for the $Na_xK_{(2-x)}FeO_4$ synthesis, when the ratio of K_2FeO_4:NaOH was 1 to 10. A solution of 5g K_2FeO_4 in 18 ml of 6 M NaOH was added to 28 ml of 18 M NaOH, with vigorous stirring. The precipitate, obtained by filtration, was purified by dissolving once more in 14.4 ml of 6 M NaOH, and re-precipitated by addition of 22.4 ml of 18 M NaOH solution. The black crystalline precipitate obtained was filtered, and washed sequentially with the following organic solvents for removing water and alkali hydroxide salts: n-hexane 4×8ml, iso-propanol 2×8ml, methanol 8×8ml and finally 3×8ml diethylether. All solutions used in this synthesis had to be cooled, to prevent Fe(VI) decomposition. After 1 hour drying under vacuum the yield was 2.6 g of 96.3% purity $Na_{1.1}K_{0.9}FeO_4$ with the Fe(VI) oxidation state determined by chromite analysis, and the relative rubidium to potassium ratio determined by ICP.

Figure 2.1 compares the x-ray powder diffraction spectra measured for the pure K_2FeO_4 and Cs_2FeO_4, compounds synthesized as described above. The crystalline structure for each of these two compounds has been previously determined to be consistent with the Pnma orthorhombic system, by similar x-ray diffraction powder analysis of the structure by Herber and Johnson and others (*33-35*). Computer simulation based on these parameters yields a simulated x-ray diffraction spectrum with the same diffraction as in Figure 2.1.

Figure 2.2 compares the FTIR spectra, measured in a KBr pellet, for the synthesized alkali ferrate salts K_2FeO_4, Cs_2FeO_4 and Ru(K)FeO_4, as well as the alkali earth ferrate compounds $BaFeO_4$ and $SrFeO_4$. In each-sample, $BaSO_4$ has been added as a quantitative standard, in the fixed mass ratio of 1:10 to the sample. As can be seen, the Cs_2FeO_4 compared to K_2FeO_4 spectra are similar,

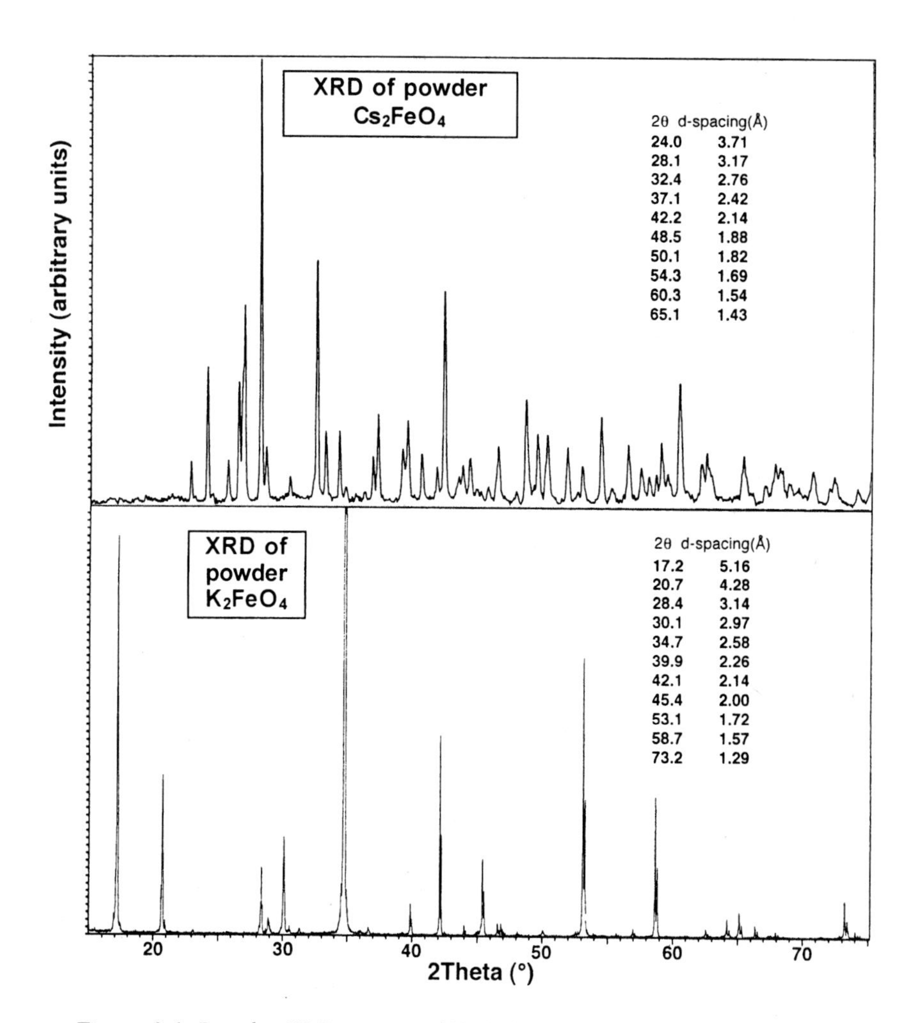

Figure 2.1. Powder XRD pattern of K_2FeO_4, and Cs_2FeO_4. (Reference 14)

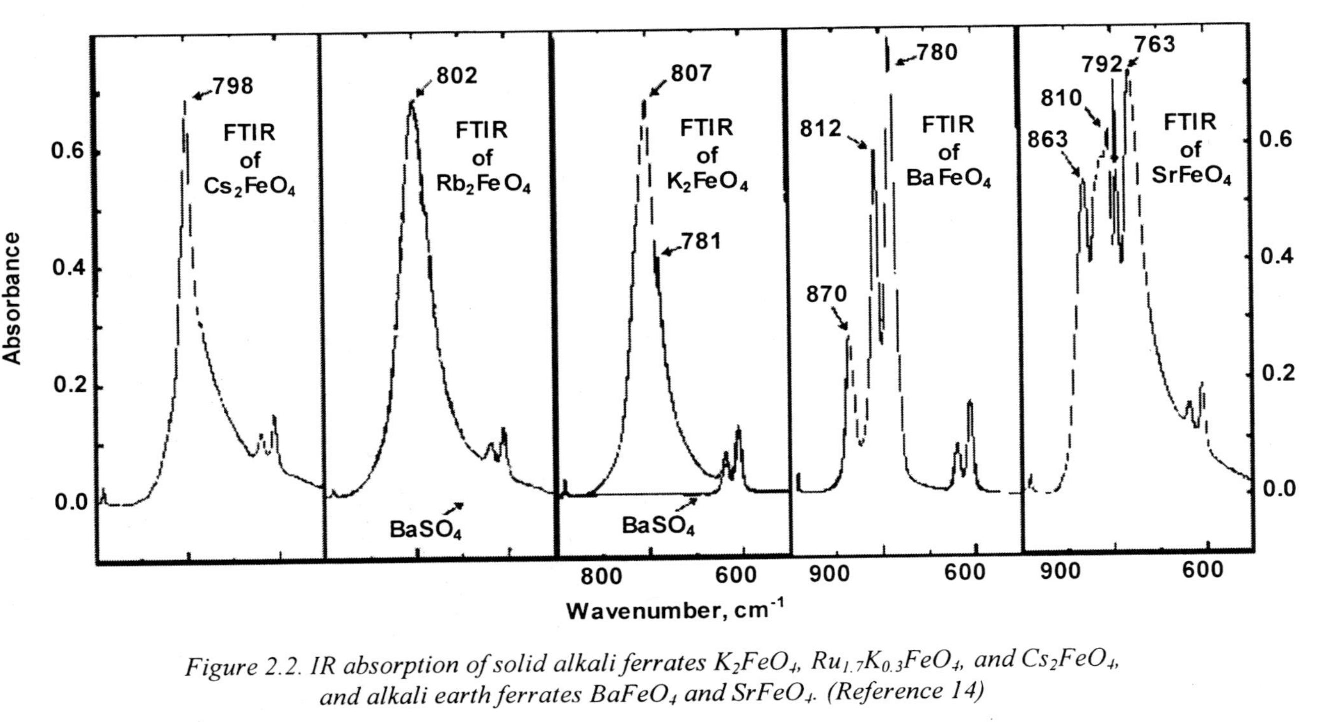

Figure 2.2. IR absorption of solid alkali ferrates K_2FeO_4, $Ru_{1.7}K_{0.3}FeO_4$, and Cs_2FeO_4, and alkali earth ferrates $BaFeO_4$ and $SrFeO_4$. (Reference 14)

with the principal difference between these two in the 798 compared to 807 cm^{-1} absorption peak for the Cs_2FeO_4 compared to K_2FeO_4 compound. The FTIR results of $BaFeO_4$ and $SrFeO_4$ will be detailed discussed in section 3.

As with the cesium, the preparation of the rubidium, Fe(VI) salt from K_2FeO_4, results in retention of a high degree of purity of the Fe(VI) valence state. However, unlike the cesium salt, conversion to rubidium from potassium was incomplete, resulting in a salt containing 85 equivalent percent Rb and 15 equivalent percent K, or $Rb_{1.7}K_{0.3}FeO_4$. FTIR of this mixed rubidium potassium salt results in IR absorption peaks which are largely indistinguishable from the pure K_2FeO_4 compound, providing a challenge for de-convolution to obtain the IR spectrum of a pure Ru_2FeO_4 compound, or to distinguish it as a composite or single compound. As seen in the figure, the spectrum appearance is intermediate to that observed for either the pure K_2FeO_4 or Cs_2FeO_4 salts, with a peak absorbance at 802 nm and a shoulder that occurs at approximately 780 nm. The location and magnitude of these near lying K_2FeO_4, Ru_2FeO_4, and Cs_2FeO_4 Fe(VI) absorption peaks differ from the spectra we, or others, have measure for the lower valence Fe state salts including Fe_3O_4, Fe_2O_3, or $Fe(OH)_3$ (*36, 37*). In the pure potassium salt case, the IR stretching frequencies of K_2FeO_4 had been interpreted as evidence of the equivalence, symmetric and tetrahedral distribution of the oxygen atoms surrounding the iron center (*38*).

Attempts to synthesize Na_2FeO_4 by aqueous oxidation of Fe(III) salts had led to salts containing high levels of Fe(III) impurity [> 40% Fe(III)], evidently due to rapid decomposition of the Fe(VI) species in the requisite alkaline sodium hypochlorite synthesis solution. Alternately, we retain high Fe(VI) purity using a direct synthesis, by reaction of K_2FeO_4 with sodium salts, as described above. However, this preparation leads to a mixed Na/K salt, rather than the pure sodium salt. ICP analysis shows that the salt contains 1.1 equivalents of sodium per 0.9 equivalent of potassium. As evidenced by the FTIR spectra in Figure 2.3, rather than a pure compound of $Na_{1.1}K_{0.9}FeO_4$, the compound appears to act as a composite mixture of 55 equivalent percent Na_2FeO_4 and 45% K_2FeO_4. As also seen in the figure, the spectra of the pure Na_2FeO_4 compound can be deduced by weighted difference of the spectra. This deconvolution reveals principal IR absorption peaks for Na_2FeO_4 occurring at 868, 802, and 784 cm^{-1}.

The degree of synthesis optimization to obtain the sodium and rubidium Fe(VI) salts is dissimilar. The mixed sodium potassium salt summarized represents our best attempt to maximize the Na to K ratio in the Fe(VI) salts. However, we believe that further optimization of the rubidium to potassium level beyond 0.85 to 0.15 is possible, although the high expense of the rubidium reactants precluded its extensive, further optimization.

The main portion of Figure 2.4 compares the 71° stability of Cs_2FeO_4, K_2FeO_4, $Rb_{1.7}K_{0.3}FeO_4$, and $Na_{1.1}K_{0.9}FeO_4$. It is seen that the highest stability is observed for K_2FeO_4, and that there is a modest trend of a drop in purity in time for either the lighter ($Na_{1.1}K_{0.9}FeO_4$) or heavier alkali ($Rb_{1.7}K_{0.3}FeO_4$, and Cs_2FeO_4) salts.

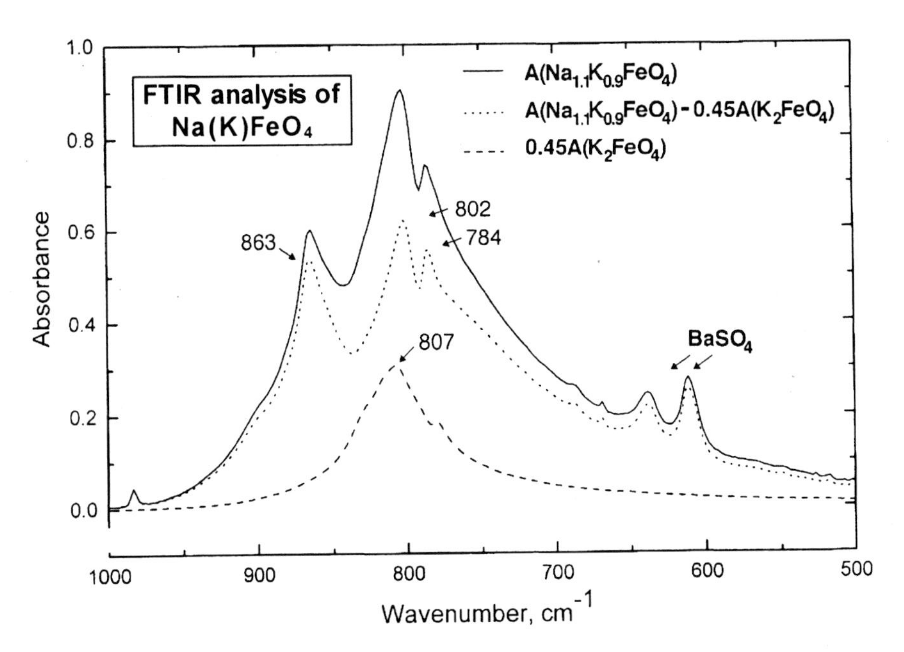

Figure 2.3. IR absorption of solid $Na_{1.1}K_{0.9}FeO_4$ and K_2FeO_4, as well as the computed spectra of pure Na_2FeO_4 (by deconvolution of these spectra). (Reference 14)

Inset of Figure 2.4 compares the measured solublities of Cs, and the mixed Rb and Na/K super iron salts in KOH electrolytes. Of interest is the domain of the high concentration KOH electrolytes. This domain includes the electrolyte used for the (zinc and metal hydride) anode in conventional alkaline batteries. The higher solubility observed for the sodium Fe(VI) salt in KOH electrolyte is consistent with the observed high solubility for the potassium Fe(VI) salt in NaOH electrolyte. Hence it is consistent to express this simply as FeO_4^{2-} dissolution as limited by either separate sodium or potassium alkali hydroxide electrolyte. As seen in the figure, for each of the alkali Fe(VI) salts, soluble Fe(VI) concentration decreases with increasing KOH concentration, and solubility is in the millimolar domain in a saturated KOH electrolyte. Saturation decreases down the alkali column, that is the solubility, S, varies in the order $S(Na_{1.1}K_{0.9}FeO_4) > S(K_2FeO_4) > S(Rb_{1.7}K_{0.3}FeO_4) > S(Cs_2FeO_4)$.

Traditionally, the preparation of pure Li_2FeO_4 has been a technical challenge and rarely studied. We have marginally improved on the synthesis of Gump and Wagner, (*39*) by the addition of a final acetonitrile wash, to produce Li_2FeO_4 of up to 20%, rather 15%, purity as determined by chromite analysis (*15*). ICP analysis averaged over three syntheses determines the average sample is comprised of 39±2 % by weight of Fe, 9±1% Li, and 0.8±0.5% K. The analysis indicates a small amount of potassium retained from the synthesis, and

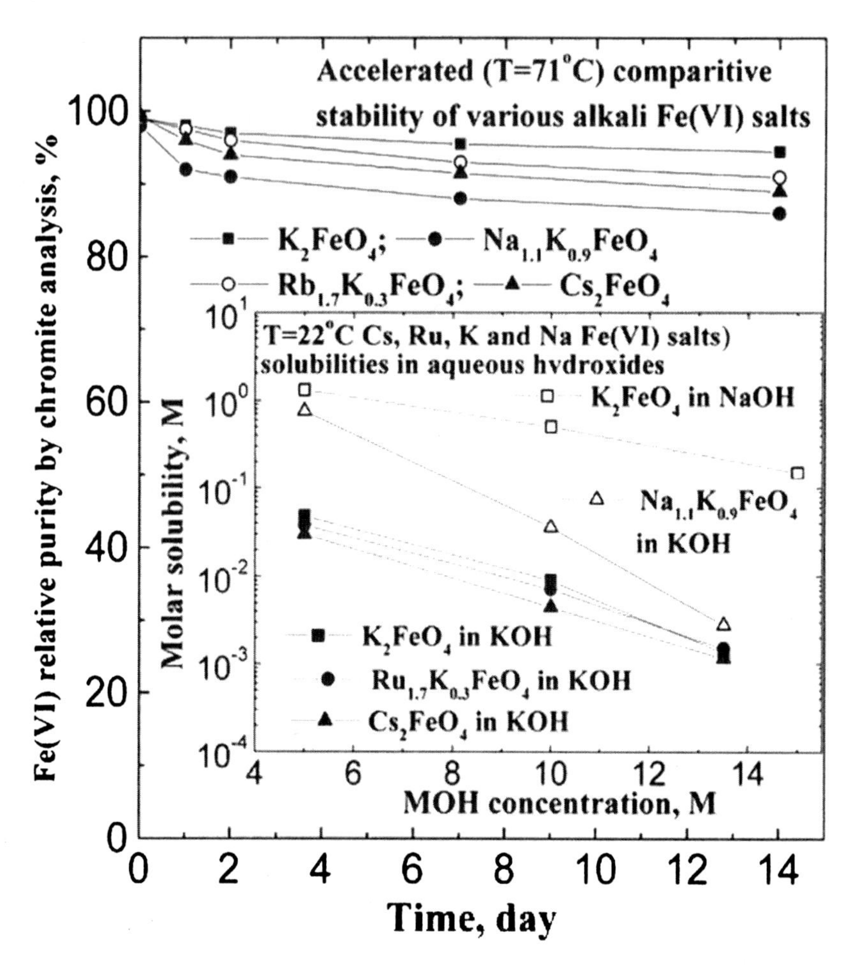

Figure 2.4. Solid state stability of Cs_2FeO_4, K_2FeO_4, $Rb_{1.7}K_{0.3}FeO_4$ and $Na_{1.1}K_{0.9}FeO_4$ at 71°C. Inset: Room temperature solubility of Cs_2FeO_4, K_2FeO_4, $Rb_{1.7}K_{0.3}FeO_4$ and $Na_{1.1}K_{0.9}FeO_4$ in various hydroxide solutions. (Reference 14)

indicates that the large impurities consist of iron salts, presumably hydrated Fe(III) oxides. Specifically in the synthesis, 1.6 g of AR grade $LiClO_4 \cdot 3H_2O$ was added to an aqueous 3°C solution of 1 g 97.5% purity K_2FeO_4 in 5 ml H_2O. After stirring at 3°C, the $KClO_4$ precipitate was removed by filtration. The filtrate was dried under vacuum for 90 minutes. The paste obtained was washed 4 times with 25 ml of ACN, and dried under vacuum. The Li_2FeO_4 obtained was a black powder with Fe(VI) purity determined by chromite analysis.

3. Solution precipitation synthesis of $SrFeO_4$, and $BaFeO_4$

As the synthesis of alkali Fe(VI) salts, the dried, solid state stable K_2FeO_4 may be used for alkaline earth Fe(VI) salts, $BaFeO_4$ and $SrFeO_4$ synthesis, directly or after storage. The salts were synthesized by utilizing the higher alkaline insolubility of strontium or barium ferrate(VI) compared to that of potassium ferrate(IV) (*5, 16*). We have observed effective Fe(VI) precipitates occur starting with barium nitrate, chloride, acetate or hydroxide salts. This section summaries the synthesis of $BaFeO_4$ from barium hydroxide, and the synthesis of $BaFeO_4$ and $SrFeO_4$ from barium acetate.

$BaFeO_4$ may be synthesized from barium hydroxide. In this synthesis, 0.210 kg $Ba(OH)_2 \cdot 8H_2O$ (98%, Riedel-de-Haen) was dissolved in 5 liter deionized water, with CO_2 removed by argon flow, at 0°C, and the solution is filtered through GF/A filter paper (solution A). In a second solution, 0.08kg K_2FeO_4 was dissolved at 0°C in 1.6 liter 2% KOH solution (37.6 g KOH in 1.6 liter water, with CO_2 removed by argon flow), and then filtered through GF/A filter paper (150 mm) into the solution A with stirring at 0°C (using an ice bath). Stirring is continued in the mixture for 20 minutes. The mixture obtained was filtered on a single funnel with GF/A glass microfiber paper, diameter of 230 mm, and then, the residue of $BaFeO_4$ was washed with 10 liter cold distilled water without CO_2, until the $BaFeO_4$ reached pH=7. The resultant $BaFeO_4$ is dried for 16-24 hours under room temperature vacuum (at 2-3 mbar) and yields 90-93 g of 96-98% purity $BaFeO_4$ as determined by chromite analysis.

$SrFeO_4$ and $BaFeO_4$ may also be prepared from the respective acetate salts in accord with the reaction:

$$K_2FeO_{4\text{-aq}} + M(C_2H_3O_2)_{2\text{-aq}} \rightarrow MFeO_{4\text{-solid}} + KC_2H_3O_{2\text{-aq}}; \quad M{=}Sr \text{ or } Ba \qquad (3.1)$$

Specifically, 96 g of $Sr(C_2H_3O_2)_2$ (Aldrich ®) and 1 g $Sr(OH)_2$ were dissolved in 240 ml deionized water, with CO_2 removed by argon flow, at 0°C, then filtered through 150 mm diameter Whatman glass microfiber GF/A paper (solution A). Separately, into 320 ml of aqueous 0.50 wt% KOH solution, with CO_2 removed, is dissolved 80 g K_2FeO_4 at 0°C. This solution is filtered through 150 mm GF/A paper into solution A with stirring at 0°C (using an ice bath). $SrFeO_4$ will precipitate immediately. Stirring is continued in the mixture for 3 min. The mixture obtained was filtered on a Buchner, sintered glass filter, and then washed 6 times with 100 ml of chilled ethanol, followed by six times with 50 ml of chilled ether. The $SrFeO_4$ powder was dried under vacuum (at 2-3 mbar), at room temperature, for 3 hours, yielding ~75g of 94-95% purity $SrFeO_4$, determined by chromite analysis. To prepare the analogous $BaFeO_4$ salt, 120 g of $Ba(C_2H_3O_2)_2$ (Aldrich ®) and 1 g $Ba(OH)_2$ were dissolved in 1.2 L deionized water, with CO_2 removed, at 0°C, then filtered through 150 mm diameter GF/A paper (solution A'). Separately, into 1.6 L of aqueous 2.0 wt% KOH solution, without CO_2, is dissolved 80 g K_2FeO_4 at 0°C. This solution is filtered through 230 mm GF/A paper into solution A' with stirring at 0°C. Stirring is continued in the mixture for 40 min. The mixture obtained was filtered through 230 mm GF/A paper, and then washed with ~10 liter of cold deionized water, without CO_2, to reduce the pH to 7. The $BaFeO_4$ paste was dried under vacuum (at 2-3 mbar), at room temperature, for 26 hours, yielding ~90g of 98-99% purity $BaFeO_4$, determined by chromite analysis.

The FTIR spectra of synthesized $SrFeO_4$ and $BaFeO_4$ are shown in Figure 2.2. As can be seen, the spectra are readily distinguishable, and the latter is consistent with a previous qualitative IR determination of $BaFeO_4$ (*38*). The $SrFeO_4$ spectrum is more complex than observed for either $BaFeO_4$or K_2FeO_4, whose IR stretching frequencies had been interpreted as evidence of the equivalence, symmetric and tetrahedral distribution of the oxygen atoms surrounding the iron center (*37*).

Figure 3.1 shows the x-ray powder diffraction spectra of $BaFeO_4$ and $SrFeO_4$ prepared via the synthesis described in this section. Note that $BaFeO_4$ and $SrFeO_4$ powders were each synthesized from the K_2FeO_4 (XRD spectrum is shown as Figure 2.1), and in both cases exhibit a broader diffraction pattern. We, interpret this as evidence of a more amorphous structure of these latter salts, although as also evident in the XRD, each is still significantly crystalline. The $BaFeO_4$ spectrum is consistent with an orthorhombic crystal system (Pnma) (*35*), and the structural features interpreted from the distinctive $SrFeO_4$ XRD need future studies.

Figure 3.2 compares the measured solubilities of Sr, K and Ba super iron salts in various alkaline electrolytes. As seen in the figure, at all concentrations through KOH saturation (KOH saturation occurs at ~13.5 M at 22°C), $SrFeO_4$ is more soluble, than the minimum solubility of $BaFeO_4$'s in a $Ba(OH)_2$ containing KOH electrolyte. Furthermore at low KOH concentrations, $SrFeO_4$ is more soluble than K_2FeO_4. However of interest is the domain of the most concentrated KOH electrolytes. This domain is of significance, as it includes the electrolyte used in conventional alkaline batteries. It is in this domain that the $SrFeO_4$ solubility is minimal. In this medium, its very low solubility is similar to K_2FeO_4 and $BaFeO_4$ [without added $Ba(OH)_2$]. $SrFeO_4$ solubility will further diminish in the case in which $Ba(OH)_2$ is also added to this electrolyte.

4. Solution precipitation synthesis of Ag_2FeO_4

The silver ferrate compounds may be chemically synthesized as converted from potassium or barium ferrate, Fe(VI), salts via a substitution reaction with silver nitrate. Best results were obtained when potassium ferrate, K_2FeO_4, was used for this Ag_2FeO_4 synthesis. ICP analysis of Ag_2FeO_4 synthesized from $BaFeO_4$ indicated traces present of $BaFeO_4$ in the silver ferrate obtained, due to its competitively low, solution phase solubility (*17*).

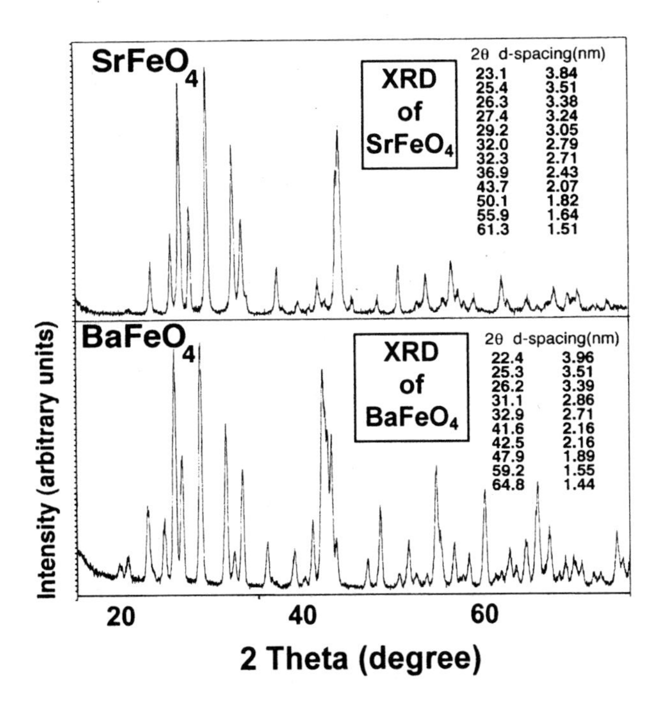

Figure 3.1. Powder XRD pattern of $SrFeO_4$, and $BaFeO_4$. (Reference 16)

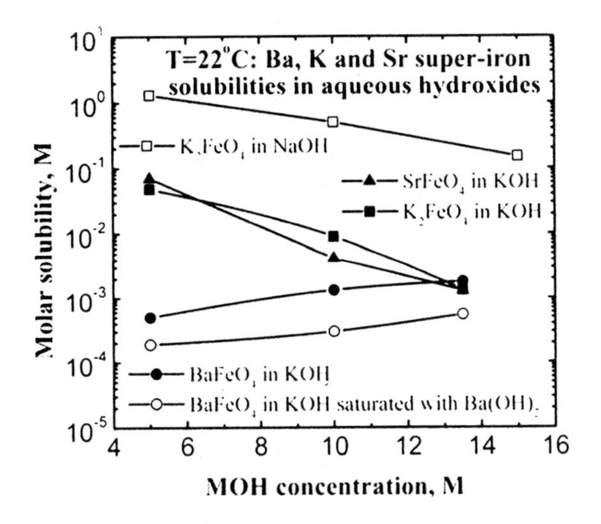

Figure 3.2. Room temperature solubility of $BaFeO_4$, K_2FeO_4, and $SrFeO_4$ in various aqueous hydroxide electrolytes. (Reference 16)

Upon dissolving the potassium ferrate salt in an aqueous solution of silver nitrate, a spontaneous, immediate black precipitate of silver salt was obtained:

$$K_2FeO_4 + 2AgNO_3 \rightarrow Ag_2FeO_4 + 2KNO_3 \quad (4.1)$$

Due to the high sensitivity of silver compounds to light, aluminum foil protected glassware or brown glassware were used in this synthesis. Specifically, to a solution of 5 g $AgNO_3$ (from Fruitarom 5553260) in 50 ml triply deionized water, chilled to –3 °C, 2 g, high purity, K_2FeO_4 (98.2%) was added. After 5 minutes of stirring, the black precipitate was filtered on GF/A filter paper, and washed three times with 50 ml of triply deionized water. The resultant product, Ag_2FeO_4, was dried for 6 hours under room temperature vacuum (at 2-3 mbar) yielding 2.85 g Ag_2FeO_4.

Purity of the product Ag_2FeO_4 was determined by chromite analysis and ICP. Inductively coupled plasma analysis provides evidence that the conversion from potassium Fe(VI), to the silver Fe(VI), salt has been almost complete. Specifically, the material contains less than 1% equiv. of potassium relative to silver. The ICP determined molar ratio of iron to silver is greater than the expected molar ratio of 0.5:1 (±0.03), and measured to be 0.57:1. Presumably, the measured excess iron is due to a lower valence iron oxide such as Fe_2O_3 or its more hydrated salt, $Fe(OH)_3$ or to a mixed Ag(I)Fe(III) oxide. Chromite analysis, probes the total oxidation of the synthesized salt, and compares it to the intrinsic maximum 5 electron/equivalent capacity, based on the combined $3e^-$ alkaline reduction of Fe(VI) and $2e^-$ alkaline reduction of each Ag(I):

$$5Cr(OH)_4^- + 3Ag_2FeO_4 + 5OH^- + H_2O \rightarrow 6Ag + 3Fe(OH)_3(H_2O)_3 + 5CrO_4^{2-} \quad (4.2)$$

This chromite analysis determines that the material is 86.7% pure based on redox state. The remaining iron is in a lower valence state, although at these relatively low concentration levels, the specific nature of this ferric impurity is difficult to distinguish. Presumably the excess iron exists as several amorphous ferric salts, which, consistent with the ICP measured excess iron and chromite analyses, can be generalized as a 13% ferric oxide impurity.

The observed IR spectra of Ag_2FeO_4 is not similar to previously observed Fe(VI) compounds. Figure 4.1, compares the infrared spectra, measured in a KBr pellet, of solid Ag_2FeO_4 with the Ag(I) compound, Ag_2O, and the Fe(VI) compound K_2FeO_4. Spectra are presented over two frequency ranges from 600 to 900 cm^{-1}, and from 1250 to 1800 cm^{-1}. The K_2FeO_4 compound does not display the IR absorption peaks found for the Ag(I) compounds in the 1300 to 1500 cm^{-1} range. In this range the absorption of the synthesized Ag_2FeO_4 compound is broad compared to that observed for Ag_2O, and does not exhibit the distinct Ag_2O peaks observed at 1477, 1449, 1376 and 710 cm^{-1}. In the figure also note that the Ag_2FeO_4 compound does not display the significant IR Fe(VI) absorption peaks in the 750 to 850 cm^{-1} range, found for a range of

Fe(VI) compounds, including K_2FeO_4, and Cs_2FeO_4, and for the alkali earth Fe(VI) salts $SrFeO_4$ and $BaFeO_4$.

As previously demonstrated, the synthesized K_2FeO_4 salt does not decompose and exhibits a full solid stability at room temperature. As shown in Figure 4.2, the observed stability of Ag_2FeO_4 is considerably less than that of K_2FeO_4. As seen in the figure, only when stored in the freezer (at 0 °C) is the material moderately stable over a period of 1 week, but does decrease in purity from 87% to 83% during this period. At higher temperatures, over the same time period, the Ag purity falls to 73% at 25 °C, and falls to 27% at 45 °C, and falls more rapidly at 71 °C, over a period of three days to 34%. Light considerably accelerates instability of the Ag_2FeO_4. After 5 h exposure to ambient fluorescent light, FTIR reveals formation of the distinctive Ag_2O IR absorption peaks 1477, 1449, 1376 and 710 cm^{-1}, and after 1 daylight exposure, the FTIR spectra is characteristic of a salt substantially converted to Ag_2O.

5. Solid synthesis of $BaFeO_4$

Solid phase reaction preparation of $BaFeO_4$ is generalized as the center of Scheme 1. Specially, in this synthesis, 1 equivalent (45.6 wt%) of solid K_2FeO_4 is mixed per 0.5 equivalent (36.5 wt%) of solid $Ba(OH)_2{\cdot}8H_2O$ and 0.5 equivalent (17.8 wt%) of solid BaO. The mix is ground for 15 minutes in a Fritsch Pulversette 0, yielding a viscous dough-like paste. The paste is transferred to a GF/A filter paper under suction, and washed with CO_2 free water until the wash decreases to pH=7; then transferred to a drying flask, reaching a constant mass, under vacuum (at 2-3 mbar) for 4-5 hour, to yield the product $BaFeO_4$. 12 g of K_2FeO_4 yields 13 g of 97% Fe(VI) purity salt containing 96% $BaFeO_4$ and 3% K_2FeO_4. Three fold larger syntheses proceed to similar purity, and with improved relative yield due to smaller transfer losses.

Probing and optimizing the above solid phase reaction was accomplished in several steps. Solid K_2FeO_4 reacts with a suspension (a supersaturated aqueous solution) of $Ba(OH)_2$ to yield a mixture of pure $BaFeO_4$ and pure K_2FeO_4. For example, a suspension is prepared of 81.2 g (0.26 mol) of $Ba(OH)_2$ in 2 liter of 10 M KOH. To this suspension solid 51.0 g K_2FeO_4 (0.26 mol) is added, and stirred 30 minutes. The K_2FeO_4 is highly insoluble in the solution, and is converted towards $BaFeO_4$. The resultant powder, still undissolved, is removed by filtration, and the precipitate washed with organic solvents, as previously described for similar purification step in K_2FeO_4 preparation. The reaction yields a pure mixture of Fe(VI) salts (as determined by chromite, FTIR and ICP analysis) containing approximately a 4:1 ratio of $BaFeO_4$ to K_2FeO_4.

Using solid state reactants instead, no room temperature reaction was observed for a 1:1 mole ratio of BaO to K_2FeO_4, when ground together for 3 hours. However, spontaneous conversion to $BaFeO_4$ is achieved by replacing the BaO with conventional solid $Ba(OH)_2{\cdot}8H_2O$. In this case, a 1:1 mole ratio of $Ba(OH)_2{\cdot}8H_2O$ to K_2FeO_4 yields upon grinding an immediate reaction to

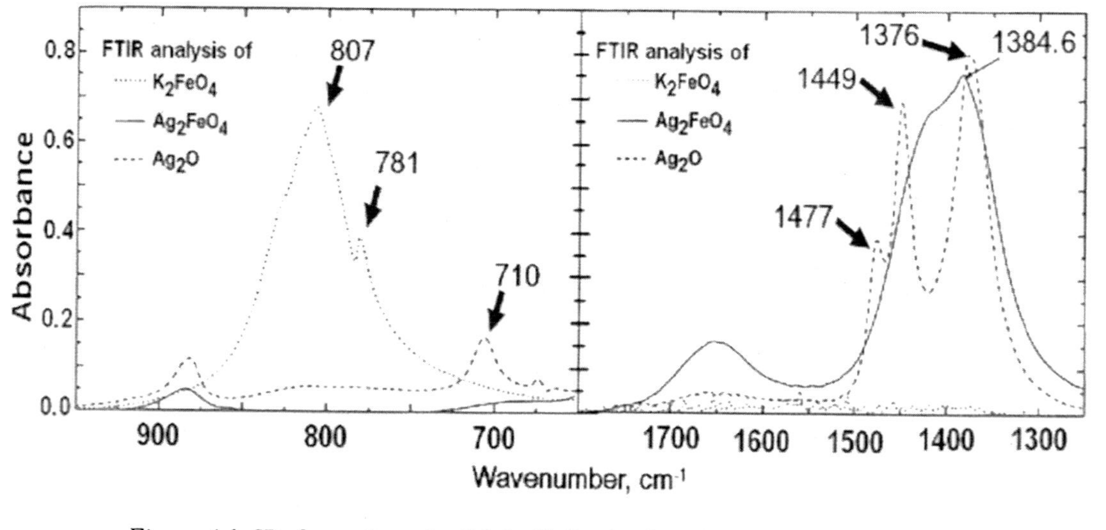

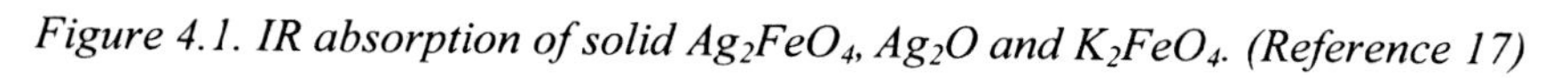

Figure 4.1. IR absorption of solid Ag_2FeO_4, Ag_2O and K_2FeO_4. (Reference 17)

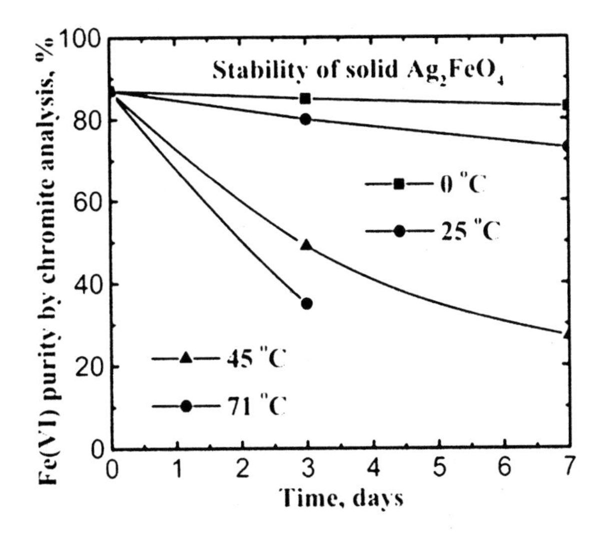

Figure 4.2. Solid state stability of Ag_2FeO_4, at 0, 25, 45, 71°C. (Reference 17)

$BaFeO_4$. Hence, FTIR analysis of the ground solid BaO/ K_2FeO_4 mixture yields the spectra of pure K_2FeO_4 (a single absorption at 807 cm^{-1}), without any of the three $BaFeO_4$ identifying absorptions which occur in the same region. At room temperature, the presence of bound water, included within the hydrated solid $Ba(OH)_2$ salt, clearly facilitates the reaction of the ground mixture to yield $BaFeO_4$. However, this solid K_2FeO_4/solid $Ba(OH)_2{\cdot}8H_2O$ reaction yields a wet paste, a suspension of solid $BaFeO_4$ in 13.9 M aqueous KOH, due to the dissolution product of 2 moles of KOH per 8 moles (0.14 kg) H_2O generated, in accord with:

$$K_2FeO_4 + Ba(OH)_2{\cdot}8H_2O \rightarrow BaFeO_4 + 2KOH + 8H_2O \qquad (5.1)$$

Intermediate syntheses demonstrated solid BaO could drive the reaction to $BaFeO_4$, when combined with as little as 50 mol% of $Ba(OH)_2{\cdot}8H_2O$. The resultant mix, equivalent to the tetrahydrate $Ba(OH)_2{\cdot}4H_2O$, are sufficient to support a substantially complete (96-97%) room temperature conversion of the K_2FeO_4 to $BaFeO_4$, and generate a viscous dough-like blend of solid $BaFeO_4$ mixed with supersaturated KOH in accord with generation of only 4 moles of H_2O for 2 moles of KOH:

$$K_2FeO_4 + 0.5BaO + 0.5Ba(OH)_2{\cdot}8H_2O \rightarrow BaFeO_4 + 2KOH + 4H_2O \qquad (5.2)$$

6. Direct Electrochemical Synthesis of $BaFeO_4$

Direct Electrochemical synthesis of $BaFeO_4$ can proceed through the route illustrated on the right side of Scheme 1 (*8*). The synthesis was carried out with an electrolysis cell configured with a two compartment cell; the two compartment cell is separated by a Raipore R1010 anion impermeable membrane (surface area 192 cm^2) into two 80ml compartments. High surface area iron anodes were prepared from iron wire using Fluka (0.1% Cu, 0.1% Ni and 0.7% Mn) iron wire, of diameter, d = 0.2 mm, coiled at the lengths (L) of 128 m, with surface area, $A= L\pi d = 800\ cm^2$. The cathode consisted of high surface area platinum gauze (200 mesh per cm^2). Electrode pretreatment included 3 minutes sonication in 1:3 HCl, followed by triply deionized water washing (Nanopure water system) to pH=7. The electrolytes were prepared from analytical grade reagents and triply deionized water. The experiments were conducted under N_2 bubble induced convection. Electrolysis cell temperature was controlled by a large ethanol bath (Heto) with a cooling/heating control of $\pm 1^{\circ}C$. Constant direct charging currents were applied by Pine AFRDE5 bipotentiostat ($i_c \geq 1.6A$) and EP613 DC power source ($i_c \leq 150mA$).

The direct (*in-situ*) electrochemical synthesis is accomplished through oxidizing a iron wire anode, in a $NaOH/Ba(OH)_2$ co-electrolyte for a fixed time. During the electrolysis, $BaFeO_4$ is spontaneously formed and precipitated during iron oxidation in the chamber of the electrolysis compartment. Then the precipitation is vacuum filtered on a GF/A filter paper, and washed with triply deionized water to pH=7. The $BaFeO_4$ product is dried at room temperature for 16-18 hours under 2 Torr to a constant mass. In principle, this direct electrochemical synthesis has several advantages: i) Fe(VI) synthesis is simplified to a one step process. ii) Fe(VI) instability is avoided through the direct formation of the solid product. iii) On a volumetric basis, the product is formed at several orders of magnitude higher concentration than for solution phase generation. iv) The system avoids the consumption of chemical oxidants used in chemical Fe(VI) synthesis. v) The electrolytic generation of solid super-iron salts may provide pathways for more electroactive Fe(VI) compounds.

Fundamentals of the electrochemical alkaline solution phase generation of FeO_4^{2-} are further detailed in reference 8. Parameters affecting the solution phase Fe(VI) (as FeO_4^{2-}) generation including the electrolyte concentration, anodic current density, temperature and separator effects have been studied and evaluated (*8*). The direct electrochemical synthesis of solid $BaFeO_4$ is based on the optimized conditions for electrochemically formation of solution phase FeO_4^{2-}. At sufficiently anodic potentials in alkaline media, an iron anode is directly oxidized to the Fe(VI) species FeO_4^{2-}, in accord with the oxidation reaction:

$$Fe + 8OH^- \rightarrow FeO_4^{2-} + 4H_2O + 6e^- \qquad (6.1)$$

This process occurs at potentials > 0.6 V vs the standard hydrogen electrode, SHE (*8*). The *in-situ* electrochemical $BaFeO_4$ synthesis uses NaOH/$Ba(OH)_2$ co-electrolyte. The electrochemically generated FeO_4^{2-} reacts with $Ba(OH)_2$ to precipitate $BaFeO_4$. The spontaneous $BaFeO_4$ formation reaction may be generalized as:

$$Fe + 6OH^- + Ba(OH)_2 \rightarrow BaFeO_4 + 4H_2O + 6e^- \quad (6.2)$$

Combined this reaction with the corresponding H_2 evolution reaction at the cathode side:

$$Fe + 2OH^- + 2H_2O \rightarrow FeO_4^{2-} + 3H_2 \quad (6.3)$$

the net cell reaction becomes:

$$Fe + Ba(OH)_2 + 2H_2O \rightarrow BaFeO_4 + 3H_2 \quad (6.4)$$

As previously observed, NaOH electrolytes support higher solution Fe(VI) generation rates and efficiencies than KOH electrolyte (*40-42*). It was also found that little Fe(VI) is generated electrochemically in a pure (NaOH-free) aqueous $Ba(OH)_2$ electrolyte, evidently limited by the comparatively low concentration of hydroxide sustainable in such solutions. As previously observed, the current efficiency of Fe(VI) synthesis strongly depends on the solution-phase hydroxides concentration (*8*) due to the effects of the solution's activity and conductivity (*43, 44*) on the kinetics of Fe(VI) formation. In the pure aqueous $Ba(OH)_2$ electrolyte, the maximum concentration of solution-phase hydroxide ($[OH^-] < 0.5$ M) is limited by the saturation of $Ba(OH)_2$, which is insufficient to sustain high rates of Fe(VI) formation. However, the synthesis progresses rapidly in an aqueous NaOH/$Ba(OH)_2$ co-electrolyte, forming at rates comparable to the pure NaOH electrolyte, but directly forming the solid Fe(VI) salt. Table 6.1 summarizes a series of *in-situ* electrochemical syntheses performed in 10 or 14 M NaOH electrolytes. Attempts to conduct analogous measurements in $[NaOH] \leq 8$ M, resulted in unsuitably low yields (< 8%). At the high end of the concentration range, that is in electrolytes more concentrated than 14 M, problems were encountered due to insufficient solubility of the $Ba(OH)_2$ co-electrolyte. As can be seen in the table, increasing the NaOH concentration from 10 to 14 M, leads to improvements of both purity and yield.

Figure 6.1 probes the influence of the internal cell temperature on the *in-situ* $BaFeO_4$ synthesis. The slight improvement in $BaFeO_4$ purity with the cell temperature is consistent with the higher solubility of $Ba(OH)_2$ in NaOH at elevated temperatures, retarding the formation of a solid $Ba(OH)_2$ contamination of the product. Also presented in Table 6.1 is the effect of varying the molar ratio between the FeO_4^{2-} and $Ba(OH)_2$ reactants. As can be seen, increasing the FeO_4^{2-} content leads to minor improvements of purity and yield, apparently due

Table 6.1. *In-situ* Electrochemical Syntheses of $BaFeO_4$ Under a Variety of Physical and Chemical Conditions.
(Reference 8)

[NaOH] (M)	*$[Ba(OH)_2]$ (M)*	*Electrolysis time (min)*	*Internal temp. (°C)*	*n_{Ba}/n_{Fe} in reactants*	*Theoretical $BaFeO_4$ produced (mmol)*	*$BaFeO_4$ purity (%)*	*Yield (%)*	*Synthesis efficiency (%)*
10	50	80	25	1.0:1.0	4.00	74.6	22.6	16.9
14	45	50	25	1.0:1.0	3.60	85.7	25.2	21.6
14	45	45	35	1.0:1.0	3.60	91.6	75.5	69.2
14	45	250	35	1.0:5.5	3.60	74.5	87.6	65.3
14	45	30	45	1.5:1.0	2.40	92.8	60.5	56.1
14	45	45	45	1.0:1.0	3.60	93.8	61.7	57.9
14	45	70	45	1.0:1.5	3.60	94.5	61.9	58.5
14	45	135	45	1.0:3.0	3.60	85.8	74.9	64.3
14	45	45	55	1.0:1.0	3.60	93.4	42.2	39.4

NOTE: Cell conditions are described in the Fig. 6.1 legend. The indicated yield is determined as 100% × (synthesized $BaFeO_4$/ theoretical $BaFeO_4$. Synthesis efficiency is calculated as (purity × yield)/100%.

to the minimization of traces of insoluble $Ba(OH)_2$ in the precipitate. In a solution with $Ba(OH)_2/FeO_4^{2-}$ molar ratio of 1.50, the $BaFeO_4$ purity is 94.5%, which approaches the $BaFeO_4$ purities previously obtained for chemical (hypochlorite driven oxidative) syntheses of Fe(VI) salts. Increasing this ratio above 50%, as evident in the table for the 35-45°C measurements, induces a decrease in the purity, although accompanied by an increase in the yield. The cause for these observations lies in the relatively longer time required for such experiments. After a lengthy electrolysis, Fe(III) can accumulate. This directly affects the purity and the total mass of the contaminated product.

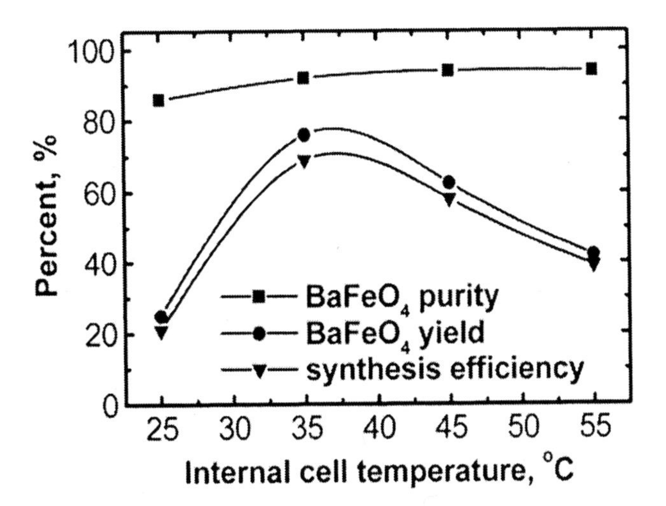

Figure 6.1. In-situ electrochemically synthesized $BaFeO_4$ purity, yield, and synthesis efficiency dependence on the internal cell temperature. Related measurements are detailed in Table 6.1. (Reference 8)

Optimized *in-situ* electrochemically synthesized $BaFeO_4$ was obtained during 70 min of 1.6 A (S_{anode}=800 cm^2, J=2 mA/cm^2) electrolysis in a co-electrolyte of 14 M NaOH and 45 mM $Ba(OH)_2$ at 45°C. Seven concurrent syntheses generated products of purity 93.3, 93.2, 93.6, 94.8, 95.0, 94.2, and 94.0%, indicating a high degree of reproducibility of the optimized *in-situ* electrochemical syntheses. Figure 6.2 compares FTIR and XRD spectra of the high purity chemical, compared to *in-situ* electrochemical, synthesis of $BaFeO_4$. As can be seen, both analyses present nearly identical crystalline diffraction patterns [consistent with the space group *D*2h (*Pnma)*] (*10, 33*) and IR absorption spectra (including the well-defined, typical $BaFeO_4$ triplet at 750-800 cm^{-1}). These measurements also verify that high purity $BaFeO_4$ was obtained employing the electrochemical optimized preparation procedure.

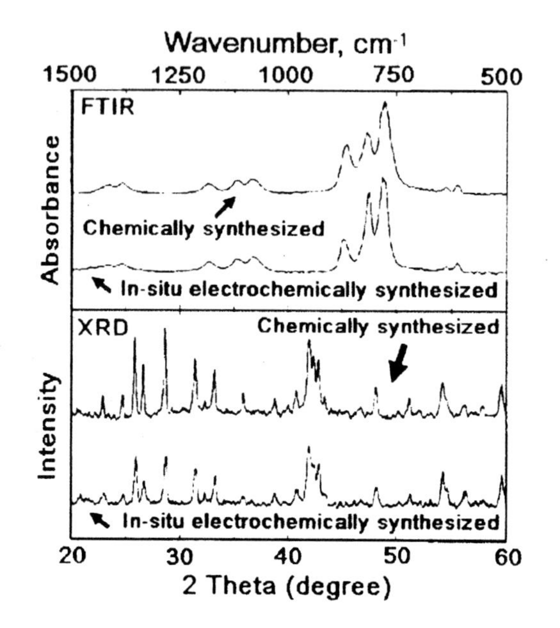

Figure 6.2. XRD (bottom) and FTIR (top) spectra of chemically and direct electrochemically synthesized $BaFeO_4$. (Reference 8)

Refinements of cell size and electrolysis conditions can be expected to lead to further purity and yield improvements. Figure 6.3 explores the effect of membrane variation on the *in-situ* electrochemical synthesis of $BaFeO_4$. In each case the same synthesis electrolyte [80 ml 14 M NaOH, 0.045 M $Ba(OH)_2$] and conditions (T = 40°C, 1.6 A synthesis current, 128 m of 0.2 mm diam Fe wire electrode) were used, but the separation material between the iron electrode synthesis chamber and the counter electrode compartment were modified, from the Raipore 1010 membrane used in the previous experiments, to either an alternate alkaline stable, cation selective membrane (Nafion 350), or a nonselective separator (either 1 or 5 layers of Whatman glass microfiber paper) or an open compartment (no separator). As seen in the figure, the separator choice has little effect on the purity of the synthesized salt, with the Nafion separator cell providing a marginally higher 96% purity of synthesized $BaFeO_4$. However, the separator choice has a substantial effect on the synthesis yield. As expected, both a nonselective separator and no separator lead to lower yield, as a portion of the synthesized Fe(VI) can be lost via reduction at the counter electrode. The cation selective separator impedes anionic Fe(VI) diffusion to the counter electrode, preventing this loss, and as seen in the figure, yield is highest with Nafion membrane separation.

Initial syntheses were also conducted, demonstrating a several fold scale-up, using the same surface area membrane separation, but with a 580 mL, rather

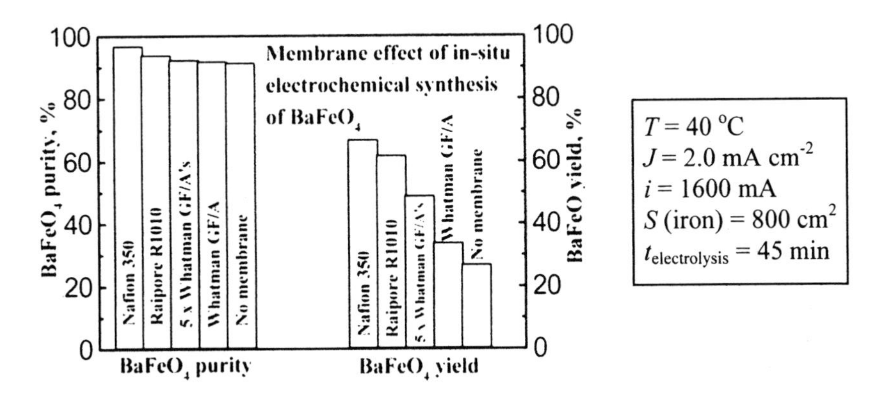

Figure 6.3. Membrane effect on the electrochemical synthesis of in-situ electrochemically synthesized $BaFeO_4$. With the exception of various different separators located between the synthesis and counter electrode compartments, the anode, cathode and cell configuration used for this synthesis are as described in the text. (Reference 8)

than 80 mL 14 M NaOH, 0.045 M $Ba(OH)_2$ electrolyte, and using a longer 640 m wound Fe (0.2 nm diameter) wire anode, to synthesize at a current of 2.88 A and temperature of 50°C for 120 min. Following a wash with 7 L water (CO_2-free), this yielded a low yield of 1.97 g (or 29.3% conversion of the available barium), but very high purity of 95.9% $BaFeO_4$.

In Table 6.2 are presented analyses and summaries of the chemical and direct electrochemical synthesis of $BaFeO_4$. One major difference in the methodologies is in the synthesis temperature. Whereas the chemical synthesis of K_2FeO_4 and its conversion to $BaFeO_4$ are performed at low temperatures to avoid decomposition, direct electrochemical syntheses require higher temperature (25-45°C) to facilitate the Fe(VI) electrolysis. Under such conditions, the electrochemical syntheses are one to two orders of magnitude faster than the chemical synthesis, and $BaFeO_4$ purity obtained approaches the purity of the chemical synthesis. Also presented in the table is the IR absorption ratio between carbonates/hydroxide (1420 cm^{-1}) to ferrate (780 cm^{-1}) measured for the $BaFeO_4$ produced by the different methods, indicating a strong correlation between these impurities and the total Fe(VI) purity detected by the chromite titration. ICP analysis results support a $BaFeO_4$ stoichiometry with a small excess of Ba that increases with decreased purity. Traces of K and Sr are due to trace amounts of these elements in the NaOH and $Ba(OH)_2$ reagents while Mn, Ni, and Cu are due to their initial presence in the iron anode.

An interesting aspect of the direct electrochemically synthesized $BaFeO_4$ is a substantially enhanced stability of the salt as summarized in Figure. 6.4. Rather than the higher stability observed at 25°C, the experiments are conducted

Table 6.2. Properties of Direct Electrochemical and the Chemical Synthesized $BaFeO_4$. *(Reference 8)*

	Direct Electrochemical synthesized $BaFeO_4$				Chemically synthesized $BaFeO_4$			
Optimal synthesis conditions	[NaOH]=14M, T=45°C $0.67 < n_{Ba}/n_{Fe} < 1.00$ J=2mA cm^{-2}				$3°C < T < 5°C$ $n_{Ba}/n_{Fe} = 1.5$			
Synthesis Time	45 - 75 minutes				17 hours (1000 minutes)			
Purity	93-95%				95-98%			
Yield	61-62%				90-97%			
FTIR analysis A_{1420}/A_{780}	0.10-0.16				0.08-0.12			
ICP analysis Molar Ratio	K 0.027	Na 0.004	Ba 1.02	Fe 1.00	K 0.081	Na 0.02	Ba 1.02	Fe 1.00
	10^{-3} Cu < 4	10^{-3} Ni < 4	Mn 0.007	Sr 0.003	10^{-3} Cu < 2	10^{-3} Ni < 3	Mn 0.003	Sr 0.001

NOTE: Comparison includes synthesis conditions, yield, and the following analyses of the ferrate product: chromite Fe(VI) purity analysis, FTIR ratio of the 1420 and 780 absorption bands, and ICP elemental analyses normalized to Fe.

at higher temperature (45°C) to accelerate any decomposition, and to complete the analysis in a reasonable time period (1 month).

As can be seen, comparing the time-dependent purity of a similar initial purity chemically and electrochemically synthesized $BaFeO_4$, stored under air at 45°C, reveals a substantially higher stability for the direct electrochemically synthesized $BaFeO_4$. It was previously observed that the solid-state stability of the chemically synthesized $BaFeO_4$ salt may be improved by a variety of factors, including sonication during synthesis to increase the Ba^{2+} and K_2FeO_4 solution reaction rate, rapid drying at elevated temperature (50°C), and retaining 2-6% H_2O in the dried salt (*10*). A principle factor observed for improving stability of the chemically synthesized $BaFeO_4$ salt is minimizing the exposure time of the salt to water during the Ba^{2+}/K_2FeO_4 conversion step. K_2FeO_4 is insoluble in acetonitrile, ACN, and a wide variety of nonaqueous salts (*22*), and a slurry of K_2FeO_4 and ACN does not react with $Ba(CH_3COO)_2$ or other anhydrous barium salts. However, the conversion reaction does progress quickly when water is introduced to the slurry, resulting in the salt with stability improved compared to the pure aqueous phase reaction as seen in Figure. 6.4. As seen in the figure, the highest stability is observed for the direct electrochemical synthesized $BaFeO_4$.

The improved stability characteristic of the electrochemically synthesized material is unexpected considering the high similarity in XRD and FTIR spectra

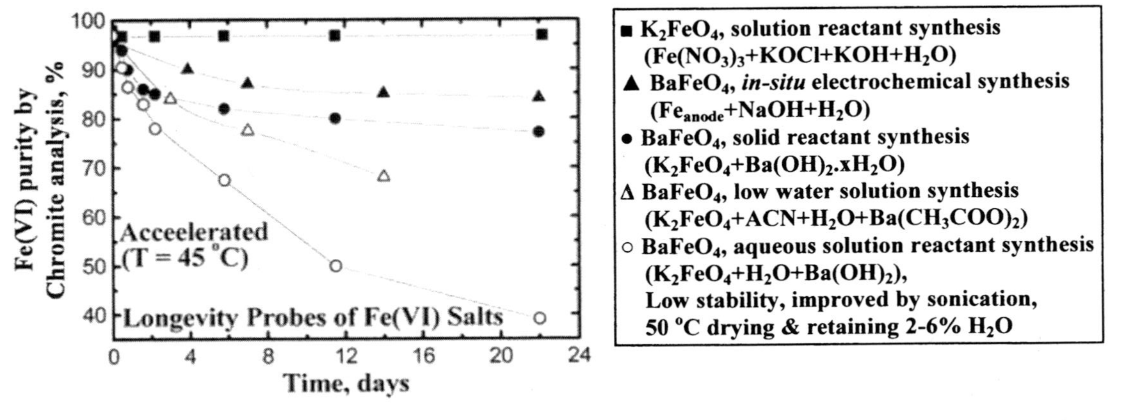

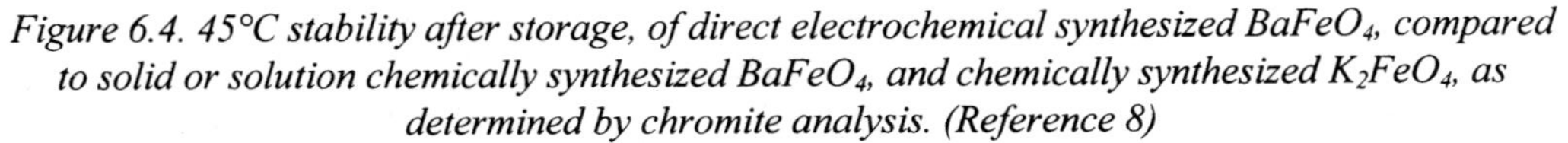

Figure 6.4. 45°C stability after storage, of direct electrochemical synthesized $BaFeO_4$, compared to solid or solution chemically synthesized $BaFeO_4$, and chemically synthesized K_2FeO_4, as determined by chromite analysis. (Reference 8)

of the differently prepared substances. However, SEM results indicate that the grain size of the electrochemical synthesized $BaFeO_4$ is substantially smaller, as shown in Figure 6.5 (*11*). One interesting observation toward an interpretation of the different stabilities evident is the higher trace concentrations of Mn for the *in-situ* electrochemically compared to chemically synthesized $BaFeO_4$ determined from the ICP elemental analyses of the different products (Table 6.2). Manganese can be concurrently oxidized with the iron to form high valence manganese oxide salts. Based on our previous experience with manganese additives (*19, 23*), they can stabilize Fe(VI) salts by the renewal of decomposed Fe(III), according to reactions such as:

$$3MnO_4^{2-} + H_2O + Fe_2O_3 \rightarrow 3MnO_2 + 2OH^- + 2FeO_4^{2-} \qquad (6.5)$$

Figure 6.5. SEM of chemical (a) and electrochemical (b) synthesized $BaFeO_4$. Grain size is stable before and after 45°C, 30 day storage. (Reference 11)

7. Preparation of Fe(III/VI) Thin Films

Fe(VI) salts, synthesized by solution phase, solid phase and electrochemical method, have been introduced as a novel series of charge storage super-iron salts (*4-32*). These salts exhibit up to three electrons of charge storage occurring at a single, electropositive cathodic potential. Whereas, facile, primary charge transfer has been extensively demonstrated, reversible charge transfer of these salts has been problematic. In principle, a sufficiently thin film Fe(III/VI) cathode should facilitate electronic communication with a conductive substrate to sustain cycled charge storage or reversible deposition. However, a variety of Fe(VI) thin films, formed by pressure and/or mix with a granular conductor such as small grain carbons, (*26*) had passivated upon charge cycling. We observed that rechargeable ferric films can be generated, formed by electrodeposition onto conductive substrates from solution phase Fe(VI) electrolytes (*11, 12*).

Preparation of Fe(VI/III) Reversible Film on Smooth Pt Substrate

A thin film, conducive to Fe(VI) charge cycling, is generated from micropipette controlled, microliter volumes of dissolved Fe(VI) in alkaline solution. The solution is placed on a Pt foil electrode, and isolated (by cation selective, alkali resistant, Nafion 350 membrane) from a larger volume, which contains an immersed reference electrode and a Pt gauze counter electrode in alkaline solution. Figure 7.1 presents two representative examples of the cathode formation, including deposition of either a 410 or 110 nM film, each with a surface area of 0.3 cm^2 on Pt foil, and with respective film thicknesses of approximately 0.8 μm and 0.2 μm. As illustrated in the bottom and top of the figure, either potentiostatic, or galvanostatic depositions approach 100% coulombic efficiency. The potentiostatic process for cathode film formation is preferred, as the galvanostatic deposition necessitates multiple constant current steps, and longer time for film completion. The film may be cycled *in-situ*, or removed, and washed with de-ionized water, and returned to the cell. It has been observed that the film is also equally active, whether formed from a solution containing dissolved, chemically synthesized Fe(VI) salts, or an electrochemically formed solution of dissolved Fe(VI) (*11*). The choice of alkaline Fe(VI) solution (*e.g.* Li_2FeO_4, Na_2FeO_4 or K_2FeO_4, in LiOH, NaOH or KOH) and concentration, permits variation of the thickness and composition of the generated film.

Whereas a one micron Fe(III/VI) layer, deposited on Pt foil by this technique, can support > 1000 reversible charge/discharge cycles to 30% DOD, the film passivates within several cycles at deeper discharge levels. However a thinner film delays the onset of passivation. A 0.1 μm Fe(VI) film sustains 200 cycles of 50% DOD, or 20 cycles of deeper (80%) DOD, after which the film passivates.

It was also observed that a nanofilm (for example, a 3 nm thickness Fe(VI) film), formed by the same technique is highly reversible. This Fe(III/VI) film on Pt foil is formed by reduction of 5 mM Na_2FeO_4 in 10 M NaOH at 100 mV *vs.* Ag/AgCl, followed by film inspection and galvanostatic cycling. The film is rigorous, and when used as a storage cathode, exhibits charging and discharging potentials characteristic of the Fe(VI) redox couple, and extended, substantial reversibility (can sustain over 100 charge/discharge cycles of 80% DOD).

Fe(VI/III) Reversible Film Preparation on Extended Conductive Matrixes

As demonstrated above, ultra-thin (eg. 3-nm-thick) super-iron films can sustain an extended reversibility. However, thicker films were not rechargeable due to the irreversible buildup of passivating (resistive) Fe(III) oxide, formed during film reduction (as illustrated in Scheme 7.1). Hence, thicker films had been increasingly irreversible (eg. either a 100 nm or a 1000 nm super-iron film

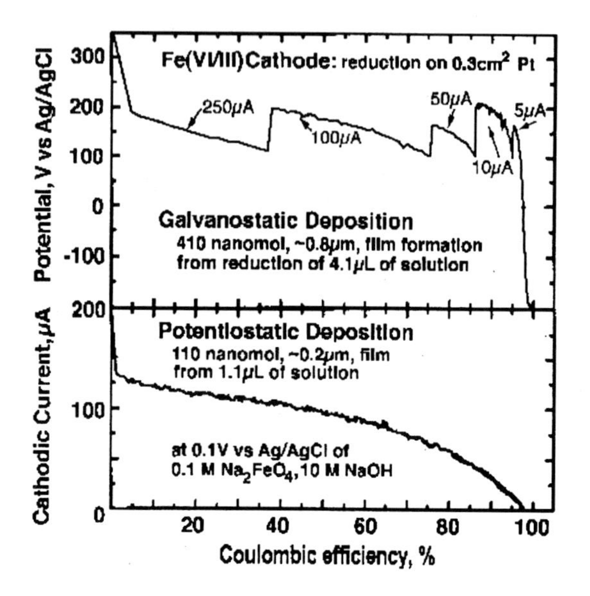

Figure 7.1 Electrochemical, galvanostatic or potentiostatic deposition of a thin Fe(VI/III) film onto Pt foil. Coulombic efficiency is integrated current during the deposition, normalized to the 3F available 0.1 M Fe capacity. (Reference 11)

had passivated after only 20 cycles or 2 cycles, respectively). We probed that preparation of Fe(III/VI) thin films on an extended conductive matrix can facilitate the film's reversibility (*12*).

Substrates for the film preparation are Pt or Ti foil. The extended conductive matrixes are prepared through depositing platinum (or Pt-Au co-deposition) on the substrates. Prior to deposition of platinum, platinum substrates were polished using aluminum oxide cloth (600 grit), etched in aqua regia (HCl/HNO_3 (3:1)) for 10-20 min, sonicated in distilled water for 20 min, and then electrochemically cleaned by cycling between -0.2 V and -1.5 V vs Ag/AgCl for 50 cycles at a scan rate of 500 mV s^{-1} in 1 M H_2SO_4. Titanium substrates were polished with 320, and then 600, grit aluminum cloth, then sonicated in 6 M HCl, followed by a deionized water rinse prior to other treatments.

On platinum substrates, Pt was potentiostatically deposited in a three-electrode cell at 0.2 V vs Ag/AgCl from aqueous 0.1 M H_2PtCl_6. The working electrode was a Pt foil with an exposed geometrical area of 8 cm^2, and circular Pt foil was used as a counter electrode. Pt was deposited onto Ti in a similar manner, but the deposition potential was -0.1 V vs Ag/AgCl, and a more concentrated, more conductive, 0.2 M solution of Na_2PtCl_6 in 0.1 M $HClO_4$ was

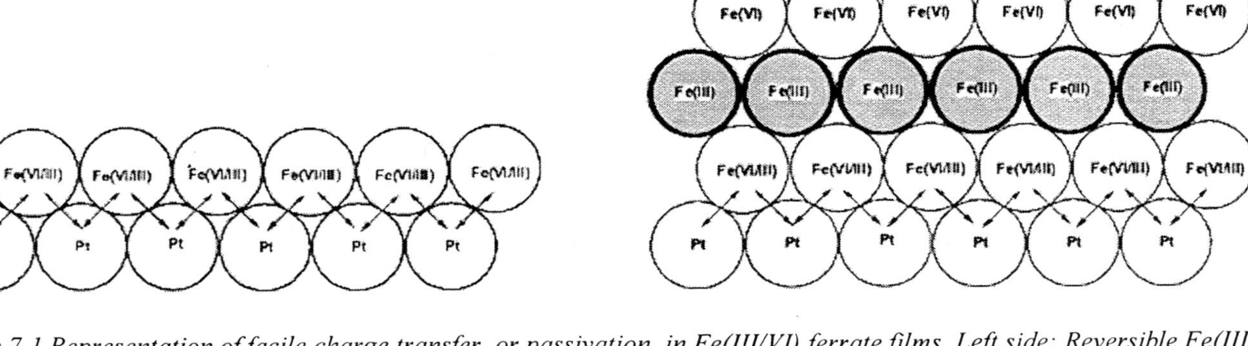

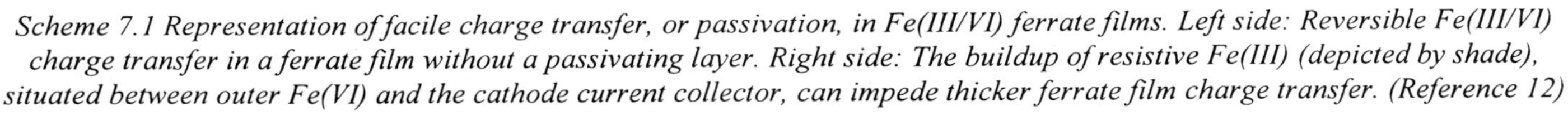

Scheme 7.1 Representation of facile charge transfer, or passivation, in Fe(III/VI) ferrate films. Left side: Reversible Fe(III/VI) charge transfer in a ferrate film without a passivating layer. Right side: The buildup of resistive Fe(III) (depicted by shade), situated between outer Fe(VI) and the cathode current collector, can impede thicker ferrate film charge transfer. (Reference 12)

used to obtain higher coverage (>5 mg cm^{-2}) of Pt deposit. Gold and platinum were codeposited onto Ti from a mixed solution of H_2PtCl_6 and $HAuCl_4$ (0.1:0.1 M) in 0.1 M $HClO_4$, again at -0.1 V vs Ag/ AgCl.

Super-iron films were electrodeposited from 30 mM K_2FeO_4 as dissolved in 10 M NaOH. This was chosen as the electrolyte due to the high solubility of super-iron salts in NaOH (K_2FeO_4 is highly insoluble in concentrated KOH electrolytes). This (millimolar level) K_2FeO_4/(molar level) NaOH, is effectively equivalent to an Na_2FeO_4/NaOH electrolyte containing 0.3% potassium. Electrodeposition was conducted in a cell formed from clamped, alkaline-resistant polypropylene squares. One square contained a cylindrical well, machined through the square. A well was formed by covering the horizontal bottom of the hole with the substrate and clamping the second square below the substrate. The 30 mM K_2FeO_4 in 10 M NaOH electrodeposition electrolyte was added to the well, covering the substrate, a working electrode with an exposed geometric area of 4 cm^2 of either Pt or Ti. A nickel counterelectrode was positioned just above the working electrode, and a Ag/AgCl reference electrode was immersed in the electrolyte. A deposition potential of 0.1 V vs Ag/AgCl was potentiostatically applied, which initiates ferrate deposition onto the platinum, or platinized platinum, substrates. A deposition potential of -0.01 V vs Ag/AgCl was required when platinized Ti was used as the substrate to obtain a durable super-iron film. Prior to use, the film electrode was cleaned with a 10 M NaOH (super-iron free) solution.

A substantial improvement to sustain thick film charge transfer is obtained when an extended conductive matrix was utilized as the film substrate. Specially, in a half-cell configuration, a 100 nm Fe(VI) cathode, electrodeposited on the extended conductive matrixes, sustained 100-200 reversible three-electrode galvanostatic charge/discharge cycles, and a 19 nm thin film cathode sustained 500 such cycles. Full-cell storage (anode/cathode) was also probed. In conjunction with a metal hydride anode, a 250 nm super-iron film cathode film sustained 40 charge/discharge cycles, and a 25 nm film was reversible throughout 300 cycles. The 2 orders of magnitude increase, up to 250 nm, in the rechargeable Fe(III/VI) cathode thickness. The facilitated super-iron charge transfer, upon platinization, as a result of the expanded conductive matrix to facilitate charge transfer, is represented in Scheme 7.2. Without direct contact with the substrate, the shaded Fe(III) centers in Scheme 7.1 had posed an impediment to charge transfer. This was partially (Scheme 7.2, left side) and fully alleviated (right side) by intimate contact with the enhanced conductive matrix, which maintains extended direct contact with the substrate.

Using Ti foil as the substrate, a 50 nm Fe(III) film on 7.5 mg Pt cm^{-2} platinized Ti can sustain over 200 charge/discharge cycles. The one drawback encountered with the platinized Ti substrate is the difficulty in obtaining larger Pt deposits, more than 7.5 mg cm^{-2}, which resulted in an unstable powdery surface, and because of this, it was difficult to deposit more than a 70 nm super-iron film on platinized Ti. However, a substantial improvement in the stability

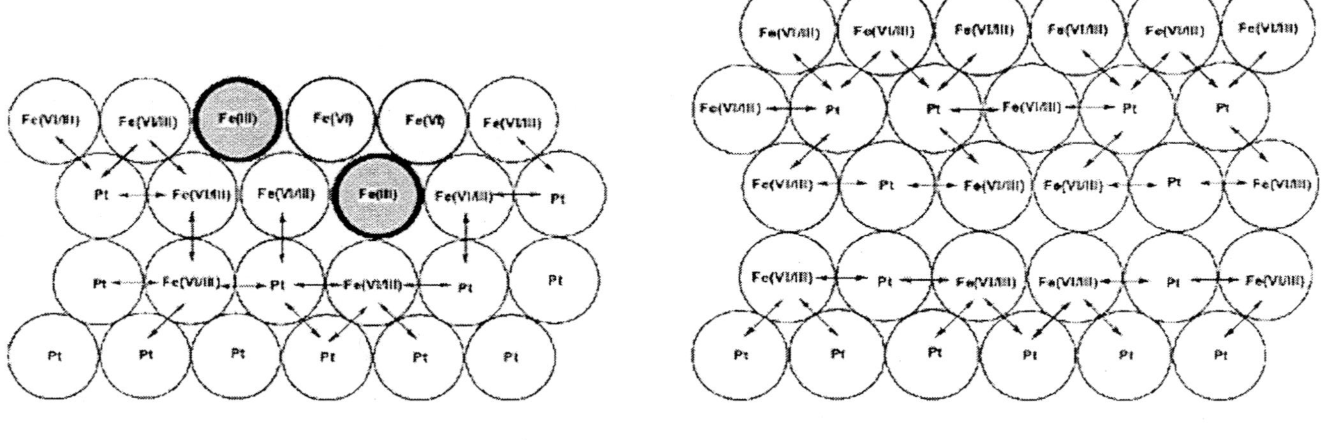

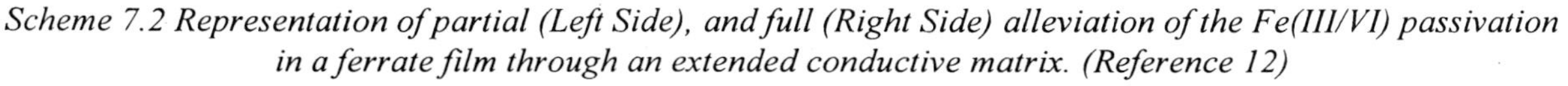

Scheme 7.2 Representation of partial (Left Side), and full (Right Side) alleviation of the Fe(III/VI) passivation in a ferrate film through an extended conductive matrix. (Reference 12)

and upper limit of the thickness of the film was observed when Pt-Au codeposited Ti surface was used as the substrate. A 300 nm super-iron film displayed a moderate charge/discharge cycle life of 20 (*12*).

8. Preparation of Zr or Mn Coated Fe(VI) Salts

We have developed a novel zirconia coating methodology, and only 1% of this zirconia coating dramatically improves the cathodic storage capacity of K_2FeO_4 (*13*). The charge transfer chemistry of synthesized super-iron salts has been probed in detailed (*4-32*). Among the super-iron cathodes, K_2FeO_4 exhibits higher solid state stability and higher intrinsic $3e^-$ capacity than pure $BaFeO_4$, but the rate of charge transfer is higher in the latter. At low current densities the cathode approaches the intrinsic, over 400 mAh/g storage capacity (*4*). However, the Fe(VI) forms a ferric overlayer (*11, 12*), upon storage the bulk Fe(VI) remains active, but the overlayer passivates the alkaline cathode towards further discharge. Whereas the fresh pure K_2FeO_4 discharges well, the capacity of K_2FeO_4 decreases seriously after storage (*13*).

The novel zirconia coating is derived from an organic soluble zirconium salt. A variety of coating solvents were studied, and of these, ether was chosen due to its facile evaporation (bp = 34 °C), $ZrCl_4$ solubility, and no reaction or solubility with the cathode materials. A 1 w% zirconia coating, prepared with 30 min coating time, is observed to have the best effect on charge retention of a coated cathode. 0.3 to 5% zirconia coatings were prepared. Excess coating is observed to the cathode overpotential, whereas, a lesser coating is insufficient for maximum charge retention. Specially, 1% ZrO_2 coating on K_2FeO_4 is prepared: 8 mg $ZrCl_4$ (AR grade, ACROS®) is dissolved in 8 ml ether (Fisher®), and stirred with 0.8 g solid (insoluble) K_2FeO_4 in air for 30 min, followed by vertex suction, then vacuum removal of the remaining solvent, and drying overnight. This ZrO_2 methodology is also available for coating other cathode materials, such as MnO_2, NiOOH, $NaBiO_3$, KIO_4 and AgO. ATR/FT-IR (Attenuated Total Reflectance Fourier Transform Infrared) spectrometry analysis results of several uncoated and coated cathode materials are shown in Figure 8.1. Pure ZrO_2 is also analyzed for comparison. The prominent 1608 cm^{-1} peak of the commercial $ZrCl_4$ fully disappears (not shown), and as seen in the figure, new 1396 and 1548 cm^{-1} peaks on the coated material coincides with the absorption spectra of pure $ZrO_2/Zr(OH)_4$ depending on extent of hydration: (*45*)

$$ZrCl_4 + 2O_2 \rightarrow ZrO_2 + 2Cl_2; \quad ZrO_2 + 2H_2O \leftrightarrow Zr(OH)_4 \qquad (8.1)$$

Stabilized zirconia was introduced as a pH sensor for high temperature aqueous systems, (*46*) and $Zr(OH)_4$ has long been known as a hydroxide ion conductor which will readily exchange between solution phase hydroxide, phosphate, fluoride, and sulfate (*47–50*). The insoluble Zr centers provide an

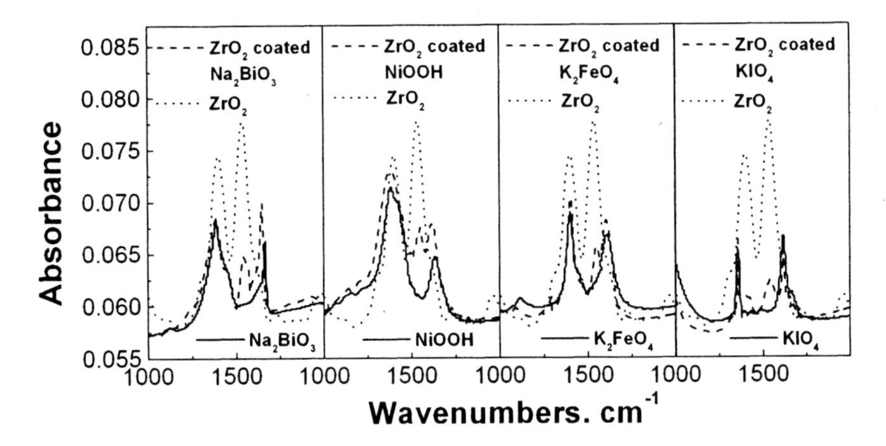

Figure 8.1 ATR/FT-IR spectra of zirconia coated and uncoated cathode materials $NaBiO_3$, NiOOH, K_2FeO_4 and KIO_4. Spectra of 5% coating included for emphasis; a 1% zirconia coating exhibits evident, but proportionally smaller, 1396 and 1548 cm^{-1} peaks. (Reference 13)

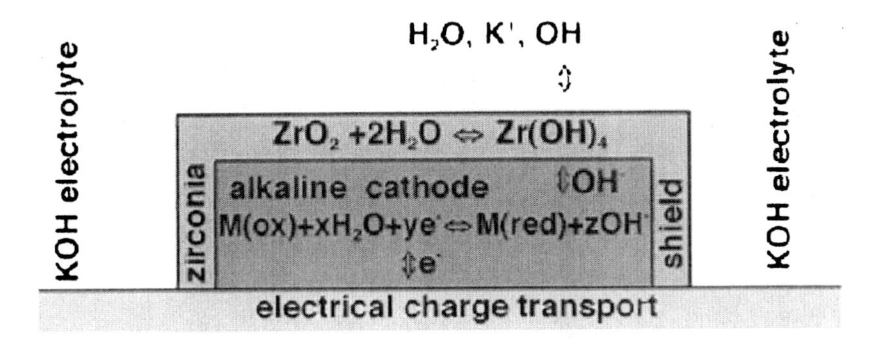

Scheme 8.1 Representation of zirconia alkaline cathode protection. (Reference 13)

intact shield, as represented in Scheme 8.1, and with eq. (8.1) right, a necessary hydroxide shuttle to sustain alkaline cathode redox chemistry.

The less soluble $BaFeO_4$ salt is expected to be intrinsically more stable than K_2FeO_4, but as shown in Figure 6.4, the chemically synthesized $BaFeO_4$ is somewhat less stable. We have found that a five percent coating of $KMnO_4$ improves the $BaFeO_4$ robustness. Specially, the coating is prepared as: 4.74 g $KMnO_4$ (30.0 millimoles) was dissolved by stirring in 0.33 liter of acetonitrile. 90.0 g (0.348 moles) $BaFeO_4$ powder is added. $BaFeO_4$ is insoluble in this solution and the suspension was stirred for 30 minutes. Acetonitrile is removed under vacuum, initially with stirring for 60 minutes to remove the majority of the acetonitrile. This is continued without stirring for 3 hours to fully dry the 5% $KMnO_4$ coated $BaFeO_4$.

9. Fe(VI) Analysis

This section focuses on the analysis of synthesized Fe(VI) compounds. Analytical Fe(VI) methodologies summarized are FTIR, ICP, titrimetric (chromite), UV/VIS, XRD, potentiometric, galvanostatic, cyclic voltammetry, constant current electrochemical discharge and Mössbauer analyses.

FTIR Fe(VI) Analysis

FTIR can provide a quantitative tool for the determination of Fe(VI) compounds when a suitable standard is added as a constant fraction to the Fe(VI) sample. Characteristics of a suitable Fe(VI) FTIR standard are: (i) that it is inert towards Fe(VI) compounds and (ii) has a clear, intrinsic IR spectra isolated from the Fe(VI) absorption bands. The small sample size, comprising only ~1% by weight of a typical FTIR KBr pellet, as well as the precise placement of the pellet in the spectrometer, provides challenges to the quantitative analysis of the spectra. These challenges are overcome by the use of such an added standard, prepared as a fixed concentration mixed with the sample, prior to extracting a segment of the sample to prepare the KBr pellet.

We have found barium sulfate is a suitable standard for Fe(VI) FTIR determination, which provides a reproducible, inert, and distinctive, but isolated, IR absorption. In this procedure for quantitative Fe(VI) analysis, detailed in reference *6*, a KBr pellet is formed which contains a known mass of $BaSO_4$, as well as a known mass of the sample to be analyzed, and the FTIR spectrum measured. As seen in Figure 9.1, and in the figure insert, the absorbance of $BaFeO_4$ at 780 cm^{-1}, relative to the $BaSO_4$ absorbance at 1079 (or 1183) cm^{-1}, of a fixed concentration or $BaSO_4$, grows in linear proportion to the $BaFeO_4$ concentration, and provides a route for quantitative analysis of the $BaFeO_4$

concentration. The analysis utilizes k, the $BaSO_4$ to $BaFeO_4$ conversion constant, defined by the absorptivity of $BaSO_4$ compared to that of $BaFeO_4$, in a standard sample containing equal $BaSO_4$ and $BaFeO_4$ weight fractions:

$$k\ (\text{wt\% } BaFeO_4 = \text{wt\% } BaSO_4) \equiv A_{BaSO_4}(1079\ \text{cm}^{-1}) / A_{BaFeO_4}(780\ \text{cm}^{-1}) \quad (9.1)$$

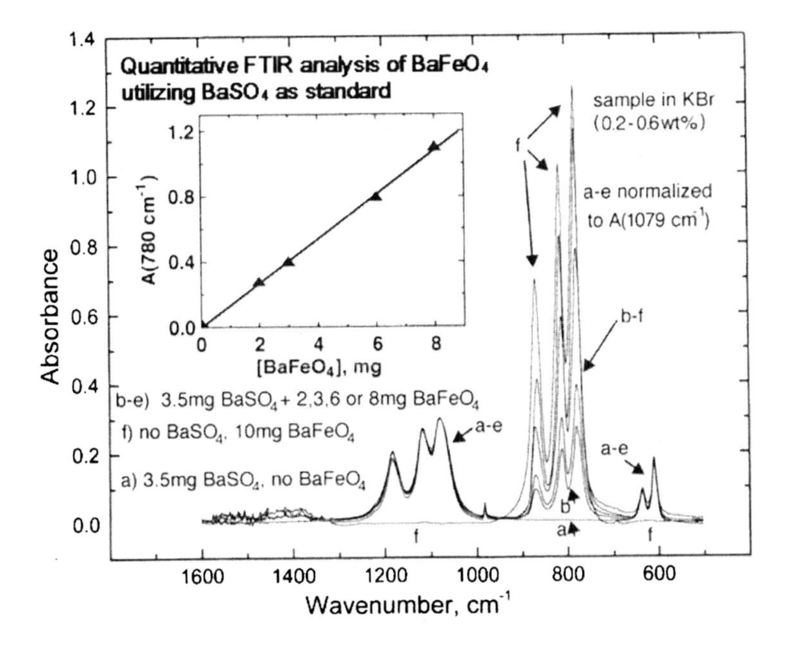

Figure 9.1. FTIR analysis of BaFeO4 utilizing a BaSO4 standard. (Reference 6)

From the magnitude of the absorption spectra in Figure 9.1, an appropriate methodology for FTIR analysis of an unknown in $BaFeO_4$ is suggested: (i) prepare a 100 mg of sample comprising two parts of the unknown with one part by weight of $BaSO_4$; thereby forming a sample with a constant fraction $f = 1/3$ in $BaSO_4$; (ii) prepare a KBr pellet containing ~1 wt.% of this admixture; (iii) measure and baseline correct the FTIR spectrum; (iv) calculate the relative $BaFeO_4$ to $BaSO_4$ absorption peaks, as R from the measured absorption peaks,

$$R = A(780\ \text{cm}^{-1}) / A(1079\ \text{cm}^{-1}) \quad (9.2)$$

which determines the $BaFeO_4$ concentration in the $BaSO_4$ sample admixture

$$[BaFeO_4,\ \text{wt:\% in } BaSO_4 \text{ admixture}] = 100\%\ Rkf \quad (9.3)$$

(v) convert this ratio, by $f/(1-f)$, to the $BaFeO_4$ composition in the unknown as

$$[BaFeO_4, \text{wt:\% determined in unknown}] = 100\% \, Rk \,/\, (f^1 - 1) \qquad (9.4)$$

where k is the $BaSO_4$ to $BaFeO_4$ conversion constant, which may be determined from a mixture containing equal weight fractions of $BaSO_4$ and $BaFeO_4$ and measured as $1/R$. More precisely, k, may be determined using the inverse slope of a calibration curve of $[BaFeO_4]/[BaSO_4]$ versus R measured in $f = 1/3$ $BaSO_4$ mixes, with KBr and also containing 0, 4, 10, 20, 33, 50, or 66 wt.% known, pure $BaFeO_4$ (e.g. 100 mg samples containing 33.3 mg $BaSO_4$, and either, 0, 4, 10, 20, 33.3, 50 or 66.7 mg $BaFeO_4$, with the remainder as added KBr). Calibration of k, will compensate for path length, spectrometer, $BaSO_4$ and Fe(VI) material composition variations (*e.g.*. a sample which contains additional Fe(VI) salts or other interferences). However, in the absence of a pure $BaFeO_4$ sample, a $k = 1.15$ is appropriate.

ICP Fe(VI) Analysis

Inductively coupled plasma analysis of K_2FeO_4 and $BaFeO_4$ samples was conducted with a ICP Perkin-Elmer Optima 3000 DV) to determine the relative elemental weight percent and mole percent compositions of the principal cations, and possible impurities, in the sample. Such conventional inductively coupled plasma analytical methodologies provide elemental composition information, but not information regarding a compound's valence state. Hence, these methodologies are convenient, but not specific, to Fe(VI) analysis, and are only briefly described in this section. These methodologies, or related atomic absorption or emission and x-ray fluorescence techniques, are important to determine total iron relative to other elements in a super-iron compound. From these values, the mole ratio of principal cations, the mass percent of the principal cations, and the maximum contribution of alternate cation impurities is determined.

Titrimetric (Chromite) Fe(VI) Analysis

Following the quantitative determination of total iron in a compound, for example as described by ICP analysis in the previous section, the extent of the iron existing in the Fe(VI) valence state can be determined by titrimetric chromite analysis. Alternately, if the type of Fe(VI) compound is known (for example as K_2FeO_4 or $BaFeO_4$), then the sample's mass yields the theoretical oxidation capacity (calculated as three equivalents per Fe), which in turn is compared with the chromite analyzed oxidation capacity.

The chromite analysis methodology varies with the specific Fe(VI) compound to be ascertained. For example, the highly insoluble $BaFeO_4$

compound must be heated during dissolution, and alternate competing oxidants must be removed. In this section are presented our optimized chromite methodologies for several Fe(VI) related materials. In each case, the Fe(VI) sample is dissolved into solution as FeO_4^{2-} to oxidize chromite, Cr(III) to chromate Cr(VI):

$$Cr(OH)_4^- + FeO_4^{2-} + 3H_2O \rightarrow Fe(OH)_3(H_2O)_3 + CrO_4^{2-} + OH^- \quad (9.5)$$

The generated chromate is then titrated with a standard ferrous ammonium sulfate solution, using a (0.5 g per 100 ml) aqueous sodium diphenylamine sulfonate indicator solution. Each of the examples of Fe(VI) chromite analysis methodologies described below utilizes analytical reagents in doubly deionized water. Further details, including the chromite analysis of a coated Fe(VI) salts is presented in reference 6.

K_2FeO_4 Chromite Analysis

Dissolution of the solid K_2FeO_4 sample into solution: A concentrated NaOH solution is prepared from 720 g NaOH in 1 liter of solution, and pretreated to remove potential interferences by reaction (boiling) with 0.1 g K_2FeO_4 until the characteristic purple/black FeO_4^{2-} color disappears. A Cr(III) solution is prepared from 16.66 g $CrCl_3 \cdot 6H_2O$ per 100 ml solution. A 20 ml of the pretreated, concentrated NaOH solution is combined with 5 ml of the Cr(III) solution, a further 5 ml of water is added, and the solution is cooled to room temperature in an ice bath. A 150 - 200 mg K_2FeO_4 sample is added to this solution, which is stirred for 30 minutes, until full K_2FeO_4 dissolution.

Preparation of the solution to be titrated: Prior to titration, to the K_2FeO_4 solution (or 25 ml of a known Cr(VI) solution) sequentially is added: 150 ml water, followed by 65 ml of 1/5 H_2SO_4 (prepared from a 1/5 dilution of 95-97% H_2SO_4), then 15 ml of a H_2SO_4 /H_3PO_4 solution (prepared from 240 ml water, 150 ml 85% H_3PO_4, and 60 ml 95-97% H_2SO_4), and 7-8 drops of the sodium diphenylamine sulfonate indicator solution.

Preparation of the 0.085 N ferrous ammonium sulfate titrant: A 250 ml solution is prepared with 8.34 g $Fe(NH_4)_2(SO_4)_2 \cdot 6H_2O$. The precise normality of the ferrous ammonium sulfate solution is determined by using this solution to titrate a Cr(VI) known sample solution. The Cr(VI) solution is prepared from 25 ml of 0.085 *N* $K_2Cr_2O_7$ solution (prepared as 1.042 g $K_2Cr_2O_7$ per 250 ml solution), and is treated as described above prior to titration. From the concentration of the $K_2Cr_2O_7$ solution, the normality, *N*, of the ferrous ammonium sulfate is then determined from *V*(ml), the volume of $Fe(NH_4)_2(SO_4)_2 \cdot 6H_2O$ titrant needed to yield a color change from purple to green, as $N = 0.085 \times 25/V$.

Determination of the K_2FeO_4 purity: The K_2FeO_4 is inserted into solution and the solution prepared for titration as described above. From *V*(ml), the volume of $Fe(NH_4)_2(SO_4)_2{\cdot}6H_2O$ titrant needed in the titration to a color change from purple to green, and the normality, *N*, of the ferrous ammonium sulfate, is then determined the K_2FeO_4 purity, P(%,) as:

$$P(\%) = 100\ VN\ Fw\ /3\ m \qquad (9.6)$$

using the Fw = 198.04 g/mole formula weight of K_2FeO_4, n=3 equivalents per Fe(VI), and the sample mass m(mg).

$BaFeO_4$ chromite analysis

As with the K_2FeO_4 sample, a concentrated NaOH solution is prepared from 720 g NaOH in 1 liter of solution, and pretreated by boiling with 0.1 g K_2FeO_4 until the solution color disappears. A Cr(III) solution is again prepared from 16.66 g $CrCl_3{\cdot}6H_2O$ per 100 ml solution. Now however, 20 ml of the pretreated, concentrated NaOH solution is combined with 3 ml of the Cr(III) solution, and the solution is cooled to room temperature in an ice bath. A 100-110 mg $BaFeO_4$ sample is added to this solution, which is stirred for 30 minutes. Following addition of 50 ml distilled water, the mixture is heated to 90-95°C (using a water bath) for 1 hour. After cooling, to the solution is added a cooled mixture of 240 ml water, 150 ml 85% H_3PO_4, 60 ml 95-97% H_2SO_4, and 7-8 drops of sodium diphenylamine sulfonate.

The $BaFeO_4$, inserted into solution and prepared for titration, is titrated with a ferrous ammonium sulfate solution prepared and standardized as described in the previous K_2FeO_4 chromite analysis section. From *V*(ml), the volume of $Fe(NH_4)_2(SO_4)_2{\cdot}6H_2O$ titrant needed to yield a color change from purple to green, and the normality, *N*, of the ferrous ammonium sulfate, is then determined from equation 9.6 using the 257.11 g/mole formula weight of $BaFeO_4$, and the sample mass m(mg).

UV/VIS Fe(VI) Analysis

Fe(VI), dissolved as FeO_4^{2-}, has a distinctive UV/Vis spectrum. However, the quantitative analysis of solid Fe(VI) salts by dissolution and UV/Vis absorption spectroscopy is limited by (i) the relative insolubility of Fe(VI) salts such as $BaFeO_4$ in aqueous solutions (*18*), the general tendency of Fe(VI) salt insolubility in a wide variety of organic solvents (*15*), and the tendency of dissolved aqueous Fe(VI) salts to decompose in aqueous solutions other than specific electrolytes, such as highly concentrated KOH electrolytes, and electrolytes specifically excluding Ni(II) and Co(II) catalysts (*4*). The decomposition with water takes the form of:

$$2FeO^{2-} + 3H_2O \rightarrow 2FeOOH + 3/2O_2 + 4OH^- \tag{9.7}$$

When occurring, the decomposition tends to lead to the formation of colloidal ferric(III) oxide, and to limit the time available for Fe(VI) analysis. Colloidal ferric oxide interference is minimized by a 385 nm baseline correction, and/or solution centrifugation prior to spectroscopic analysis. The visible absorption spectrum of Fe(VI) in highly alkaline solution exhibits a maximum at 505 nm, an absorption shoulder at 570 nm and two minima at 390 nm and 675 nm. The molar absorptivity measured at 505 nm is 1070±30 M^{-1} cm^{-1}. The molar absorptivity remained constant up to 200 mM. Similarly, at a fixed ferrate concentration, the measured absorbance was independent of alkali hydroxide cation and concentration. Hence, to within better than 5%, the 505 nm absorbance of 2 mM K_2FeO_4 is the same in 5 M Li, Na, K and Cs hydroxides, and also the same in 5-15 M NaOH, 5-13.5 M KOH and 5-15 M CsOH.

Solution pre-treatment (such as inducing decomposition of 2mM K_2FeO_4 in KOH solution at 60° to 80°C, followed by removal of the decomposition products) can improve the subsequent Fe(VI) stability. The solubility of Fe(VI) salts is low in this solution, but the stability of the dissolved FeO_4^{2-} is high (*4*). A useful procedure for UV/Vis analysis of a solid Fe(VI) sample is to first prepare a solution of this pre-treated saturated KOH. Dissolve a precise mass (2 to 3 mg) of the sample in 10 ml (~1mM Fe(VI)) of the pre-treated saturated KOH solution. Measure the UV/Vis spectrum, and baseline correct to avoid 385 nm colloidal ferric oxide interference. Determine the absorbance at 505 nm and convert to FeO_4^{2-} concentration. Note, that the presence of $Ba(OH)_2$ will further diminish FeO_4^{2-} solubility. This increases the challenge to dissolve a solid Fe(VI) sample. $BaFeO_4$ is insoluble in aqueous saturated $Ba(OH)_2$. In KOH solutions also containing saturated $Ba(OH)_2$, $BaFeO_4$ dissolves only at the submillimolar level (*18*).

XRD Fe(VI) Analysis

We have measured Fe(VI) powder XRD data using a Philips Analytical X-ray, B. V. diffractometer, operating with CuKα radiation, with a flat sample holder mounted on a SPG spectrogoniometer (*6*). Measured powder x-ray diffraction spectra exhibit little variation using salts ranging averaging in particle size from 35 to 100 μm, or measured with a wide range of (2θ) scan rates. The obtained XRD spectra are often consistent with an orthorhombic crystal system with the spaces group D_{2h} (Pnma) (*6*). Although often used as a tool for qualitative rather than quantitative, analysis, we also use XRD to distinguish between coated barium ferrate from pure barium ferrate. For example, a several percent coating from $KMnO_4$ is distinguishable as a low level of the expected, known $KMnO_4$ XRD pattern superimposed on the regular $BaFeO_4$ pattern. Graphites and carbon blacks added to an Fe(VI) mix in the preparation of a cathode do not significantly interfere with the observed Fe(VI) XRD patterns.

Fe(VI) Electrochemical Analyses

Electroanalytical techniques to probe Fe(VI) compounds can be conveniently categorized as either solution phase (dissolved Fe(VI)) or solid cathode techniques. Solution electrochemical techniques such as potentiometric, galvanostatic and cyclic-voltammetry methods are convenient, and are carried out using a conventional potentiostat (AFCBP1 Pine bipotentiostat) in a three-electrode electrochemical cell. The working and counter electrodes are made of bright platinum. The reference electrode was saturated calomel electrode (SCE).

The solid cathode methodologies are largely free of the challenges associated with decomposition challenges which may occur with the solution phase techniques, and provide a direct probe of the kinetic (as a function of fixed load, current or power densities) and thermodynamic (as a function of cell potential) oxidizing capacity of the Fe(VI) compounds. Various different Fe(VI) solid cathode discharge methodologies have been described (*4-32*) before, in which the potential was measured in time via LabView Data Acquisition on a PC, and cumulative discharge, as ampere hours, Ah, and watt hours, Wh, determined by incremental integration. A constant current discharge mode will be demonstrated in this section.

Fe(VI) potentiometric analysis

The measured redox potentials, at platinum electrode, of K_2FeO_4 in various NaOH solutions was measured yielding $E°(\text{in NaOH})_{FeO_4^{2-}} = 0.66 \pm 0.01$ V (SHE). As expected, redox potential shifts to more positive values with log of the increase in ferrate concentration and was analyzed in detail in reference 6. In aqueous alkali and alkali earth hydroxide electrolytes, we have observed that a variety of Fe(VI) compounds, over a wide concentration range, exhibit a potential varying from 0.55 to 0.75 V vs SHE.

Fe(VI) solution phase galvanostatic analysis

Galvanostatic reduction of dissolved ferrate yields a direct electrochemical measurement of the oxidation state of iron in ferrate. Figure 9.2 presents the time evolution of the potential, during ferrate reduction. A $c_{initial}$ = 2 mM potassium ferrate solution in v = 3 mL of 15 M NaOH is reduced at a current density J = 10 μA/cm^2. Integration of the charge transferred yields the relative oxidation state, $\Delta q'$, where F is the Faraday constant and t is time:

$$\Delta q' = t \times J/(c_{initial} \times v \times F) \tag{9.8}$$

and which may be compared to the intrinsic charge of the insoluble Fe(III) product. A planar platinum electrode does not provide the optimum surface to probe solution phase Fe(VI) reduction. The Pt surface tends to passivate in time as the reduced Fe(III) layer builds on the surface. This passivation is alleviated by (i) minimizing the thickness of the layer, employing low volumes and low concentrations of dissolved Fe(VI) salts, (ii) using low initial current densities, such as 10 μA cm^{-2}, and (iii) further diminishing the current by an order of magnitude as the overpotential increases towards the end of the reduction. Under these conditions, and as seen in the curve in Figure 9.2, the oxidation state of the starting material approaches Fe(VI) in accord with equation 9.8.

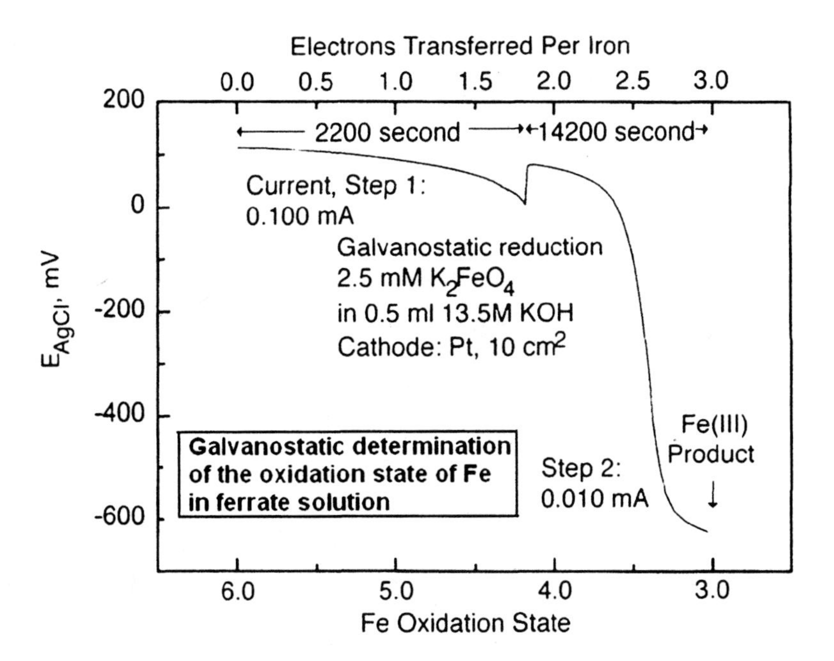

Figure 9.2. Galvanostatic determination of the measurement of the oxidation state of iron in ferrate(VI) solutions. (Reference 6)

Fe(VI) cyclic voltammetry analysis

Representative voltammetric curves for the reduction of K_2FeO_4 dissolved in 15 M NaOH at a Pt electrode are shown in Figure 9.3 and are further detailed in reference 6. The negative potential scan reveals a cathodic current of ferrate(VI) reduction at potentials less positive than 200 mV, and cathodic reduction of ferrate(VI) proceeds with an overvoltage of approximately 150 mV. O_2 evolution in the oxidation sweep interferes with oxidation of ferrate(III). The peak cathodic current density increased linearly with ferrate concentration and was proportional to (scan rate)$^{1/2}$, indicating diffusion limitation of ferrate(VI) reduction.

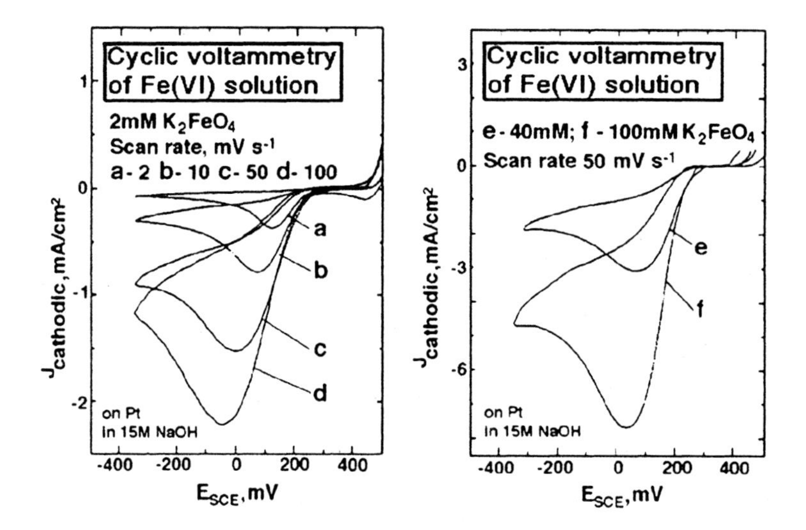

Figure 9.3. Cyclic voltammetry of ferrate(VI) solutions. Left: Variation with potential scan rate. Right: Variation with ferrate(VI) concentration. (Reference 6)

Fe(VI) solid cathode discharge analysis

Using a cathode mixed with a high level of (30% by weight) of an effective conductive matrix (1μm graphite), the full 406 mAh/g storage capacity of K_2FeO_4 is obtained during discharge, as seen in Figure 9.4 (and compared to an MnO_2 cathode reduction). The discharge is in accord with the Fe(VI → III) $3e^-$ reduction described in equation 9.9. Coulombic efficiency will be further affected by other additives than graphite, and control of packing, electrolyte and particle size (*6*).

$$FeO_4^{2-} + 5/2H_2O + 3e^- \rightarrow 1/2Fe_2O_3 + 5OH^- \quad (9.9)$$

Figure 9.4 summarizes constant current discharge results obtained in a relatively small button configuration cell. Useful analytical information may also be obtained with larger cells packed with a Fe(VI) compound cathode mix, using cell configurations such as the conventional AAA dimensioned cylindrical cell, and again using a KOH electrolyte, and the remaining components removed from a commercial alkaline cell (including the Zn gel anode, the separators, the anode current collector, and the cathode current collector (the outer case).

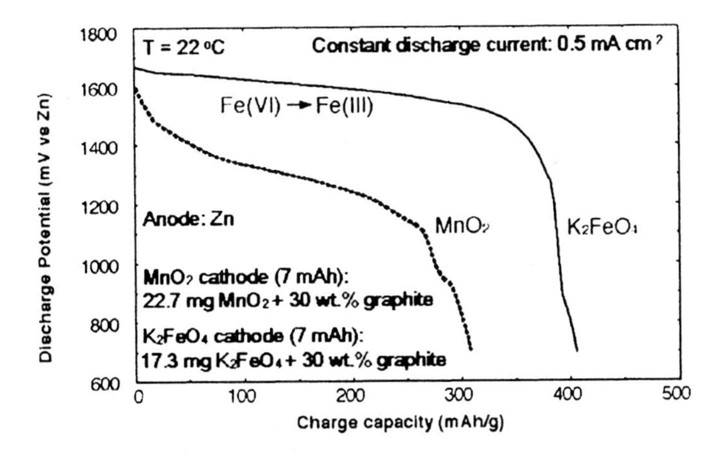

Figure 9.4. Constant current discharge of a K_2FeO_4 cathode compared to a MnO_2 cathode-limited Zn cell. Cell discharge is measured with Labview Software interfaced Pine model AFCBP1 bipotentiostat. In each case, a commercial 1.1 cm diameter button cell, containing excess Zn, is opened, the cathode removed, and replaced with the 7 mAh cathode. (Reference 6)

Fe(VI) Mössbauer Analysis

As shown in Figure 9.5, Mössbauer analysis is an excellent means to distinguish between the trivalent and hexavalent states of iron. As recently reported (*29*), Mössbauer studies were performed using a conventional constant acceleration Mössbauer drive and a 50 mCi^{57} Co:Rh source. The velocity calibration and isomer shift reference are those of a thin foil of α-iron. The spectra were analyzed by a least-squares fit program to several quadrupole doublets. The relative areas of the doublets were taken as the relative abundances of the Fe(VI) and Fe(VI) components. *In-situ* electrochemical Mössbauer studies have also been performed using a specially modified cell in a synchrotron spectroelectrochemical (*26*). The source was 25 mCi^{57} Co in a Rh lattice. Velocity calibration of the spectrometer was based on the hyperfine splitting of α-Fe at room temperature (*26*).

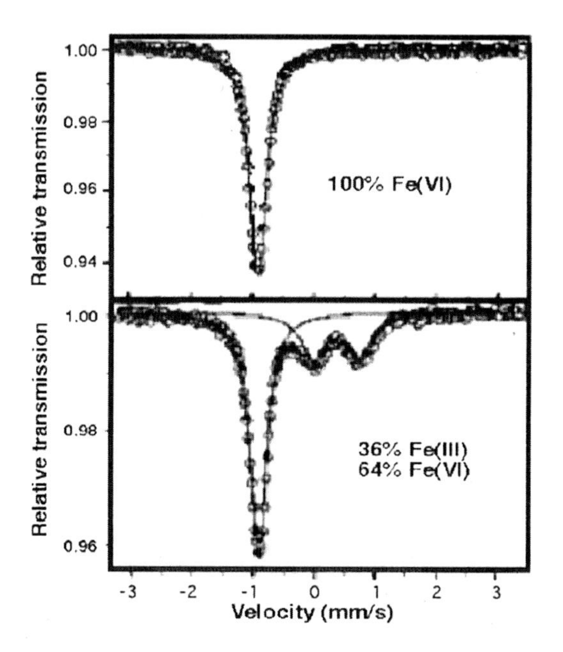

Figure 9.5. Mössbauer spectra of a sample containing pure K_2FO_4 (Fe(VI), or K_2FO_4 partially reduced to the Fe(III) valence state (Reference 29).

Conclusions

In this review, two conventional, as well as two improved Fe(VI) synthetic routes are presented. The conventional syntheses include solution phase oxidation (by hypochlorite) of Fe(III), and the synthesis of less soluble super-irons by dissolution of FeO_4^{2-} and precipitation with alternate cations. The new synthetic pathways include a solid synthesis route for Fe(VI) salts, and the direct (*in-situ*) electrochemical synthesis of Fe(VI) salts.

High purity (96.5-99.5%) and high solid state stable K_2FeO_4 salts was synthesized through chemically oxidation of Fe(III) salts in alkaline solution by hypochlorite. Then the solid K_2FeO_4 was used to synthesize other Fe(VI) salts through the ion exchange with alternate cations to precipitate less soluble ferrates in the alkaline solution medium. High purity alkali ferrates Cs_2FeO_4, Rb_2FeO_4, and $K_xNa_{(2-x)}FeO_4$, low purity Li_2FeO_4, as well as high purity alkali earth Fe(VI) salts $BaFeO_4$, $SrFeO_4$, and also Ag_2FeO_4, have been successfully synthesized through this solution phase precipitation synthesis way. Unlike solution phase synthesis, an alternate rapid synthesis of Fe(VI) salts using all solid state room temperature reactants can eliminate synthesis steps. High purity $BaFeO_4$ salt generated by means of solid phase synthesis was summarized. High purity $BaFeO_4$ was efficiently synthesized electrochemically, directly as a solid phase, from concentrated alkaline media. Direct electrochemically synthesized $BaFeO_4$ exhibits smaller grain size and enhanced stability towards prolonged thermal decomposition. Direct electrochemical synthesis of Fe(VI) compounds has several advantages of shorter synthesis time, simplicity, reduced costs (no chemical oxidant is required) and providing a possible pathway towards more electroactive and thermal stable Fe(VI) compounds.

Ultra-thin (3-nm) Fe(III/VI) films exhibited a high degree of three-electron reversibility (throughout 100-200 charge/discharge cycles), However, thicker films had been increasingly passive toward the Fe(VI) charge transfer. Extended conductive matrix facilitates a 2 orders of magnitude enhancement in charge storage for reversible Fe(III/VI) thin films. High-capacity (Fe(III/VI) super-iron films were electrochemically deposited by electrochemical reduction of Na_2FeO_4 with an intrinsic 3 e^- cathode storage of 485 mAh g^{-1}. Films were alternatively deposited on either smooth conductive substrates or on extended conductive matrixes, composed of high-surface-area Pt, Pt on Ti, and, Pt/Au on Ti. Using extended conductive matrixes, a 100 nm Fe(VI) cathode sustained 100-200 reversible charge/discharge cycles, and a 19 nm thin film cathode sustained 500 such cycles. Building a full-cell storage (anode/cathode) system in conjunction with a metal hydride anode, a 250 nm super-iron cathode film sustained 40 charge/discharge cycles, and a 25 nm film was reversible throughout 300 cycles.

A low level (1%) zirconia coating derived from $ZrCl_4$ through organic solvent medium significantly stabilized the high capacity super-iron cathode, K_2FeO_4, and enhanced the stability and electrochemical capacity of super-iron batteries. A 5% $KMnO_4$ coating prepared through acetonitrile solvent can improve the robustness of chemical synthesized $BaFeO_4$.

Various Fe(VI) analytical methodologies, include FTIR, titrimetric (chromite), UV/VIS, XRD, ICP, potentio-metric, galvanostatic, cyclic voltammetry, constant current electrochemical discharges and Mössbauer analysis, were summarized. The FTIR methodology becomes quantitative with use of an internal standard such as added barium sulfate. Electrochemical techniques which utilize a solid cathode and spectroscopic techniques which utilize a solid sample are preferred over solution phase techniques. The chromite methodologies can be modified to determine the extent of Fe(VI $\rightarrow$ III) oxidation power in both unmodified or coated Fe(VI) compounds. Mössbauer analysis was an excellent means to distinguish between the trivalent and hexavalent states of iron.

References

1. Carr J. D.; Kelter P. B.; Tabatabai A.; Splichal D.; Erickson J.; Mclaughlin C. W. *Proceedings of the 5th Conference, Water Chlorination;* Jolly R. L., Eds.; Lewis: Chelsea, MI, 1985; p. 1285.
2. Sharma V. K.; Smith J. O.; Millero F. J. *Environ. Sci. Tech.* **1997**, *31*, 2486.
3. Licht S.; Yu X. *Environ. Sci. Technol.* **2005**, *39*, 8071.
4. Licht S.; Wang B.; Ghosh S. *Science* **1999**, *285*, 1039.
5. Licht S.; Naschitz V.; Ghosh S.; Liu B.; Halperine N.; Halperin L.; Rozen D. *J. Power Sources* **2001**, *99*, 7.
6. Licht S.; Naschitz V.; Lin L.; Chen J.; Ghosh S.; Lui B. *J. Power Sources* **2001**, *101*, 167.
7. Licht S.; Tel-Vered R.; Halperin L. *Electrochem. Com.* **2002**, *4*, 933.
8. Licht S.; Tel-Vered R.; Halperin L. *J. Electrochem. Soc.* **2004**, *151*, A31.
9. Tel-Vered R.; Rozen D.; Licht S. *J. Electrochem. Soc.* **2003**, *150*, A1671.
10. Licht S.; Naschitz V.; Wang B. *J. Power Sources*, **2002**, *109*, 67.
11. Licht S.; Tel-Vered R. *Chem. Comm.* **2004**, 628.
12. Licht S.; DeAlwis C. *J. Phys. Chem. B* **2006**, *110*, 12394.
13. Licht S.; Yu X.; Zheng D. *Chem. Comm.* **2006**, 4341.
14. Licht S.; Naschitz V.; Rozen D.; Halperin N. *J. Electrochem. Soc.* **2004**, *151*, A1147.
15. Licht S.; Wang B. *Electrochem. Solid- State Lett.* **2000**, *3*, 209.
16. Licht S.; Naschitz V.; Ghosh S.; Lin L.; Lui B. *Electrochem. Com.* **2001**, *3*, 340.
17. Licht S.; Yang L.; Wang B. *Electrochem. Comm.* **2005**, *7*, 931.
18. Licht S.; Wang B.; Ghosh S.; Li J.; Naschitz V. *Electrochem. Comm.* **1999**, *1*, 522.
19. Licht S.; Wang B.; Xu G.; Li J.; Naschitz V. *Electrochem. Comm.,* **1999**, *1*, 527.
20. Licht S.; Wang B.; Li J.; Ghosh S.; Tel-Vered R. *Electrochem. Comm.* **2000**, *2*, 535.
21. Licht S.; Ghosh S.; Dong Q. *J. Electrochem. Soc.* **2001**, *148*, A1072.
22. Licht S.; Naschitz V.; Ghosh S. *Electrochem. Solid-State Lett.* **2001**, *4*, A209.
23. Licht S.; Ghosh S.; Naschitz V.; Halperin N.; Halperin L. *J. Phys. Chem., B* **2001**, *105*, 11933.
24. Licht S.; Ghosh S. *J. Power Sources* **2002**, *109/2*, 465.
25. Licht S.; Naschitz V.; Ghosh S. *J. Phys. Chem. B* **2002**, *106*, 5947.
26. Ghosh S.; Wen W.; Urian R. C.; Heath C.; Srinivasamurthi V.; Reiff W. M.; Mukerjee S.; Naschitz V.; Licht S. *Electrochem. Solid-State Lett.* **2003**, *6*, A260.
27. Nowik I.; Herber R. H.; Koltypin M.; Aurbach D.; Licht S. *J. Phys. Chem. Solids* **2005**, *66*, 1307.
28. Koltypin M.; Licht S.; Tel-Vered R.; Naschitz V.; Aurbach D. *J. Power Sources* **2005**, *146*, 723.

29. Koltypin M.; Licht S.; Nowik I.; levi E.; Gofer Y.; Aurbach D. *J. Electrochem. Soc.* **2006**, *153*, A32.
30. Lee J.; Tryk D.; Fujishima A.; Park S. *Chem. Comm.* **2002**, 486.
31. Yang W.; Wang J.; Oan T.; Xu J.; Zhang J.; Cao C. *Electrochem. Com.* **2002**, *4*, 710.
32. De Konnick M.; Belanger D. *Electrochim. Acta* **2003**, *48*, 1435.
33. Herber R. H.; Johnson D. *Inorg. Chem.* **1979**, *18*, 2786.
34. Audette R. J.; Quail J. W.; Black W. H.; Robertson B. E. *J. Solid State Chem.* **1973**, *8*, 49.
35. Saric A.; Music S.; Nomura K.; Popvic S. *J. Mol. Struct.* **1999**, *481-481*, 633.
36. Wang Y.; Muramatsu A.; Sugimoto T. *Colloids Surf.* **1998**, *134*, 281.
37. Tripathi A. K.; Kamble V. S.; Gupta N. M. *J. Catal.* **1999**, *187*, 332.
38. Audette R. J.; Quail J. W.; *Inorg. Chem.* **1972**, *11*, 1904.
39. Gump J. R.; Wagner W. F. *Trans. Kent. Acad. Sci.* **1954**, *15*, 112.
40. Bouzek K.; Rousar I. *J. Appl. Electrochem.* **1997**, *27*, 679.
41. Bouzek K.; Schmidt M.; Wragg A. *Electrochem. Commun.* **1999**, *1*, 370.
42. Licht S. Manassen J. *J. Electrochem. Soc.* **1987**, *134*, 1064.
43. Licht S. *Anal. Chem.* **1985**, *57*, 514.
44. Licht S. *Electroanalytical Chemistry*, Vol. 20, A. Bard, I. Rubinstein, ed., Marcel Dekker, NY, **1998**, *20*, 87.
45. Fang X.; Fang C.; Chen J. *J. Chin. Ceram. Soc.* **1996**, *6*, 732.
46. Hettiarachchi S.; Kedzierzawski P.; Macdonald D. *J. Electrochem. Soc.*, **1985**, *132*, 1866.
47. Chitrakar R.; Tezuka S.; Sonoda A.; Sakane K.; Ooi K.; Hirotsu T. *J. Colloid Interface Sci.*, **2006**, *297*, 426.
48. Parks G.; De Bruyn P. *J. Phys. Chem.*, **1962**, *66*, 967.
49. King R.B..; *Ency. Inorg. Chem,* **1995**, *8*, 4480.
50. Clearfield A.; Nancollas G.; Blessing R. *Ion Exchange and Solvent Extraction,* Marinsky J.A., Eds.; Decker: NewYork, 1973, vol. 5, p 1–120.

Chapter 2

The Role of the Electrode and Electrolyte Composition in the Anode Dissolution Kinetics

Comparison of the Ferrate(VI) Synthesis in the Solutions of NaOH and KOH of Various Ratio

Zuzana Mácová[1], Karel Bouzek[1,*], and Virender K. Sharma[2]

[1]Department of Inorganic Technology, Institute of Chemical Technology Prague, Technicka 5, 166 28 Prague 6, Czech Republic
[2]Chemistry Department, Florida Institute of Technology, 150 West University Boulevard, Melbourne, FL 32901
[*]Corresponding author: bouzekk@vscht.cz

This study deals with the influence of the anode material and the electrolyte composition on the kinetics of the electrode dissolution and on the Ferrate(VI) formation. The cast iron rich in iron carbide and the silicon rich steel are compared as perspective materials that provide the highest current efficiency of the electrochemical synthesis process. Pure NaOH, KOH solutions and their mixtures were used as an electrolyte. Electrochemical impedance spectroscopy was used to quantify the influence of the individual parameters on the anode dissolution kinetics. The results were compared with the pure iron study results, in order to assess impact of the individual anode material components on the anode's behavior. The theory, derived on the basis of the electrochemical impedance spectroscopy and cyclic voltammetry data, was confirmed by the batch electrolysis experiments.

Introduction

The first report of the electrochemical preparation of Ferrate (Ferrate means Ferrate(VI) throughout this work) was done by J. C. Poggendorf in 1841. The preparation was accomplished by the anodic dissolution of pure iron electrode in a strong alkaline solution. Two waves of the interest in this compound could be recognized in the past: one at the beginning of the 20th century and the other during the fifties-sixties. However, only during the past decade, the most significant increase of published papers and first attempts to commercialize this compound has happened. This is mainly due to the exacerbating environmental problems of developed societies and Ferrate's high potential to solve them.

The Ferrate can be synthesized by the chemical, thermal, or electrochemical methods. The main advantage of the electrochemical synthesis in comparison to the other two methods is the high purity of the product, and the utilization of an electron as a so-called "clean oxidant". In addition, this approach results in a substantial reduction of the amount of solvents needed to produce Ferrate of high purity.

The synthesis of Ferrate by an anodic iron dissolution proceeds in the transpassive potential region. At these conditions the surface of the iron anode is covered by a partly disintegrated (*e.g.*, containing cracks and/or pores) oxidic layer. The synthesis efficiency is strongly influenced by the protective properties of this layer. These properties can be influenced by the reaction conditions, *i.e.*, by the electrolyte concentration, composition and temperature, cell arrangement, and by the anode material composition.

Haber *(1)* and Pick *(2)* have already observed the positive effect of hydroxide concentration on the anodic dissolution of iron and on the stability of the formed Ferrate. Different authors *(2-5)* have found the optimum concentration for the Ferrate synthesis to be 14 M NaOH. One of the first detailed studies on the effect of electrolyte composition on the Ferrate synthesis was provided in a previous work *(6)*. NaOH, KOH and LiOH solutions were compared by means of batch electrolysis experiments at various temperatures (20, 30, and 40 °C). NaOH was found to be the anolyte that provides the highest current yields. According to this study, the explanation exists in different solubility of individual products (FeO_4^{2-} and its intermediates) in the electrolyte solution. Another possible explanation represents an impact of the individual cations on the structure of surface layers and their protective properties. Lapicque and Valentin *(7)* investigated mixtures of NaOH and KOH of different K^+:Na^+ ratios, in order to identify conditions that take the advantage of the reduced Ferrate solubility and allow direct electrochemical synthesis of solid K_2FeO_4. They observed that a suitable K^+ content in the anolyte causes a decrease in the Ferrate solubility without causing significant decrease in the process efficiency.

Pick *(2)* was the first who addressed the issue of the anode material composition impact on the electrochemical Ferrate synthesis efficiency. He found that the current efficiency of the electrochemical Ferrate synthesis increases with the increase of the used carbon content in the iron anode material. Denvir and Pletcher (*8,9*) confirmed this observation. In the previous studies *(3-5)*, different iron materials were investigated. The researchers concluded that only the presence of carbon in the form of iron carbide enhances deterioration of the oxidic layer protective properties. In the last systematic study published in this field, Lescuras - Darrou et al. *(10)* claimed silicon to be responsible for the depression of anode surface deactivation.

The first objective of this study is to compare the dissolution kinetics of the two materials referred to as the most efficient for the Ferrate synthesis, *i.e.* white cast iron with high iron carbide content, and steel with high silicon content. The pure iron anode will be used as a reference material. Impact of the two studied anode components (iron carbide and silicon) on the properties of the surface oxo-hydroxide layers will be assessed by comparing the results of the materials that contain these compounds to the reference. In the next step, the effect of a second cation, namely K^+, introduced into the solution of NaOH on the kinetics of the anode surface processes was followed. Solutions of different Na^+:K^+ molar ratio will be compared in order to clarify the solution composition effect on the structure and properties of the surface layers developed on iron material. The results will be verified using the batch electrolyses.

Experimental

All chemicals used in this work were of analytical grade purity. The studied anode materials were white cast iron (WCI) of composition: 3.16 wt. % C in the form of Fe_3C, 0.44 wt. % Mn and 0.036 wt. % Ni; silicon rich steel (SRS) containing 3.17 wt. % Si, 0.47 wt. % Cu, 0.23 wt. % Mn and 0.03 wt. % Ni; and pure iron with the degree of purity 99.95 wt. % containing less than 0.005 wt. % C, 0.0048 wt. % Ni and 0.0003 wt. % Mn.

The OH^- concentration of all studied solutions was 14 M. Pure NaOH, pure KOH, and solutions of composition varying in Na^+:K^+ molar ratios (1:3, 1:1 and 3:1) were used as an electrolyte.

Cyclic voltammetry (CV) and electrochemical impedance spectroscopy (EIS) were used to characterize the electrode processes taking place on the surface of the individual anode materials under study. Related experiments were accomplished in classical three-electrode arrangement with an appropriate anode material as a working electrode. Platinum foil served as a counter electrode. HgO/Hg electrode in 14 M NaOH solution connected with the cell using a Haber-Luggin capillary served as a reference. All potentials mentioned in this study refer to this electrode.

PINE potentiostat AFCBP 1 (U.S.A.), controlled by PC, was used to obtain cyclic voltammetric curves. SI 1287 Electrochemical Interface and SI 1250 Frequency Response Analyser by SOLARTRON (U.K.), also controlled by PC, were used in EIS experiments. Perturbation signal of frequency 65 kHz to

100 mHz and amplitude of 10 mV was applied. The anode potential range that corresponds to the Ferrate formation was investigated.

The galvanostatic electrolyses performed at current density ranging from 1.0 to 28.0 mA cm^{-2} took place in the batch electrolytic cell with anode and cathode compartments separated by a PVC diaphragm with the main pore diameter of 28 μm. The effect of temperature was investigated in the range of 20 °C to 60 °C. Every electrolysis was run for 180 minutes. After the electrolysis termination, the resulting anolyte solution was analyzed with respect to the Ferrate and total iron contents. On the basis of these data, the current efficiency of the studied systems was evaluated.

The analysis of the Ferrate content was performed by standard chromite method *(11)*. The total iron content was determined using atomic absorption spectroscopy (AAS).

Results

Cyclic voltammetry

The CV has been used to provide basic information on the processes taking place on the electrode surface. The potential region of the Ferrate synthesis was identified for each particular electrolyte composition and electrode material structure. A typical example of a polarization curve of the pure iron and the two studied materials is shown in Figure 1.

As demonstrated, the WCI electrode generally shows the highest electrochemical activity. This has been explained previously *(12)* in terms of the preferential dissolution of the iron carbide phase that results in continuous disintegration of the protective oxo-hydroxide layer covering the anode surface. Also, the cathodic current peak, located at the potential of –70 mV and corresponding to the Ferrate reduction, is very pronounced at this material. It corresponds to the reported high efficiency of the WCI in the Ferrate synthesis.

Surprisingly, SRS shows the lowest electrochemical activity from all the materials studied. This does not correspond to the published literature data *(10)* which claims that the Ferrate synthesis efficiency obtained using SRS anode can be equal, or even higher than the efficiency obtained using WCI. Nevertheless, both materials (SRS and pure iron) exhibit well-defined anodic shoulders in the potential region of the oxygen evolution commencement (620 mV) corresponding to the Ferrate synthesis. At the same time, zooming in both curves of the pure iron as well as of the SRS, shows well-defined Ferrate reduction peaks at the potential of approx. –120 mV. The current density of the reduction peaks is comparable for the SRS and for the pure iron. The shift of the Ferrate reduction peak to the more cathodic potential corresponds to the stronger passivated electrode surfaces at these two materials.

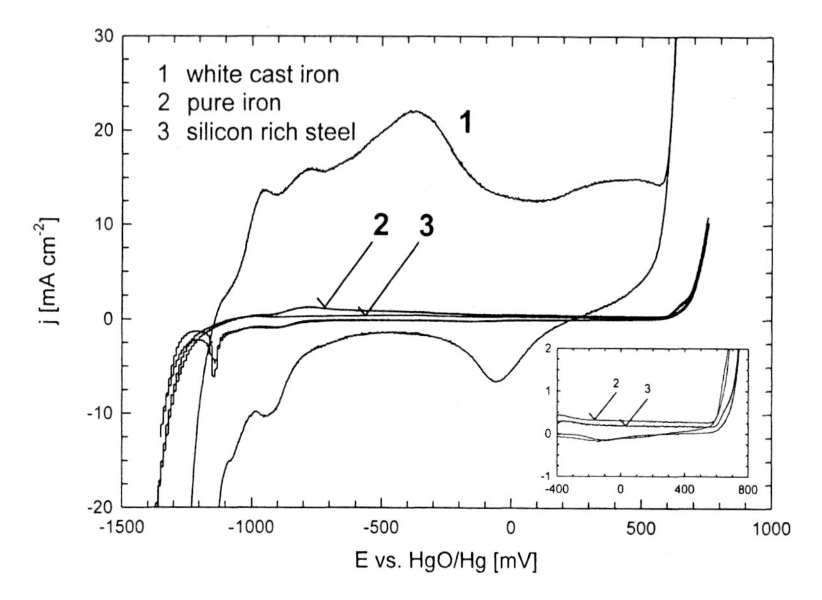

Figure 1. Cyclic voltammetric curves of studied anode materials, lines belonging to the particular materials are indicated in graph, temperature 20 °C, potential scan rate 10 mV s^{-1}, electrolyte 14 M NaOH; inset shows the curve zoomed to the Ferrate reduction region.

From this point of view, the properties of the surface oxo-hydroxide layer covering the anode in the transpassive potential region are the most important. The layer maintaining high protective properties until high anodic potentials are achieved causes low efficiency of the Ferrate synthesis. This is because anodic current flows through the system first when the oxygen evolution becomes the anode reaction kinetically preferred over the Ferrate synthesis. On the other hand, the absence of surface passivation may result in extremely low Ferrate synthesis current efficiency as well. This apparently paradoxical behavior is due to the catalytical influence of the iron compounds dissolved in the electrolyte near the anode surface on the oxygen evolution reaction *(13)*.

The CV, however, doesn't allow a more detailed insight into the behavior of these three materials. It provides just qualitative characterization of the intensity of material dissolution. This is because the efficiency of the Ferrate production may be assessed on the basis of the current density of the cathodic current peak corresponding to its reduction *(9)*.

In the next step, the influence of the electrolyte composition was investigated. For the sake of clarity, the changes in the CV curves caused by the introduction of the second cation, namely K^+, into the NaOH basic anolyte will first be documented on the pure iron anode example. The increasing amount of the K^+ ion shows a well detectable impact on the shape of the polarization curve, especially in the transpassive potential region. At 20 °C, the current density in the transpassive potential region increases most rapidly with the increasing anode potential for pure KOH solution. Already, the first addition of the Na^+ ions into the solution causes a significant decrease of the current density in the identical potential region. At the ratio of K^+:Na^+=1:3, the current density reaches in the potential range of 625 to 700 mV a stationary value of approximately 25 % of a value observed for the pure KOH solution. With increasing temperature, the situation becomes less apparent. However, the general trend could be derived easily. For the pure NaOH solution, the trend is characterized by the current density increasing with temperature and reaching a value, comparable to KOH at 40 °C. However, at 60 °C, the current density of the pure NaOH solution even exceeds the value observed for KOH. Surprisingly, all electrolytes containing both cations at various ratios provide current densities lower or comparable to that of the pure electrolytes.

For both remaining materials, the observed behavior can be summarized as follows. In the case of SRS, the NaOH provides higher current density of the polarization curves in the transpassive potential region for the entire temperature range under study than does KOH. In the case of WCI, the situation is just the opposite of that in the pure iron, *i.e.*, NaOH provides higher current density at 20 °C. At 60 °C, the higher current density was reached in KOH. It is clear that these differences are the consequences of the anode material composition. The situation seems to be relatively clear for the pure iron. Here the important aspect represents the chemical reaction of the hydroxyl anion with the surface layer, which is homogeneous in composition. As it follows from the results of Åckerlöf and Kegeles *(14)* and Åckerlöf and Bender *(15)*, at 20 °C, the mean electrolyte activity coefficient of NaOH in a 14 M solution corresponds to a value of 13.5. In the case of KOH, the mean electrolyte activity coefficient for the same molar concentration reaches a value of 26.9. At 60 °C, the situation changes dramatically. The activity coefficient of the NaOH solution reduces to 5.9. But for KOH, the decrease is even more significant. It reduces to 4.8. The reactivity of the hydroxyl anion with the passivating oxo-hydroxide layer depends on its activity. It is thus clear that with increasing temperature, the depassivating ability of NaOH prevails over KOH, which predominates at 20 °C. This is in agreement with the above described behavior of the pure iron electrode during CV experiments.

In the case of additional components or phases in the iron anode structure, the situation changes significantly. The decisive factor is the interaction of the electrolyte with each particular component, *e.g.* silicon or iron carbide, being present on the anode surface.

Electrochemical Impedance Spectroscopy

EIS is a powerful tool allowing the sensitive analysis of the electrochemical processes' kinetics. Its applicability to the characterization of the anode material properties related to Ferrate synthesis has been proven already during our previous study focusing on the behavior of pure iron and WCI in concentrated NaOH solution *(16)*. In agreement with the previous results, EIS spectra of all materials under study are characterized by two time constants. A typical example of the impedance spectra obtained is shown in Figure 2.

The electrode whose surface is covered by a layer (or layers) of more or less insulating properties represents a specific case in the EIS experiments. A typical example is passivated metal surface, which is the present system. Under these circumstances, the electrode reaction kinetics (iron dissolution and/or oxygen evolution reaction) obtained from the EIS spectra of the system is, to a certain extent, determined by the surface layer properties. This is because the transport of the electrical charge across the layer represents the rate determining step. This is documented by the fact that the iron dissolution rate in the case of pure iron is nearly zero; even the overvoltage for this reaction in the passivity region has reached more than 1 V. In the present case, the two time constants of the spectra therefore indicate duplex surface oxo-hydroxide layer and not two electrode reactions *(17)*. A duplex structure of the surface layer was further confirmed previously by an independent study using Mössbauer spectroscopy *(18)*.

As indicated by the shape of the spectra for the pure iron anode, one of the layers is characterized by a low resistance against the charge transfer. On the other hand, the second layer represents, in certain electrode potential limits, an efficient barrier against charge transfer and thus against the anode material dissolution. With increasing anode potential, the protective properties of the latter layer practically disappear. In order to allow the comparison of properties of the individual materials, it is necessary to evaluate the charge transfer properties of these layers quantitatively. It is known from EIS theory that an analytical mathematical model allowing the direct fitting of kinetic constants of the reaction mechanism steps on the base of the experimentally determined spectra can be derived only for simple systems with well defined electrode process mechanisms *(19)*. This is clearly not the case in Ferrate synthesis. Therefore, an approximate approach based on the definition of the suitable equivalent circuit was used in the present case. The proposed equivalent circuit was based on the physical model of the two surface layers. The circuit consists of two parallel R-CPE (Constant Phase Element) elements connected in series with an electrolyte resistance *(17)*. The capacitance used in the original literature was replaced in the present work by the CPE in order to satisfy nonidealities of the surface layers, especially in the transpassive potential region. The equivalent circuit used is shown in Figure 3. As it was found, this simple model shows surprisingly good agreement with experimental data in passive and early transpassive potential region, where the latter one corresponds to the Ferrate formation conditions. This is confirmed in Figure 4, which shows a typical example of the impedance spectra together with the related fit of the equivalent circuit. At the higher anode potentials corresponding to the intensive oxygen evolution, the model fails. This is because the conditions do not correspond to the simplification assumptions made, which consider mainly the integrity of the both surface layers.

As with cyclic voltammetry, the results will first be discussed on the example of the pure iron anode in NaOH solution. For this system, the resistance of both sublayers decreases with increasing anode potential. Similar behavior was observed for increased electrolyte temperature.

Whereas the value of the CPE element of the internal surface layer remains independent of the anode potential at 20 °C, the value for the external layer exhibits a significant increase in the potential range of 650 to 725 mV, followed by sudden decrease down to nearly zero. This corresponds to the reduction of the external layer's thickness and its disintegration above the upper limit of this electrode's potential range. Important information is that the internal (protective) layer does not disintegrate (CPE remains constant), and the increase in the charge transfer kinetics corresponds more to the oxygen evolution reaction. In comparison, at 60 °C, the increase in the value of the internal layer CPE is more significant and takes place already in the potential range of 575 to 650 mV. At these conditions, the internal layer is also clearly damaged and loses its integrity. Both these observations strongly support the theory of the chemical reaction of the surface layers with hydroxyl anions under formation of the soluble products. In agreement with the general theory on the chemical reaction kinetics, the rate of this reaction is significantly enhanced by increased temperature.

In the case of SRS, the situation is similar regarding the individual layers' resistance. It decreases as the anode potential increases. The dependence of the internal layer capacitance on the electrode potential at both temperatures shows similar characteristics as well.

The WCI anode also exhibits similar behavior. With increasing temperature, the resistivities of both surface films are reduced. Also, in regards to the films' capacitance, the disintegration proceeds more rapidly than in the case of the pure iron.

Interesting was also the development of the EIS spectra with the changing content of the K^+ ion in the anolyte solution. The anode potentials of 625 mV at 20 °C and 600 mV at 60 °C were chosen for the comparative study. The Ferrate synthesis occurs with these conditions. However, the parasitic reaction represented by the oxygen evolution does not yet reach a significant extent.

For the pure iron anode, the resistance of the internal protective oxide layer decreases at 20 °C from the value of almost 600 Ω found for pure NaOH, with increasing content of the K^+ ion to about 100 Ω reached for the pure KOH. At 60 °C, the value of the resistance grows from the value of 11 Ω in pure NaOH to

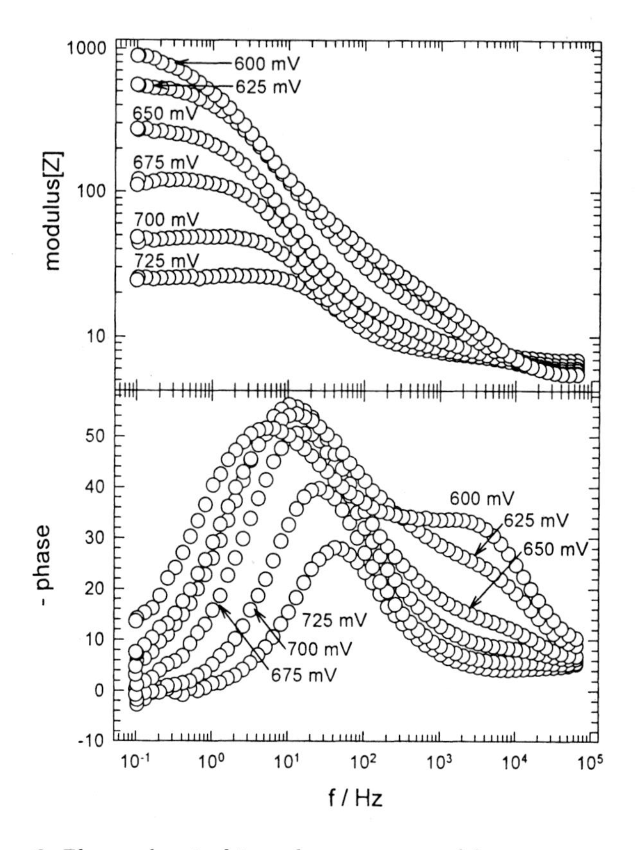

Figure 2. Electrochemical impedance spectra of the pure iron anode in dependence on the electrode potential; corresponding anode potentials are indicated in the graph. Temperature 20 °C; electrolyte 14 M NaOH.

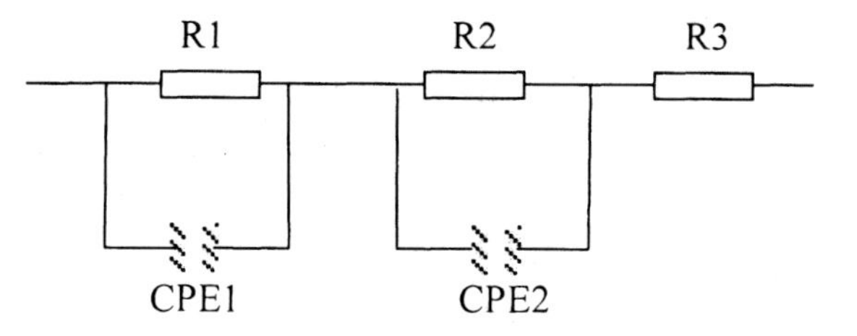

Figure 3. Equivalent electrical circuit used to evaluate EIS data; index 1 indicates the internal sublayer, index 2 is the external sublayer, and index 3 is the electrolyte.

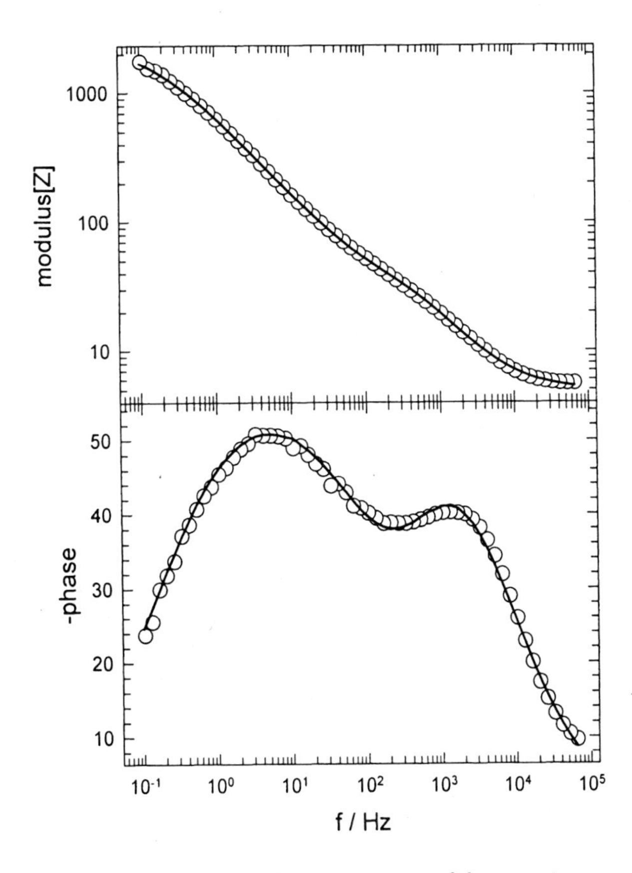

Figure 4. Electrochemical impedance spectra of the pure iron anode at the potential of 575 mV, temperature 20 °C, electrolyte 14 M NaOH; the circles indicate experimental points, the line fit of corresponding equivalent circuit (see Figure 3) with optimized parameters.

the limiting value of 32 Ω found for the K^+:Na^+ ratio 3:1. These values indicate that the barrier film is at the potential studied substantially disintegrated. An exception is NaOH solution at 20 °C. The layer still keeps its basic protecting properties at this temperature.

In the case of the external porous layer, the situation is qualitatively similar at 20 °C. Its resistance value is decreasing with the increasing K^+ content. At 60 °C, it remains approximately constant at the value of 0.6 Ω, which is more or less just the resistance of the electrolyte film at the anode surface.

The CPE element related to the internal layer shows at 20 °C, the lowest value of 0.31 mF for the pure NaOH solution. The first addition of the K^+ ion into the electrolyte has already resulted in the significant increase of the capacitance to the stationary value of 0.85 mF. It confirms previous data that the first portion of K^+ results in a significant change of the internal layer structure. Raising the temperature to 60 °C tripled the increase of the barrier film capacitance value. But in this case, its value remained relatively independent of the K^+ content.

These results confirm the indications of the CV experiments. The kinetics of the charge transfer indicated by the optimized equivalent circuit parameters is at 20 °C faster in KOH and vice versa at 60 °C in NaOH as an anolyte. It is connected with the mean activity coefficients of NaOH and KOH in concentrated water solutions. However, it is difficult to find a reliable explanation of the observed CPE value dependence on the anolyte composition and temperature based only on this data. Dependence of the capacitance value on the anode potential has to be determined as well. It shows that the capacitance of the internal layer doesn't decline in the solutions of higher K^+ content as rapidly in the region of the upper limit of the anode potential range studied as it was observed for the pure NaOH solution. It was continuously growing at 20 °C in the entire potential range under study and it had attained a value of 0.39 mF for KOH at 750 mV. In NaOH, at identical conditions, a value of 0.66 mF was reached at 700 mV. Afterwards, its value was falling down to 0.44 mF at 750 mV. At 60 °C, KOH reaches a capacitance value of 2.2 mF at 600 to 625 mV and then falls down to 1.8 mF at 650 mV. In the case of NaOH, it has a value of 3.0 mF at 600 mV and 2.8 mF at 650 mV. It seems that, in agreement with expectation, lower solubility of potassium salts of anode dissolution products results in the thicker anode surface layer, even at 60 °C. This is probably due to the compounds precipitated from the locally oversaturated solution at the anode surface vicinity. Comparison of the charge transfer resistance data with the capacitance values for the surface film highlights one additional important finding: the thickness of the film resulting from the evaluation of the capacitance data doesn't have necessary ability to indicate its protective properties. The problem has to be handled in whole complexity.

The CV and EIS data discussed provide closer insight into the properties of the oxo-hydroxide film covering the anode surface and the influence of the electrolyte composition and anode structure on them. Nevertheless, complementary information is needed. This is because the significant reduction of the anode surface layer protective properties can result in two processes: (i) required anode dissolution and subsequent Ferrate production and/or (ii) initiation of the intensive oxygen evolution. Depassivation of the anode surface may result in both alternatives. Batch electrolysis will separate these two processes on the intermediate and longtime scale.

Batch electrolyses

Following from the previous results *(6)* for NaOH as an anolyte and for the pure iron anode, the temperature increase from 30 °C to 40 °C resulted in the increase of the current density providing optimal current efficiency. The temperature, however, did not influence the highest current yield obtained. In the case of the WCI anode, the temperature increase from 20 °C to 30 °C was even counter productive. The highest current efficiency achieved was reduced by more than 10 %. Using KOH as an anolyte, the situation is significantly different. Here an increase in the current yield with increasing temperature could be observed for both anode materials under study.

In the present work, this study was extended in several aspects. Beside the extended temperature range and anolyte compositions, a new material (SRS) was included in the study in order to contribute to the understanding of the anode material aspect and to optimize the operational parameters used for the Ferrate production.

Using NaOH as an anolyte, both studied materials, WCI and SRS, have shown identical dependency on the temperature, as shown in Figure 5. At the temperature of 20 °C, the current yield has reached more than 50 %. At 60 °C it falls below 10 %. The situation changes dramatically using KOH solution as an electrolyte. In such a case, clearly the highest current yield has been reached for WCI. At 20 °C, the current yield was approximately 30 %, and it further increased to more than 50 % at 60 °C. In the case of the SRS, the highest current yield of approximately 30 % could be obtained at 60 °C and was practically independent of the current density used.

These observations are highly interesting for several reasons. The first is related to the applicability of high temperature for the Ferrate production. It was believed up to now, with a few exceptions *(20)*, that high temperature would generally result in product decomposition. This was confirmed for the NaOH solution and WCI anode. As can be seen in Figure 6, the concentration of iron in the oxidation state lower than Fe(VI) in the anolyte increases substantially with the anolyte temperature. It has reached a value up to 5 g dm^{-3} for the electrolysis conditions under study. The amount of total iron dissolved remained the same, or was even slightly higher when compared to the 20 °C. It is thus possible to say that the theory on the instantaneous decomposition of the Ferrate produced at high temperatures has been supported for NaOH anolyte. As shown above, under identical conditions, the EIS data indicate significant deterioration of the barrier layer protective properties. With respect to these results, even higher iron content in the anolyte was originally expected at the elevated temperature. The observed disintegration of the passive layer, however, does not lead to significantly more intensive anode dissolution because of the competing oxygen evolution reaction.

In the case of KOH, the theory on fast Ferrate decomposition was not proven. The main reason is that Ferrate is produced directly in a solid form. It is thus not sensitive to homogeneous decomposition. Moreover, as indicated by the EIS results, protecting properties of the barrier layer remain preserved to some extent due to the limiting solubility of the potassium iron salts. This allows slowing down the oxygen evolution kinetics. Subsequently, the iron dissolution process is more efficient.

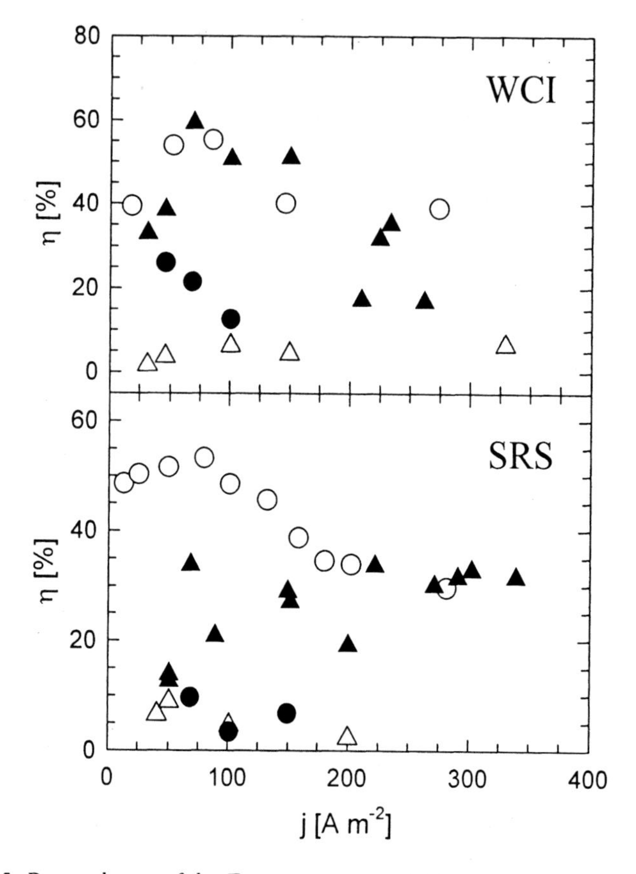

Figure 5. Dependence of the Ferrate current yield on the anode current density for 14 M electrolyte, WCI – white cast iron, SRS – silicon rich steel, ○ - 20 °C, △ - 60 °C; empty symbols – NaOH, filled symbols – KOH.

In the case of SRS, the situation is different. The temperature increase from 20 °C to 60 °C causes decrease in the Ferrate yield by approximately an order of magnitude in NaOH as an anolyte. At the same time, concentration of iron in lower oxidation states in the anolyte remains nearly the same or even decreases, as shown in Figure 6. In the KOH solution, the situation is closer to the WCI anode. With the increased temperature, the current efficiency rose by about 5 times. Also the concentration of iron in lower oxidation states increased 3 to 5 times.

This clearly indicates that using SRS anode in the NaOH solution, the kinetics of the oxygen evolution prevails with increasing temperature over the iron dissolution kinetics. This is a behavior opposite to the WCI anode. Using KOH the situation is not so clear. It is possible to assume that replacement of Na^+ ions in the anolyte by K^+ increased resistance of the barrier layer covering the anode surface

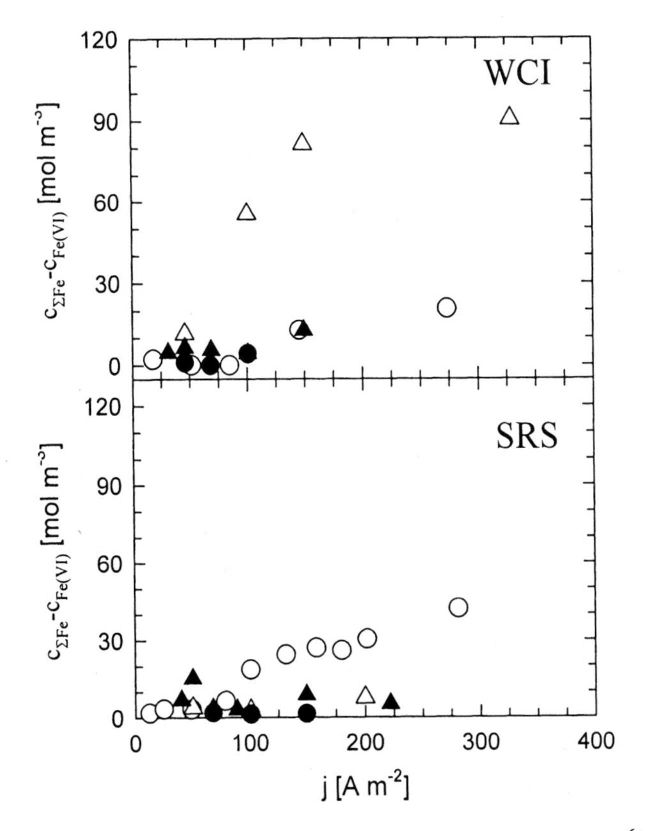

Figure 6. Dependence of the iron in oxidation state lower than Fe^{6+} content in the anolyte on the anode current density for 14 M electrolyte, WCI – white cast iron, SRS – silicon rich steel, ○ - 20 °C, △- 60 °C; empty symbols – NaOH, filled symbols – KOH.

at 60 °C, as observed by EIS. This phenomenon allows slowing down of the oxygen evolution kinetics and thus enhances the iron dissolution.

There is a partial discrepancy between the last finding and the originally proposed explanation of the silicon effect on the iron dissolution kinetics, *i.e.* gradual anodic dissolution by the strongly alkaline electrolyte. At enhanced temperature, this process is supposed to exhibit even higher kinetics than at 20 °C. Because of this, a more intensive anode material dissolution was expected. When compared to the pure iron in NaOH at 50 °C *(12)*, we may conclude that the iron dissolution kinetics is strongly hindered when using SRS. The reason for this behavior is not understood yet and additional experiments, focusing especially on the determination of the anode surface composition, have to be carried out.

Conclusions

The present work examined the impact of the anode material and electrolyte composition on the Ferrate synthesis efficiency. The results obtained represent an important contribution with respect to understanding the influence of the surface film's properties on the synthesis of Ferrate. These layers are influenced not only by the structure of the anode, but significant impact was observed also for the anolyte composition. EIS results indicate changes of the charge transfer kinetics with the change of the cation being present in the solution. It seems that the reason for such behavior is partial precipitation of the anode dissolution products in the electrolyte layer adjacent to its surface at sufficient K^+ ion concentration in the anolyte. The data obtained confirm previously published theory on the influence of the iron carbide present in the anode structure on the continuous depassivation of the anode surface. The silicon present in the structure seems to have similar properties at low electrolyte temperatures. However, as the temperature increases, the behavior deviates from the expected one. This result might be explained by the silicon layer remaining on the anode surface but the reason for such behavior is not clear yet.

The experiments performed confirmed that the superior material from the current efficiency point of view is cast iron rich in the iron carbide (WCI). However, at the lower temperatures, similar results can be obtained by means of silicon rich steel. The main advantage of this material is that it is easily machineable and it thus allows construction of more sophisticated anode structures. Three-dimensional electrodes are one example of such structures. An important aspect represents also the possibility of the electrochemical production of Ferrate directly in the solid form. When using a WCI anode, the simplest approach uses pure KOH as an anolyte at the operational temperature of 60 °C. This approach, however, is not applicable in the case of SRS. This material shows high oxygen evolution kinetics at given conditions. It is necessary to use the mixture of NaOH and KOH in an appropriate ratio and at a suitable operational temperature. The current efficiency in this case does not reach the level of the WCI in KOH. Nevertheless, this is a subject of further research, especially with regard to the contribution of the utilization of three-dimensional electrode to the overall process efficiency.

Acknowledgements

The financial support of this research by the Grant Agency of the Czech Republic under project No. 104/05/0066 and by the Institute of Chemical Technology under project No. 105/08/0016 is gratefully acknowledged. We also acknowledge the support U.S.-Czech Republic Partnership Program supported by the National Science Foundation and Czech Republic Academy.

References

1. Haber, F. *Z. Elektrochem.* **1900**, *7*, 215.
2. Pick, W. *Z. Elektrochem.* **1901**, *7*, 713.
3. Bouzek, K.; Roušar, I. *J. Appl. Electrochem.* **1993**, *23*, 1317-1322.
4. Bouzek, K.; Roušar, I.; Taylor, M. A. *J. Appl. Electrochem.* **1996**, *26*, 925-932.
5. Bouzek, K.; Roušar, I. *J. Appl. Electrochem.* **1996**, *26*, 919-924.
6. Bouzek, K.; Schmidt, M. J.; Wragg, A. A. *Electrochem. Commun.* **1999**, *1*, 370-374.
7. Lapicque, F. ; Valentin, G. *Electrochem. Commun.* **2002**, *4*, 764-766.
8. Denvir, A.; Pletcher, D. *J. Appl. Electrochem.* **1996**, *26(7)*, 815-822.
9. Denvir, A.; Pletcher, D. *J. Appl. Electrochem.* **1996**, *26(7)*, 823-827.
10. Lescuras-Darrou, V.; Lapicque, F.; Valentine, G. *J. Appl. Electrochem.* **2002**, *32(1)*, 57-63.
11. Schreyer, J. M.; Thompson, G. W.; Ockerman L. T. *Anal. Chem.* **1950**, *22(11)*, 1426-1427.
12. Bouzek, K.; Roušar, I. *J. Appl. Electrochem.* **1997**, *27*, 679-684.
13. Bouzek, K.; Lipovská, M.; Schmidt, M.; Roušar, I.; Wragg, A. A. *Electrochimica Acta* **1998**, *44*, 547-557.
14. Åkerlöf, G.; Kegeles, G. *J. Am. Chem. Soc.* **1940**, *62*, 620-640.
15. Åkerlöf, G.; Bender, P. *J. Am. Chem. Soc.* **1948**, *70*, 2366-2369.
16. Bouzek, K.; Bergmann, H. *Corr. Sci.* **1999**, *41*, 2113-2128.
17. Jüttner, K. *Electrochim. Acta* **1990**, *35(10)*, 1501-1508.
18. Bouzek, K.; Nejezchleba, M. *Collect. Czech. Chem. Commun.* **1999**, *64*, 2044-2060.
19. Gabrielli, C. *Identification of electrochemical processes by frequency response analysis – Technical report Number 004/83* **1983**, Solartron Instruments, UK.
20. He, W.; Wang, J.; Shao, H.; Zhang, J.; Cao, C. *Electrochem. Commun.* **2005**, *7*, 607-611.

Chapter 3

Electrochemical Ferrate(VI) Synthesis: A Molten Salt Approach

M. Benová [1], J. Híveš [1], K. Bouzek [2], and V. K. Sharma [3]

[1]Department of Inorganic Technology, Slovak University of Technology in Bratislava, 812 37 Bratislava, Slovakia
[2]Department of Inorganic Technology, Institute of Chemical Technology Prague, 166 28 Prague, Czech Republic
[3]Florida Institute of Technology, Melbourne, FL 32901

The electrochemical synthesis of ferrate(VI) was studied for the first time in a molten salt environment. An eutectic NaOH-KOH melt at the temperature of 200 °C was selected as a most appropriate system for the synthesis. Cyclic voltammetry was used to characterize the processes taking place on the stationary platinum (gold) or iron electrodes. The identified anodic current peak corresponding to the ferrate(VI) production was close to the potential region at which oxygen evolution begins. During the reverse potential scan, well defined cathodic current peak corresponding to the ferrate(VI) reduction appears. However, the peak was shifted to less cathodic potential than that of potential corresponding to the electrolysis in aqueous solutions. This indicates less progressive anode inactivation in a molten salts environment.

The interest in ferrate(VI) has increased substantially for the last decade because of the development of a high capacity battery (so called super iron battery) and its potential as a multipurpose chemical for water and wastewater treatment (*1-9*). Ferrate(VI) is able to decompose rapidly many toxic pollutants, including chemical weapons. It is also known to be an efficient disinfecting agent (*10,11*).

Potassium salt of Fe(VI) (K_2FeO_4) can be produced by thermal, chemical, and electrochemical techniques (*4-16*). All of these synthetic techniques have disadvantages in producing ferrate(VI) for its intended uses. The thermal technique uses high temperature (>800 °C) and also gives lower yields (<50%) due to potential decomposition of ferrate(VI) at temperatures above 250 °C (*5*). The chemical technique requires large amounts of chemicals and several steps are involved in the production of K_2FeO_4 (*16*). Electrochemical synthesis applies electrolysis of iron (or iron salt) under concentrated hydroxide solution, which creates problems because ferrate(VI) is reduced by water ($2FeO_4^{2-}+5H_2O \rightarrow 2Fe^{3+}+3/2O_2+10OH^-$) (*1,10*).

Electrochemical ferrate production in a molten hydroxide environment represents a promising alternative approach. The most important advantage of this method consists in absence of water in the electrolyte. Therefore, ferrate(VI) produced is, after reaction mixture cool down, in a solid dry form and thus stable.

In order to minimize thermal product decomposition the system with lowest melting temperature was used. From this point of view, the eutectic mixture of the NaOH-KOH (51.5 mol. % NaOH) was found to be the most attractive. It is characterized by the relatively low eutectic melting point of 170 °C and high conductivity of $\kappa_{200\,°C} = 0.588\ \Omega^{-1}\ cm^{-1}$ (*17*).

Experimental

Cyclic voltammetry was conducted using the AUTOLAB electrochemistry system (ECO Chemie). The cell consists of a platinum crucible, which served at the same time as a counter electrode (CE). A platinum, gold or iron rod was used as the working electrode (WE). The iron used as a WE was of 99.1 % purity. The main impurity was Mn (0.6 wt. %). No suitable reference electrode (RE) has been established for molten hydroxide system (*18*). Therefore Pt quasi-RE was attempted in this work. Its potential is supposed to be sufficiently stable in the experimental conditions of the present study.

The major problem in the preparation of fused hydroxide melt consists in the control of level of the water being present in the electrolyte. Water is the only significant impurity in reagent grade hydroxides beside carbonate. Carbonate does not disturb studied process do to its electrochemical and chemical inertness at the conditions used, however water may influence results significantly. Unfortunately, water cannot be quantitatively removed by simple heating the fused hydroxides at 500 °C (*18*). A special drying procedure was adopted. NaOH and KOH (both p.a. grade, Merck) were dried in a vacuum drying oven in the presence of P_2O_5 for several days at gradually increasing temperature up to 200 °C. Despite this special drying procedure, melt contained approx. 1.5 mol % of water according to differential thermal analysis (DTA) and thermal gravimetry (TG).

During experiment the crucible containing 60-150 g of dried eutectic hydroxide mixture is placed in an open vertical laboratory furnace. Thus experiments are performed under the air atmosphere. The melt temperature is measured by means of a Pt-Pt10Rh thermocouple.

Results and Discussion

Figure 1 shows the cyclic voltammogram for platinum electrode in eutectic NaOH-KOH melt in absence and presence of Fe_2O_3 at 200 °C. The anodic limit of the system is due to oxidation of melt and likely corresponds to the one or both of the following reactions

$$2OH^- \rightarrow \tfrac{1}{2}O_2(g) + H_2O + 2e^- \qquad (1)$$

$$4OH^- \rightarrow O_2^- + 2H_2O + 3e^- \qquad (2)$$

The cathodic limit is due to the reduction of H_2O, evolved on the anode surface, according to the reaction

$$H_2O + e^- \rightarrow \tfrac{1}{2}H_2(g) + OH^- \qquad (3)$$

The small cathodic current peak (C_{PtOx}) in Figure 1 (dashed line) located at 0.0 V corresponds to the reduction of PtO_x film formed during anodic potential scan on the Pt working electrode surface. The cyclic voltammogram of the Pt electrodes in a molten eutectic NaOH-KOH system with addition of Fe_2O_3 at temperature of 200 °C is given as a solid line in Figure 1. In comparison to the system without Fe_2O_3 addition voltammogram is characterized by a sharp oxidation peak occurring at 0.42 V (A_3) which corresponds to the formation of ferrate(VI) according to the following reaction (*19*).

$$FeO_2^- + 4OH^- \rightarrow FeO_4^{2-} + 2H_2O + 3e^- \qquad (4)$$

The FeO_2^- ion originates from the reaction of Fe_2O_3 with the fused electrolyte according to Eq. (5).

$$Fe_2O_3 + 2OH^- \rightarrow 2FeO_2^- + H_2O \qquad (5)$$

Two current peaks occurred during the cathodic potential scan. The first one (C_{3a}) corresponds to the reduction of the FeO_4^{2-} to FeO_2^- according to the reaction (6).

$$FeO_4^{2-} + 2H_2O + 3e^- \rightarrow FeO_2^- + 4OH^- \qquad (6)$$

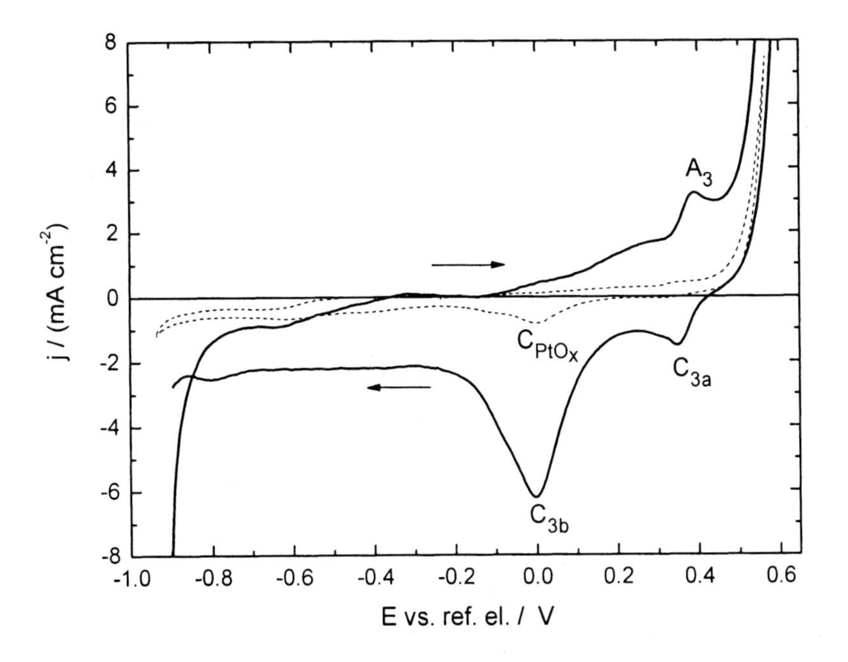

Figure 1. Cyclic voltammogram of the platinum working electrode in the NaOH-KOH eutectic system at 200 °C. Platinum crucible was used as a counter electrode, RE: Pt. Potential scan rate 100 mV s^{-1}. Dash line - voltammogram in the electrolyte without addition of Fe_2O_3, solid line - voltammogram in the electrolyte with addition of Fe_2O_3 (0.5 wt. %).

The peak C_{3b} likely corresponds to the reduction of iron-oxo species produced at high anodic potentials. However, the potential of this peak is also similar to the potential of the PtO_x layer reduction peak, C_{PtOx} (see Figure 1), which suggests that C_{3b} peak at 0.0 V may also corresponds to PtO_x reduction. These two possibilities were distinguished by replacing Pt electrode with a noble metal, gold. Gold does not form surface oxide layer. As shown in Figure 2, the gold electrode also gave C_{3b} peak though it was approximately 100 mV less cathodic when compared to Pt. Origin of peak C_{3b} is therefore not directly related to the reduction of PtO_x layer. It suggests that the reduction of PtO_x is a prerequisite of the rapid iron-oxo species reduction, which takes place only on the bare Pt surface. The inhibiting role of the surface oxidic layer for the selected Fe redox electrode reactions reactions is confirmed by the significant variation of the peak A_3 current density observed for the Pt and Au electrodes. Whereas on the Pt electrode has current density of this peak reached just 3.2 mA cm^{-2}

(Figure 1), on the Au electrode it rose to 50 mA cm^{-2} (Figure 2). This difference is caused by the passivation of Pt in the anodic potentials region. According to Damjanovic et al. (*20*) in water environment the oxides film formation on the Pt surface is completed at the potential of about 1.0 V below the oxygen evolution commencement. It is therefore possible to consider, that also in the molten salts environment is in the potential range of the ferrate(VI) formation the Pt surface already fully passivated. In contrary to this, Au surface does not cover by the oxides layer. Thus, it provides significantly higher activity towards ferrate(VI) formation.

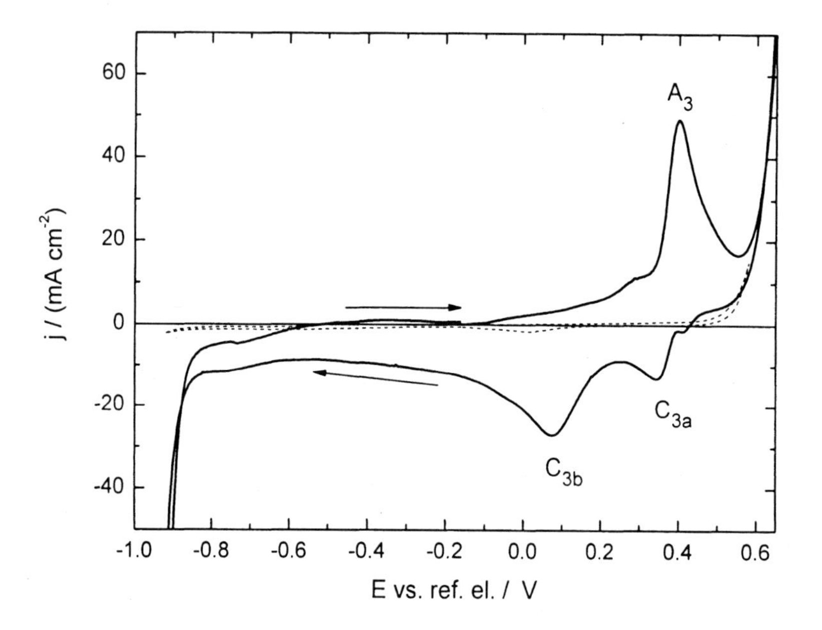

Figure 2. Cyclic voltammogram of the gold electrode in the NaOH-KOH eutectic system at 200 °C. Platinum crucible was used as a counter electrode, RE: Pt. Potential scan rate 100 mV s^{-1}. Dash line - voltammogram in the electrolyte without addition of Fe_2O_3, solid line - voltammogram in the electrolyte with addition of Fe_2O_3 (0.5 wt. %).

The cyclic voltammograms were recorded at various scan rates (50 - 900 mV/s) in order to characterize more precisely individual current peaks. Resulting polarization curves are presented in Figure 3. The course of the curves indicates the electrode reaction corresponding to the current peak A_3 to be reversible. This is in agreement with the observation made for the water environment (*19*).

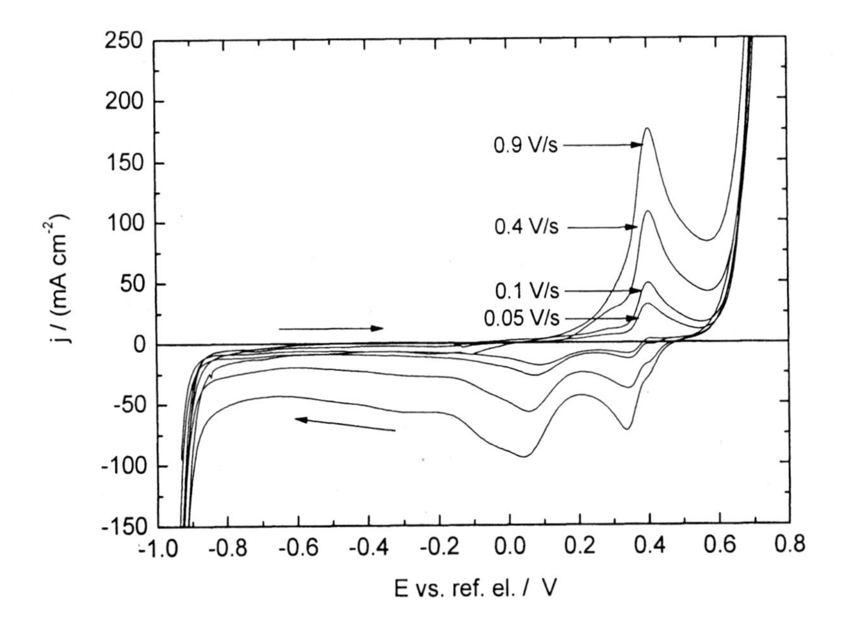

Figure 3. Cyclic voltammogram of the gold working electrode in the NaOH-KOH eutectic system with addition of Fe_2O_3 (0.5 wt. %) at 200 °C. Platinum crucible was used as a counter electrode, RE: Pt. Potential scan rates used are indicated in the figure.

The dependence of the A_3 peak current density was plotted against the square root of the potential scan rate, see Figure 4. Linear dependence was found; indicating electrode process controlled by the semiinfinitive linear diffusion of the electroactive specie to the electrode surface. From the analysis of $|E_p - E_{p/2}|$ the number of electrons exchanged at the reaction corresponding to the peak A_3 was evaluated as $n = 2.99 \pm 0.08$. The same number of electrons exchanged were also confirmed for the cathodic peak (C_{3a}). Both these findings confirm assignment of the peak A_3 to the oxidation of ferric ion to ferrate(VI) according to reaction (4). Identical results were obtained also for the Pt working electrode.

In order to confirm origin of current wave in the potential region of the hydrogen evolution commencement, an alternative approach was chosen. Before

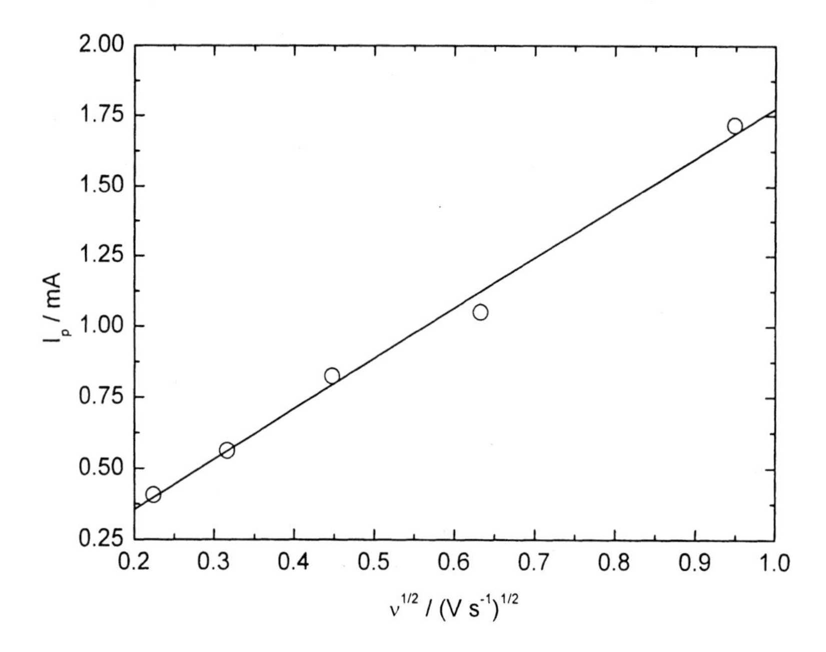

Figure 4. Dependence of the anodic peak (A_3) current on the square-root of the potential scan rate. Gold working electrode in the molten eutectic NaOH-KOH system with addition of Fe_2O_3 (0.5 wt. %) at 200 °C. Platinum crucible was used as a counter electrode, RE: Pt. ○ experimental data, line – linear regression analysis.

the start of the potential scan, the electrode (Pt or Au) was cathodically polarized for 10 minutes at –1.0 V in order to cover the electrode surface with reproducible amount of the cathodically deposited iron. An example of resulting voltammogram is shown in Figure 5.

A sharp oxidation peak A_1 occurred at -0.58 V. Comparison to the electrolysis in alkaline water electrolytes (*28*) suggests that it corresponds to the formation of FeO_2^{2-} ions according to reaction (7).

$$Fe + 4OH^- \rightarrow FeO_2^{2-} + 2H_2O + 2e^- \quad (7)$$

The obtained number of exchanged electrons $n = 1.99 \pm 0.05$ from $|E_p - E_{p/2}|$ analysis of several voltammograms recorded at different scan rate confirms this peak assignment.

The anodic current peak (A_2) corresponds either to the oxidation of FeO_2^{2-} ions to FeO_2^- according to reaction (8) or to the changes in the structure and composition of the anode surface hydroxide layer.

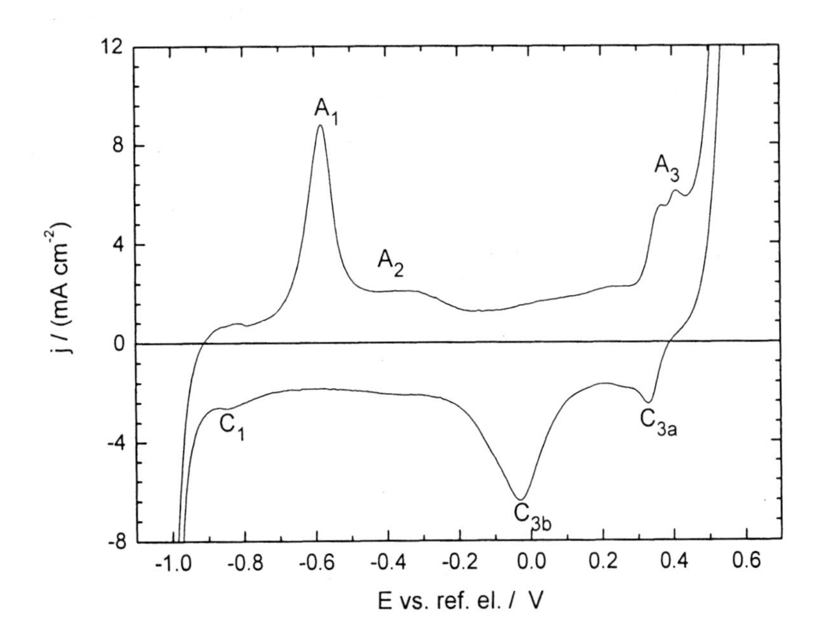

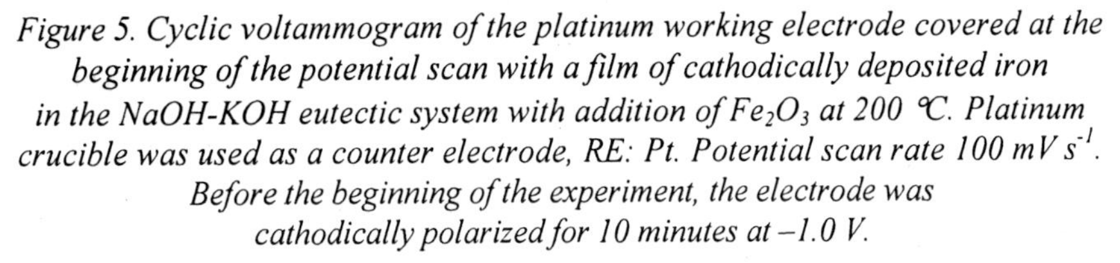

Figure 5. Cyclic voltammogram of the platinum working electrode covered at the beginning of the potential scan with a film of cathodically deposited iron in the NaOH-KOH eutectic system with addition of Fe_2O_3 at 200 °C. Platinum crucible was used as a counter electrode, RE: Pt. Potential scan rate 100 mV s^{-1}. Before the beginning of the experiment, the electrode was cathodically polarized for 10 minutes at –1.0 V.

$$FeO_2^{2-} \rightarrow FeO_2^{-} + e^{-} \qquad (8)$$

Peak A_3 corresponds to the ferrate(VI) production and is in the potential range of the oxygen evolution commencement.

Three current peaks occurred during the cathodic potential scan. The first one (C_{3a}) corresponds to the reduction of the FeO_4^{2-} to FeO_2^{-} according to the reaction (6). The second cathodic current peak (C_{3b}) corresponds to the further reduction of iron oxide film. The third cathodic current peak (C_1) corresponds to the reduction of the iron species to the metallic iron. It is important to note the shape of the CV curve at the electrode potentials higher than -0.2 V. It is identical with the Pt electrode in the melt with Fe_2O_3 addition. This, together with the shape of the bulk iron electrode polarization curve shown in Figure 6 indicates that the metallic iron film was completely dissolved from the Pt electrode surface by the reactions corresponding to the peaks A_1 and A_2. These reactions provide corresponding iron species necessary for the ferrate(VI) formation.

Formation of ferrate(VI) was further studied by conducting cyclic voltammetric experiments with an iron rod as working electrode. Typical polarization curve obtained is characterized by the sharp oxidation peak (A_1) (Figure 6). As already discussed, current peak A_1 corresponds to the formation of FeO_2^{2-} ions according to Eq. (7). Analysis of the difference $|E_p - E_{p/2}|$ of several volammograms recorded at different scan rates gave n=1.99±0.04; thus confirming this peak assignment.

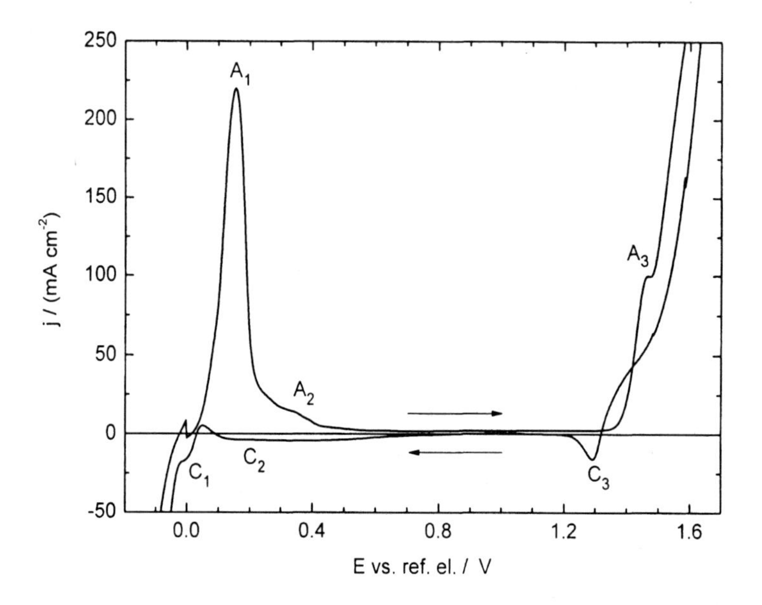

Figure 6. Cyclic voltammogram of the iron working electrode, in the NaOH-KOH eutectic system at 200 °C. Platinum crucible was used as a counter electrode, RE: Fe. Potential scan rate 50 mV s^{-1}.

A further anodic oxidation of FeO_2^{2-} resulted in FeO_2^- (peak A_2). Peak A_3 corresponds to the ferrate(VI) production. Furthermore, this is accompanied during the cathodic potential scan by a current peak C_3 (Figure 7). The analysis of C_3 peak showed 2.97±0.05 electrons reduction *i.e.* reduction of Fe(VI) to Fe(III) ion. Unfortunately, a similar analysis for peak A_3 could not be conducted because of partial overlap of the peak with the oxygen evolution reaction.

The anodic vertex potential was varied in order to confirm interconnection between the peaks A_3 and C_3. In the first potential scan the vertex potential was set to 1.4 V, *i.e.* to the potential lower than peak A_3 commences. In such a case no C_3 peak was observed (Figure 8). When vertex potential was increased to 1.6 V, both current peaks (A_3 and C_3) appeared on the voltammetric curve. This, together with n corresponding to peak C_3 confirms an assignment of peak A_3 to the ferrate(VI) production.

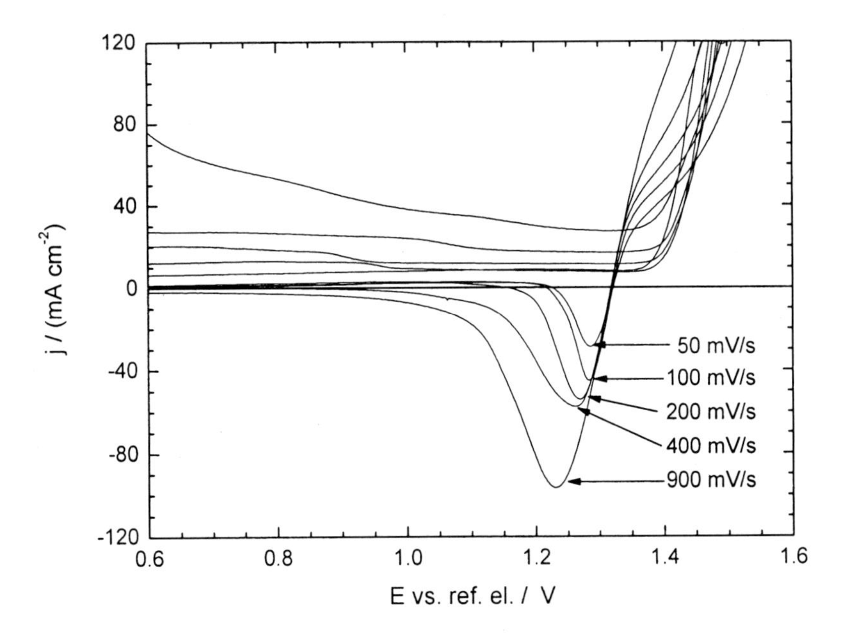

Figure 7. Cyclic voltammograms of the iron working electrode in the NaOH-KOH eutectic system at 200 °C. Detail of the cathodic current peak C_3. Scan rate is marked in the figure, anodic vertex potential 1.5 V. Platinum crucible was used as a counter electrode, RE: Fe.

Importantly, the potential of peak C_3 in Figure 6 and Figure 8 is shifted to the substantially less cathodic potentials (<0.20 V) relative to peak A_3 than reported for the aqueous solution environment in which the shift was ~0.40 V (*21*). This indicates that the Fe electrode is not covered by the compact passive layer after polarization at a strong anodic potentials. This is a critical finding with respect to the possible fast deterioration of the anode process efficiency in the electrochemical synthesis of ferrate(VI) in the presence of water. Thus, continuous degradation of passivating film with high temperature together with high concentration of OH^- increases solubility of the electrode surface layer.

As shown in Figure 9, the dependence of the C_3 peak current density on the anodic vertex potential exhibits a clear maximum. It appears at the vertex potential of 1.65 V.

Integrated cathodic charge corresponding to the peak C_3 shown in Figure 10 confirms the highest amount of the ferrate(VI) being reduced at the electrode surface for the anode vertex potential of 1.65 V. According to Denvir and Pletcher (*4*), this charge is proportional to the amount of ferrate(VI) produced in

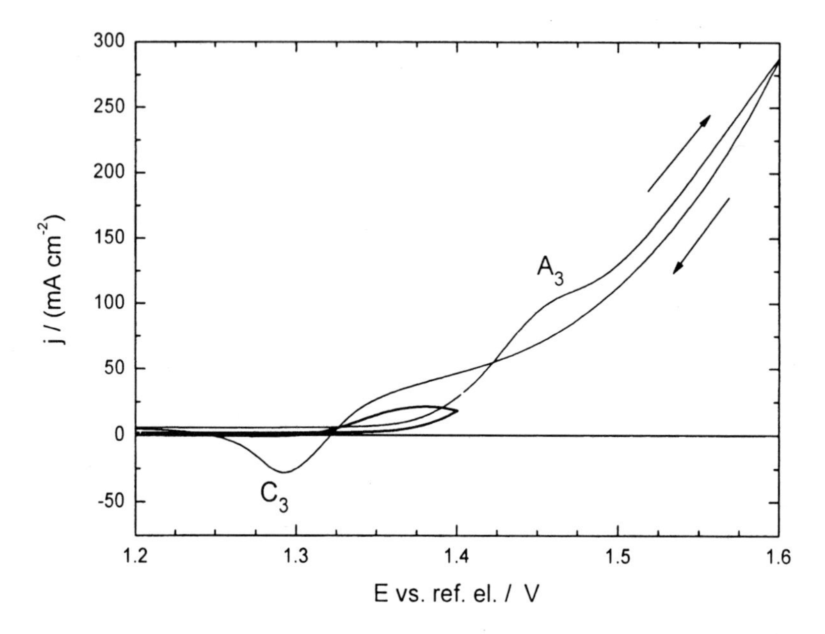

Figure 8. Cyclic voltammograms of the iron in the NaOH-KOH eutectic system at 200 °C. Detail of the cathodic current peak C_3. Anodic vertex potential 1.4 V and 1.6 V, scan rate 50 mV s^{-1}. Platinum crucible was used as a counter electrode, RE: Fe.

the anodic period of the process. This indicates decrease in the production of ferrate(VI) if the anode potential is to high. Oxygen bubbles are produced with a very high intensity at >1.65 V, which causes the substantial decrease in the anode active surface area. This phenomenon was also observed in synthesis under aqueous environment (*5, 6*).

Conclusions

Electrochemical investigations of iron behavior in the molten hydroxide system based on the eutectic mixture of KOH and NaOH were carried out. The voltammetric experiments performed demonstrate the appearance of the three-electron reversible electrode process of the FeO_4^{2-}/FeO_2^- redox couple. The ferrate(VI) can thus be synthesized electrochemically by anodic oxidation of the iron electrode as well as iron species present in a molten eutectic NaOH-KOH mixture. This synthetic method does not require separation steps to obtain solid

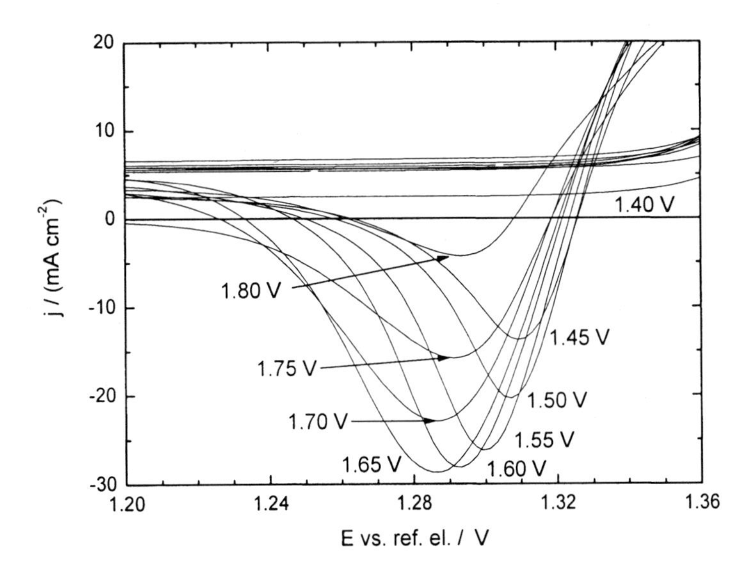

Figure 9. Cyclic voltammograms of the iron working electrode in the NaOH-KOH eutectic system at 200 °C. Detail of the cathodic current peak C_3. Anodic vertex potential is marked in the figure, scan rate 50 mV s^{-1}. Platinum crucible was used as a counter electrode, RE: Fe.

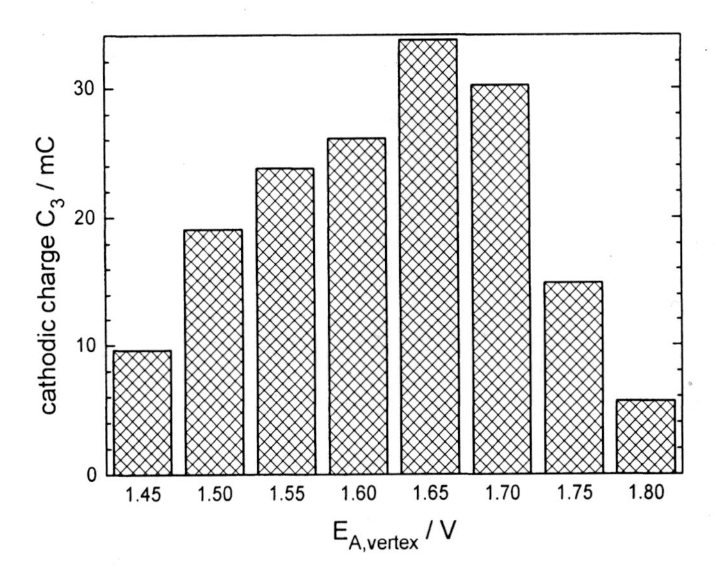

Figure 10. Cathodic charge corresponding to the peak C_3 area vs. the anodic vertex potential in the NaOH-KOH eutectic system at 200 °C, WE: Fe, CE: Pt, RE: Fe.

ferrate(VI) from the solution. It would thus facilitate applications of a ferrate(VI) as a reagent in the green chemistry and in the treatment of toxins and pollutants.

Acknowledgment

This work was financially supported by the Ministry of Education of the Slovak Republic project VEGA No. 1/2476/05 and by the Grant Agency of the Czech Republic under project No.104/05/0066. The partial support of NATO collaborative grant (EST.CLG project no. 979931) is also acknowledged.

References

1. Bouzek, K. ; Roušar, I. *J. Appl. Electrochem.* **1993**, *23*, 1317.
2. Bouzek, K. ; Roušar, I. *Electrochim. Acta* **1993**, *38*, 1717.
3. Denvir, A.; Pletcher, D. *J. Appl. Electrochem.* **1996**, *26*, 815.
4. Denvir, A.; Pletcher, D. *J. Appl. Electrochem.* **1996**, *26*, 823.
5. Bouzek K.; Roušar, I. *J. Appl. Electrochem.* **1997**, *27*, 679.
6. Bouzek, K. ; Schmidt, M.J.; Wragg, A.A. *Coll. Czech. Chem. Comm.* **2000**, *65*, 133.
7. Lescuras-Darrou, V.; Lapicque, F.; Valentin, G. *J. Appl. Electrochem.* **2002**, *32*, 57.
8. Lapicque, F.; Valentin, G. *Electrochem. Commun.* **2002**, *4*, 764.
9. Licht, S.; Naschitz, V.; Wang, B. *J. Power Sources* **2002**, *109*, 67.
10. Jiang, J.Q.; Lloyd, B. *Water Res.* **2002**, *36*, 1397.
11. Sharma, V. K. *Advances in Environmental Research* **2002**, *6*, 143.
12. Stahl, G.E. referred according to Gmelins Handbuch, Eisen, No.: 59, 1932, p. 912.
13. Poggendorf, J.C. *Pogg. Ann.* **1841**, *54*, 372.
14. Haber, F. *Z. Elektrochem.* **1900/1901**, *7*, 215.
15. Scholder, R. ; Bunsen, H. ; Kindervater, F.; Zeiss, W. *Z. anorg. allgem. Chem.* **1955**, *282*, 268.
16. Ninane, L.; Kanari, N.; Criado, C.; Jeannot, C.; Evrard, O.; Neveux, N. *Ferrates: Synthesis, Properties and Aplications in Water and Wastewater Treament*, Proceedings of the 232-nd ACS National Meeting, San Francisco, 2006, Vol.46, p. 558.
17. Dauby, C.; Glibert, J.; Claes, P. *Electrochimica Acta* **1979**, *24*, 35.
18. Miles, M.H. *J. Appl. Electrochem.* **2003**, *33*, 1011.
19. Bouzek, K.; Roušar, I.; Bergmann, H.; Hertwig, K. *J. of Electroanalytical Chemistry* **1997**, *425*, 125.
20. Damjanovic, A.; Yeh, L.-S.R.; Wolf, J.F. *J. Electrochem Soc.* **1980**, *127*, 1945.
21. Beck, F.; Kaus, R.; Oberst, M. *Electrochmica Acta* **1985**, *30*, 173.

Chapter 4

Electrochemical Behavior of Fe(VI)–Fe(III) System in Concentrated NaOH Solution

Cun Zhong Zhang[1,2,*], HongBo Deng[1], Tingting Zhao[1], Feng Wu[1,2,*], Wei Liu[3], Shengmin Cai[4], Kai Yang[1,2], and Virender K. Sharma[5]

[1]School of Chemical Engineering and The Environment, Beijing Institute of Technology, Beijing 100081, People's Republic of China
[2]National Development Center of Hi-Tech Green Materials, Beijing 100081, People's Republic of China
[3]School of Environment Sciences and Engineering, Sun Yat-Sen University, Guangzhou 510275, People's Republic of China
[4]College of Chemistry and Molecular Engineering, Peking University, Beijing 100871, People's Republic of China
[5]Chemistry Department, Florida Institute of Technology, 150 West University Boulevard, Melbourne, FL 32901

The electrochemical behavior of a Fe(VI)/Fe(III) system was investigated with both a SnO_2-Sb_2O_3/Ti electrode and a powder microelectrode. Results showed that ferrite ion (FeO_2^-), formed in solution and solid phase, was a suitable Fe(III) species for the generation of FeO_4^{2-} and for construction of a Fe(VI)/Fe(III) redox system under strong basic conditions. Solid ferrite was made by the method of molten melting a mixture of Fe_2O_3 and NaOH at 973 ± 2K. The product was confirmed to be $NaFeO_2$ by X-ray diffraction (XRD) measurements. Ferrate (Fe(VI)) was also synthesized from ferrite (solid), $Fe(OH)_3$, and Fe_2O_3 using an hypochlorite oxidation method. In this method, it was demonstrated that ferrite was the most suitable material among these Fe(III) compounds for ferrate synthesis. Results also indicated: a) FeO_4^{2-}/FeO_2^- is a suitable redox system under studied experimental conditions, b) rate and the reversibility of the redox reaction between FeO_4^{2-} and FeO_2^- were promoted markedly with the increase of temperature from 293 K to 333 K, c) the electro-oxidation of FeO_2^- is a fast two-electron transfer reaction on both the SnO_2-Sb_2O_3/Ti electrode and the $NaFeO_2$ powder microelectrode, and d) open-hearth dust of iron and steel industry could be used as a promising and valuable raw material for ferrate synthesis.

Introduction

Ferrate (Fe(VI), $Fe^{VI}O_4^{2-}$) has been proposed in wastewater treatment in the past few decades because of its dual quality of oxidant and flocculent in a single dose (*1,2*). Ferrate has also been used in organic synthesis (*3*) since it has a strong but controllable oxidative potential. A new application of ferrate in a super-iron battery was first reported by Licht in 1999 (*4*) and since then many investigations have been carried out on this subject (*5,6*). Fabricating a safe and rechargeable super-iron battery is a potential application of ferrate because this battery is environmentally-friendly with high specific energy. Until recently, chemical, electrochemical, and high temperature methods (*3,5-7*) have been developed for ferrate synthesis.

There are some variations on the chemical composition and formal potential (or standard potential) of the Fe(VI)/Fe(III) system in previous reports (*8-10*). Additionally these results showed dependence on experimental conditions. Moreover, the results indicated that the Fe(VI)/ Fe(III) system is not a simple redox system. Besides, it also indicated that the published Latimer profile of Fe(VI)/Fe(0) (*9*), especially the Fe(VI)/Fe(III) part, is incomplete. First of all, such information might play an illusive role in the development of selective electrochemical synthesis of ferrate and even in the manufacture of safe and rechargeable batteries. Furthermore, insufficient clarity may limit the communication between electrochemical research and research of other fields and also affects the application of the Marcus theory on the Fe(VI)/Fe(III) redox system (*11*).

Thermodynamic data of different valence state Fe species indicate that Fe(III) is the most stable species under most of the environmental conditions.

Therefore, synthesis of ferrate from pure iron by electrochemical methods did not completely exhibit the actual charge/discharge process of the positive electrode material of a super-iron battery. It should be noticed that the most published potential of ferrate generation is higher than that of *Oxygen Evolution Reaction* (*OER*). *OER* thus induce a sharp increase of the inner pressure of the battery and will possibly result in a dangerous blast during charging process of the secondary super-iron battery. This will also reduce the high efficiency electro-generation of ferrate in simple unsealed electrochemical equipment. Hence, the study of the selective electro-chemical synthesis of ferrate is critically needed. In addition, the electro-generation of ferrate will broaden our knowledge of the behaviors of such a high-valent oxidant under extreme conditions.

This paper addresses issues in selective electro-generation of ferrate using different Fe(III) compounds. Special electrode materials with high over-potential of *OER* could be used as an effective tool for the exploration of electro-generation of strong oxidants in aqueous solution. In the last century, dimensionally stable anode (DSA) has played important role in many fields for its excellent properties. The electrochemical behavior of the molecule on different electrode materials provides useful and meaningful information. Therefore DSA should be used as an effective and helpful tool for extreme reaction conditions.

The open-hearth dust in the iron and steel industry in Inner Mongolia of China has been widely used (*12,13*). Uses include manufacturing of catalysis and magnetic materials. However, the electrochemical property of the main ingredient, Fe_2O_3, of the dust has not been investigated. In this paper, we have used SnO_2–Sb_2O_3/Ti electrodes and powder microelectrode techniques (*14*) to understand the mechanism of the Fe (VI)/Fe (III) system.

Experimental

Electrochemical measurements were carried out using a CHI660A electrochemical workstation (CH Instruments). The conventional three-electrode electrochemical cell was used for electrochemical experiments. A Hg/HgO (in 14M NaOH) electrode as a reference electrode (RE) was set in an independent compartment and introduced into electrolytic cell by a salt bridge having layers of basic-resistant semi-permeable membranes and Luggin capillary. A sintered Pt/Ti sheet electrode was used as a counter electrode. Two types of microelectrodes, SnO_2–Sb_2O_3/Ti electrodes and powder microelectrode (*14*), were used as working electrodes. The coating number used was 20 times in order to keep the stability of the SnO_2–Sb_2O_3/Ti electrodes in concentrated NaOH aqueous solution. A piece of platinum wire with diameter of 0.1mm was used to fabricate powder microelectrode. Each kind of investigated sample powder, Fe_2O_3, Fe $(OH)_3$, and $NaFeO_2$, was loaded in the micro-hole at the very tip of powder microelectrode. A $NaFeO_2$ powder was made from the mixture of Fe_2O_3 and NaOH at a molar ratio of 1.2 : 1 (Na : Fe) at 973 ± 2 K. The $Fe(OH)_3$ was prepared by mixing of $Fe(NO_3)_3$ and NaOH solutions at pH = 7 - 8. The $Fe(OH)_3$ deposition was washed thoroughly by redistilled water. Concentrated NaOH aqueous solutions having no FeO_2^- ion and 0.02 M FeO_2^- ion (*15*) were used as blank and working electrolyte solutions, respectively. Saturated hypochlorite solution was prepared by dissolving Cl_2 gas into 14.0 -14.5 M NaOH solution. All of the reagents were of analytic grade. The XRD analysis was carried out on a Dmax-2500 diffractometer using Cu Kα radiation. The UV-Visible spectroscopic measurements were carried out with the Hitachi U-2800 Spectrophotometer.

Results and Discussion

Chemical Behavior of Fe(III) Compounds

Figure 1 shows the cyclic voltammograms (CVs) on SnO_2–Sb_2O_3/Ti electrodes in blank and working solutions at 293 K. The blank solution showed no peak and the value of the onset potential of *OER* was ~1 V. It demonstrated that the over-potential of *OER* on SnO_2–Sb_2O_3/Ti electrodes was considerably high in 14 M NaOH solution. In working solution, one anodic peak (P_a) and one cathodic peak (P_c) appeared in the CV curves. It is difficult to produce the integrated anodic peak in strong basic solution on an ordinary electrode due to the limitation of electrochemical windows. According to previous reports (*5,15-17*), the integrated anodic peak and cathodic peak are related to electro-generation and electro-reduction of ferrate, respectively.

Addition of Fe_2O_3 powder did not change the CV behavior in NaOH solution, possibly due to the low solubility of this powder in alkaline

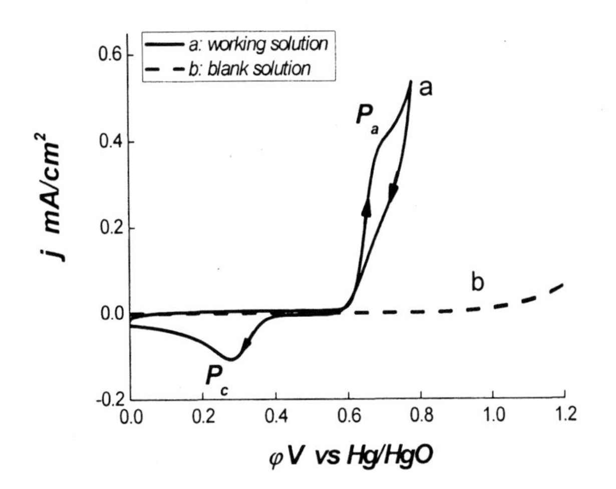

Figure 1. Cyclic voltammograms of SnO_2–Sb_2O_3/Ti electrodes in: (a) 14 M NaOH containing 0.02M FeO_2^- , (b) 14 M NaOH blank solution, scan rate: 10 mV/s, the arrows indicate the scan direction.

solution. The soluble FeO_2^- may thus be a suitable species for the electro-generation of ferrate. Results using FeO_2^- have reported that the onset potential and peak potential of the anodic peak, P_a, were almost kept at the same position, and no change occurred due to the surface component of electrode materials and scan rate (*5,15,16*). However, the peak potential of the cathodic peak, P_c, was easily affected by these variables (*5,6,15-17*). This indicates that the electro-reduction reaction of ferrate differs from the electro-oxidation of FeO_2^-. In order to find a suitable ferric ions source for the electro-generation of ferrate, different Fe(III) compounds were investigated. The XRD profile of $NaFeO_2$ is shown in Figure 2.

The reactions of Fe(III) compounds, Fe_2O_3, $Fe(OH)_3$, and $NaFeO_2$ with saturated hypochlorite solution were investigated by a UV-Visible spectrophotometer at 293 ± 0.5K. The results are shown in Figure 3. Both $Fe(OH)_3$ and $NaFeO_2$ showed a maximum absorption peak at 505 nm and an absorption shoulder at 570 nm. However, the peak intensity obtained in the use of $NaFeO_2$ is more remarkable than that of $Fe(OH)_3$. Meanwhile, there was no absorption peak obtained in using Fe_2O_3. It appears that the Fe_2O_3 could not be dissolved in saturated hypochlorite basic aqueous solution to have any reaction with the hypochlorite ion. The spectral results of Fe_2O_3 in saturated hypochlorite solution are in agreement with the result of electro-

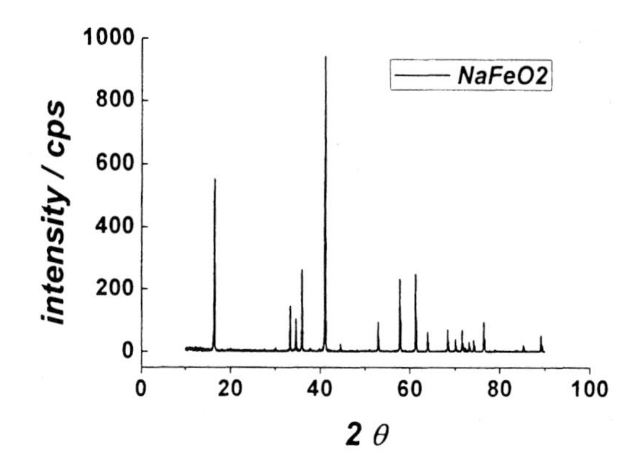

Figure 2. The XRD results of a $NaFeO_2$ sample

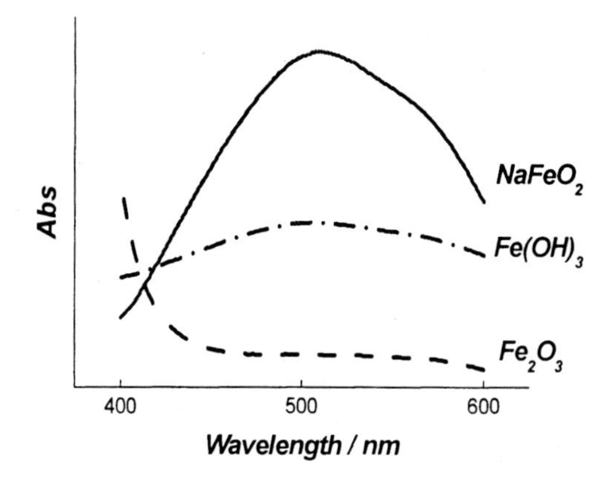

Figure 3. Spectra of solution obtained in the mixture of solid state Fe(III) compounds (equi-molar of Fe atom) with saturated hypochlorite solution.

chemical investigation (see Figure 1). This comparative result demonstrated that FeO_2^- ion is the most active Fe(III) species for the generation of ferrate in concentrated basic solution. The use of $Fe(OH)_3$ showed a weak acidity to form FeO_2^- in a basic solution.

The use of molten basic electrolyte may have a suitable to ensure activity of Fe_2O_3. The reaction (1) may occur at 973 ± 2 K to give sodium ferrite.

$$2\ NaOH + Fe_2O_3 \rightarrow 2\ NaFeO_2 + H_2O \qquad (1)$$

The equation (1) indicates that the ferrite could be generated by thermo-chemical reaction.

The electrochemical behavior of different Fe(III) compounds was also investigated by using a powder microelectrode technique in 14 M NaOH solution at 293 K (Figure 4). The results indicate that no reaction occurred on the Fe_2O_3 electrode. *OER* was a unique obvious phenomenon until the potential exceeded 0.75 V. Under the same conditions, a weak hump and an obvious current peak in the potential range from 0.6 to 0.7 V, appeared on the segment of positive-going sweep of CV curves for both $Fe(OH)_3$ and $NaFeO_2$. However, obvious *OER* could also be found after the weak hump on the curve for $Fe(OH)_3$. There were weak and strong cathodic peaks from ~ 0.4 to 0.2 V for $Fe(OH)_3$ and $NaFeO_2$, respectively, during the segment of negative-going sweep on CV curves. Results clearly demonstrate that FeO_2^- ion was the most suitable Fe(III) compound for the electro-generation of ferrate in strong basic solution.

The electro-generation of ferrate from $Fe(OH)_3$ and $NaFeO_2$ were also investigated by the cyclic voltammetry method in 14 M NaOH solution at different temperatures. The results are shown in Figure 5. The generation rate and reversible degree of the electrochemical redox reaction on either $NaFeO_2$ or $Fe(OH)_3$ were remarkably promoted with the increase in temperature. In addition, electro-generation of ferrate using $NaFeO_2$ could be separated from *OER* in comparison with $Fe(OH)_3$ at all studied temperatures. Thus, $NaFeO_2$ is the most suitable material to form ferrate.

With the increase in temperature, an anodic peak also grew gradually on the $Fe(OH)_3$ electrode. The acidic property and acidic ionization of $Fe(OH)_3$

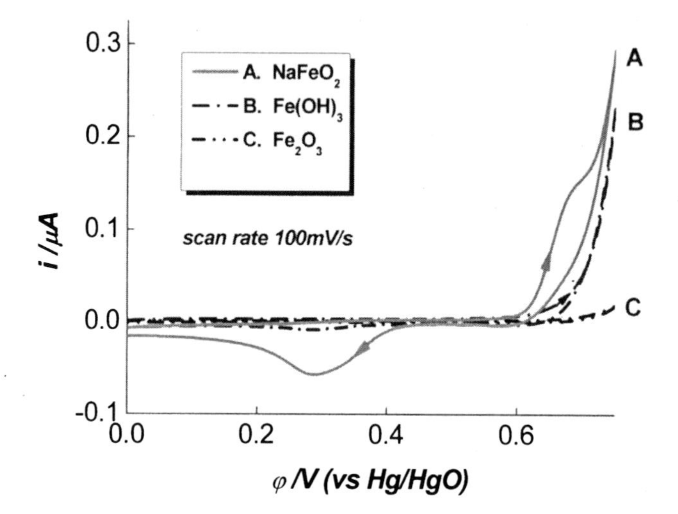

Figure 4. The cyclic voltammetric curves for Fe(III) compounds scan rate: 100mV/s, arrows indicate scan direction (Reproduced with permission with J. Functional Material.*)*

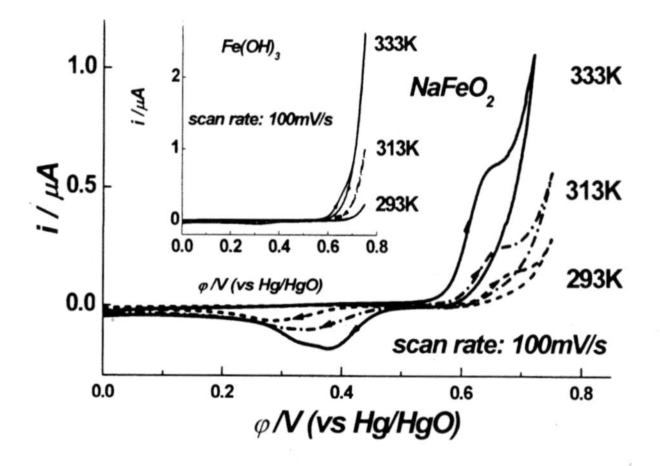

Figure 5. The cyclic voltammetric curves for $Fe(OH)_3$ (inset profile) and $NaFeO_2$ at three temperatures, scan rate: 100mV/s, arrows indicate scan direction (Reproduced with permission with J. Functional Material.*)*

became stronger at higher temperature. This phenomenon is consistent with earlier reported results (*15,17*).

Mechanism of Electro-Generation of Ferrate

In general, the apparent potential of the Fe (VI)/Fe(III) system (*3,8,10*) is higher than that of *OER*. This inhibits the selective electro-generation of ferrate in aqueous solution. It is thus necessary to understand the electrochemical mechanism for the reorganization of the Fe (VI)/Fe (III) system for selective electro-generation of ferrate. The parameter of $\left|\varphi_{3/4} - \varphi_{1/4}\right|$ of the anodic peak, P_a in Figure 1 was investigated in different NaOH solutions and results are given in Table 1.

Table. 1. $\left|\varphi_{3/4} - \varphi_{1/4}\right|$ in different concentrated NaOH solutions

C_{NaOH}(M)	10	12	14	16	18
$\left|\varphi_{3/4} - \varphi_{1/4}\right|$(mV)	28.5	28.3	28.5	29.1	29.5

Table 1 suggests that the step corresponding to the appearance of P_a should be a fast step with a two-electron transfer in the total process of CEC (*6,18*). This step can be described as follows:

$$FeO_2^- + 2e^- \rightarrow FeO_4^{3-} \qquad (2)$$

The characteristic parameter $\left|\varphi_{P_a/2} - \varphi_{P_a}\right|$ of the anodic peak, P_a on solid state ferrite was 30 ± 1 mV in 14 M NaOH aqueous solution (see Figure 4). This value is close to the corresponding parameter P_a on SnO_2–Sb_2O_3/Ti electrode having 0.02 M FeO_2^- in 14 M NaOH aqueous solution. The number of transferred electrons in an electrochemical step on solid state $NaFeO_2$ electrode was determined as approximately 2.0. This indicates the occurrence of an electrochemical reaction (2) on the solid state ferrite electrode. This also implies that the electrochemical reaction (2) is a fast electron-transfer reaction (or a facile charge transfer reaction) and can not be affected by electrode material and scan rate. This finding might be helpful to explore parameters of the Fe(V)/Fe(III) redox couple and electrochemical behaviors of the Fe(VI)/Fe(III) system. According to work reported in the literature (*2,11,19*), the redox couple of Fe(V)/Fe(III) might be a reversible redox couple and possibly participates in the redox system of Fe(VI)/Fe(III). The potential of redox couple Fe(V)/Fe(III) should be higher than that of the Fe(VI)/Fe(III) redox system (i.e. $\varphi_{Fe(V)/Fe(III)} > \varphi_{Fe(VI)/Fe(III)}$). Apparently, the detection of the corresponding potential of the FeO_4^{3-}/FeO_2^- redox couple must be necessary to obtain the precise potential value of the Fe(VI)/Fe(III) redox system. Moreover, this would help to modify the known Latimer profile of the redox system (*9*). Unfortunately, until now, the electrochemical behavior of FeO_4^{3-} (Fe(V) species) could not be successfully achieved under the experimental conditions due to the short-life of FeO_4^{3-} species (*2,19*). It should also be one of possible reasons for the variation in the potentials obtained in the Fe(VI)/Fe(III) redox system.

In this investigation, the nature of P_c and the relationship between P_c and P_a were used to learn the cause of short-life of FeO_4^{3-} and to understand the mechanism of electro-generation of ferrate. The results are shown in Figure 6. In the potential range from 0.5 to 0.0 V, the cathodic peak of the CV curve of a mixed powder micro-electrode, containing $NaFeO_2$ and K_2FeO_4 showed that P_c must arise from electro-reduction of the FeO_4^{2-} ion on the $NaFeO_2$ electrode. This suggests that the disproportionation reaction of FeO_4^{3-} may be happening according to the equation (3).

$$Fe(V) \rightarrow Fe(VI) + FeO_2^- \qquad (3)$$

Reaction (3) is different from previous results (*16*), which reported disproportionation of Fe(V) species. However, reaction (3) is a fast homogenous reaction in the electrochemical step and is not affected by the electrode material. These combined results with the UV-Visible spectral measurements suggest that the amount of FeO_2^- and electro-generated FeO_4^{3-} can be expected considerably different on $Fe(OH)_3$ and $NaFeO_2$ electrodes.

The amount of formed FeO_4^{2-} would thus be different on investigated electrodes. Thus weak and strong Faradaic current signals on $Fe(OH)_3$ and $NaFeO_2$ electrodes, respectively, could be used as a reliable evidence to demonstrate the disproportionation reaction and mechanism on both used electrodes. Interestingly, the formal potential of ClO^-/Cl^- is larger than the $\varphi_{1/2}$ of P_a, which may be used as a promising path to explain the property and process of FeO_4^{3-} from the oxidation reaction of FeO_2^-.

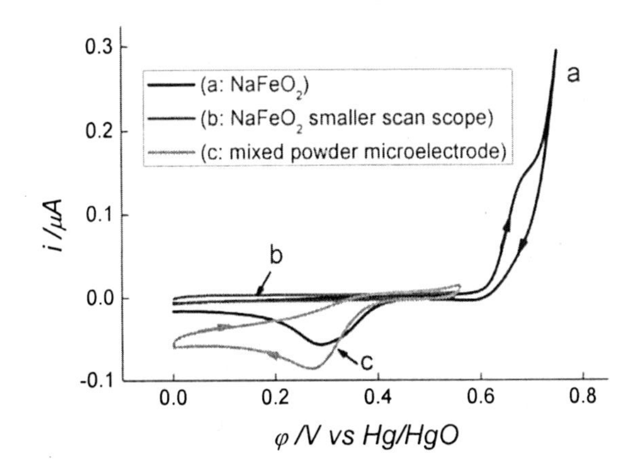

Figure 6. The cyclic voltammetric curves for the $NaFeO_2$ and mixed powder microelectrode, containing $NaFeO_2$ and K_2FeO_4; scan rate: 100mV/s, arrows indicate scan direction

It should be pointed out that the peak potential of the cathodic peak varied significantly with different electrode materials (*5,6,8,15-17*). The difference should be induced by the surface components of electrode materials. In addition, there was no opposite anodic peak of the cathodic peak on CV curves in previous work. All these results confirm that electro-reduction of FeO_4^{2-} ion is an irreversible electrochemical reaction. Based on the knowledge of electrode kinetics and previous results (*2,11,19*), the possible product of the first step of electro-reduction of FeO_4^{2-} should be

Fe(V). Due to the presence of over-potential, the formal potential of FeO_4^{2-} /Fe(V) should be situated in the potential scope from 0.3 to 0.6 V (vs. Hg/HgO in 14 M NaOH). Although the mechanism of electro-reduction is still unclear, the peak potential of electro-reduction of ferrate on $NaFeO_2$ is almost equal to that of ferrate on Pt (*9,15,17*).

Acknowledgements

This work was supported by the National Key Research and Development Program (Grand No. 2002CB211800), the National Key Program for Basic Research of China (2001CCA05000), Teaching and Researching Fund of Beijing Institute of Technology (05120404), and Excellent Young Scholars Research Fund of Beijing Institute of Technology (000Y05-19). Authors thank anonymous referees for the knowledgeable advice and valuable guidance.

References

1. Waite, T.D.; Gilbert, M. *J. Water Pollution Control Fed.*, **1978**, *50*, 543
2. Sharma, V.K. *Adv. Environ. Res.* **2002**, *6*, 143.
3. Delaude, L.; Laszlo, P. *J. Org. Chem.*, **1996**, *61*, 6360.
4. Licht, S.; Wang, B.H.; Ghosh, S. *Science,* **1999**, *285*, 1039.
5. Koninck, M.D.; Bélanger, D. *Electrochim. Acta.,* **2003**, *48*, 1425.
6. Zhang, C.Z.; Liu, Z.; Wu, F.; Qi, F. *Electrochem. Comm.*, **2004**, *6*, 1104.
7. Kopelev, N.S.; Perfiliev, Y.D.; Kiselev, Y.M. *J. Radioanal. Nucl. Chem,* **1992**, *162*, 235.
8. Licht, S.; Naschitz, V.N.; Halperin, L.; Haperin, N.; Halperin, L.; Lin, L. Chen, J.; Ghosh, S.; Liu, B. *J. Power Sources.*, **2001**, *101*, 167.
9. Bard, A.J.; Parsons, R. *Standard Potentials in Aqueous Solution*, Marcel Dekker, INC. NewYork and Basel. **1985**, p391.
10. Wood, R.H. *J. Am. Chem. Soc.*, **1958**, *80*, 2038.
11. Johnson, M.D.; Sharma, K.D. *Inorg. Chim. Acta,* **1999**, *293*, 229.
12. Xu, A.J.; Liu, S.C. *J. Inner Mongolia Normal University, (Natural ScienceEdition),* **2003**, *32(3)*, 388.

13. Xu, A.J.; Liu, S.C.; Zhang, Q.; Liu, S.T. *J. Inner Mongolia Normal University, (Natural Science Edition)*, **2002**, *31(4)*, 365.
14. Cha, C.S.; Li, M.C.; Yang, H.S. *J. Electroanal. Chem.* **1994**, *368*, 47.
15. Bouzek, K.; Roušar, I.; Bergmann, H.; Hertwig, K. *J. Electroanal. Chem.*, **1997**, *425*, 125.
16. Bouzek, K.; Roušar, I. *J. Appl. Electrochem.*, **1993**, *23*, 1317.
17. Beck, F.; Kaus, R.; Oberst, M. *Electrochim.Acat.*, **1985**, *30(2)*, 173.
18. Bard, A.J.; Faulkner, L.R. *Electrochemical methods: Fundamentals and applications*, New York: John Wiley, **2001**, Chapter 5.
19. Rush, J.; Bielski, B.H.J. *J. Am. Chem. Soc.*, **1986**, *108*, 525.

Chapter 5

Preparation of Potassium Ferrate by Wet Oxidation Method Using Waste Alkali: Purification and Reuse of Waste Alkali

Jiang Chengchun[1], Liu Chen[2], and Wang Shichao[2]

[1]School of Civil and Environmental Engineering, Shenzhen Polytechnic, Shenzhen, China 518055
[2]Department of Urban and Civil Engineering, Shenzhen Graduate School, Harbin Institute of Technology Shenzhen, China 518055

A new method of preparing potassium ferrate using waste alkali is developed in this report. After preparation of potassium ferrate by the wet oxidation method, the waste alkali was purified and reused for a further preparation runs. The purification of waste alkali and the temperature for the furification were studied. The results indicated that the waste alkali can be used for preparing potassium ferrate, and the purity and yield of potassium ferrate product were steadily higher than 90% and 60%, respectively after ten recycles of the waste alkali. Therefore, due to the use of waste alkali, the cost is reduced sharply, and a green synthesis for potassium ferrate is achieved.

Introduction

Ferrate(VI) ion of FeO_4^{2-} is a very strong oxidant. Under acidic conditions, the redox potential of ferrate is greater than ozone and is the highest of all the oxidants practically used for water and wastewater treatment (*1*). Moreover, during the oxidation process, ferrate is reduced to Fe(III) ions or ferric hydroxide. This suggests that the ferrate is a dual-function chemical reagent, which has the potential to perform both oxidation and coagulation in a single treatment step (*2*). Thus, in recent years considerable attention has been paid to the aspects of ferrate treatment such as the inactivation of micro-organisms (*3*), its reactivity with a wide range of aqueous contaminants, such as ammonia and heavy metals (*4*) and treatment of industrial and municipal wastewaters (*5-7*). Yet, owing to the instability and the preparation economics, ferrate is not available commercially. Currently, there is a need for further studies to improve the method of ferrate preparation, such as decreasing the cost and increasing the yield and stability.

In general, there are three methods for the synthesis of ferrate (*8*): (1) the electrochemical method by anodic oxidation of iron in a KOH electrolyte solution. The production yield is revealed to be strongly dependent on the electrolyte concentration and current density; (2) the dry method by which various iron-oxide-containing minerals are melted under extremley alkaline and aerobic conditions. This method proves to be quite dangerous and difficult, since the synthesis process could cause detonation at elevated temperatures; (3) the wet method by which a Fe(VI) salt is oxidized under extremely alkaline conditions by either hypochlorite or chlorine. Among the three approaches, the wet oxidation method has been well developed. However, owing to the complication of the procedure, high cost and harmful environmental impact, application of this method in a large scale has not been realized.

In view of the disadvantages suffered by the wet oxidation method for preparing potassium ferrate, a cost-effective and green preparation method was evaluated. In the present work, efforts have been exerted to explore the possibility of preparing potassium ferrate salt using waste alkali, which resulted in serious waste of chemicals and pollution if discharged directly after the preparation of potassium ferrate. In the first part of the study, the method of purifying waste alkali was investigated. In the second part, the operation condition which avoids the co-precipitation of impurities with potassium ferrate when using purified waste alkali was studied.

Materials and Methods

Chemicals

The main chemicals used were ferric nitrate (from Guangzhou tanshan yueqiao chemical factory), potassium hydroxide (from Guangzhou taishan chemical reagent and plastic Ltd.), potassium permanganate (from Guangzhou zhuhai chemical reagent factory.), and hydrochloric acid (from Guangzhou donghong chemical plant.). All chemicals used in this work were analytical reagent grade and used without any further purification. The solutions were prepared with water that had been distilled and then passed through an 18MΩ Milli-Q water purification system.

Experimental procedures

Potassium ferrate preparation

The procedure of solid potassium ferrate preparation was followed by modifying the method of reaction between OCl^- and $Fe(NO3)_3$ in strongly basic media and isolating the as-prepared ferrate from the saturated KOH solution(*5*). Two aspects of modification have been conducted: (1) After preparation of potassium ferrate by the wet oxidation method, the waste alkaline solution was purified and reused for further preparation runs instead of fresh KOH solution; (2) The temperature of purifed waste alkaline solution used for re-precipitation (isolating) of as-prepared ferrate was at ambient temperature (23±2 °C) instead of chilling the KOH solution.

Waste alkali purification

The waste alkaline solution produced during the process of precipitation and re-precipitation of potassium ferrate was mixed and cooled to 0 °C in an ice bath. Next, the waste alkaline solution was stirred rapidly while a certain amount of KOH was added slowly into the solution until it was saturated by the KOH under cooling conditions. Stirring was continued for additional 30 min at 0 °C. Subsequently, the resultant suspension was filtered with a glass filter (P-4), then the precipitate was discarded, and the filtrate was collected for subsequent use in further preparation of potassium ferrate.

Analytical methods

The purity of potassium ferrate was determined by the Arsenite-Bromate method (*9*), which is based on the reduction of the ferrate to ferric ion in alkaline arsenite solution. A weighted sample of ferrate is added to a standard alkaline arsenite solution, in which the amount of arsenite is larger than that required for the reduction of ferrate ions, then the excess arsenite is back-titrated with standard bromate and the equivalent of consumed bromate is calculated.

The concentration of the waste alkali was determined by the conventional volumetric titration method, however, the violet color of potassium ferrate in the waste alkali must be eliminated by alcohol rather than acid.

The concentration of chloride and nitrate ions was determined by ion chromatography (ICS — 1500, DIONEX Co.). Samples of 24 μL were injected into the separating column, where the ions were separated by ion exchange according to their size and charges. The elute is 9 m mol sodium carbonate solution whose signal was eliminated by suppressor column. The column and cell temperature were 30 °C and 35 °C, respectively.

Result and Discussion

Purification of waste alkali

Some authors (*10*) conducted the experiments of using the waste alkaline solution without any purification to prepare potassium ferrate. They found that the waste alkaline solution could be used directly in the potassium ferrate synthesis, but the impurities accumulate in the waste alkaline solution, and the purity of prepared potassium ferrate decreased to 12% after recycles of the waste alkaline solution, According to the research by Thompson et al. (*11*), the impurities in the waste alkaline solution were mainly potassium nitrate and potassium chloride, thus it was crucial to remove these impurities before reusing the waste alkaline solution.

Figure 1 shows the solubility of potassium chloride at different concentrations of potassium hydroxide. Evidently, there was a pronounced decrease in the solubility of potassium chloride as the potassium hydroxide molarity increased, indicating the feasibility to remove the potassium chloride impurities by adding potassium hydroxide into the waste alkaline solution until it was saturated. However, adding potassium hydroxide into the waste alkaline solution directly at ambient temperature (herein room temperature) to purify the waste alkaline solution was not a cost-effective choice. As we know, higher

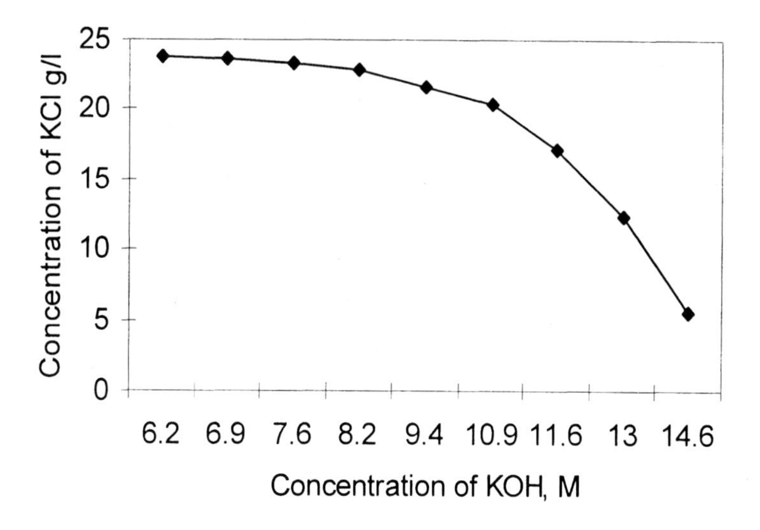

Figure 1. Solubility of KCl in aqueous solution of potassium hydroxide (Data from Reference 11).

temperatures result in higher addition of potassium hydroxide when potassium hydroxide is used to saturate the waste alkaline solution. In addition, higher temperatures also cause the increase of the impurity solubility. The equilibrium concentration of chloride and nitrate ions in 11 M waste alkaline solution under different temperatures are shown in Table 1. It can be seen that the concentration of chloride and nitrate ions appears to decrease with the decreasing temperature of waste alkaline solution, and the precipitation of the chloride and nitrate ions is accomplished by making the temperature of waste alkaline solution decrease from 25 °C to 0 °C .

The method of purification of waste alkaline solution in this study includes three steps. First, the waste alkaline solution produced from preparation of potassium ferrate was cooled to 0 °C in an ice-bath. Second, a certain quantity of potassium hydroxide was added continuously by until the solution was saturated by potassium hydroxide at 0 °C. Last, the precipitated impurities of potassium chloride and potassium nitrate were removed by filtration through a fritted glass (P-4).

Table 1. The Concentrations of Chloride and Nitrate Ions in 11 M Waste Alkaline Solution under Different Temperatures

Temperature °C	*25*	*20*	*15*	*10*	*5*	*0*
Cl^- g/l	9.35	8.41	8.15	7.44	6.74	5.83
NO_3^- g/l	19.87	17.00	15.80	14.16	12.20	10.14

Re-precipitation of potassium ferrate using purified waste alkali

Although the purification of the waste alkaline solution was carried out, there was still a portion of impurities of potassium chloride and potassium nitrate remained in the purified waste alkaline solution. It was also crucial to avoid co-precipitation of impurities when using purified waste alkali to re-precipitate potassium ferrate. The method of purifying the crude potassium ferrate by Thompson et al. (*11*) and Delaude et al. (*5*) was to re-precipitate the crude potassium ferrate using chilled saturated potassium hydroxide solution. It is feasible for fresh saturated potassium hydroxide solution, but for purified waste alkali, low temperature will result in the co-precipitation of impurities in some extent.

Figure 2 displayed that the solubility of both potassium ferrate and potassium chloride decreased with the increase in the concentration of potassium hydroxide, and the solubility of potassium ferrate changed insignificantly but the potassium chloride changed sharply when the concentration of potassium hydroxide changed from 10 M to 13 M. Considering the yield and purity of potassium ferrate, the final solution of around 11 M in the potassium hydroxide was suitable for the re-precipitation of crude potassium ferrate when using the purified waste alkali.

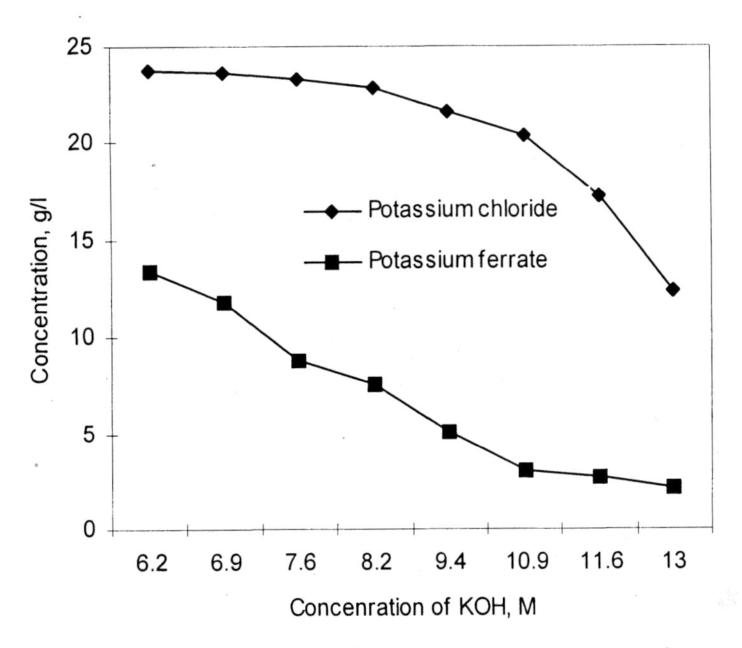

Figure 2. Solubility of KCl and K_2FeO_4 in aqueous solution of potassium hydroxide (Data from Reference 11).

The equilibrium concentrations of potassium ferrate in the 11 M waste alkaline solution under different temperaturse are shown in Table 2. It can be seen that the concentration of potassium ferrate changed very little (from 1.61 g/L to 2.78 g/L) when the temperature of waste alkaline solution changed from 0 °C to 25 °C . Whereas, the concentration of both chloride and nitrate ions increased dramatically from 5.82 g/L to 9.35 g/L and from 10.14 g/L to 19.87 g/L, respectively (Table 1). The data indicated that the re-precipitation of the bulk of the potassium ferrate was accomplished by making the ambient temperature (23±2 °C) of purified waste alkaline solution feasible, but the major part of the potassium chloride impurity as well as the potassium nitrate would still be in solution.

Table 2. The Concentrations of Potassium Ferrate in 11 M Waste Alkaline Solution under Different Temperatures

Temperature °C	*25*	*20*	*15*	*10*	*5*	*0*
K_2FeO_4 g/l	2.78	1.91	1.88	1.73	1.67	1.61

Further, we carried out the re-precipitation of the crude potassium ferrate by using the solution of purified waste alkaline solution at ambient temperature (23±2 °C) rather than under chilled conditions, and the final solution is ca. 11M in potassium hydroxide. By this method of re-precipitation operation, the purity and yield of potassium ferrate product were steadily higher than 90% and 60% respectively after ten recycles using the purified waste alkali.

Conclusion

The results presented here indicated that the waste alkali can be used for preparing potassium ferrate, and the purity and the yield of potassium ferrate product were steadily higher than 90% and 60% respectively after ten recycles using the purified waste alkali. Therefore, due to the use of waste alkali, the cost is reduced sharply, and a green preparation for potassium ferrate is achieved.

Acknowledgement

The National Nature Science Foundation of China (No. 50478049; 50678104) and Nature Science Foundation of Guangdong Province (No.04011215) are acknowledged for the financial support. The valuable advice of Dr. Liu, Hong in the preparation of this report is also appreciated.

References

1. Jiang, J. Q.; Lloyd B. *Water Res.* **2002**, *36*, 1397-1408.
2. Deluca, S. J. ; Cantelli, M. *Water Sci. Technol.* **1992**, *26*, 2077-2080.
3. Kazama, F. *Water Sci. Technol.* **1995**, *31*, 165-168.
4. Murmann, R. K. ; Robinson, P. R. *Water Res.* **1974**, *8* ,167-176.
5. Delaude, L.; Laszlo, P. *J. Org. Chem.* **1996**, *61*, 6360-6370.
6. Johnson, M. D.; Read, J. F. *Inorg. Chem.* **1996**, *35*, 6795-6799.
7. Sharma, V. K.; Rivera, W.; Joshi, V. N.; Millero, F. *J. Environ. Sci. Technol.* **1999**, *33*, 2645-2650.
8. Li, C.; Li, X. Z.; Graham, N. *Chemosphere*, **2005**, *61*, 537-543.
9. Graham, N.; Jiang, C. C.; Li, X. Z. et al. *Chemosphere*, **2004**, *56*, 949-956.
10. Tian, B. Z.; Qu, J. H. *Environmental Chemistry*, **1999**, *18*, 173-177 (in Chinese).
11. Thompson G. W.; Ockerman L.T. and Schreyer J.M. *J. Am. Chem. Soc.* **1951**, *73*, 1379-1381.

Chapter 6

New Processes for Alkali Ferrate Synthesis

L. Ninane[1,*], N. Kanari[2], C. Criado[1], C. Jeannot[3], O. Evrard[3], and N. Neveux[4]

[1]Solvay SA, 1 rue Gabriel Péri 54110 Dombasle sur Meurthe, France
[2]Institute National Polytechnique Lorraine, Nancy, France
[3]University Henry Poincare, Nancy, France
[4]NanciE, Nancy, France
[*]Corresponding author: Leon.ninane@solvay.com, leon.ninane@wanadoo.fr

A new process for the synthesis of the potassium ferrate(VI) salt was developed at the University of Nancy, France under an EEC program; started in 2001. This program had an objective of synthesizing a large quantity of ferrate in order to feed large scale applications in the field of water treatment such as drinking water, municipal waste water, and industrial waste water treatment. The raw materials used were ferrous sulphate, potassium hydroxide, and calcium hypochlorite (or chlorine). In this process, mixing of three solids took place in a mixer in which the potassium ferrate(VI) salt was stabilized. Another objective of the program was to develop synthesis of solid sodium ferrate(VI), which is cheaper to produce because its preparation requires less expensive materials: caustic soda instead of potassium hydroxide and sodium hypochlorite (or chlorine gas) instead of calcium hypochlorite. The final objective was to develop a better technology, which could be cheaper and easier to scale-up. This chapter also describes successful results of lab and small pilot tests.

Introduction

Iron(VI) has been known for a long time; although it has been observed in 1841 by Fremy, it is still not well known (*1,2*). Iron salts are known only in the ferrous (Fe(II)) and the ferric forms (FeIII); the higher oxidized forms Fe(IV, Fe(V), Fe(VI) give unstable compounds designated under the general name of ferrate. Sometimes all iron salts from Fe(II) to Fe(VI) are called ferrate. Ferrate(VI) is not registered at the European inventory of chemicals (Einecs). The FeO_4^{2-} ion or ferrate (Fe(VI)) has raised considerable interest because of potentially interesting applications in the treatment of water and wastewater.

Physicochemical Properties of Ferrate(VI)

The dry crystalline potassium ferrate(VI) is stable if protected against humidity. When this salt is heated, it decomposes at temperature higher than 250 °C with a release of oxygen. This release is not accompanied by immediate reduction of Fe(VI) to Fe(IV). Ferrate(V) may appear as intermediate state of reduction of Fe(VI) depending on temperature. The red-violet aqueous solution of ferrate(VI) is unstable and gradually decomposes according to the equation:

$$2\,FeO_4^{2-} + 5\,H_2O \;=\; 2\,Fe(OH)_3 + 4\,OH^- + 3/2\,0_2 \qquad (1)$$

The rate of decomposition depends on the temperature and the alkalinity. The ferrate is more stable at high alkalinity and low temperatures. The decomposition of ferrate(VI) ion in solution is also accelerated by the presence of metal ions or impurities in solution (e.g. Ni, Co). Ferrate(VI) is a powerful oxidant, often superior in oxidizing power than permanganate.

Oxidation potentials of ferrate at several pH values:

- 2.2 volt at pH = 0 — Species: $H_3Fe0_4^+$
- 1.4 volt at pH =7 — Species: $HFe0_4^-$
- 0.7 volt at pH =14 — Species: FeO_4^{2-}

This potential is always higher than water over the entire pH range. Therefore water is a potential reductor of ferrate in the whole range of pH.

Comparison with other oxidants

MnO_4/MnO_2	1.68 V in acidic conditions
	0.59 V in alkaline conditions
Cl_2/Cl^-	1.51 V in acidic conditions
$HClO/Cl^-$	1.3 V in neutral conditions
ClO^-/Cl^-	<1 V in alkaline pH
O_3/O_2	2.07 V

The degradation product of ferrate is $Fe(OH)_3$, which acts as a coagulant like ferric chloride to precipitate impurities.

Effect on Bacteria and Viruses

Ferrate has a strong effect on bacteria; in preliminary work (*9*), the reduction of bacteria population in municipal wastewater has been reported: a reduction of 99.9 - 99.8% of the coli bacteria forms has been observed.

Processes Described in Literature

Until now, production of solid ferrate has been possible only at a small scale in the lab, in the form of potassium ferrate K_2FeO_4; the synthesis is difficult, the yield is low, and the stability of the pure ferrate solution is low. The existing synthesis processes are based on the principle of chemical or electrochemical oxidation of a ferric salt in strongly aqueous alkaline solution; this oxidation leads to the production of an alkaline solution of potassium ferrate(VI); solid potassium ferrate is then precipitated by addition of solid potassium hydroxide in a large quantity. Several successive washings by organic solvents can give a pure potassium ferrate.

Synthesis of ferrate(VI) can be carried out using three methods:

- Dry methods
- Wet methods
- Electrochemical methods

A brief description of these methods is given below and a new process is described.

Dry Methods

Several dry processes are described in literature (*1,2*). The oldest method is the combustion of iron scraps with potassium nitrate at high temperature (*1,2*). Several other methods involve the mixing of an alkali peroxide (Na_2O_2 or K_2O_2) with iron salt, most often an iron oxide, at high temperature. These methods were developed in laboratory using expensive chemicals and are difficult to scale-up.

Wet Methods

The oxidation of iron salt (iron hydroxide or iron sulfate) is carried out in strong alkaline solution with a chlorine gas or strong soluble oxidizer (hypochlorite of sodium or potassium) (*1,2*).

Example: $2\ Fe(OH)_3 + 3\ NaOCl + 4\ NaOH = 2\ Na_2Fe0_4 + 3\ NaCl + 5\ H_2O$

The addition of excess solid potassium hydroxide can precipitate out the solid potassium ferrate(VI).

These methods do not produce solid sodium ferrate (VI). The ferrate(VI) solutions are not very stable and decomposition is very sensitive to temperature and to the presence of impurities. The solid potassium ferrate(VI) is not very pure and presence of impurities such as KCl and $KClO_3$ have been detected.

Electrochemical Method

The electrochemical method involves the oxidation of iron anode in an electrochemical cell (*1*). Old researchers describe this method using diaphragm cells. Recently patents have been filed for the production of ferrate(VI) by membrane cells (*3-5*).

New Dry Process

A new dry process has been designed at the University of Nancy, France and the results of the work have been patented. Four doctoral thesis (*6*) have extensively studied this process including properties and stability of the produced ferrate, the analytical methods of its identification, and its potential applications (*7*). A new process produced a solid iron(VI) derivative, the potassium ferrate(VI). This process used raw materials of ferrous sulphate, potassium hydroxide, and hypochlorite or chlorine. The potassium ferrate(VI) was stabilized by sulfate and the by-product was the non-toxic iron (III) salt, which acts as a coagulant in the water treatment processes.

The work consists of designing a new dry process that uses ferrous sulphate (a residue in metallurgy and titanium dioxide production processes) as a raw material. This process works at room temperature. The general formula of the product is $M(Fe, X)O_4$ where M can be K_2 or Na_2 , X is the stabiliser in solid solution. The formation of solid solution of this ferrate(VI) product depends on the crystal structure of K_2FeO_4 and K_2XO_4 that is isomorphic with similar crystal parameters. $K_2(Fe,S)O_4$ was synthesized from the oxidation of $FeSO_4$ in potassium hydroxide. The evaluation of the properties of this ferrate(VI) product has been made during the first phase of the EEC programme; supported by CEE (Reference: BR PR – CT-97-0392). The $K_2(Fe,S)O_4$ was more stable than the pure K_2FeO_4.

$$FeSO_4 .nH_2O_s + Ca(OCl)_2.Ca(OH)_{2\,s} + 6\,KOH =$$
$$2\,K_2(Fe_{0.5},S_{0.5})O_{4\,S} + 2\,Ca(0H)_{2\,s} + 2\,KCl + (n+2)H_2O$$

The new process involved mixing a set of powders in a rotating reactor at room temperature (Figure 1). A mixing powder contained an iron salt (e.g. powder of ferrous sulphate), a strong alkali (pellets of potassium hydroxide), and an oxidant in the solid form (powder of calcium hypochlorite). The reaction started at room temperature and used small quantity of water, which came from the reagents. Crystallisation of potassium ferrate(VI) started at the surface of potassium hydroxide pellets and progressed to the core of these pellets. The colour of potassium hydroxide changed from white to a deep purple and the progress of the oxidation was monitored by Mossbauer spectroscopy.

Objectives of Recent EEC Program

A consortium of eight Universities and industrial partners with the following objectives has completed a new EEC project:

1. Production of a large quantities of potassium ferrate(VI) for applied research; the pilot plant used the mixing technology of powders described earlier.

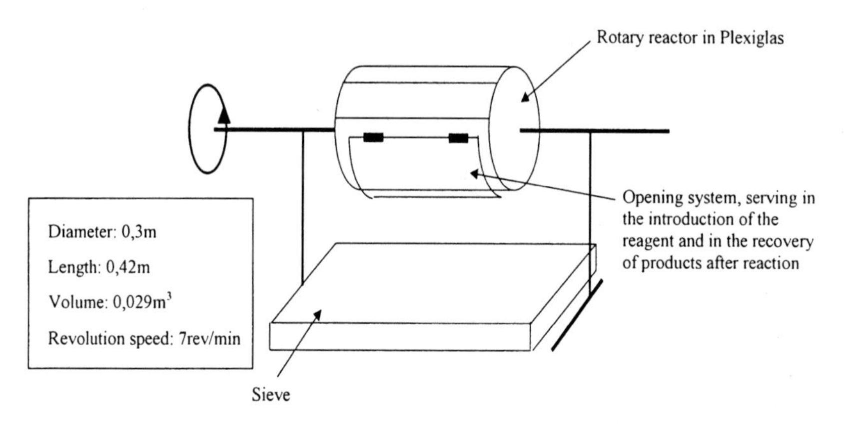

Figure 1. Experimental set-up

2. Development of a new fluid bed technology, which should be easier and safer to scale-up to a large capacity than the existing technology.
3. Development of the process using cheaper raw materials like caustic soda and chlorine or hypochlorite.
4. Carrying out a large range of investigations for applications in drinking water treatment, in municipal waste water treatment, in industrial waste water, and in other fields where a strong oxidant can be used.
5. Collection of the relevant data on the alternative chemicals or systems actually used in water treatment.

This paper describes the significant progress made in the synthesis of solid sodium and potassium ferrate(VI).

Results of the mid-term program (*2*)

The EEC program that started at the end of 2001 synthesized a large quantity of ferrate(VI) to feed a large scale applications in the field of water treatment: drinking water, municipal waste water and industrial waste water. A description of pilot is shown in Figure 2.

This process allowed the production of large quantities of ferrate(VI) needed for applied research but has limitations due to difficulty in mixing and controlling the heat of the reaction. Continous operation improved the process but the yield of Fe(VI) was still too low ($Fe(VI)/Fe_{tot}$ = 50 %). Research on other process has been going on during the second half of the program

Analysis of Fe(VI)

The reference method for the analysis of Fe(VI) is the Mössbauer spectrometry, however, this method is long and is not very adaptable for the production and quality control purposes. An example of a Mössbauer spectroscopic analysis is shown in Figure 3.

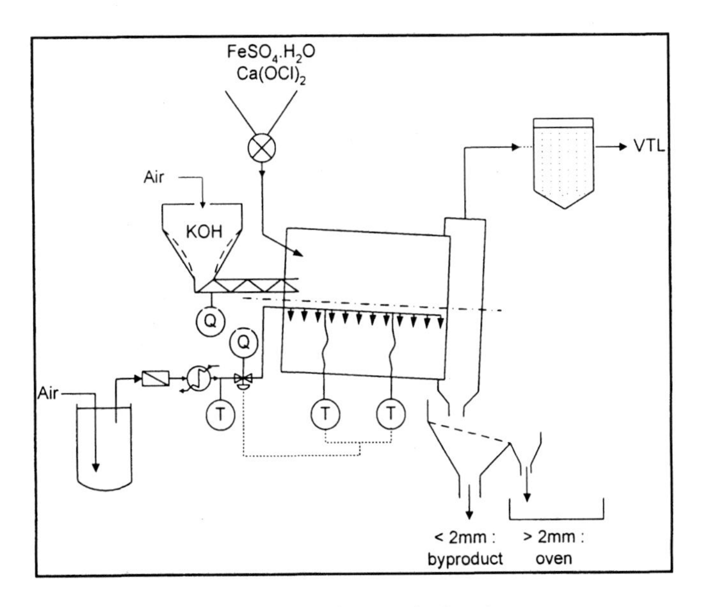

Figure 2. Pilot 150 kg batch

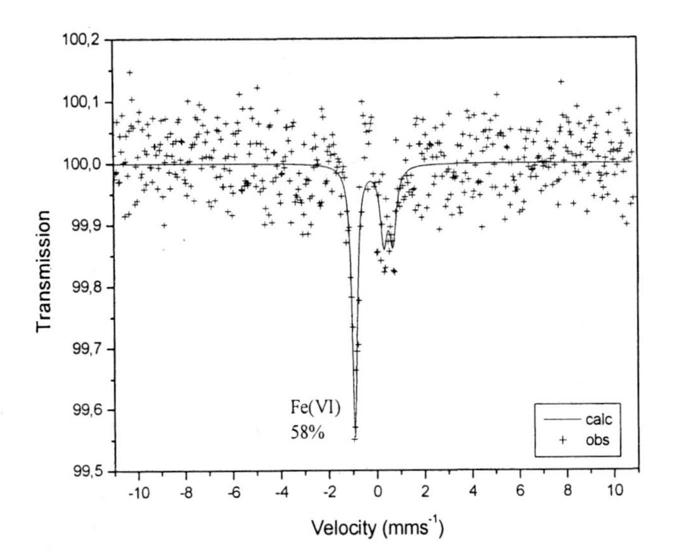

Figure 3. Mössbauer results

Typical results were:

Fe total: 90 g/kg
Fe(VI): 50 g/kg

Several method of titration used for testing the ferrate:

- Chemical titration with As (III) in alkaline solution
- Chemical titration with Cr(III):Cr(III) oxidized to Cr(VI)
- Visible absorption spectroscopy of purple color of ferrate ion (λ = 508 nm); The method by absorption is very difficult, due to the fast decomposition of ferrate ion in aqueous phase
- These methods do not seem appropriate for the titration of Fe(VI) in potassium sulfate solution

Considering the difficulty of analysis, a new method involving reduction of Fe(VI) by Fe(II) and titration of the excess Fe(II) with dichromate was investigated. This method was performed in very strict conditions of inert atmosphere in order to avoid oxidation of excess Fe(II) by oxygen. This method gave reliable results and was much faster and easier.

New Process at the End of Program

Semi-dry synthesis in rotating drum was very difficult to scale-up at industrial size because of following reasons:

- temperature is difficult to control
- Severe problems of segregation are encountered
- Scale-up is inherently difficult.

Two new processes using fluid bed technology instead of mixing powders have been explored to eliminate these problems. Patents have been filed on both processes.

Process 1: Granulation Fluid Bed by spraying miscellaneous solutions

Granulation Fluid bed

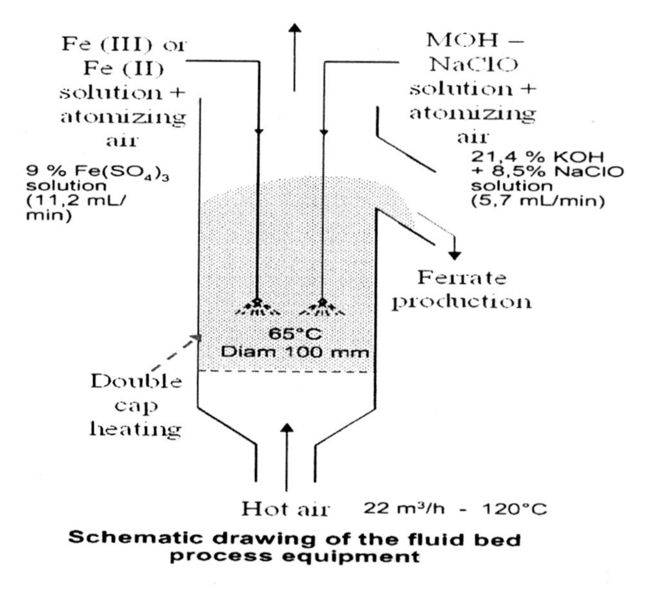

Schematic drawing of the fluid bed process equipment

Tests were made with:

{NaClO – KOH} or {NaClO –NaOH} solution
$Fe_2(SO_4)_3$, or $FeSO_4$, or $Fe(NO_3)_3$,or $FeCl_3$,or $FeCl_2$ solution

Two solutions were atomized inside a fluid bed; the small drops covered the surface of solid particles; and reacted together, evaporated on the bed particles surface, and made the particle grow. The produced ferrate continuously exited the bed by a lateral outlet.

Parameters were:

- Raw materials (KOH, NaOH, Fe(II) or Fe(III) salts).
- Relative humidity of air.
- Bed temperature
- Initial load of inert particle for start-up.
- Particle size : best range of 300 to 1000 μm
- Fe6+/ Fe yield : >75 % in best tests

Conclusions made were

- Fe(VI)/ Fe_{tot} yield ratio was significantly improved
- Crystal bed temperature was easier to control than in rotating drum
- Ferrate obtained in the Fluid Bed was already dried (no after- extra drying) and no sieving needed

Process 2: Fluid bed with chlorine

A new process using fluid bed technology and chlorine was also disclosed at the ACS meeting.

Synthesis with Chlorine: The overall reaction with chlorine, caustic soda and ferrous sulfate is given below:

$$FeSO_4 .H_2O + 2\ Cl_2 + 8\ NaOH \quad = Na_2FeO_4\ (?) + Na_2SO_4 + 4\ NaCl + 5\ H_2O$$

The process can be divided into individual steps:

– Premixing of solid sodium hydroxide and solid ferrous sulphate in a mixer. During this operation, ferrous sulphate reacted with potassium hydroxide and the reaction progressed inside the pellets.

$$2\ NaOH + FeSO_4 .\ H2O \rightarrow Na_2SO_4 + Fe(OH)_2 + H_2O$$

– This premixed solid, which had almost the same particle size as the original solid caustic soda introduced in the mixer. More consistent information was revealed by SEM investigation.

- The external part of the caustic pellets was essentially composed of Na,O,S and Fe; indicating a reaction of NaOH with iron sulphate.
- NaOH pellets became porous; facilitating the diffusion of reactive gas during the Sodium ferrate(iv) synthesis
- The core of the pellet showed the presence of unreacted NaOH

– The premixed solid was fluidised by air with a small content of chlorine gas. Visual observation indicated a good fluidisation without dust formation. The solid after chlorination was examined by optical microscope, which confirmed that the shape after treatment was similar to the shape of the original caustic soda pearls. The solid ferrate(VI) colour was purple; indicating the presence of sodium ferrate(VI).

Comparison of the Two Fluid Bed Processes

Process 1	Miscellaneous iron salts	Yield $Fe(VI)/Fe_{Tot}$=75 %	K or Na Ferrate
Process 2	NaOH, Fe^{2+}, Cl_2	Yield $Fe(VI)/Fe_{Tot}$=50%	Na Ferrate only

Main Results in Applications

Results in drinking water treatment

- From surface water resource: little technical and economical advantage of ferrate(VI) or ferrate(VI) in combination with ferric chloride compared to traditional solution
- From sub surface resource: these water are often contaminated with As and/or Mn
- Ferrate(VI) used in combination with ferric chloride is a good technical and economical alternative to existing treatment of As with a target level of 10 µg/L
- Ferrate(VI) used in combination with ferric chloride is a good technical and economical alternative to existing treatment of Mn with a target of 50 µg/L

Urban waste water

- Ferrate(VI) used at small dosage in combination with ferric chloride
- Few specific applications are possible in tertiary treatment

Industrial waste water

- Some specific applications have been found in treatment of waste water from textile industry allowing direct discharge of treated water in municipal plant
- Some specific applications have been found in treatment of waste water from tannery industry allowing direct discharge of treated water in municipal plant

- Some specific applications have been found in removing aromatic residue in water from coke.

Acknowledgements

This work was performed in the frame of contract No G5 RD-CT-2001-03011 and we want to thank the support of the EEC.

References

1. Evrard, O.; Geradin, R.; Schmitt, Evrard, L.L. Ferrate of Alkali or Alkaline Earth Metals, Their Preparation and Their Industrial Applications. International patent WO 91/07352, **1991**.
2. Ninane, L.; Veronneau, C.; Neveux N.; Jeannot C.; Dupre B. Manufacture of super oxidants for the treatment of industrial wastes. In proceedings of environmental clean technologies for sustainable production and consumption, Vancouver, **2003**.
3. Kanari, O.; Ninane, L.; Neveux, E. Synthesis of Alkali Ferrate using a waste as Raw Material, *JOM*, **2005**.
4. Gmelin Handbuch der anorg. Chemie, Eisen Teil B, **1932**.
5. Kokarovstseva, B. Russ. *Chem. Rev.* **1972**, *41(11)*, 929-936
6. Asahi Glass. Japanese patent 57/19827 of 27/5/1981
7. Olin Corp, US 4435256 of 23/3/1981
8. Olin Corp, US 4435257 of 23/3/1981
9. Four Ph.D. Thesis. N. Neveux, N. Aubertin, A. Lechaudel, C. Jeannot at the University of Nancy, France.
10. Sharma, V.K. Potassium Ferrate (VI): an environmentally friendly oxidant *Adv. Environ. Res.* **2002**, *6*, 143-156.
11. Neveux, N. Le sulfato-ferrate de potassium, un nouvel agent oxydant coagulant. *Informations Cimie no 386*, **1997**.

Chapter 7

Higher Oxidation States of Iron in Solid State: Synthesis and Their Mössbauer Characterization

Yurii D. Perfiliev[1] and Virender K. Sharma[2]

[1]Faculty of Chemistry, Moscow Lomonosov State University, Leninskii gory, Moscow 119992, Russia
(email: perf@radio.chem.msu.ru)
[2]Chemistry Department, Florida Institute of Technology, 150 West University Boulevard, Melbourne, FL 32901

The solid state synthesis of iron ion in high oxidation states (≥+4) is briefly reviewed. Reactions in solid state are mainly considered through some unusual cases of obtaining iron ions with oxidation states of Fe(V), Fe(VI), Fe(VII), and Fe(VIII). The significance of the Mössbauer spectroscopic technique for identifying valence forms of elements in complex oxygen-containing compounds is emphasized. Sodium ferrate(IV), Na_4FeO_4, is highly hygroscopic and decomposes to Fe(VI) and Fe(III) upon contact with water. The basic Mössbauer parameter, isomer shift, δ, for alkali and alkaline earth metal ferrate(VI) changes in a very narrow range of –0.87 to –0.91 mm s^{-1} (with respect to α-Fe). This suggests a weak influence of the outer ions on iron bound in an oxygen tetrahedron. The Mössbauer spectra of unknown oxoferrate ions with oxidation states of Fe^{5+} and Fe^{8+} are reported. The isomer shift, which is a fundamental parameter decreases with increasing valence state. This tendency persists for all possible high oxidation states of iron.

Introduction

There has been increasing interest in higher oxidation states of iron (Fe(IV), Fe(V), and Fe(VI)) because they are involved as an alternate for battery cathodes, green oxidants for organic synthesis, environmental-friendly oxidants in pollution remediation processes, and as intermediates in Fenton-type reactions and biological transfer processes (*1-7*). Investigations on the synthesis conditions, stability, reactivity, and physical chemical properties of iron compounds in higher oxidation states (over +3) are critical to understanding their properties and how they can be used in various applications. Of the higher oxidation states of iron, Fe(VI) has special interest because of its high oxidation power in removing pollutants in wastewater with the formation of non-hazardous products (*1,2*). Other higher oxidation states, Fe(V) and Fe(IV) are stronger oxidants in solutions than Fe(VI) (*8*-10). The oxidation states higher than +6, Fe(VII) and Fe(VIII), are also of interest and such iron compounds have been suggested (*11*).

It is often difficult to identify the oxidation state of iron in compounds using chemical methods. However, the nuclear γ-resonance method based on the Mössbauer effect can easily identify different valence states of iron in multiphase systems and complex compounds concurrently containing several differently charged ions (*12*). This is particularly true for iron oxo derivatives containing iron in high oxidation states.

The basic parameter that suffices to identify the oxidation states of iron is the isomer shift, which tends to decrease with an increase in oxidation states below six (*13*). This trend may be extended to the possible oxidation states +7 and +8 (*12*). This is supported by analogy with the isomer shifts of ruthenium (*14*), the nearest iron homologue in the group (Figure 1).

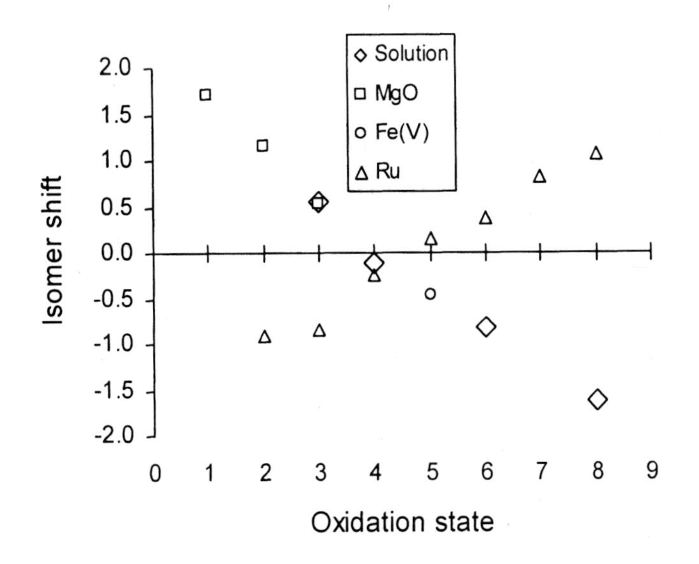

Figure 1. Correlation between ^{57}Fe isomer shift (relative to standard substance, α-Fe) and oxidation state of iron in oxo derivatives at 77 K (12). The isomer shift of Fe(V) in K_3MnO_4 is taken from (27). The isomer shifts (relative to Ru in Rh) for ionic (with oxygen and fluorine) ruthenium compounds at 4 K (14).

Ruthenium shows the opposite trend in which the isomer shift increases with an increase in oxidation state (Figure 1). The opposite change is due to the positive ΔR/R (ΔR - the change of the nuclear radius upon excitation) ratio for the ^{99}Ru nucleus. This ratio for ^{57}Fe is negative. For simplicity, Figure 1 shows data for systems in which several forms of iron can coexist and the isomer shift of the iron line is little affected by the crystal lattice or by the character of chemical bonding. This can happen in emission Mössbauer experiments where the isomer shift of a "nucleogenic" ^{57}Fe is measured. Thus the emission spectra of ^{57}Co introduced into MgO (*15*) and frozen solutions containing different oxidation states of iron (*16*) were used. Other details are provided elsewhere (*12*).

Different approaches have been applied to synthesize higher oxidation states of iron in the solid state. These approaches attempt to create physicochemical conditions favorable to the formation of a new species with higher-valent iron states than found in the starting material. Variable parameters are usually temperature, pressure, concentration, and the reactivity of the components. This paper briefly summarizes synthesis of oxo compounds of iron in oxidation states $\geq$+4. The Mössbauer parameters of these compounds are also discussed.

Iron(IV) Compounds

Solid sodium metaferrate(IV) (Na_2FeO_3) has been obtained by heating Na_2O_2 and Fe_2O_3 (the molar ratio Na : Fe = 2 : 1) in oxygen at 370 °C (*17*). Importantly, the trituration of mixture components with carbon tetrachloride was mandatory for preparing it. Its Mössbauer spectrum gave a single line with the isomer shift -0.08 mm/s at 294 K, which was consistent with the shifts of Fe^{4+} oxidation state (*13*). The effective magnetic moment of the iron ion at 293 K was determined to be 4.65 μB, which corresponds to the $3d^4$ configuration of the tetravalent high-spin iron ion.

The black powder of sodium orthoferrate(IV)has been synthesized using solid-state reaction of sodium peroxide Na_2O_2 and $Fe_{1-x}O$ in a 4:1 molar ratio of Na to Fe at 400 °C (exposition 15 h) (*18*). Powder X-ray and neutron diffraction studies suggest that Na_4FeO_4 is in the triclinic system P-1 with the following cell parameters:

a = 8.48(1) Å	α = 124.7(1)°
b = 5.76(1) Å	β = 98.9 (3)°
c = 6.56(1) Å	γ = 101.8 (3)°

Na_4FeO_4 is isotypic with other known solid phases of the form Na_4MO_4 (where M = Ti, Cr, Mn, Co, Ge, Sn, and Pb). The structure of Na_4FeO_4 is characterized as a three-dimensional network of isolated FeO_4 tetrahedra connected by Na atoms. This compound exhibits Jahn-Teller distortions due to their high spin d^4 configurations in the tetrahedral FeO_4 coordination (*18*). Crystals of Na_4FeO_4 are highly hygroscopic and dissociate to Fe^{3+} and $Fe^{VI}O_4^{2-}$ in water (as shown in Eq. 1).

$$3\ Na_4FeO_4 + 8\ H_2O \rightarrow 12\ Na^+ + Fe^{VI}O_4^{2-} + 2\ Fe^{3+} + 16\ OH^- \qquad (1)$$

The hydrolysis of Fe^{3+} then gives amorphous $Fe(OH)_3$ or FeO(OH). The Mössbauer spectrum of Na_4FeO_4 gave one doublet (δ = -0.218(5) mm/s, Δ = 0.407(5) mm/s at 295 K (Figure 2). A sextet at 6 K was observed, which characterizes Fe(IV) in a high spin tetrahedral FeO_4 coordination. Figure 2 shows an increase in isomer shift with decreasing temperature.

The reaction of iron(III) oxide with potassium peroxide did not yield potassium ferrate(IV), presumably, due to the extremely favorable kinetics for the formation of ferrate(VI). However, potassium ferrate(IV) could be prepared upon thermal decomposition of K_2FeO_4 at 500°C in dry oxygen, and its spectrum is a singlet with δ = -0.16(2) mm/s and Γ_{exp} = 0.35(4) mm/s (*19*). The same procedure resulted in cesium ferrate(IV) with an indefinite chemical composition; this ferrate has the isomerl shift -0.11(2) mm/s, which is typical of Fe(IV) ions in an octahedral environment of oxygen atoms (*20*).

Crystals of barium iron(IV) oxides, Ba_2FeO_4, and Ba_3FeO_5 were first prepared by dehydrating the barium iron(III) hydroxides with the appropriate Ba/Fe ratios at 700 °C under a dynamic atmospheric pressure of oxygen (*21*). A molten KOH-$Ba(OH)_2$ flux synthesis has also been reported (*22*). The low-temperature isomer shifts obtained were –0.152 and –0.142 mm/s (relative to α-iron) for Ba_2FeO_4 and Ba_3FeO_5, respectively. These isomer shifts, together with magnetic susceptibility measurements, confirmed the valence state of +4 for iron in these compounds (*22*).

Finally, efficient monitoring of iron oxidation states allowed improving their preparation procedures by a solid-phase synthesis and many iron salts, oxides, and different oxidants were tested. The reaction of Fe_2O_3 with MO_x (M = Na, K, Cs) as the oxidant gave the best results. In all other cases, the total ferrate(IV) content did not exceed 25 mol % (*17*).

Iron(V) Compounds

Though, iron(V) is an uncommon oxidation state, the preparation of its compounds has been reported (*23,24*). The oxidation states of iron in these compounds were determined by chemical analysis and magnetic susceptibility measurements. Some workers thus do not rule out the possibility that these compounds are represented by a mixture of equal amounts of ferrate(IV) and ferrate(VI) (*25*). The structure of K_3FeO_4 crystals has been recently studied (*26*), but Mössbauer spectroscopic characterization of this compound has not been carried out.

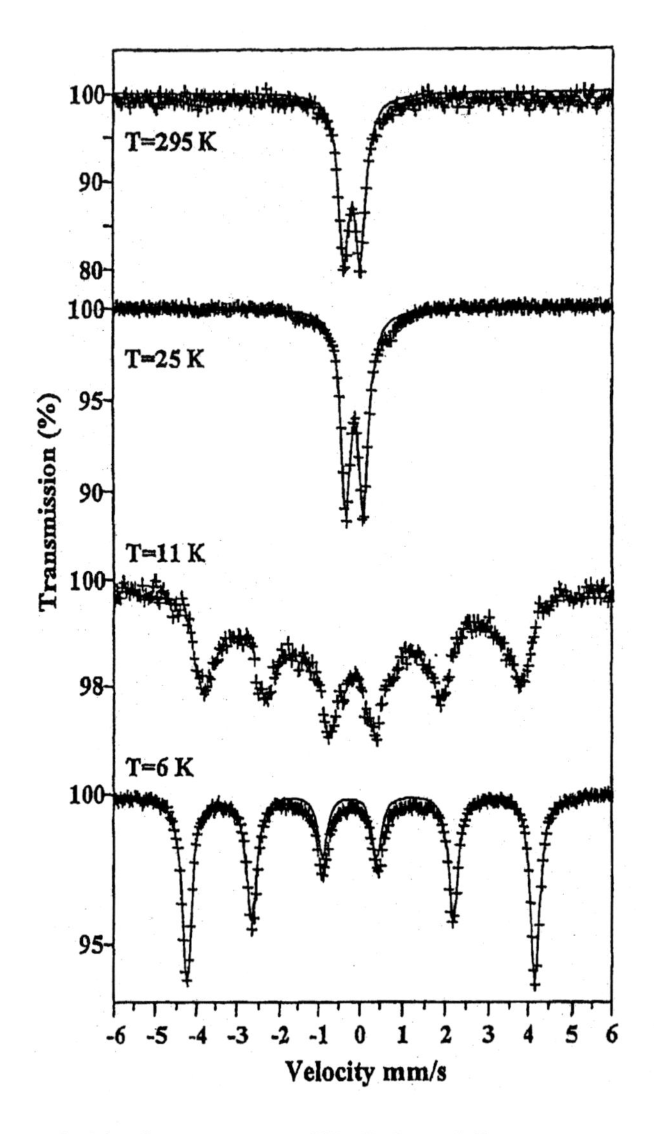

Figure 2. Mössbauer spectra of Na_4FeO_4 at different temperatures. (Reproduced from reference18. Copyright 2002.)

The +5 oxidation state of iron was stabilized by introducing iron ion into potassium manganate(V) (K_3MnO_4) (*27*). The K_3MnO_4 compound was separately synthesized by the reaction of Mn_2O_3 with KO_2 at 800 °C. The synthesized compound was amorphous to X-ray diffraction. Figure 3 shows the Mössbauer spectra of iron embedded in K_3MnO_4 at room temperature (*27*).

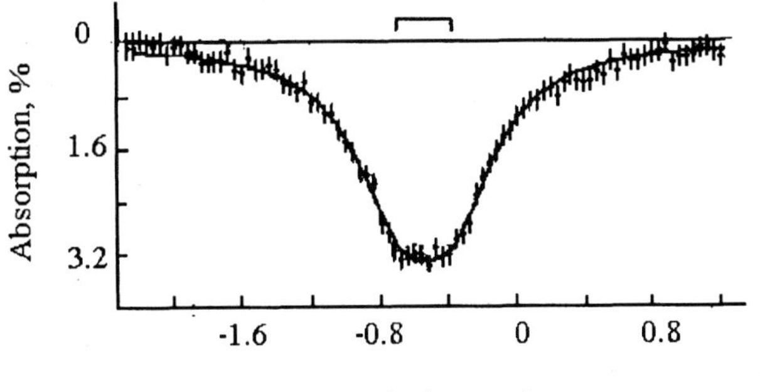

Figure 3. Mössbauer spectrum of K_3MnO_4 doped with iron at 294 K.

The room-temperature spectrum shows a single broadened line (δ =-0.549(1) mm/s). Using Figure 1, this line was assigned to the pentavalent iron ion. As expected, this δ value is intermediate between the δ values for Fe^{6+} (-0.92 to -0.88 mm/s for K_2FeO_4 (*24*)) and Fe^{4+} (-0.13 to -0.08 mm/s for metal ferrates (IV) (*13*) and is close to the isomer shift -0.45 mm/s; assigned to the +5 iron ion in an octahedral position (*28*). However, δ = -0.549(1) mm/s in Figure 2 is 0.14 mm/s lower than the isomer shift for octahedral +5 iron, which suggests tetrahedral coordination of the iron ion in K_3MnO_4. The isomer shifts of Fe^{2+} or Fe^{3+} ions occupying tetrahedral or octahedral positions in oxygen iron compounds differ by the same value (*13*).

Recently, Fe^{5+} iron in manganates was obtained by a similar procedure by using the initial molar composition $[KO_2]$:$[MnO_2]$=8:1 at 800 °C (*29*). The Mössbauer parameters obtained were δ = -0.538(2) mm/s and Δ = 1,067(10) mm/s, which are consistent with the +5 oxidation state of iron. The introduction of +5 valent iron has also been demonstrated in V_2O_5 and K_2RuO_4 (*27*).

Iron(VI) Compounds

A number of alkali and alkaline earth ferrates of iron(VI) have been synthesized using both dry and wet techniques. These techniques have been recently summarized (*30*). Dry techniques are generally carried out using a thermal technique whereas chemical and electrochemical procedures are applied in wet techniques (*30*). The wet techniques are given in other chapters, hence are not reviewed here. Ferrate(VI) of the M_2FeO_4 composition (where M = Na, K, Rb, Cs, and Ag), two alkali earth metal ferrates(VI) ($SrFeO_4$ and $BaFeO_4$), and two mixed cation ferrates(VI) ($K_3Na(FeO_4)_2$ and $K_2Sr(FeO_4)_2$) have been prepared successfully (*30*).

When thermal techniques are used, a substantial yield of sodium ferrate(VI) (Na_2FeO_4) can be reached only by using the multistage temperature program and a special prior treatment of the mixture of oxides (*30*). A 4:1 molar ratio of Na to Fe (370 °C, exposition >12 h), results in a nearly 100% yield of ferrate(VI). This compound was formulated as Na_4FeO_5 based on the chemical analysis data (*31*). The main advantages of dry methods are high yields and a one-step process.

The characteristics of ferrates(VI) using Mössbauer spectroscopy technique are given in Table 1 (*32*).

In wet techniques, iron(VI) was generally produced by oxidizing a basic solution of Fe(III) salt by hypochlorite ion (*38*). However, the use of chlorine creates chlorinated by-products; hence it is not an environmentally friendly procedure. Moreover, there is also an emphasis on green chemistry; therefore, chlorine alternates are being sought. Recently (*39*), the formation of iron(VI) in ozonalysis of iron(III) in alkaline solution was demonstrated (eq 2).

$$2Fe(OH)_4^- + 3O_3 + 2OH^- \rightarrow 2FeO_4^{2-} + 3O_2 + 5H_2O \qquad (2)$$

The UV-visible spectrum of iron(VI) was compatible with a tetrahedral geometry of other high-valent metal oxoanions such as CrO_4^{3-} and MnO_4^-. Furthermore, iron(VI) in the synthesis was confirmed by Mössbauer spectroscopic techniques (Figure 4). The δ in Figure 4 is similar to values of δ for various salts of the iron(VI) ion (see Table 1).

Oxoferrates(VI) with molecular formula $M_2Fe^{VI}O_4$ with M= Li, Na, $(CH_3)_4$, $N(CH_3)_3BzI$, and $N(CH_3)_3Ph$ were also synthesized using cation exchange reactions with K_2FeO_4, followed by freeze-drying of the resulting aqueous solution (*40*). Crystals of lithium ferrate(VI) (Li_2FeO_4), which decompose at -10 ± 3 °C, were monohydrated.

Recently, an Fe(VI)-nitrido complex was prepared photochemically (*41*). Interestingly, the δ values of FeO_4^{2-} ion are considerably lower than those of the iron(VI)-nitrido complex ($[(Me_3cy\text{-}as)FeN](PF_6)_2$) ($\delta$ = 0.40 mm s^{-1}) . This is not surprising because both have different geometries and electronic structures. The coordination number influences the isomer shift, as has been demonstrated clearly by comparing salts of the $[FeCl_4]^-$ anion with $[FeCl_6]^{3-}$ trianion (*42*), in which the isomer shift of the latter is larger by 0.23 mm s^{-1}. In case of +6 oxidation states of iron, the iron(VI)-nitrido has an octahedral coordination, while the iron(VI) ion has a coordination number of four. Another significant difference is that the iron(VI)-nitrido complex has one strong iron-nitrogen multiple bond, whereas the iron(VI) ion has four strong and covalent Fe=O

Table 1. Mössbauer characteristics of ferrate(VI).

Formula	δ (mm s^{-1}) 298 K	Δ (mm s^{-1}) 298 K	H (T)	T_N (K)
$K_3Na(FeO_4)_2$	**-0.89**[a]	**0.21**	**no magnetic ordering down to 4.2 K**	-
K_2FeO_4	-0.88 **-0.90**	0	14.4±2 at 2.8 K[e] 14.7 at 0.15 K[e,f]	4.2[e] 3.6[f]
Rb_2FeO_4	-0.89	0	14.9±2 at 2.8K[e]	2.8-4.2[e]
Cs_2FeO_4	**-0.87**	0	15.1±2 at 2.8K[e]	4.2-6.0[e]
$K_2Sr(FeO_4)_2$	-0.91[c] **-0.90**	**0.14**	8.7[e] at 2 K unresolved sextet	~3[c]
$BaFeO_4$	-0.90[d] -.90	**0.16**	11.8±2 at 2.8 K[e]	7.0-8.0[e]

[(a)] The results from *(32)* are printed in bold; errors in these data do not exceed ±0.01 mm·s^{-1}; [(b)] Shinjo *et al.* (*33*); [(c)] Ogasawara *et al.* (*34*); [(d)] ±0.02 mm·s^{-1}, Ladriere *et al.* (*35*); [(e)] Herber and Johnson (*36*); [(f)] Corson and Hoy (*37*).

Table 1 demonstrates that isomer shifts, δ, of different iron compounds are for +6 valent state, which change very little (–0.87 to –0.91 mm/s). This indicates a weak influence of the outer ions on iron bound in an oxygen tetrahedron, which is the main structural unit of all ferrates(VI).

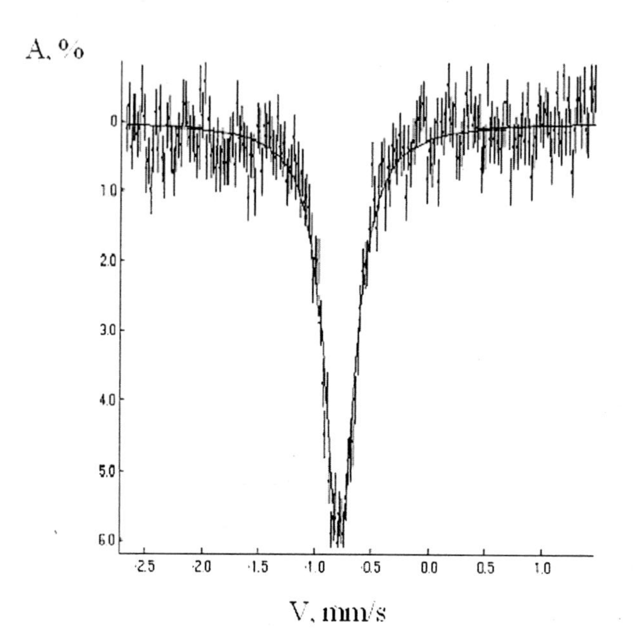

Figure 4. Mössbauer spectrum of frozen solution of Fe(VI) in 5 M NaOH (77 K). (Reproduced from reference 39. Copyright 2007.)

double bonds. This may also influence the variation in the isomer shifts of two species.

Iron(VII) Compounds

The existence of iron(VII) compounds is still not verified though the possibility of iron(VII) in iron-doped compounds of heptavalent elements has been suggested. Iron-doped sodium and cesium ruthenates were studied by Mössbauer spectroscopy (*43*). The possibility of substituting iron for ruthenium was proven by synthesizing iron-containing sodium ruthenate(VI). The valence state of iron in this compound was +6, which was expected from simple isomorphous miscibility considerations. The synthesis of cesium ruthenate(VII) gave a mixture of two phases, $CsRuO_4$ and $Cs_3(RuO_4)_2$. In this composition, doped iron ions occurred in two forms. Spectrally, one of them was easily identified as the Fe(VI) state (δ = -0.76(1) mm/sat 77 K). The other form, manifested in the spectrum as a weaker line at the left slope of the major line of Fe^{6+}, was characterized by the isomer shift -1.03(2) mm/s . This value identified this form to a higher oxidation state.

Iron(VIII) Compounds

An alkaline solution containing octavalent iron, Fe(VIII), was prepared by anodic dissolution of iron metal in an alkaline medium (*12,16*). The Mössbauer spectrum of a frozen solution showed five lines of various intensities (Figure 5).

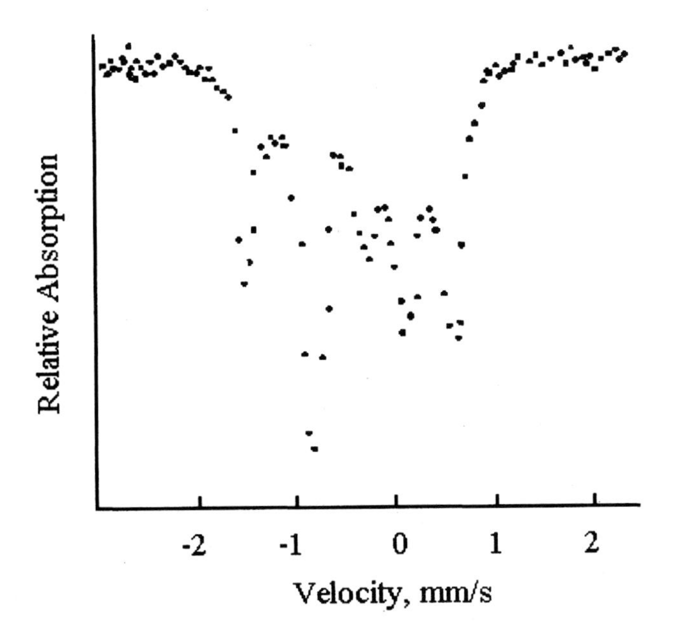

Figure 5. Mössbauer spectrum of a sample containing iron in different oxidation states. (Reproduced from reference 12. Copyright 2002 American Chemical Society.)

The line positions corresponded to velocities v_1 = -1.62, v_2 = -0.82, v_3 = -0.10, v_4 = 0.18, and v_5 = 0.92 mm/s. Peaks 4 and 5 had equal intensities in all experiments, whereas their contributions to the total resonance absorption could change considerably. Thus it was deduced that these peaks arise from the same oxidation state. That, along with the corresponding Mössbauer parameters, (δ = 0.56(4) mm/s, Δ = 0.71(4) mm/s), made it possible to assign peaks 4 and 5 to the doublet from Fe(III) ions. The positions of peaks 2 and 3 corresponded to Fe(VI) and Fe(IV). The remaining peak, whose intensity did not correlate with other absorption lines, was assigned to octavalent iron. The isomer shift of this peak, in contrast to the isomer shift of Fe(VII), fits the general trend of a decrease in the isomer shift with an increase in the iron oxidation state. The Fe(VIII) state is unstable. It should be pointed out that separate spectra recorded within the first hours after electrolysis were combined in the spectrum shown in Figure 5.

Acknowledgment

We wish to thank Professor Zoltan Homonnay for useful comments.

References

1. Sharma, V.K. *Adv. Environ. Res.* **2002**, *6*, 143-156.
2. Sharma, V.K.; Kazama, F.; Jiangyong, H.; Ray, A.K. *J. Water Health* **2005**, *3*, 45-58.
3. Sharma, V.K.; Mishra, S.K.; Nesnas, N. *Environ. Sci. Technol.* **2006**, *40*, 7222-7226.
4. Sharma, V.K. *Water Sci. Technol.* **2007**, *55*, 225-230.
5. Groves, J.T. *J. Inorg. Biochem.* **2006**, *100*, 434-447.
6. Oliveria, F.T.D.; Chanda, A.; Benerjee, D.; Shan, X.; Mandal, S.; Que, Jr. L.; Bominaar, E.L.; Munck, E.; Collins, T.J. *Science* **2007**, *315*, 835-838
7. Delaude, L.; Laszlo, P. *J. Org. Chem.* **1996**, *61(18)*, 6360-6370.
8. Sharma, V.K. *Rad. Phys. Chem.* **2002**, *65*, 349-355.
9. Sharma, V.K.; O'Connor, D.; Cabelli, D.E. *J. Phys. Chem. B* **2001**, 11529-11532.
10. Sharma, V.K.; Burner, C.R.; Yngard, R.; Cabelli, D.E. *Environ. Sci. Technol.* **2005**, *39*, 3849-3855.
11. Atanasov, M. *Inorg. Chem.* **1999**, *38*, 4942-4948.
12. Perfiliev, Yu.D. *Russ. J. Inorg. Chem.* **2002**, *47*, 611-619.
13. Menil, F. *J. Phys. Chem. Solids.* **1985**, *45(7)*, 763-789.
14. Good, M.L. In: A review of the Mössbauer spectroscopy of ruthenium -99 and ruthenium-101, in Mössbauer Effect Data Index, Stevens, J.G. and Stevens, V.E., Eds., New York: IFI/Plenum, **1973**, 51-63.
15. Chappert, J.; Frankel, R.B.; Misetich, A.; Blum, N.A. *Phys. Rev.* **1969**, *149*, 578-589.
16. Perfiliev, Yu.D., Kopelev, N.S., Kiselev, Yu.M., Spitsyn, V.I. *Proc. Acad. Sci. USSR, Phys. Chem. Sect.(Eng. Transl.)* **1987**, *296*, 1028-1031.
17. Kopelev, N.S., Perfiliev, Yu.D., Kiselev, Yu.M. *J. Radioanal. Nucl. Chem.* **1992**, *162(2)*, 239-251.
18. Jeannot, C., Malaman, B., Gérardin, R., Oulladiaf, B. *J. Solid State Chem.* **2002**, *165*, 266-277.
19. Kiselev, Yu.M., Kopelev, N.S., and Perfil'ev, Yu.D. *Zh. Neorg. Khim.(in Russian)*, **1987**, *32 (12)*, 2982-2986.
20. Kopelev, N.S., Popov, A.I., and Val'kovsky, M.D. *J. Radioanal. Nucl. Chem. Lett.* **1994**, *188*, 99-108.
21. Scholder, R.; Vonbunsen, H.; Zeiss, W. Z. *Anorg. Allg. Chem.* **1956**, *283*, 330-337.

22. Delattre, J.L.; Stacy, A.M.; Young, V.G.; Long, G.L.; Hermann, R.; Grandjean, F. *Inorg. Chem.* **2002**, *41*, 2834-2838.
23. Kokarovtseva, I.G.; Belyaev, I.N.; Semenyakova, L.V. *Russ. Chem. Rev.* **1972**, *41(11)*, 928-937.
24. Kopelev, N.S. In Mössbauer Spectroscopy of Sophisticated Oxides, Vertes, A. and Hommonnay, Z., Eds., Budapest: Akademiai Kiadó, **1997**, 305-332.
25. Klemm, W. and Wahl, K. *Angew. Chem.* **1963**, *65*, 261-265.
26. Hoppe, R. and Majer, K. *Z. Anorg. Allg. Chem,* **1990**, *586*, 115-124.
27. Perfiliev, Yu.D. *J.Radioanal.Nucl.Chem.* **2000**, *246(1)*, 21-25.
28. Demazeau, G.; Buffat, B.; Ménil, F.; Fournès, L.; Pouchard, M.; Dance, J.M.; Fabritchnyi, P.; Hagenmuller, P. *C. R. Acad. Sci., Ser. II* **1981**, *16*, 1465-1472.
29. Perfiliev, Y.D.; Alkhatib, K.E.; Kulikov, L.A. *Vestnik MGU(Bulletin of Moscow University), ser. 2, Chem,* **2007**, *48(2)*, 139-142.
30. Perfiliev Yu.D.; Sharma V.K. In: Ferrate(VI) Synthesis: Dry and Wet Methods, Proceedings of Int. Symp, "Innovative Ferrate (VI) Technology in water and Wastwater Treatment" May 31, 2004, Prague, Czech Republic, pp 32-37
31. Kiselev, Y.M.; Kopelev, N.S.; Zav'yalova, N.A.; Perfiliev, Y.D.; Kazin, P.E. *Russ. J. Inorg. Chem.* **1989**, *34*, 1250-1253.
32. Dedushenko, S.K.; Perfiliev, Y.D.; Goldfield, M.G.; Tsapin, A.I. *Hyperfine Interactions*, **2001**, *136(3)*, 373-377.
33. Shinjo, T.; Ichida,T.; Takada, T. *J. Phys. Soc. Japan* **1970**, *29(1)*, 111-.
34. Ogasawara, S.; Takano, M.; Bando, Y. *Bull. Inst. Chem. Res. Kyoto Univ.* **1988**, *66*, 64-65.
35. Ladriere, J.; Meykens, A.; Coussement, R.; Cogneau, M.; Boge, M.; Auric, P.; Bouchez, R.; Benabed, A.; Godard, J. *J. de Phys. Colloque C2* **1979**, *40*, C2-20.
36. Herber, R.H., Johnson, D. *Inorg. Chem.* **1979**, *18*, 2786-2790.
37. Corson, M.R.; Hoy, G.R. *Phys. Rev.* **1984**, *B29*, 3982.
38. Schreyer, J.M., Thompson, G.W., Ockerman, L.T., Potassium ferrate(VI). Inorg. Synthesis **1953**, 4164-4168.
39. Perfiliev, Y.D.; Benko, E.M.; Pankratov, D.A.; Sharma, V.K.; Dedushenko, S.K. *Inorg.Chim.Acta*, **2007**, *360*, 2789-2791.
40. Malchus M.; Jansen M. *Z anorg allg Chem* **1998**, *624*, 846-1854.
41. Berry, J.F.; Bill, E.; Bothe, E.; George, S.D.; Mienert, B.; Neese, F.; Wieghardt, K. *Science* **2006**, *320*, 1937-1941.
42. Greenwood, M.N.; Gibb, T.C. Mossbauer Spectroscopy. Chapman and Hall, London, **1971**.
43. Perfiliev Y.D.; Kholodkovskaya L.N.; Kiselev Y.M.; Kulikov L.A. *ICAME-95, Conf. Proceed (*ed. I. Ortalli, Sif, Bologna), **1996**, *50*, 517-520.

Chapter 8

Thermal Stability of Solid Ferrates(VI): A Review

Libor Machala[1], Radek Zboril[1], Virender K. Sharma[2,*], Jan Filip[1], Oldrich Schneeweiss[1,3], János Madarász[4], Zoltán Homonnay[5], György Pokol[4], and Ria Yngard[2]

[1]Nanomaterial Research Centre, Palacky University, Svobody 26, 771 46 Olomouc, Czech Republic
[2]Chemistry Department, Florida Institute of Technology, 150 West University Boulevard, Melbourne, FL 32901
[3]Institute of Physics of Materials AS CR, Žižkova 22, 61662 Brno, Czech Republic
[4]Department of Inorganic and Analytical Chemistry, Budapest University of Technology and Economics, Szt. Gellért tér 4, H–1521 Budapest, Hungary
[5]Laboratory of Nuclear Chemistry, Eötvös Lorand University, H–1117 Budapest, Pázmány P. s. 1/A, Budapest, Hungary
[*]Corresponding author: vsharma@fit.edu

This review critically summarizes currently known results concerning the thermal decomposition of the most frequently used ferrate(VI) salts (K_2FeO_4, $BaFeO_4$, Cs_2FeO_4). Parameters important in the thermal decomposition of solid ferrates(VI) include the initial purity of the sample, a presence of adsorbed and/or crystal water, reaction atmosphere and temperature, crystallinity, phase transitions, and secondary transformation of the decomposition products. The confirmation and identification of metastable phases formed during thermal treatment can be difficult using standard approach. The *in-situ* experimental approach is necessary in some cases to understand better the decomposition mechanism. Generally, solid ferrates(VI) were found to be unstable at temperatures above 200 °C as one-step reduction accompanied by oxygen evolution usually proceeds. The most known and used ferrate(VI) salt, potassium ferrate(VI) (K_2FeO_4), decomposes at high temperatures to potassium ortho-ferrate(III), ($KFeO_2$), and potassium oxides. The resulting phase composition of the sample heated in air can be affected by accompanying secondary reactions with the participation of CO_2 and H_2O in air. However, the thermal decomposition of barium ferrate(VI) ($BaFeO_4$) is not sensitive to constituents of air and is mostly reduced to non-stoichiometric $BaFeO_x$ ($2.5 < x < 3$) perovskite-like phases stable under ordinary conditions. Such phases contain iron atoms with oxidation state +4; exhibiting the main difference in the decomposition mechanisms of K_2FeO_4 and $BaFeO_4$.

Introduction

Iron, generally known in the +2 and +3 oxidation states, can also be obtained in higher oxidation states such as +4, +5, and +6 under in a strong oxidizing environment (*1-4*). In recent years, there has been an increasing interest in the +6 oxidation state of iron, ferrate(VI) ($Fe^{VI}O_4^{2-}$), due to its potential use in high energy density rechargeable batteries, in cleaner ("greener") technology for organic synthesis, and in treatment of contaminants and toxins in water and wastewater (*5-11*). Chemical, electrochemical, and thermal techniques are usually applied to prepare solid ferrate(VI) salts (*12-21*).

The synthesis of the ferrate(VI) salts by the chemical technique requires several synthesis steps and large amounts of chemicals (*16,17*). Electrochemical synthesis applies electrolysis of iron (or iron salt) in concentrated hydroxide solution followed by a separation step in order to obtain the solid K_2FeO_4 product. The formation of passive iron oxide on the electrode reduces the ferrate(VI) yield (*18*). Possible reduction of ferrate(VI) in water ($2FeO_4^{2-} + 5H_2O \rightarrow 2Fe^{3+} + 3/2O_2 + 10OH^-$ (*22*)) lowers the product yield of K_2FeO_4. A dry technique is thus attractive as it can avoid difficulties associated with the wet techniques. Dry thermal techniques are relatively simple and are generally based on the reaction between iron(III) oxide and MO_x (M = Na, K, Cs,...; x = 1 or 2) under a stream of dried oxygen (*12-15*). However, the decomposition of ferrate(VI) occurs simultaneously at the elevated temperatures used in the thermal synthesis technique, which results in a usually less than 60 % yield of ferrate(VI). Prevention of the decomposition of ferrate(VI) might be accomplished by optimization of the temperature conditions which could lead to an increase in the ferrate(VI) yield. This would thus require a profound understanding of the mechanism of the thermal decomposition of ferrate(VI) salts.

In the literature, the results obtained by different authors on thermal decomposition of ferrate(VI) salts are in disagreement, particularly regarding observation and identification of hypothetic intermediate oxidation states of iron,

Fe(V) and Fe(IV), and observed mass loss in the thermal decomposition of ferrate(VI). This review provides a summary of the present knowledge with respect to the thermal decomposition of the most frequently used ferrates(VI) salts (K_2FeO_4, $BaFeO_4$ and Cs_2FeO_4). A critical discussion of reasons for discrepancies in the results obtained by different authors is also given.

Thermal Decomposition of K_2FeO_4 in Air

Scholder (*21*) was the first who studied the thermal behavior of K_2FeO_4 under an oxygen stream. The decomposition of ferrate resulted in a strong oxygen evolution between 200 and 350 °C. Microscopic images obtained from samples heated between 350 and 550 °C in this study showed a mixture of two crystalline phases of dark and light green particles. The light green phase was assumed to be potassium orthoferrite(III) ($KFeO_2$) while the darker phase was calculated to be a solid solution of K_2FeO_4 and $K_3Fe^{V}O_4$ in a 1:2 molar ratio. The overall mean oxidation number of iron species was measured to be +4.4 and eq 1 was suggested to explain the decomposition process.

$$5\ K_2FeO_4 \rightarrow K_2FeO_4 \cdot 2K_3FeO_4 + 2\ KFeO_2 + 2\ O_2 \tag{1}$$

Finally, the pure +3 oxidation state of iron in the form of $KFeO_2$ was obtained at 1000 °C after complete evaporation of K_2O.

Ichida (*23*) applied Mössbauer spectroscopic and X-ray diffraction techniques to determine the decomposition products of K_2FeO_4 in air. Heating the ferrate sample for about 90 days below 200 °C resulted in an X-ray amorphous Fe^{3+} compound. In this process Fe^{6+} ions were directly reduced to Fe^{3+} and none of the intermediate valence states of iron, Fe^{5+} or Fe^{4+}, were observed during the decomposition process. Above 250 °C, potassium ortho-ferrite(III) was identified as the only crystalline compound and the decomposition process was described by chemical equation 2,

$$K_2FeO_4 \rightarrow KFeO_2 + KO_x + (2\text{-}x/2)\ O_2 \tag{2}$$

where x stands for an uncertainty in the chemical form of a poorly crystalline potassium oxide.

Fätu and Schiopescu (*24*) used simultaneous thermogravimetry (TG) and differential thermal analysis (DTA) to investigate the thermal behavior of K_2FeO_4 in air (Figure 1). A continuous 14.3% decrease of sample weight recorded between 50 and 320 °C (with a heating rate of 10 °C/min) was ascribed to the release of 3/4 moles of oxygen per one mole of decomposed K_2FeO_4 (eq 3).

$$2\ K_2FeO_4 \rightarrow K_2O \cdot Fe_2O_3 + K_2O + 3/2\ O_2. \tag{3}$$

However, the theoretical mass loss calculated using eq 3 is 12.1 % and is significantly smaller than that observed in TG experiments (Figure 1). A simultaneously obtained DTA curve displayed a narrow endothermic peak with the minimum at 620 °C, interpreted as a phase transition of Fe_2O_3 by the authors. It is known however that polymorphous transformations of Fe_2O_3 (*e.g.* maghemite to hematite) take place at considerably lower temperatures (< 500 °C) and exhibit an exothermic effect on the DTA (DSC) curve (*25*).

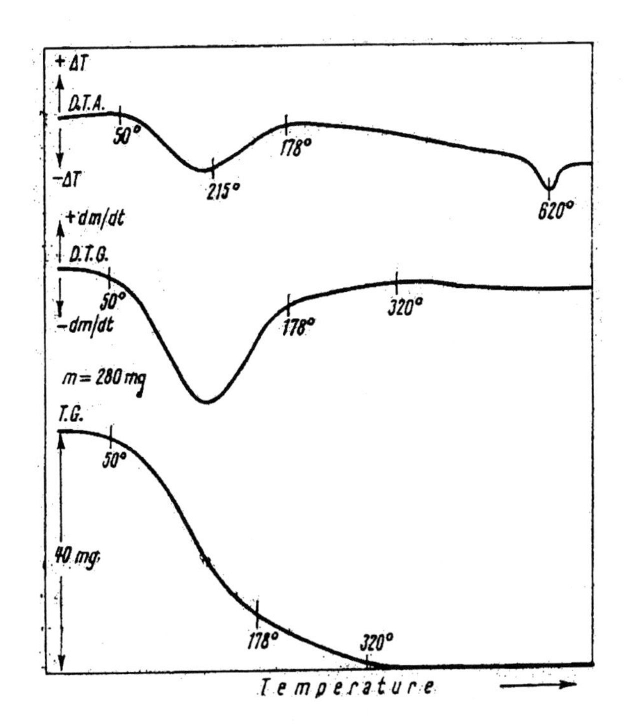

Figure 1. TG/DTG/DTA curves of K_2FeO_4 simultaneously measured in air. Reproduced with permission from Fätu and Schiopescu (24). Copyright 1974.

More recently, Machala *et al.* (*26*) reinvestigated the mechanism of the thermal decomposition of K_2FeO_4 in static air using *in-situ* techniques, including thermal analysis (TG/DSC, 5 °C/min rate), high-temperature Mössbauer spectroscopy and variable temperature X-ray powder diffraction (VT XRD). This approach has the advantage that it allows direct monitoring of the phase composition during the thermally induced process including identification of reaction intermediates. In addition, secondary chemical transformations of the decomposition products due to the interaction with air humidity could be prevented.

TG and DSC analyses performed under static air showed a thermal stability of K_2FeO_4 up to 230 °C (Figure 2). The slight weight loss of 0.3 % below 230 °C can be ascribed to the release of adsorbed water, which was also observed in the DSC curve by an endothermic minimum at 80 °C. Between 230 and 280 °C, a weight loss of 8.0 % was observed, which suggests a release of oxygen. This weight loss is however significantly lower compared to the value reported by Fătu and Schiopescu (14.3 %) (*24*). The main decomposition step is related to an endothermic effect in the DSC curve with a minimum at 256 °C (see Figure 2b). This endo-effect is immediately followed by a broadened exo-effect with a maximum at 270 °C. Importantly, previous studies conducted in an inert atmosphere did not report such an exothermic peak (*27,28*). Between 280 and 750 °C, no significant change in the sample weight was recorded. Above 750 °C, the mass loss progressively proceeded due to melting and evaporation of the decomposition products.

In-situ high temperature Mössbauer spectra were collected at four different temperatures (190, 300, 420 and 590 °C) to observe the transformation process

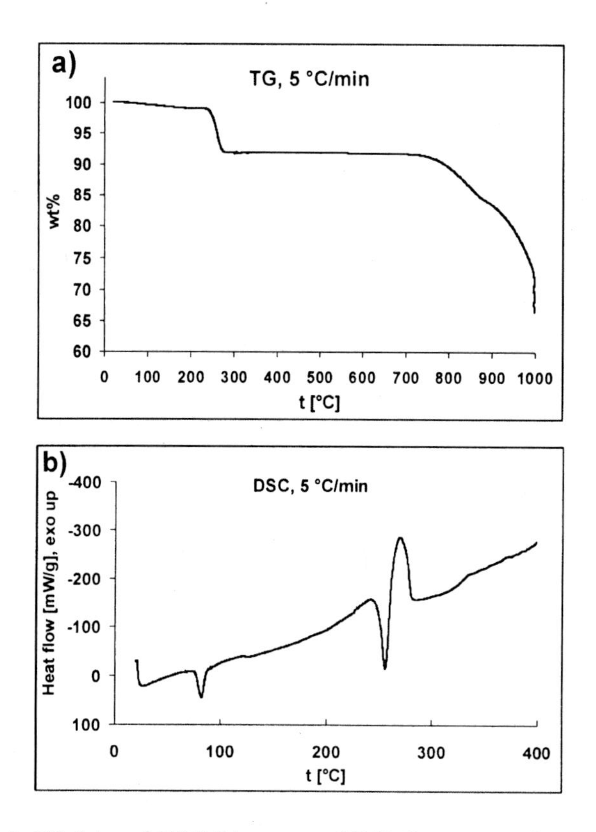

Figure 2. TG (a) and DSC (b) curves of K_2FeO_4 measured in static air. Reproduced from Machala et al. (26).

of Fe(VI) and to identify possible iron-bearing conversion intermediates (Figure 2). The spectrum of the sample heated at 190 °C (see Figure 3) consisted of two components including a singlet (δ = -1.01 mm/s) and a sextet with δ = 0.06 mm/s, ε_Q = 0.06 mm/s and B_{hf} = 45.8 T. The singlet corresponded clearly to non-transformed K_2FeO_4. The isomer shift value of the latter sub spectrum, despite of the second-order Doppler shift, is much lower than expected for an octahedral high-spin iron(III) compound (*29*). Based on the hyperfine parameters values, the sub spectrum was assigned to potassium iron(III) oxide, $KFeO_2$, where Fe^{3+} ions are tetrahedrally coordinated (*26,30*). The sextet of $KFeO_2$ represents the only one spectral component in the Mössbauer spectra measured at 300, 420, and 590 °C (Figure 3). Overall, Mössbauer spectroscopy revealed potassium iron(III) oxide, $KFeO_2$, to be the only iron-bearing phase formed during the thermal decomposition of K_2FeO_4 in air. Contrary to some earlier postulations (*21*), intermediates containing Fe(V) or Fe(IV) or other Fe(III) oxides (*e.g.* Fe_2O_3, FeO(OH)) were not identified during *in-situ* measurements.

VT XRD measurements on K_2FeO_4 samples were also carried out (Figure 4) to identify the crystalline decomposition products at high temperatures. These measurements provide information on additional besides iron phases. The VT XRD spectra demonstrate that the K_2FeO_4 incompletely transforms to potassium iron oxide ($KFeO_2$) upon heating at 190 °C (Figure 4). This process is in agreement with the Mössbauer measurements (see Figure 3). The decomposition of the ferrate(VI) into the $KFeO_2$ phase was completed at 300 °C, where new additional phases including monoclinic potassium carbonate (K_2CO_3) and potassium oxide (K_2O), clearly appeared in the XRD pattern (Figure 4). However, $KFeO_2$ and hexagonal high-temperature K_2CO_3 were the only phases detected in the XRD patterns recorded at 420 and 590 °C, without any indications of K_2O. The hexagonal high-temperature K_2CO_3 structure that appeared at the expense of the monoclinic K_2CO_3 indicates the thermally induced polymorphous transformation of potassium carbonate.

Based on the described results from the *in-situ* measurements, a new model for the decomposition of K_2FeO_4 in static air was suggested by the authors (*26*). It was postulated that the primary formation of the mixture of potassium oxide and super oxide together with $KFeO_2$ is followed by the rapid secondary reaction of carbon dioxide in air with KO_2 (eqs 4 and 5):

$$K_2FeO_4 \rightarrow KFeO_2 + 1/3\ K_2O + 1/3\ KO_2 + 1/2\ O_2 \qquad (4)$$

$$1/3\ KO_2 + 1/6\ CO_2 \rightarrow 1/6\ K_2CO_3 + 1/4\ O_2 \qquad (5)$$

This assumption is in agreement with the high affinity of KO_2 to CO_2 (*31*). Additionally, the presence of slightly overlapping endo- and exo-effects appearing in the DSC curve between 250 and 280 °C indicate a two-step formation mechanism of potassium carbonate (see Figure 2). This reflects the principal difference in the K_2FeO_4 decomposition mechanisms performed in

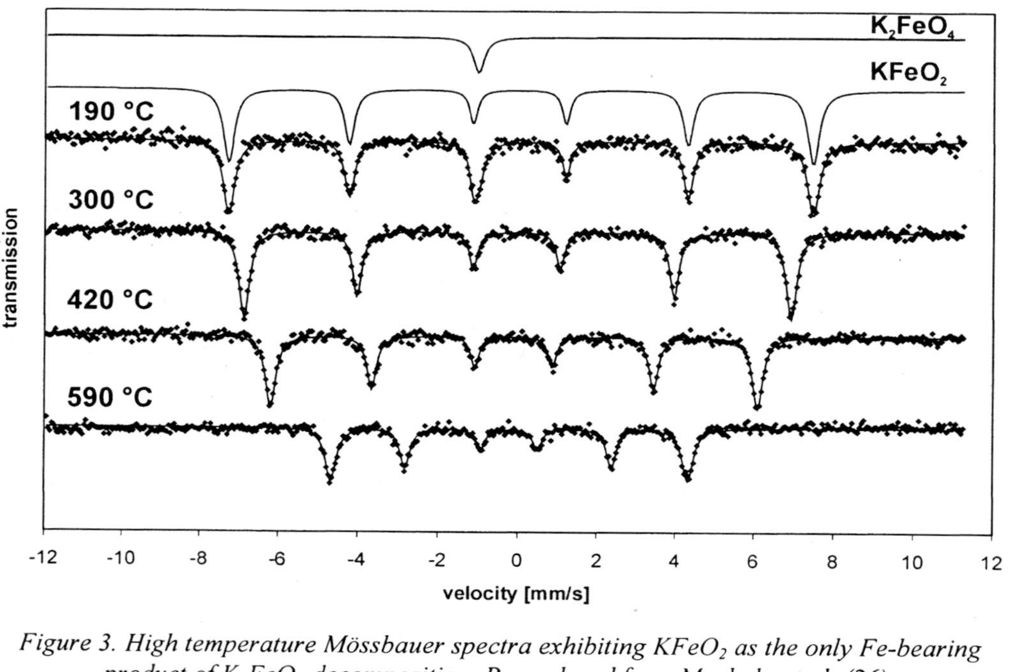

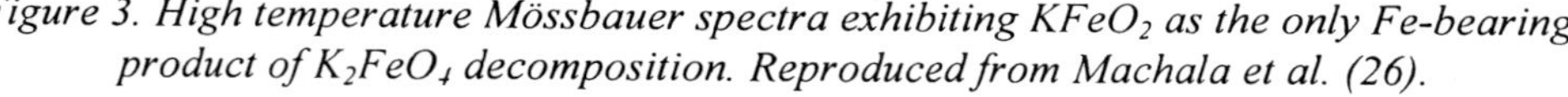

Figure 3. High temperature Mössbauer spectra exhibiting $KFeO_2$ as the only Fe-bearing product of K_2FeO_4 decomposition. Reproduced from Machala et al. (26).

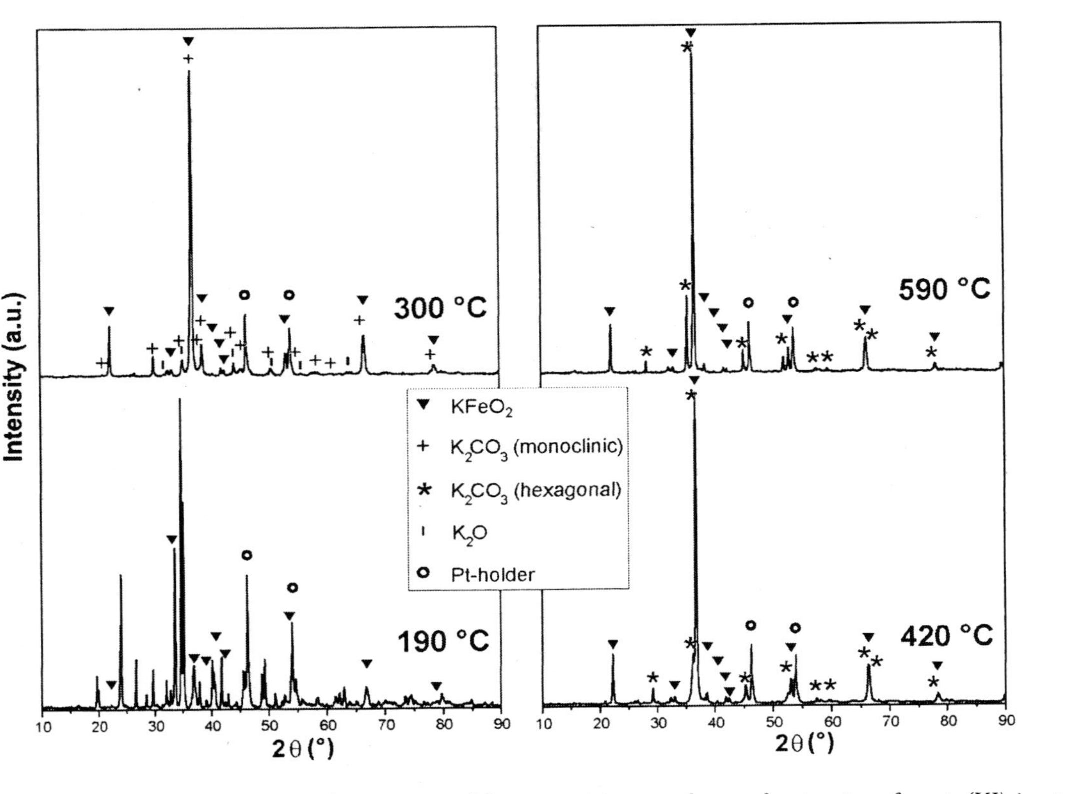

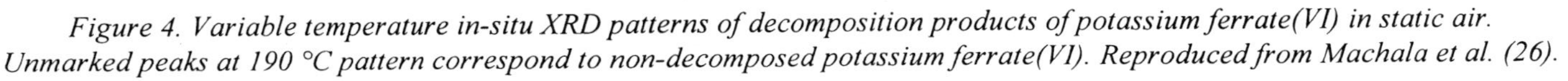
Figure 4. Variable temperature in-situ XRD patterns of decomposition products of potassium ferrate(VI) in static air. Unmarked peaks at 190 °C pattern correspond to non-decomposed potassium ferrate(VI). Reproduced from Machala et al. (26).

static air and inert atmospheres, where the latter displayed no secondary exo-effect in the DSC curve (see the next section).

No other chemical reactions were observed at temperatures above 300 °C. However, the primary decomposition products undergo various phase transitions as demonstrated by VT-XRD data. Thus, potassium oxide, clearly identified at 300 °C, is absent in the XRD pattern, recorded at 420 °C, due to its melting (the melting point of K_2O is ~350 °C). Similarly, the thermally induced polymorphous transition as observed in VT XRD patterns (*30*) changed the potassium carbonate structure at the higher temperatures from monoclinic to hexagonal. Above 750 °C, melting and evaporation of decomposition products occur as documented by a drastic decrease in the sample weight (see TG curve in Figure 2).

Thermal Decomposition of K_2FeO_4 in Inert Atmosphere

Tsapin *et al.* (*27*) studied the thermal decomposition of K_2FeO_4 under nitrogen atmosphere by TG/DSC. The thermogravimetric curve exhibited two main steps with an overall mass loss of 16.2 % (see Figure 5). Evidently, only the second step (125 – 230 °C) with 8% mass loss could be ascribed to the release of oxygen from the ferrate structure, while the first one (50 – 125 °C) was related to desorption of water from the sample surface. Such significant content of water in the initial potassium ferrate(VI) (7.2 wt%) could affect its decomposition as it readily reacts with water. As a result, a complex DSC curve was obtained reflecting a multi-step decomposition of the ferrate sample (see Figure 5).

In recent work performed by Madarasz et al. (*28*), the thermal decomposition of solid $K_2FeO_4 \cdot 0.088\ H_2O$ in an inert atmosphere (N_2 or He inert purge gas, heating rate of 10 °C/min) was studied using simultaneous TG/DTA, in conjunction with *in-situ* analysis of the evolved gases by an online coupled mass spectrometer (EGA-MS). Two decomposition steps were observed in the TG curve up to 500 °C (see Figure 6). The first one below 100 °C corresponds to the evolution of water loosely adsorbed on the sample and the second step between 210 and 310 °C is related to the decomposition of K_2FeO_4 accompanied by a release of oxygen gas as confirmed by EGA-MS. Both decomposition steps are reflected by two endothermic heat effects in the DTA curve (Figure 6). Mössbauer spectroscopic characterization performed on the decomposed sample indicated potassium orthoferrite(III), $KFeO_2$, as the only iron containing compound, which is however metastable in air. The TG curve showed a mass loss of 6.8% occurred during the decomposition (Figure 6), which was ascribed to a mixture of potassium oxide, peroxide and super oxide.

Stability of K_2FeO_4 at Room Temperature – Sample Aging

The stability of the K_2FeO_4 was studied in detail by Nowik et al. (*32*). The phase composition of the samples sealed, exposed to air, or exposed to moist air

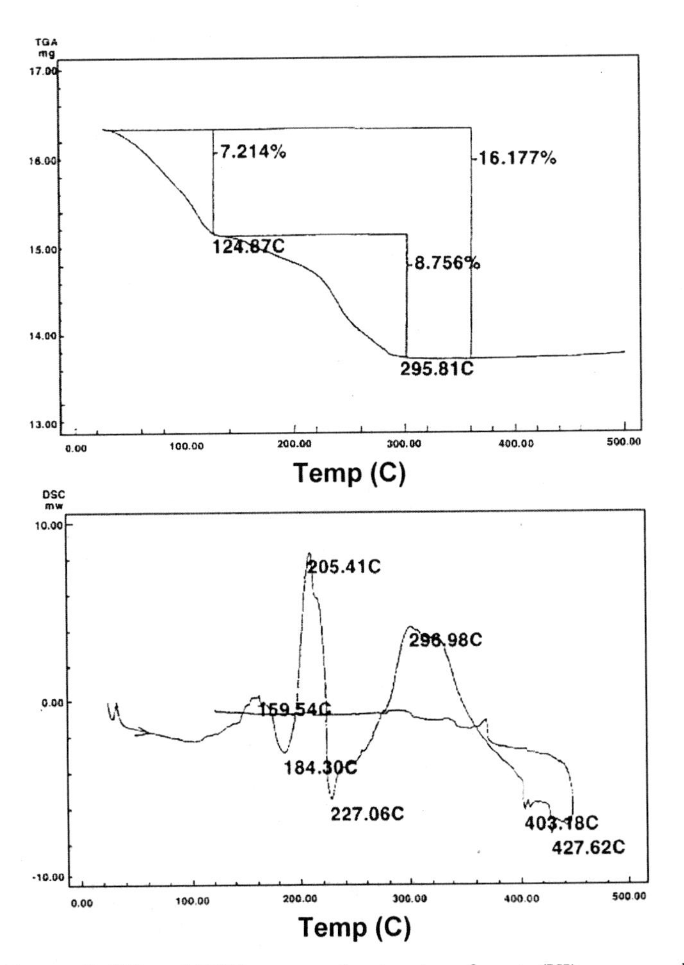

Figure 5. TG and DSC curves of potassium ferrate(VI) measured in nitrogen atmosphere with heating rate of 5 °C/min. Reproduced with permission from Tsapin et al. (27). Copyright 2000.

was determined by Mössbauer spectroscopy as a function of time. Two sub spectra (Figure 7a) appeared in the Mössbauer spectrum of a K_2FeO_4, sample that was stored for 14 months in a closed, but not well-sealed container at room temperature (RT). The minor singlet component belongs to the original K_2FeO_4, while the major doublet sub spectrum shows hyperfine parameters typical for octahedrally coordinated high-spin Fe(III) atoms. To elucidate the nature of the trivalent iron component, low temperature Mössbauer spectra of the K_2FeO_4 sample aged for 15 months were also measured (Figure 7b). While the Fe(VI) absorption sub spectrum does not exhibit any magnetic ordering down to 4.2 K, the evolution of the Fe(III) component with temperature shows the typical

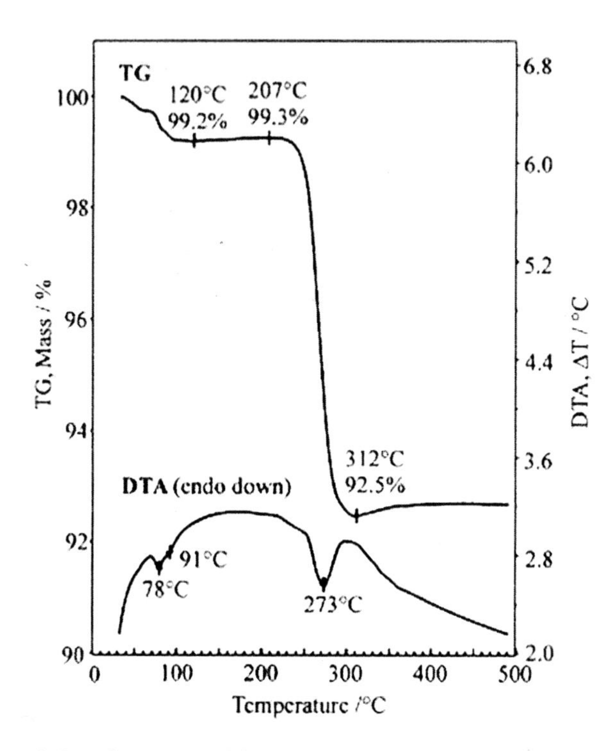

Figure 6. Simultaneous TG/DTA curves started from K_2FeO_4 sample and measured in nitrogen atmosphere (130 ml/min, 10 °C/min). Reproduced with permission from Madarasz et al. (28).

magnetic nanoparticle behavior. A relatively broad temperature range of coexistence of a super paramagnetic and magnetic ordering phase reflecting the particle size distribution is usually the case. This paper concluded that Fe_2O_3 nanoparticles are formed during aging of K_2FeO_4 based on the results of the low temperature Mössbauer measurements. However, authors did not investigate their properties such as crystal structure, particle size and morphology, and type of magnetic ordering at low temperatures. Evidently, XRD, TEM and in-field Mössbauer spectroscopy would be the complementary techniques useful for this investigation.

Barium Ferrate(VI), $BaFeO_4$

Synthesis and Thermal Decomposition of $BaFeO_4$

$BaFeO_4{\cdot}xH_2O$ salt can be easily prepared by the reaction of barium chloride with a basic solution of potassium ferrate(VI) at 0 °C (*21*). Rapid filtration of the

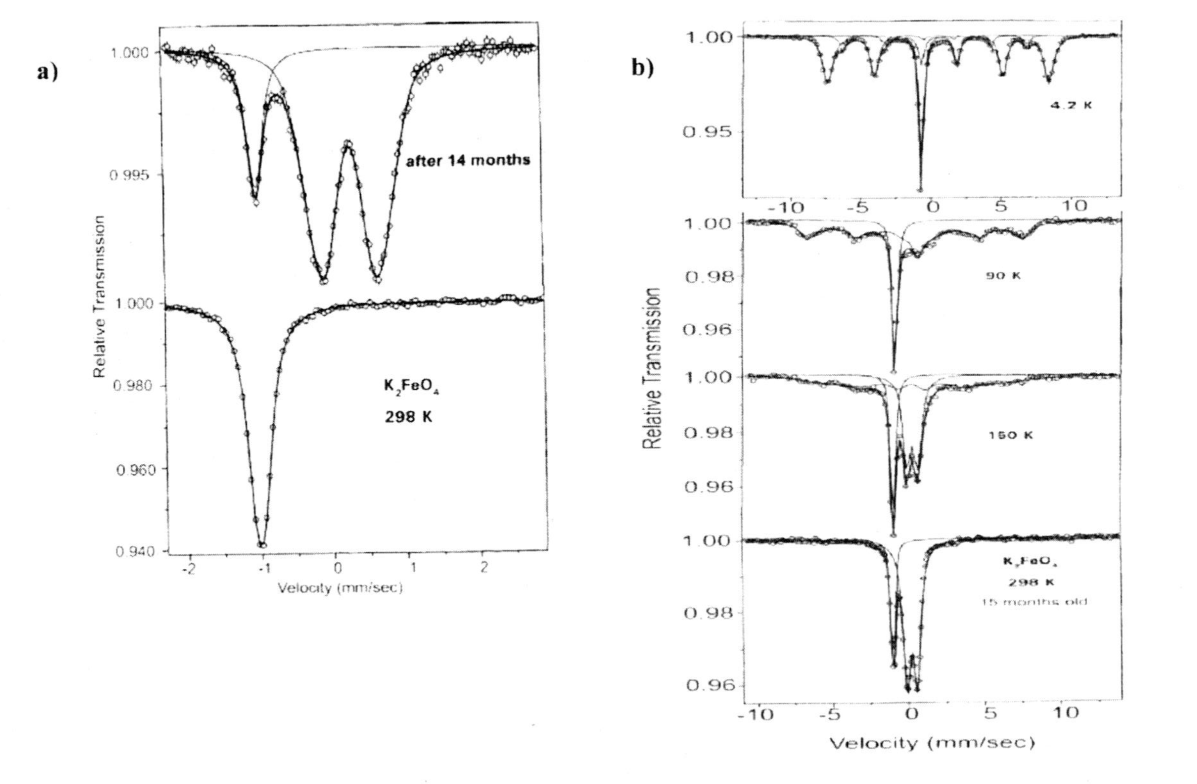

Figure 7. a) RT Mössbauer spectra of the fresh K_2FeO_4 sample (down) and the same sample aged for 14 months (up). b) RT and low temperature Mössbauer spectra of the K_2FeO_4 sample aged for 15 months. Reproduced with permission from Nowik et al. (32).

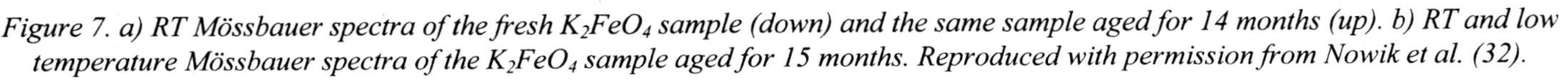

reaction mixture gives a pure product, $BaFeO_4$. Barium carbonate impurities in the prepared samples can be avoided by working in an inert atmosphere and using carbon dioxide free water. Scholder (*21*) observed that an aqueous solution of barium ferrate(VI), $BaFeO_4 \cdot xH_2O$, is unstable even at room temperature as it decomposes to $BaFeO_3 \cdot xH_2O$. In the temperature range of 200-350 °C, the thermal decomposition of vacuum dried $BaFeO_4 \cdot xH_2O$ yielded a product with an overall mean oxidation state of +3.2 for iron.

Ichida (*35*) studied the thermal decomposition of pure $BaFeO_4$ at different temperatures (up to 1200 °C) and oxygen pressures (0.2-1500 atm). Four $BaFeO_x$ phases ($2.5 < x < 3.0$) were found under various temperatures and oxygen pressures including low temperature and high temperature triclinic $BaFeO_{2.5}$, hexagonal $BaFeO_{2.63\text{-}2.95}$, and tetragonal $BaFeO_{2.61\text{-}2.71}$ (Figure 8a). The phase transformations between adjacent $BaFeO_x$ phases were examined by heating each phase under conditions favoring the formation of adjacent phases. Figure 8b schematically illustrates that some of the phase transitions were not reversible. For example, no reversed phase transformation was detected by subjecting the hexagonal phase to any condition in the formation range of either low-temperature phase.

Nanocrystalline $BaFeO_4$ with a purity above 99 % was prepared by the reaction between K_2FeO_4 and aqueous solution of $Ba(C_2H_3O_2)_2$ (*36*). The $BaCO_3$ impurity was removed by a reaction with glacial acetic acid after desiccation of the sample. The excess CH_3COOH was completely evaporated by heating, followed by rinsing with distilled water. The purity of barium ferrate(VI) was determined by performing simultaneous TG/DTA in flowing air (Figure 9). After the desorption of water (0.9 wt%), a two-step decomposition (200-250 °C: 4.8 wt%; 250-850 °C: 4.6 wt%) resulting in a Fe(III) phase was observed. The first step was accompanied by an endothermic effect in the DTA curve followed immediately by an exothermic one. The endothermic peak with a minimum at 800 °C indicated the second decomposition step. Unfortunately, identifications of the decomposition products were not performed in this study.

Recently, Yang et al. (*37*) studied the thermal decomposition and electrochemical behavior of a $BaFeO_4$ sample, prepared by the same route as in the previously mentioned work. The TGA (DTA) measurements were performed in nitrogen atmosphere (Figure 10). A mass loss of 0.99 % below 200 °C corresponds to the release of weakly bounded water. The main decomposition step was seen at ~ 230 °C and observed mass decrease of 6.36 % was consistent with the formation of a Fe(IV) phase (eq 6).

$$2\ BaFeO_4 \rightarrow 2\ BaFeO_3 + O_2 \qquad (6)$$

$BaFeO_3$, exposed to dry air interacts with CO_2 (eq 7). The formations of product phases were proven by XRD.

$$4\ BaFeO_3 + 4\ CO_2 \rightarrow 4\ BaCO_3 + 2\ Fe_2O_3 + O_2 \qquad (7)$$

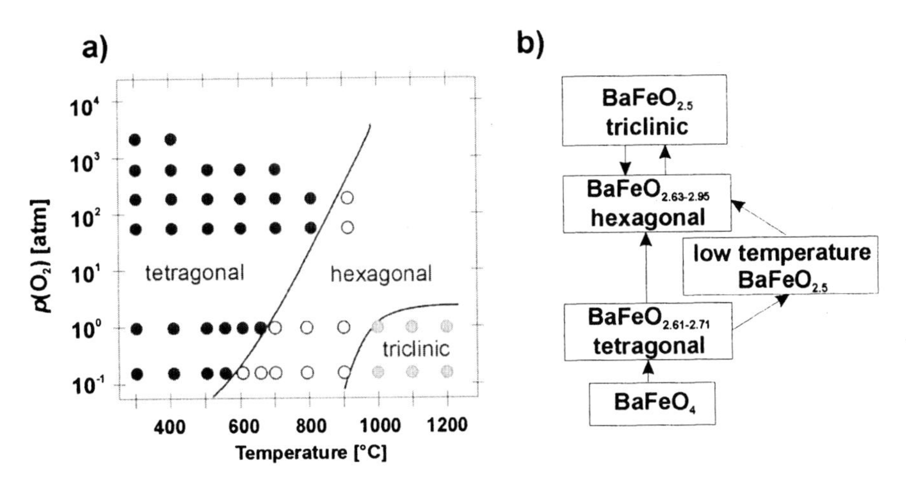

Figure 8. a) Phase diagram of the products obtained by heating $BaFeO_4$ under various temperatures and oxygen pressures. b) Possible phase transformations among $BaFeO_x$ phases. Each arrow shows the direction of phase transition. Adapted with permission from Ichida (35).

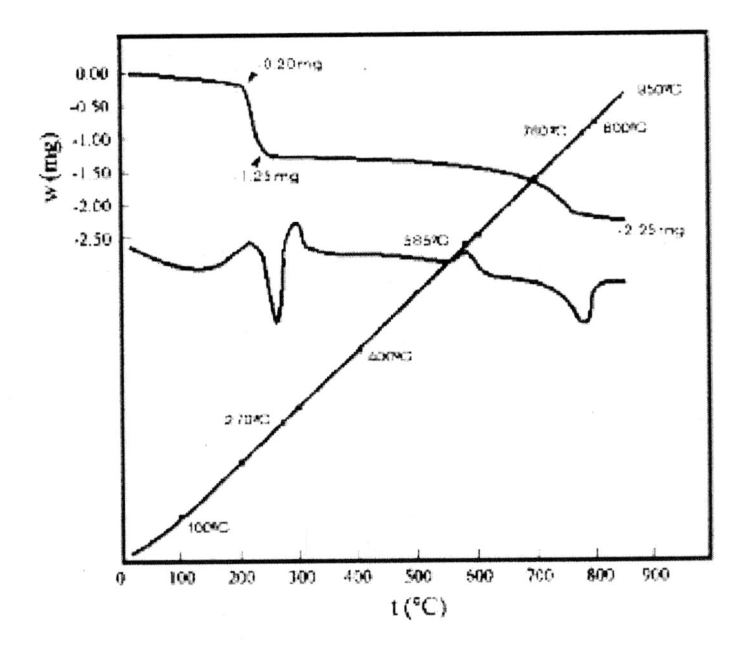

Figure 9. TG/DTA curves of $BaFeO_4$ measured in air (the original sample mass: 21.89 mg). Reproduced with permission from Ni et al. (36).

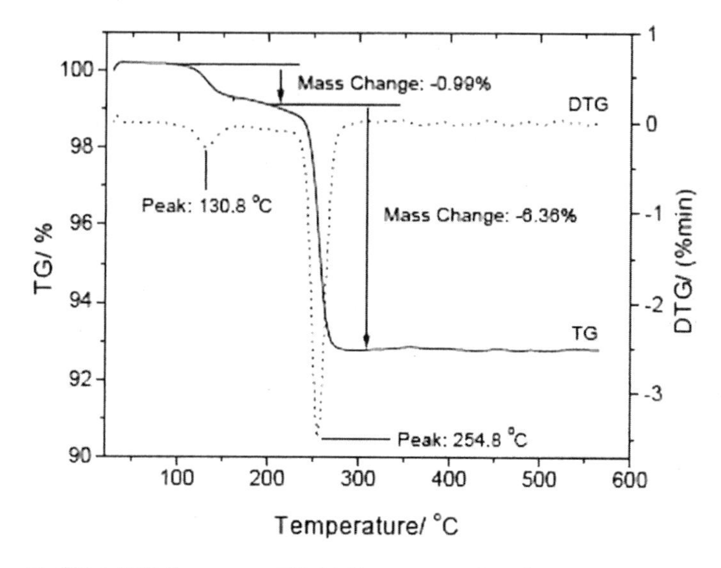

Figure 10. TGA (DTA) curves of $BaFeO_4$ measured under nitrogen atmosphere (8 °C/min). Reproduced with permission from Yang et al. (37).

Madarász *et al.* (*28*) studied the thermal behavior of a $BaFeO_4 \cdot 0.25H_2O$ by using simultaneous TG/DTA in nitrogen atmosphere (10 °C/min, flow rate of 130 ml/min) in combination with *in-situ* analysis of the evolved gases (EGA). Mössbauer spectroscopy was applied to verify the possible existence of Fe(IV) or Fe(V) intermediates and also to identify the composition of the final iron-containing decomposition products. The TG curve gave three well resolved steps including desorption of weakly bonded water (20-122 °C: 0.4 wt%), evolution of strongly bonded water molecules (122-184 °C: 1.3 wt%), and the release of oxygen (184-270 °C: 6.3 wt%) from the ferrate structure (Figure 11). The decomposition steps in the DTA curve were observed in the corresponding endothermic heat effects (Figure 11).

The Mössbauer spectroscopy analysis carried out on the fully decomposed barium ferrate(VI) sample in inert atmosphere revealed that a non-stoichiometric $BaFeO_{3-\delta}$ phase with a relative amount of Fe atoms in the intermediate valence state (III-IV) and paramagnetic at $T = 5$ K is formed. Indeed, the RT Mössbauer spectrum (see Figure 12a) consists of two singlet components, with isomer shifts of 0.37 and 0.07 mm/s typical for regular octahedrally coordinated high-spin Fe(III) atoms, and for low-spin Fe atoms with an intermediate valence state between III and IV, respectively. The low temperature ($T = 5$ K) Mössbauer spectrum of the $BaFeO_{3-\delta}$ phase (see Figure 12b) could be evaluated with the assumption of containing four spectral components including three sextets and one doublet. The magnetically split components with isomer shift of 0.48-0.50 mm/s and hyperfine magnetic fields of 46.7, 49.7 and 52.4 T correspond to high-

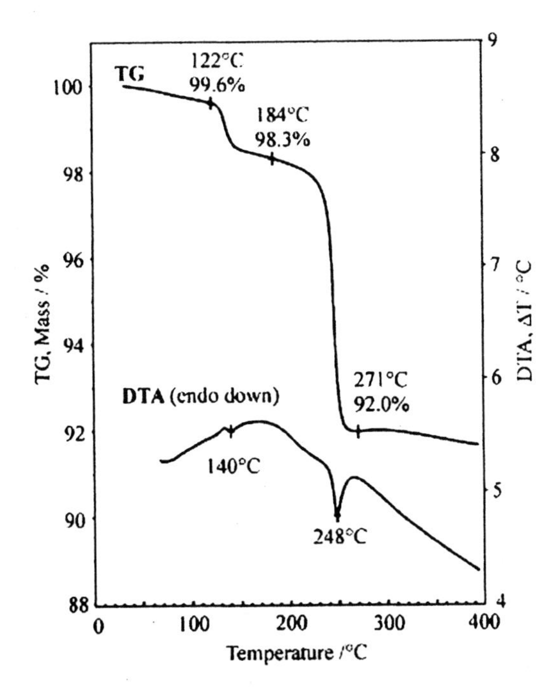

Figure 11. TG/DTA curves of a $BaFeO_4{\cdot}0.25H_2O$ sample measured in nitrogen atmosphere (130 ml/min, 10 °C/min). Reproduced with permission from Madarász et al. (28).

spin iron(III) in three different octahedral environments. The paramagnetic component with lower isomer shift of 0.30 mm/s and quadrupole splitting of 0.38 mm/s were assigned to Fe(III-IV) atoms. The formation of stoichiometric $BaFe^{IV}O_3$ or $BaFe^{III}O_{2.5}$ phases by thermally induced decomposition of barium ferrate(VI) in inert atmosphere could not be confirmed by this study.

Stability of $BaFeO_4$ at Room Temperature – Sample Aging

Similar to K_2FeO_4 study, Nowik et al. (*32*) investigated the disintegration of barium ferrate(VI) over a lengthy period (aging) under different conditions. First, the $BaFeO_4$ sample, prepared by the reaction of $Ba(OH)_2$ with K_2FeO_4, was stored in a plastic Mössbauer absorber holder (lightly sealed from air) at room temperature so the interaction with moist air was minimal. The RT Mössbauer spectra of the sample developed over time showed that the disintegration rate of $BaFeO_4$ under the given conditions is ≈ 1 % per day (Figure 13a). The spectrum of the sample, aged for 27 days (see Figure 14), was modeled by three doublet sub spectrum ascribed to the initial $BaFeO_4$ (δ= -0.88

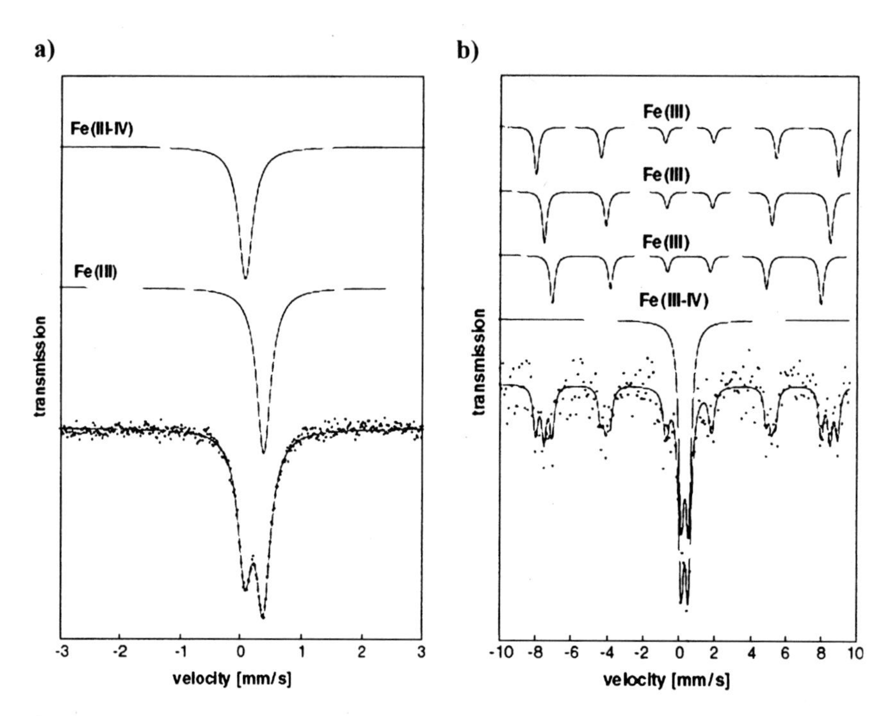

Figure 12. Room temperature a) and low temperature b) Mössbauer spectrum of a completely decomposed $BaFeO_4$ sample (N_2 atmosphere, 300 °C). Reproduced with permission from Madarász et al. (28). Copyright 2006.

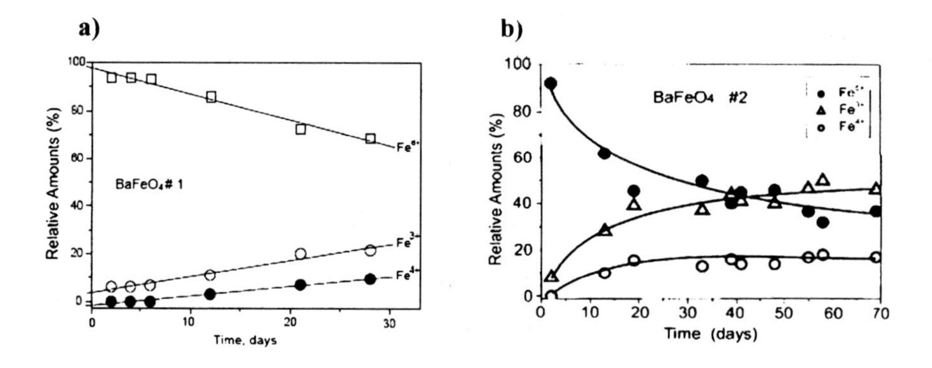

Figure 13. The relative amount of Fe ions of 3+, 4+ and 6+ valences as a function of time for a $BaFeO_4$ sample sealed from air a), and kept in a moist air b). Reproduced with permission from Nowik et al. (32). Copyright 2005.

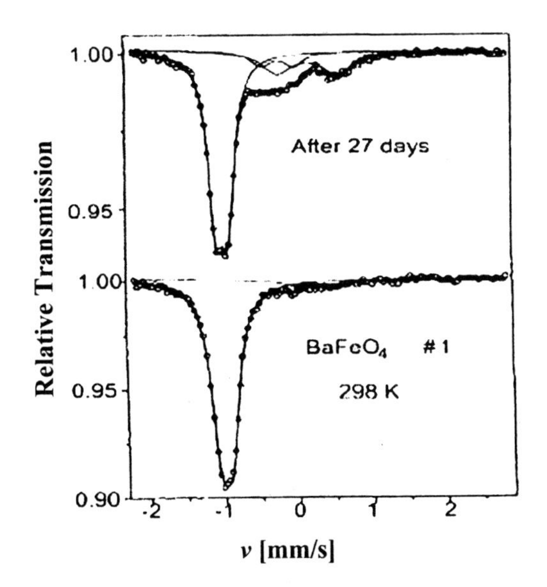

Figure 14. RT Mössbauer spectra of the as-prepared $BaFeO_4$ sample and the same sample lightly sealed in air for 27 days. Reproduced with permission from Nowik et al. (32).

mm/s, ΔE_Q = 0.16 mm/s), Fe(IV) phase (δ = -0.11 mm/s, ΔE_Q = 0.45 mm/s), and a high spin Fe(III) phase (δ = 0.33 mm/s, ΔE_Q = 0.60 mm/s). The identified Fe(III) phase was found to be in the form of Fe_2O_3 nanoparticles.

Figure 13a shows that there is a linear relationship between the time and the number of Fe ions. The decrease in the number of Fe(VI) ions is proportional to an increase in the number of Fe(III) and Fe(IV) ions.

There is however no such linear dependence detected in the case of a $BaFeO_4$ sample prepared by the same synthetic route but kept in a moist air (see Figure 13b). The disintegration process was significantly faster and not linear with time. RT Mössbauer spectra measured after different times of aging showed three-components indicating the presence of Fe(III), Fe(IV), and Fe(VI) phases similar to a sample stored in a dry atmosphere. Interestingly, low-temperature Mössbauer measurement (T = 4.2 K) of the sample aged in moist air for 90 days (not shown here), gave also a three-component spectrum. However, the isomer shift (-0.18 mm/s) of the sub spectrum related to the originally expected Fe(IV) was not consistent with that in the RT spectrum (though a second order Doppler shift was taken into account), and the hyperfine magnetic field (24.2 T) was not typical for tetravalent iron compounds (*38*). Thus, Nowik et al. (*32*) suggested that $BaFeO_3$, representing the Fe(IV) phase, undergoes a disproportionation into Fe(V) and Fe(III) phases. This interpretation now fully agrees with the low temperature Mössbauer spectrum as pentavalent iron exhibits a lower isomer shift value in comparison with tetravalent iron (*12*). The considered sub spectrum of the Fe(III) phase was likely superimposed by a much more intensive Fe_2O_3 sub spectrum.

Cesium ferrate(VI)

Cesium ferrate(VI), Cs_2FeO_4, can be prepared either via the oxidation of $Fe(OH)_3$ by Cl_2 in concentrated alkaline media, or via the interaction between solid Fe_2O_3 and cesium peroxide in a dry oxygen flow at elevated temperatures. Mössbauer spectroscopy measurements performed by Kopelev et al. (*39*) showed that thermally induced decomposition of Cs_2FeO_4 at temperatures above 600 °C resulted in $CsFeO_{2.5}$ with tetravalent iron according to reaction (8) with an uncertainty in the chemical form of cesium oxide.

$$Cs_2FeO_4 \rightarrow CsFeO_{2.5} + CsO_x + (0.75\text{-}0.5x)\, O_2 \qquad (8)$$

A detailed analysis of the X-ray powder diffraction pattern showed that $CsFeO_{2.5}$ possesses a perovskite-like structure with face-centered cubic crystals with tetravalent iron atoms octahedrally coordinated by six oxygen atoms. No iron(V) phase was observed during the thermolysis of Cs_2FeO_4.

Conclusions

The *in-situ* approach was found to be very effective contrary to "standard" methods of analysis of the cooled sample after the thermal decomposition of salts of ferrate(VI). Despite the various decomposition mechanisms suggested by different authors, $KFeO_2$ was usually identified as the Fe(III) phase primarily formed from K_2FeO_4 at high temperatures independently from the reaction atmosphere, regime of sample heating or amount of adsorbed surface water of the initial ferrate sample. On the other hand, the formation of other Fe(III) oxides (Fe_2O_3, FeOOH, *etc.*) was not observed; not even during the *in-situ* monitoring of the decomposition of K_2FeO_4. It is worthwhile to mention that both phases, K_2FeO_4 and $KFeO_2$, exhibit tetrahedral coordination of the iron atom; thus the coordination environment destroys and re-builds again during the thermal treatment. $KFeO_2$ is unstable in moist air and can significantly affect the interpretation of results. It is difficult to determine the chemical form of potassium oxide (KO_x), one of the decomposition products. The best information concerning the determination of "x" was obtained from TG analysis. Additionally, possible participation of air-CO_2 can also change the overall mass loss after the decomposition of K_2FeO_4 sample.

We can conclude that thermal treatment of $BaFeO_4$ though under different atmospheres results in the formation of $BaFeO_x$ ($2.5 < x < 3$) phases with various crystal structures. Surprisingly, neither stoichiometric $BaFeO_{2.5}$ or $BaFeO_3$ compounds were confirmed at high temperatures in air. The results independently published by different authors are more consistent for $BaFeO_4$ compared to the thermal studies carried out for K_2FeO_4. Prolonged disintegration of $BaFeO_4$ in air resulted in a Fe(III) phase in the form of Fe_2O_3 nanoparticles, however, their characterization in terms of size, morphology and magnetic behavior is incomplete.

Acknowledgment

Financial supports from the Ministry of Education of the Czech Republic (MSM6198959218 and 1M6198959201) are gratefully acknowledged. We wish to thank Dr. Rudi Wehmschulte for useful comments.

References

1. Rush, J. D.; Bielski, B. H. J. *J. Am. Chem. Soc.* **1986**, *108*, 523.
2. Jeannot, C.; Malaman, B.; Gerardin, R.; Oulladiaf, B. *J. Solid State Chem.* **2002**, *165*, 266.
3. Delattre J. L.; Stacy, A. M.; Young, V. G.; Long, G. J.; Hermann, R.; Grandjean, F. *Inorg. Chem.* **2002**, *41*, 2834.
4. Kopelev, N. S.; Perfiliev, Yu. D.; Kiselev, Yu. M. *J. Radioanal. Nucl. Chem.* **1992**, *162(2)*, 239.
5. Licht, S.; Wang B.; Ghosh S. *Science* **1999**, *28*, 1039.
6. Licht, S.; Alwis, C. D. *J. Phys. Chem. B* **2006**, *110*, 12394.
7. Waltz, K. A.; Suyama, A. N.; Suyama, W. E.; Sene, J. J.; Zeltner, W. A.; Armacanqui, E. M.; Roszkowski, A. J.; Anderson, M. A. *J. Power Source* **2004**, *134*, 318.
8. Delaude, L.; Laszlo, P. *J. Org. Chem.* **1996**, *61*, 6360.
9. Sharma, V. K.; Mishra, S. K.; Nesnas, N. *Environ. Sci. Technol.* **2006**, *40(23)*, 7222.
10. Sharma, V. K; Kazama, F; Jiangyong, H; Ray, A. K. *J. Water Health*, **2005**, *3*, 45.
11. Sharma, V. K. *Water Sci. Technol.* **2004**, *49*, 69.
12. Perfiliev, Y. D. *Russ. J. Inorg. Chem.* **2002**, *47*, 611.
13 Kokarovtseva, I. G.; Belyaev, I. N.; Semenyakova, L. V. *Russ. Chem. Rev.* **1972**, *41*, 928.
14. Neveux, N. *REWAS '99-Global Symp. Recycl., Waste Treat. Clean Tech. Proc.* **1999**, *3*, 2417.
15. Neveux, N.; Kanari, N.; Gaballah, I.; Evrard, O.; U.S. Patent 7,172,748, 2007.
16. Thompson, G. W.; Ockerman, G. W.; Schreyer, J. M. *J. Am. Chem. Soc.* **1951**, *73*, 1279.

17. Schreyer J. M.; Thompson, G. W.; Ockerman, L. T. *Inorg. Synthesis* **1953**, 4164.
18. Bouzek, K.; Schmidt, M. J.; Wragg, A. A. *Coll. Czech. Chem. Commun.* **2000**, *65*, 133.
19. Lescuras-Darrou, V.; Lapicque, F.; Valetin, G. *J. Appl. Electrochem.* **2002**, *32*, 57.
20. Kiselev, Y. M.; Kopelev, N. S.; Zav'yalova, N. A.; Perfiliev, Y. D.; Kazin, P. E. *Russ. J. Inorg. Chem.* **1989**, *34*, 1250.
21. Scholder, R.; Bunsen, H. V.; Kindervater, F.; Zeiss, W. *Z. Anorg. Allgem. Chem.* **1955**, *282*, 268.
22. Goff, H.; Murmann, R. K. *J. Am. Chem. Soc.* **1971**, *93*, 6058.
23. Ichida, T. *Bull. Chem. Soc. Jpn.* **1973**, *4*, 79.
24. Fătu, D.; Schiopescu, A. *Rev. Roum. Chim.* **1974**, *19*, 1297.
25. Zboril, R.; Mashlan, M.; Petridis, D. *Chem. Mater.* **2002**, *14*, 969.
26. Machala, L.; Zboril, R.; Sharma, V. K.; Filip, J.; Schneeweiss, O.; Homonnay, Z. *J. Phys. Chem. B* **2007**, *110*, 16248
27. Tsapin, A. I.; Goldfeld, M. G.; McDonald G. D.; Nealson, K. H.; Moskovitz, B.; Solheid, P.; Kemner, K. M.; Kelly, S. D.; Orlandini, K. A. *Icarus*, **2000**, *147*, 68.
28. Madarasz, J.; Zboril, R.; Homonnay, Z.; Sharma, V. K.; Pokol, G. *J. Solid State Chem.* **2006**, *179*, 1426.
29. Maddock, A. G. *Mössbauer Spectroscopy: Principles and Applications of the Techniques*; Horwood Chemical Science Series, UK, 1998; pp. 108.
30. Becht, H. Y.; Struikmans, R. *Acta Cryst.* **1976**, *B32*, 3344.
31. Cotton, F. A.; Wilkinson, G.; Murillo, C. A.; Bochmann, M. *Advanced Inorganic Chemistry (6th edition)*; John Wiley&Sons, Inc: New York, US, 1999, pp. 461.
32. Nowik, I.; Herber, R. H.; Koltypin, M.; Aurbach, D.; Licht, S. *J. Phys. Chem. Solids* **2005**, *66*, 1307.
33. Sharma, V. K. *Adv. Envir. Res.* **2002**, *6*, 143.
34. Ayers, K. E.; White, N. C. *J. Electrochem. Soc.* **2005**, *152*, A467.
35. Ichida, T. *J. Solid State Chem.* **1973**, *7*, 308.
36. Ni, X.-M.; Ji, M.-R.; Yang, Z.-P.; Zheng, H.-G. *J. Crystal Growth* **2004**, *261*, 82.
37. Yang, W.; Wang, J.; Pan, T.; Cao, F.; Zhang, J.; Cao, C. *Electrochim. Acta* **2004**, *49*, 3455.
38. Morimoto, S.; Kuzushita, K.; Nasu, S. *J. Magn. Magn. Mater.* **2004**, *272*, 127.
39. Kopelev, N. S.; Val'kovskii, M. D.; Popov, A. I. *Russ. J. Inorg. Chem.* **1992**, *37*, 267.

Chapter 9

A Fluorescence Technique to Determine Low Concentrations of Ferrate(VI)

Determination of Micromolar Fe(VI) Concentrations for Laboratory Investigations

Nadine N. Noorhasan, Virender K. Sharma*, and J. Clayton Baum

Chemistry Department, Florida Institute of Technology, 150 West University Boulevard, Melbourne, FL 32901
***Corresponding author: vsharma@fit.edu**

A fluorescence technique to determine low concentrations of aqueous ferrate(VI), [$Fe^{VI}O_4^{2-}$], in water was developed over a wide pH range using the reaction of ferrate(VI) with scopoletin reagent. The rates of the reaction of ferrate(VI) with scopoletin as a function of pH at 25°C were determined using the stopped-flow technique to demonstrate the reaction is rapid (< 1 min). Spectral measurements on scopoletin at different pH showed that the maximum in absorption varies with the pH while the emission maximum is independent of pH. The absorbance measurements were used to determine the acid dissociation constant, $K_a = 1.55 \pm 0.01 \times 10^{-9}$ ($pK_a = 8.81 \pm 0.05$) for scopoletin. The intensity of fluorescence for scopoletin decreases linearly with increase in the concentration of ferrate(VI), which suggests the suitability of the method. Moreover, a relatively large decrease in intensity per micromolar ferrate(VI) concentration was observed, especially at low pH, which makes fluorescence a sensitive technique to determine low ferrate(VI) concentrations.

Introduction

In recent years, there has been tremendous interest in the innovative use of ferrate(VI), which has the molecular formula $Fe^{VI}O_4^{2-}$ where iron exists in the +6 oxidation state to which four oxygen atoms are bonded covalently to give a tetrahedral structure (*1*). In the "super-iron" battery, ferrate(VI) replaces the usual manganese dioxide cathode since ferrate(VI) can gain more electrons than manganese dioxide. Additionally, the "super-iron" battery does not produce toxic compounds in contrast to the manganese cathode (*2*). Ferrate(VI) has also been proposed as a green chemistry oxidant for organic synthesis (*3*). Moreover, ferrate(VI) has the highest redox potential (+2.2V in acid) of any oxidant used in water and wastewater treatment (*4,5*). The most common treatment method is chlorination, but it produces known toxic by-products (*6-8*). In comparison, ferrate(VI) has been shown to destroy pollutants and bacterial species in seconds to minutes without producing harmful by-products (*9*). Ferrate(VI) decomposition produces Fe(III), which itself is an excellent coagulant for removal of metals and radionuclides from contaminated water (*10*).

Studies of ferrate(VI) include its production, stability, oxidation, and magnetic properties, all of which require accurate knowledge of the ferrate(VI) concentration in dilute solutions. The concentration of $Fe^{VI}O_4^{2-}$ in an aqueous sample of potassium ferrate (K_2FeO_4) can be determined by titrating it with chromium(III) (*11*) (eq. 1):

$$Cr(OH)_4^- + FeO_4^{2-} + 3H_2O \rightarrow Fe(OH)_3(H_2O)_3 + CrO_4^{2-} + OH^- \qquad (1)$$

The resulting chromate(VI) solution is then acidified as dichromate and titrated with a standard solution of ferrous ions. A similar titration procedure has also been used in the reaction of ferrate(VI) with arsenic(III) (*12*). Both methods determine concentrations only at the sub-molar to molar level of ferrate(VI). In addition, the titration steps are time consuming and use toxic heavy metals. Another method is the use of cyclic voltammetry to determine low concentrations of ferrate, but this method is inconvenient to use (*13*). To determine low concentrations in micromolar to millimolar levels of aqueous ferrate(VI), one could resort to the use of simple and convenient UV-Vis spectrophotometry where the absorbance at 510 nm is measured to determine the aqueous ferrate(VI) concentration. However, the molar absorption coefficient of ferrate(VI) at 510 nm (ε_{510nm}) is not only low, but also varies with pH (1150 $M^{-1}cm^{-1}$ at pH 9.1 to 520 $M^{-1}cm^{-1}$ at pH 6.20) (*14*). Moreover, the absence of a strong chelating agent (e.g. phosphate) in solution for complexation of Fe(III), produced from the self decomposition of ferrate(VI), causes significant errors in optical monitoring of the solution (*15*). The study of the kinetics of ferrate(VI) reactions with various substrates is presently restricted to the basic region due to fast disproportionation of ferrate(VI) at neutral and acidic pH range (*15*). The self-decomposition follows second-order kinetics and can be minimized by using lower Fe(VI) concentration.

Recently, a method was developed to determine low concentrations of aqueous Fe(VI) in acidic medium. This method uses the reaction of Fe(VI) with 2,2'-azinobis(3-ethylbenzothiazoline-6-sulfonate) (ABTS) (*16*). ABTS reacts with oxidants via a single-electron transfer to give $ABTS^{\bullet +}$, a stable and intense green colored radical that absorbs in the visible region (*16*). However, this method may not be suitable if products formed from the reaction of Fe(VI) with a substrate also absorb at similar absorption wavelengths (*17*). Under these conditions, a fluorescence method would be better to study the reactions of Fe(VI) with substrates. Thus, the present study offers an alternative method to determine concentrations at the low μM range in acidic solutions.

In this chapter, a new fluorimetric technique is proposed to determine low concentrations of ferrate(VI) in water. Fluorescence analysis in general is more sensitive than UV-Vis absorption analysis, so fluorescence should give better measurements at μM concentrations of ferrate(VI). Scopoletin (7-hydroxy-6-methoxy coumarin) (Figure 5.1), a known fluorescence agent, was chosen for this method. Scopoletin has been used to determine hydrogen peroxide using a peroxidase catalyzed oxidation method in natural waters (*18*). Although species present in natural waters were found to interfere in determining the concentrations of hydrogen peroxide, the goal of the proposed method is to determine the concentration of Fe(VI) in distilled deionized (DD) water with no interferences before adding the sample to natural waters, if at all. To demonstrate that this technique is sufficiently rapid to be useful over a wide pH range, a kinetic study of ferrate(VI) reaction with scopoletin was first examined at pH values where the reaction can still be detected at $\geq$ 0.005seconds. Detailed absorption and fluorescence spectral studies of scopoletin at different pH were carried out to choose appropriate wavelengths for fluorometric measurements. In addition, calibration curves as a function of pH were constructed to demonstrate a linear decrease in the fluorescence of scopoletin with increasing concentrations of ferrate(VI) in DD water.

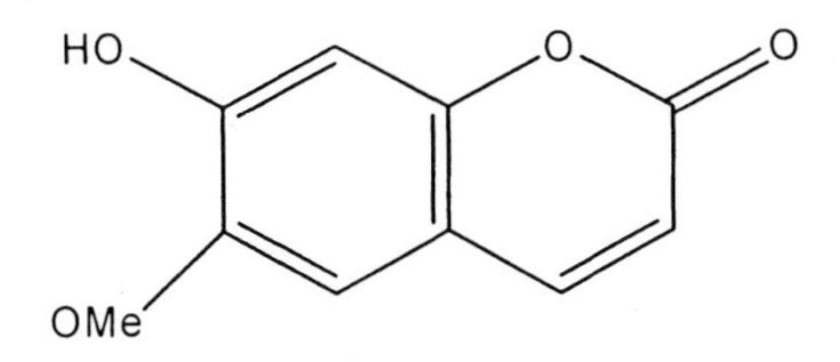

Figure 1. The molecular structure of scopoletin

Experimental

Materials

All the chemicals were purchased from Sigma-Aldrich, reagent grade or better, and were used without further purification. Solutions were prepared with water that had been distilled and passed through an 18MΩ Milli-Q water purification system. Scopoletin was prepared in 0.01M Na_2HPO_4 and it was placed in a dark bottle to prevent decomposition due to visible light (*19*). Likewise, solutions containing scopoletin were minimally exposed to light. Potassium ferrate (K_2FeO_4) of high purity (98% plus) was prepared by the method of Thompson et al. (*20*). The ferrate(VI) solutions were prepared by the addition of solid samples of K_2FeO_4 to deoxygenated 0.005M Na_2HPO_4/0.001M borate at pH 9.0. Phosphate was used in the buffer for complexing Fe(III), which would otherwise precipitate as a hydroxide to interfere with the optical monitoring of the solution (*14*). The concentrations of ferrate(VI) were determined by measuring the absorbance at 510 nm and using the molar absorption coefficient $\varepsilon_{510nm} = 1150\ M^{-1}\ cm^{-1}$ at pH 9.0 (*21*).

Apparatus

A stopped-flow spectrophotometer (SX.18 MV, Applied Photophysics, UK) equipped with a photomultiplier detector was used to make the kinetic measurements. The kinetic curves were analyzed using a non-linear least-squares algorithm within the SX.18 MV software. The temperature of this system was 25 ± 0.1°C, which was controlled by a Fischer Scientific Isotemp 3016 circulating water bath. The rate constants obtained represent the average value of six kinetic runs.

An Orion 710A ion selective electrode system equipped with a glass pH electrode was used for all pH measurements. Standard buffers of pH 4.0, 7.0, and 10.0 were used to calibrate the electrode and to determine the pH of the mixed solutions.

An HP8453 UV/Vis spectrophotometer was used for spectral studies. A 1 cm quartz cuvette was used to carry out the measurements at 25 °C. A Spex FluoroMax-3 fluorimeter was used to perform fluorescence measurements at 25 °C. The excitation and emission wavelengths were 335 nm and 460 nm, respectively. Slit widths were set at 2 nm band pass. The shutter was kept closed until the measurement was made (>3 min) in order to exclude incident radiation that can cause photobleaching (*19*).

Results and Discussion

Kinetic Experiments

In these experiments, equal volumes of 200 μM ferrate(VI) and 2000 μM scopoletin were mixed at different pH values. The reaction was followed by monitoring ferrate(VI) absorbance at 510 nm as a function of time. Excess scopoletin ensured that reactions were measured under pseudo-order conditions. The absorbance versus time profile for ferrate(VI) gave a single-exponential decay curve, indicating the reaction was first-order with respect to ferrate(VI). Reactions of ferrate(VI) with several similar compounds have also shown a first-order rate with respect to the compound (*5,22*); thus, the rate expression for the reaction of ferrate(VI) with scopoletin is assumed to be:

$$-d[Fe(VI)]/dt = k[Fe(VI)]^1[SC]^1 \quad (2)$$

where [Fe(VI)] and [SC] are the concentrations of ferrate(VI) and scopoletin, respectively, and k is the overall reaction rate constant.

The values of k were determined for different basic solutions and are given in Table 1. The rate constant of the reaction increases with a decrease in pH. At pH<11, the rates were too fast (<5 ms) under the same conditions to be measured. This agrees with previous studies on the reaction of ferrate(VI) with other compounds (*4-5,9,22,23*). Ferrate(VI) is a stronger oxidant upon protonation (eq. 3) so the reaction rates are expected to increase.

$$HFeO_4^- \Leftrightarrow H^+ + FeO_4^{2-} \qquad pK = 7.23\ (23) \quad (3)$$

The rate constants suggest that the reactions at both high and low pH are complete in less than a minute. The reaction of ferrate(VI) with scopoletin is thus fast enough for the measurement of low concentrations of ferrate(VI) over a wide range of pH in water.

Table I. The second-order rate constant, k, for the reaction of ferrate(VI) with scopoletin.

pH	*k, 10^3 $M^{-1}s^{-1}$*
13.06	1.2 ± 0.1
11.88	6.3 ± 0.1
11.02	9.3 ± 0.3

Absorption and Emission Spectra of Scopoletin

Spectral studies of the absorption of scopoletin were performed to determine the appropriate excitation wavelength for fluorescence measurements. The absorption of scopoletin at different pH values is presented in Fig. 2. A blue shift of the spectrum was observed at lower pH in agreement with previous excitation studies (*24*). The absorbance (365 nm) as a function of pH shown in the inset of Fig. 2 was used to determine the pK_a (*25*). The fit of the absorbance yields a pK_a value of 8.81±0.15 (K_a = 1.55 ± 0.01 × 10^{-9}) for scopoletin. This differs from the apparent pKa value of 7.37 reported using capillary electro-

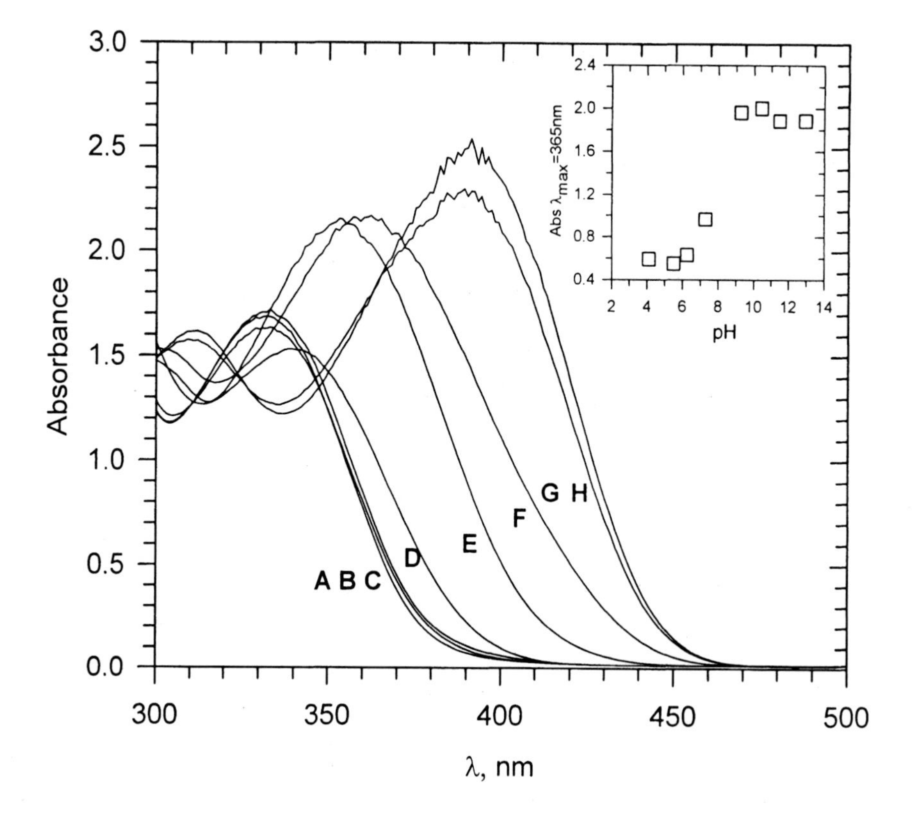

Figure 2. Absorption spectra of scopoletin at different pH values: A=4.10, B=5.49, C=6.26, D=7.28, E=9.27,F=10.44, G=11.49, & H=12.94. The inset graph shows the absorbance vs. pH at 365 nm to determine the value of pK_a.

phoresis (*26*). Although there are large absorbance changes at longer wavelengths over the pH range (factor of 4.3 at 365 nm, factor of 25 at 390 nm) at 335 nm the spectra exhibit relatively small changes in absorbance (factor of 1.4). Therefore, an excitation wavelength at 335 nm was used in this study.

The fluorescence experiments were performed over the pH range of 4.92 to 9.32. The temperature and fluorimeter settings were kept constant since these parameters can also affect fluorescence intensity (*27*). It was found that the wavelength of the emission maximum is independent of pH (Fig. 3) when scopoletin is excited at 335 nm. The fluorescence intensity decreased from pH 4.92 to 9.32 (Fig. 3) by a factor of 1.4-1.5 in agreement with the decrease in absorbance at 335 nm.

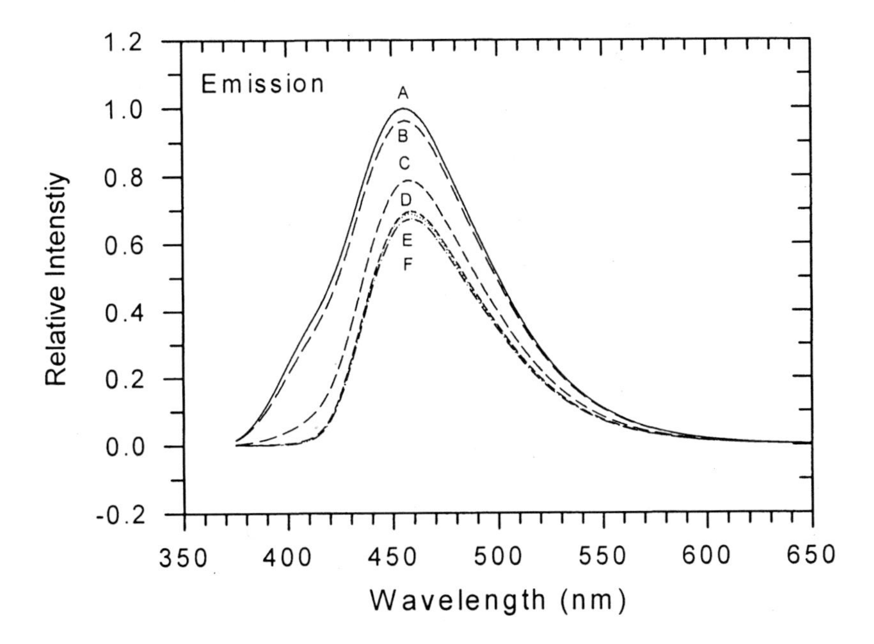

Figure 3. Emission spectra of scopoletin at different pH values: A=4.92, B=5.89, C=7.84, D=8.95, E=9.28, & F=9.32.

Calibration Curves Using Fluorescence

Calibration curves were constructed using 50 μM scopoletin with various concentrations of Fe(VI) at pH 10.10 and pH 4.95 (Fig. 4). The data points in

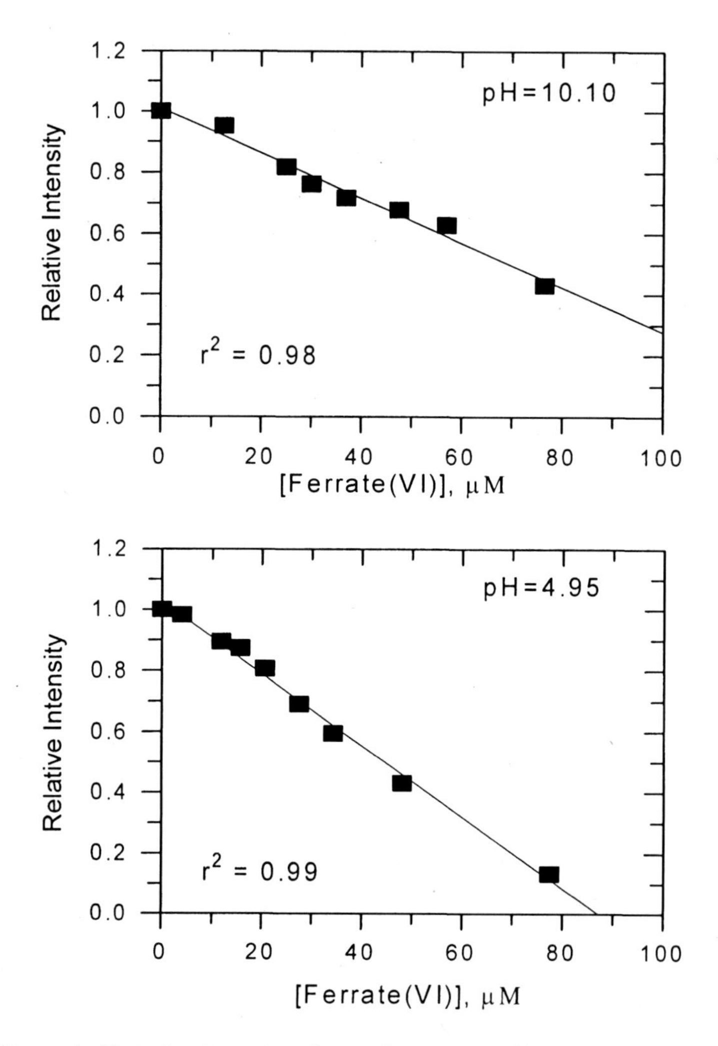

Figure 4. Emission intensity of scopoletin versus [Ferrate] at pH 10.10 and at 4.95.

Fig. 4 represent the average value of four measurements. Fig. 4 shows a linear decrease in intensity with the increase in concentration of ferrate(VI) up to 80 μM. The following slopes (±0.0004) were obtained: 0.0073 at pH 10.10, 0.0077 at pH 9.10, 0.0100 at pH 8.88, 0.0120 at pH 5.94, and 0.0120 at pH 4.95. The slope is the change in relative intensity divided by the change in concentration of Fe(VI). At the lower pH, the higher fluorescence intensity results in a greater change in intensity producing a higher slope. The steeper slope allows low concentrations of Fe(VI) to be determined even more precisely at the lower pH values, where other methods are ineffective. The linearity of the curves at low Fe(VI) concentrations clearly demonstrates the ability of the fluorimetric technique to determine low concentrations of ferrate(VI) over a wide pH range. Moreover, correlation coefficients ≥ 0.98 for the plots show that possible interfering variables do not influence the results significantly in this fluorometric method.

Additional experiments were performed at pH 9.0 to confirm that Fe(III), produced from Fe(VI) does not interfere with the fluorescence measurements. In these experiments, the fluorescence intensity was measured for solutions having 50 μM scopoletin and Fe(III) at concentrations ranging from 10 μM to 80 μM. The concentration range of Fe(III) is similar to what it would be after Fe(VI) reaction with scopoletin. The fluorescence intensity of scopoletin was unaffected by the addition of Fe(III) within experimental error. Thus, the presence of Fe(III) does not interfere with the fluorescence determination of Fe(VI) using scopoletin.

Applicability of the Method

With this technique, it is possible to determine low Fe(VI) concentrations over a wide pH range in DD water. This is important for laboratory studies involving Fe(VI), such as kinetics experiments, and for adding known concentrations of Fe(VI) to natural waters, as in water treatment. Experiments in this study showed that low concentrations of Fe(III) do not interfere in the fluorescence measurements. This fluorimetric technique to determine low Fe(VI) concentrations has several advantages. Most importantly, this fluorimetric technique is applicable over a wide pH range including acidic solutions; it is possible to measure Fe(VI) concentrations at low pH values especially below pH 6 if the absorption spectroscopy method fails due to the interferences. Finally, the simplicity of the method is that excitation and emission wavelengths can be fixed at 335 nm and 460 nm, respectively, independent of the pH and concentration in determining the standardization curves and unknown concentrations of Fe(VI).

Acknowledgment

The authors wish to thank Dr. Yunho Lee for useful comments on this manuscript.

References

1. Hoppe M.L., Schlemper E.O., Murmann R.K. *Acta. Cryst.* **1983**, *B38*, 2237-2239.
2. Stuart L., Wang B., Gosh S. *Science,* **1999**, *285*, 1039-1042.
3. Delaude L., Laszlo P. *J. Org. Chem.*, **1996**, *61*, 6360-6370.
4. Sharma V.K. *Wat. Sci. and Tech.*, **2004**, *79*, 69-74.
5. Sharma V.K., Kazama F., Jiayong H., Ray A.K. J. *Water Health*, **2005**, *3*, 42-58
6. Burrows W.D., Renner S.E. *Environ. Health Perspect.* **1999**, *107*, 975-984.
7. Craun G.F., Calderon R.L. *J. Am. Wat. Work Assoc.*, **2000**, *9*, 64-75.
8. Gunten U. Von *Water Res.*, **2003**, *37*, 1469-1487.
9. Sharma V.K. *Adv. Environ. Res.*, **2002**, *6*, 143-156.
10. Potts M.E., Churchwell D.R. *Water Environ. Res.*, **1994**, *66*, 107-109.
11. Schreyer J.M., Thompson G.W., Ockerman L.T. *J. Am. Chem. Soc.*, **1950a**, *22*, 1426-1427.
12. Schreyer J.M., Thompson G.W., Ockerman L.T. *J. Am. Chem. Soc.*, **1950b**, *22*, 691-692.
13. Venkatadri A.S., Wagner W.F., Bauer H.H. *Anal. Chem.*, **1971**, *43*, 1115-1119.
14. Rush J.D., Zhao Z., Bielski B.H.J. *Free Rad. Res.*, **1996**, *24*, 187-198.
15. Carr J. D., Kelter P. B., Tabatabai A., Spichal D., Erickson J., McLaughin C. W., in *Proc. Conf. Water Chlorin. Chem. Environ. Impact Health Effects*, R. L. Jolley Eds., New York, **1985**.
16. Lee, Y., Yoon, J., von Gunten, U. *Water Research*, **2005**, *39*, 1946-1953.
17. Rush, J.D., Cyr, J.E., Zhao, A. Bielski, B.H.J. *Free Rad. Res.*, **1995**, *22*, 349-360.
18. Li J., Dasgupta P.K. *Anal. Chem.*, **2000**, *72*, 5338-5347.
19. Donahue W. F. *Environ. Tox. & Chem.*, **1998**, *17*, 783-787.
20. Thompson G.W., Ockerman L.T., Schreyer J.M. *J. Am. Chem. Soc.*, **1951**, *73*, 1379-1381.
21. Rush J.D., Bielski B.H.J. *J. Am. Chem. Soc.*, **1986**, *108*, 523-525.
22. Sharma V.K., Burnett C.R., O'Connor D.B., Cabelli D.E. *Environ. Sci. Technol.*, **2002**, *36*, 4182-4186.
23. Sharma V.K., Burnett C.R., Yngard R., Cabelli D.E. *Environ. Sci. Technol.*, **2005**, *39*, 3849-3854.

24. Zhang L.S., Wong G.T.F. Talanta, **1999**, *48*, 1031-1038.
25. Sharma V.K., Burnett C.R., Millero F.J. *Physical Chemistry Chemical Physics*, **2001**, *3*, 2059-2062.
26. Ketai W., Huitao L., Xingguo C., Yunkun Z., Zhide H. *Talanta*, **2001**, *54*, 753-761.
27. Corbett J.T. *J. Biochem. & Biophys. Methods*, **1989**, *18*, 297-307.

Properties

Chapter 10

Aqueous Ferrate(V) and Ferrate(IV) in Alkaline Medium: Generation and Reactivity

Diane E. Cabelli[1] and Virender K. Sharma[2]

[1]Chemistry Department, Brookhaven National Laboratory, Upton, NY 11973–5000 (email: cabelli@bnl.gov)
[2]Chemistry Department, Florida Institute of Technology, 150 West University Boulevard, Melbourne, FL 32901
*Corresponding author: Cabelli@bnl.gov

This chapter reviews the generation of ferrate(V) and ferrate(IV) complexes in basic solutions. Ferrate(V) ($Fe^{V}O_4^{3-}$) is easily produced by the one-electron reduction of the relatively stable $Fe^{VI}O_4^{2-}$ ion. Comparatively, generation of a ferrate(IV) complex via one-electron oxidation of Fe(III) is rather difficult, due to the relative insolubility of Fe(III) hydroxides and the slow oxidation rate. This has resulted in limited studies of the reactivity of ferrate(IV). The most studied aquated ferrate(IV) complex is ferrate(IV)-pyrophosphate.
The reactivity of ferrate(IV) and ferrate(V) complexes with inorganic and organic substrates in alkaline solution is presented. The reactions of ferrate(IV)-pyrophosphate complex with pyrophosphate complexes of divalent metal ions are likely occurring through inner-sphere electron transfer. Ferrate(V) reacts with substrates predominantly via a two-electron transfer process to Fe(III). The only known example of one-electron reduction of ferrate(V) is its reactivity with cyanide in which sequential reduction of Fe(V) to Fe(IV) to Fe(III) was demonstrated. The reaction of Fe(V) with cyanide thus provides an opportunity for selective and unambiguous production of quantitative amounts of Fe(IV) in aqueous media.

Introduction

The higher oxidation states of iron (Fe(VI), Fe(V), and Fe(IV)) have been shown to be strongly oxidizing in enzymatic systems, where they can carry out aliphatic hydrogen abstraction (*1-5*). In addition, they have been postulated as intermediates in Fenton-type systems (*6*). Fe(VI) itself is relatively stable and has been shown to have potential as an oxidant in the so-called "green" treatment of polluted waters (*7-9*). By contrast, Fe(V) and Fe(IV) are relatively short-lived transients when produced in aqueous solution in the absence of strongly bonding ligands other than hydroxide, a feature that has limited studies of its reactivity. There has been an additional study suggesting that Fe(VI) might be useful in battery design (*10-12*). Finally, a very interesting study suggested that ferrate may be possible to oxidize insoluble chromium to chromate and thus serve to remove chromium contamination in the Hanford radioactive waste tanks (*13*). This paper summarizes the properties and reactivities of ferrate(V) and ferrate(IV) in alkaline medium.

Ferrate(V)

We have been interested in the study of the reactivity of high valent aquo iron states, ferrate(V) with inorganic and organic systems as a tool for the degradation of pollutants in water. Ferrate(V) can be generated very easily in the presence of excess ferrate(VI) through the use of reducing radicals produced in pulse radiolysis according to the following scheme (*14,15*), where ROH is

$$H_2O \text{ -/\!\backslash\!/\!\backslash\!\!\rightarrow } H(0.55),\ e_{aq}^-(2.65),\ OH(2.75),\ H_2O_2(0.72),\ H_2(0.45) \quad \text{I}$$

$$N_2O + e_{aq}^- + H_2O \rightarrow OH + OH^- + N_2 \quad (1)$$

$$H + OH^- \rightarrow e_{aq}^- + H_2O \quad (2)$$

$$OH/O^- + ROH \rightarrow H_2O/OH^- + {}^{\cdot}ROH \quad (3)$$

$$Fe(VI) + {}^{\cdot}ROH \rightarrow Fe(V) + \text{Product} \qquad k_4 = 9 \times 10^9\ M^{-1}s^{-1}\ (14) \quad (4)$$

an alcohol (e.g. ethanol, isoproponol or *tert*-butanol) that will react with the OH radical to form a simple carbon-centered radical. The reaction of ferrate(VI) with the parent alcohol used to generate the alcohol radical that reduces ferrate(VI) (reaction 3) or with some other substrate of interest requires the use of a premix device. The premix pulse radiolysis apparatus consists of two glass syringes mounted in a double syringe drive. The ferrate(VI) solution in one syringe is mixed with the alcohol solution in the other syringe. The mixed solution is promptly injected into the optical cell and exposed to an ionizing pulse.

In Figure 1, we show the visible spectral features of Fe(VI) ($Fe^{VI}O_4^{2-}$) and Fe(V) ($Fe^{V}O_4^{3-}$). The ferrate(V) spectrum has a maximum at 380 nm ($\varepsilon_{380\,nm}$ = 1460 $M^{-1}cm^{-1}$). This spectrum undergoes a blue shift with decreasing pH. Ferrate(V) absorbs much strongly in the UV region (($\varepsilon_{270\,nm} \approx 5000\ M^{-1}cm^{-1}$).

The overall decay of ferrate(V) in the acidic to basic pH range suggest three protonated forms (eqs 5-7) (*16,17*).

$$H_3Fe^VO_4 \Leftrightarrow H^+ + H_2Fe^VO_4^- \qquad 5.5 \le pK_5 \ge 6.5 \qquad (5)$$
$$H_2Fe^VO_4^- \Leftrightarrow H^+ + HFe^VO_4^{2-} \qquad pK_6 \approx 7.2 \qquad (6)$$
$$HFe^VO_4^{2-} \Leftrightarrow H^+ + Fe^VO_4^{3-} \qquad pK_7 = 10.1 \qquad (7)$$

The decay of the ferrate(V) species is strongly dependent on pH (*16,17*), where the totally deprotonated $Fe^VO_4^{3-}$ decays to a longer lived transient ($t_{1/2} \approx$ seconds) via a first-order process (reaction 8). However, as the pH is lowered, ferrate(V) disappears by second-order kinetics to form ferric ions and hydrogen peroxide (*16*). Scheme I describes the mechanism associated with the disappearance of ferrate(V) in alkaline medium, the mechanism takes deprotonated Fe(V) ($Fe^VO_4^{3-}$) all the way to Fe(III) (reactions 8-10). The second-order rate constant observed in the disappearance of Fe(V) (reactions 11 and 12) increases as the pH is lowered and is of the order of 10^7 $M^{-1}s^{-1}$ (*16*). The pH effect can be explained by reaction between $Fe^VO_4^{3-}$ and its conjugate acid, $HFe^VO_4^{2-}$ or bimolecular dimerization of the monoprotonated species reaction (*16*).

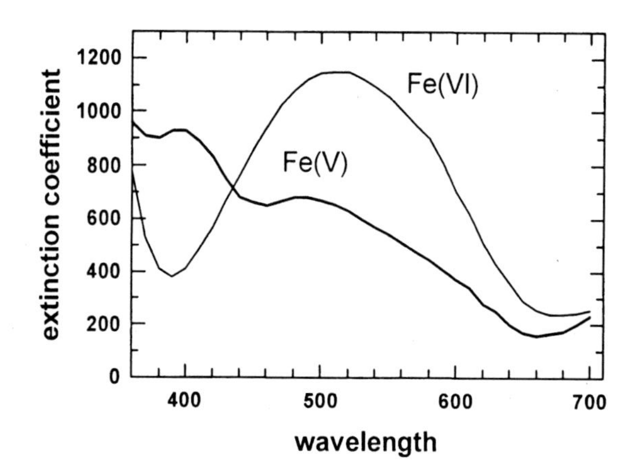

Figure 1. Visible spectra of Fe(V) and Fe(VI) in aqueous basic solution (14)

Scheme I

First-OrderDecay:

$$Fe^VO_4^{3-} + 2H_2O \rightarrow [Fe^V(OH)_4(O)_2]^{3-} \qquad (8)$$
$$[Fe^V(OH)_4(O)_2]^{3-} \rightarrow [Fe^{III}(OH)_4(O_2^{2-})]^{3-} \qquad (9)$$
$$[Fe^{III}(OH)_4(O_2^{2-})]^{3-} \rightarrow Fe^{III}(OH)_m + H_2O_2 \qquad (10)$$

Second-Order Decay:

$$2\ \text{–}Fe^V\text{=O} \xrightarrow{\text{slow}} [\text{-}Fe^{IV}(O_2^{2-})Fe^{IV}\text{-}] \xrightarrow{\text{fast}} 2Fe^{III}(O_2^{2-}) \qquad (11)$$

$$2Fe^{III}(O_2^{2-}) + H_2O \rightarrow Fe^{III}(OH)_m + H_2O_2 \qquad (12)$$

In the pH range from 3.6 to 7.0, the observed disappearance is predominantly first-order with rates that vary from 7.0 x $10^4 s^{-1}$ to 100 s^{-1} (Scheme II) (*17*). The first order process is described by the hydration of the tetrahedral ferrate(V) to a six-coordinate octahedral species prior to loss of peroxide. This aquation reaction dominates at the low and high pH range and competes with the dimerization when FeV) is either unprotonated or monoprotonated.

Scheme II

First-Order Decay:

$HFe^{V}O_4^{2-} + 2H^+ + 4H_2O \rightarrow Fe(OH)_3(H_2O)_3 + H_2O_2$ $\quad k \approx 5\ s^{-1}$ (13)

$H_2Fe^{V}O_4^- + H^+ + 4H_2O \rightarrow Fe(OH)_3(H_2O)_3 + H_2O_2$ $\quad k \approx 150\ s^{-1}$ (14)

$H_3Fe^{V}O_4(T_d) + H_2O = (Fe(OH)_5)_{aq}(O_h)$ (15)

$(Fe(OH)_5)_{aq} + H^+ = (Fe(OH)_4)^+_{aq}$ (16)

$(Fe(OH)_4)^+_{aq} + H^+ \rightarrow Fe^{III}(aq) + H_2O_2$ (17)

By using a fast premixing apparatus synchronized with the accelerator pulse, experiments can be carried out to measure the reactivity of ferrate(V) with a substrate. If the concentration of the substrate is increased such that the substrate now reacts with the OH radical, then the reaction between ferrate(VI) and an oxidized substrate radical can be measured (reaction 18). Finally, conditions can be adjusted by saturating the solution with nitrogen instead of N_2O and eliminating the addition of alcohol. Here, the electron reduces the ferrate(VI) to ferrtae(V) and the substrate is oxidized by the OH radical (reactions 18 and 19).

$$Fe(VI) + e_{aq}^- \rightarrow Fe(V) \quad (18)$$

$$S + {}^{\cdot}OH \rightarrow S_{ox} + H_2O/OH^- \quad (19)$$

Table 1 highlight the reactivity of ferrate(V), which shows the kinetic parameters in the use of high oxidation state iron as an oxidant for inorganic and organic pollutants such as cyanides, sulfur-containing, carboxylic acids, and aromatic compounds (*18*).

Gold is leached from the cyanide complex by applying soluble copper, which forms stronger complexes with cyanide than gold. The resulted copper cyanide complexes are highly toxic to aquatic life and are problematic because they are much more stable than free cyanide. The presence of copper cyanide complexes in gold industry thus presents the biggest concern in cyanide management (*19*). The destruction of cyanide by oxidation with ferrates is thus of great interest. The experiments on the reduction of ferrate(V) by cyanide have demonstrated sequential one-electron reductions of ferrate(V) to ferrate(IV) to ferrate(III) in aqueous media, where the order of reactivity is k(ferrate(V)) > k(ferrate(IV)) > k(ferrate(VI)) (*20*). Interestingly, complexation of cyanide with copper(I) enhanced the rate of the oxidation reaction (Table 1). The results demonstrate potential of ferrates in destroying cyanides in waste water.

Table 1. Ferrate(V) oxidation of compounds at pH = 12.4 in N_2O saturated 0.1 M phosphate buffer at 23-24 °C.

Pollutant (R)	Formula	k, $M^{-1}s^{-1}$
Cyanides		
Cu(I) cyanide[1]	$Cu(CN)_4^{3-}$	$1.35{\pm}0.02 \times 10^7$
Cyanide	HCN, CN^-	$1.96{\pm}\ 0.20 \times 10^4$
Thiocyanate[2]	SCN^-	$6.37{\pm}\ 0.13 \times 10^1$
Sulfur-containing compounds		
Cysteine	$HSCH_2CH(NH_2)COOH$	$4.00 \pm 0.80 \times 10^9$
Cystine	$HOOCCH(NH_2)CH_2S^-$ $SCH_2(NH_2)CHCOOH$	$1.95 \pm 0.02 \times 10^4$
Thiourea[2]	NH_2CSNH_2	$8.10 \pm 0.40 \times 10^3$
Methionine	$CH_3SCH_2CH_2CH(NH_2)COO^-$	$1.58 \pm 0.09 \times 10^3$
Carboxylic acids		
Glycine	$CH_2(NH_3^+)COO^-$	$(8.4 \pm 0.6) \times 10^6$
Alanine	$CH_3CH(NH_3^+)COO^-$	$(3.1 \pm 0.2) \times 10^6$
Aspartic	$HOOCCH_2CH_2(NH_3^+)COO^-$	$(2.6 \pm 0.1) \times 10^6$
Ketomalonic	$C(OH)_2(COOH)_2$	$(1.4 \pm 0.2) \times 10^6$
Tartaric	$HOOC(CHOH)_2COOH$	$(3.1 \pm 0.2) \times 10^3$
Glycolic	$HOCH_2COOH$	$(7.2 \pm 1.0) \times 10^2$
Malic	$HOOCCH(OH)CH_2COOH$	$(1.7 \pm 0.2) \times 10^2$
Lactic	$CH_3CH(OH)COOH$	$(1.6 \pm 0.2) \times 10^2$
Malonic	$CH_2(COOH)_2$	$(9.2 \pm 1.0) \times 10^1$
Succinic	$HOOCCH_2CH_2COOH$	$(2.0 \pm 0.2) \times 10^1$
Acetic	CH_3COOH	$(1.6 \pm 0.2) \times 10^1$
Aromatic compounds		
Histidine	$C_3H_3N_2CH_2CH(NH_2)COO^-$	$22.2 \pm 0.1 \times 10^6$
Phenylalanine	$C_6H_5CH_2CH(NH_2)COO^-$	$9.5 \pm 0.4 \times 10^6$
Tyrosine	$HOC_6H_4CH_2CH(NH_2)COO^-$	$8.1 \pm 0.2 \times 10^6$
Tryptophan	$C_8H_6NCH_2CH(NH_2)COO^-$	$9.3 \pm 0.4 \times 10^6$
Phenol[3]	C_6H_5OH	$3.8 \pm 0.4 \times 10^6$
Proline	$C_4H_7NCOO^-$	$0.1 \pm 0.01 \times 10^6$

[1] At pH = 12.2
[2] At pH = 11.2
[3] At pH = 9.0

Thiocyanate (SCN^-) is used in processes such as the manufacture of thiourea, metal separation and electroplating as well as being formed in mining wastewater (*21*). Ferrate(V) reacts with thiocyanate via a two-electron pathway (*22*) with no observable ferrate(IV) formation. As has been seen generally, the reaction of Fe(V) with SCN^- is significantly more rapid than that of Fe(VI) with SCN^-; with a rate constant that is over two orders of magnitude faster (*22*). This suggests that ferrate(VI) oxidations may be accelerated in the presence of one-electron or two-electron reducing substrates, where the reactive high valent iron species is either ferrate(V) or ferrate(IV).

Cysteine is the most reactive sulfur-containing compound (*23*) and its rate constants is three orders of magnitude higher than other compounds (Table 1). The reactivity of ferrate(VI) with sulfur compounds is determined by the nucleophilicity of the sulfur atom in compounds. The sulfur center in cysteine is most readily oxidized by ferrate(V).

The rate of oxidation of carboxylic acids by ferrate(V) varies with the nature of the substituent group at the -carbon atom of the acids, with rate constants ranging from 10^1 - 10^6 $M^{-1}s^{-1}$ (Table 1). The rate constants decrease in the order of α-NH_2 > α-C-OH > α-C-H (*24*).

$$\text{ferrate(V)} + \text{carboxylic acids} \rightarrow \text{Fe(III)} + NH_3 + \alpha\text{-keto acids} \qquad \text{II}$$

Fe(V) oxidation of aromatic compounds proceeds at rates 10^5 - 10^7 $M^{-1}s^{-1}$ (Table 1). A diamino dicarboxylic acid, histidine, reacts faster than other aromatic compounds (Table 1).

Ferrate(IV)

Iron(IV) complex with simple inorganic ligand, $P_2O_7^{4-}$ in basic medium can be generated from the corresponding parent complexe by oxidation with OH/O^- radical in aqueous solutions (reaction 20) (*25*).

$$[(P_2O_7)_2Fe^{3+}OH]^{6-} + OH \rightarrow [(P_2O_7)_2Fe^{IV}O]^{6-} + H_2O \quad k = 7.8 \times 10^7\ M^{-1}s^{-1} \qquad (20)$$

The spectrum of Fe(IV)-complex is shown in Figure 2. The spectrum has a peak at λ_{max} = 430 nm (ε = 1200 $M^{-1}cm^{-1}$). The spectra of $FeO(OH)_n{}^{2-n}$, $[(P_2O_7)_2Fe^{IV}O]^{6-}$, and $L_mFe(IV)$ were found similar. On lowering the pH, 430 nm peak undergoes blue shift. The pyrophosphate complex of iron(IV), formed at pH ≥ 10 is short lived ($t_{1/2}$ = 100-600 msec) (*25*). This complex of iron(IV) disappears by a second-order process to form a Fe(III) pyrophosphate complex and molecular oxygen (reaction 21) (*25*).

$$2\,[(P_2O_7)_2Fe^{IV}O]^{6-} + 2\,H_2O \rightarrow 2\,[(P_2O_7)_2Fe^{III}OH]^{6-} + \tfrac{1}{2}\,O_2 \qquad (21)$$

Reactions of Fe(IV) with pyrophosphate complexes of divalent transition metal ions (reaction 22) have been studied (*25*). M(II) represents a divalent

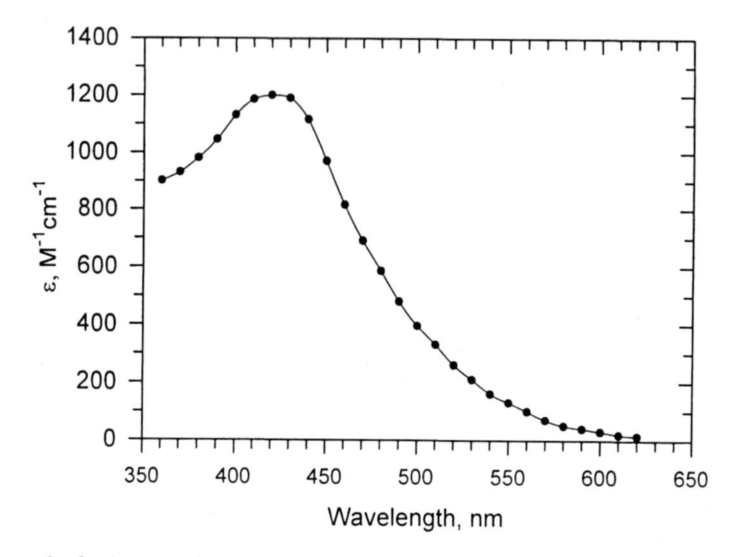

Figure 2. Spectrum of Fe(IV)-pyrophosphate complex at pH 10.0 and 25 °C (25).

pyrophosphate complex of a metal without assignment of the number of ligand L in a given complex.

$$Fe(IV) + M(II) \rightarrow Fe(III) + M(II) \qquad (22)$$

The kinetics of these reactions was investigated by pulse irradiating solutions of Fe(III) pyrophosphate, which contained pyrophosphate complexes of Mn^{2+}, Fe^{2+}, or Co^{2+}. The rate constants are listed in Table 2. The Ni(II) and Cu(II) pyrophosphate complexes showed no reactivity and their upper limit rate constants are given in Table 2.

Table 2. Reactivity of Fe(IV) pyrophosphate with some divalent pyrophosphate complexes (L_mM^{2+}; $L=P_2O_7^{4-}$) at pH 10.0 (*25*).

Metal	k, $M^{-1}s^{-1}$	Metal	k, $M^{-1}s^{-1}$
Mn^{2+}	1.2×10^6	Ni^{2+}	$<4.0 \times 10^2$
Fe^{2+}	1.6×10^6	Cu^{2+}	$<4.0 \times 10^2$
Co^{3+}	5.5×10^5		

The reduction potential of Fe(IV) is likely to be more than 1.0 V because it can oxidize Mn(II) pyrophosphate ($E^o(Mn^{III}/Mn^{II})$ = 1.0 V). Interestingly, no reaction of Fe(IV) with ferrocyanide was observed, although it is a stronger reducing agent ($E^{o}(Fe(CN)_6^{3-}/Fe(CN)_6^{4-})$=0.46V) than Mn(II) or Co(II) pyrophosphate. Because ferrocyanide is normally regarded as an outer-sphere reductant, the reactivity of Fe(IV) may not be through outer-sphere electron transfer. Moreover, the rate constants given in Table 2 are the order of 10^6 $M^{-1}s^{-1}$, which is typical for substitution reactions of Mn(II), Fe(II), and Co(II) (*25*). It is thus possible that a rate-limiting inner-sphere association between M(II) pyrophosphate and $L_2Fe^{IV}O^{6-}$ occurs and this step is followed by a rapid one-electron transfer.

The reaction of Fe(IV)-pyrophosphate complex with H_2O_2 has also been studied (reactions 23). Hydrogen peroxide is oxidized to O_2 by reactions (23) and (24).

$$[(P_2O_7)_2Fe^{IV}O^{6-} + H_2O_2 \rightarrow [(P_2O_7)_2Fe^{III}OH^{6-} + O_2^- + H^+ \quad (23)$$

$$[(P_2O_7)_2Fe^{IV}O^{6-} + O_2^- + H^+ \rightarrow [(P_2O_7)_2Fe^{III}OH^{6-} + O_2 \quad (24)$$

The rate constant for the reaction 19 is 3.9 x 10^5 $M^{-1}s^{-1}$. The reactivity of Fe(IV) pyrophosphate with metal ions as given above (Table 2) gives the estimated limits for the reduction potential at pH 10, which is $1.0 < E^{o}(Fe^{IV}_{pyrh}/Fe^{III}_{Pyrh}) < 1.35$ V.

A slope of 0.61 ± 0.11 was found between log k versus pH for the Fe(IV) reaction with cyanide. This slope is similar to slopes for reaction of Fe(VI) and Fe(V) (0.76 ± 0.07 and 0.77 ± 0.04 for Fe(VI) and Fe(V), respectively). The observed increase in reactivity of high oxidation states of iron with cyanide is probably caused by proton stabilization of the partial radical character of ferrate species ($Fe^{VI}=O \leftrightarrow Fe^{V}-O^{\bullet-}$; $Fe^{V}=O \leftrightarrow Fe^{IV}-O^{\bullet-}$; $Fe^{IV}=O \leftrightarrow Fe^{III}-O^{\bullet-}$) (*17*).

Conclusions

As is apparent, there is much more information on the formation and reactivity of ferrate(V) than ferrate(IV). This is a consequence of virtually diffusion-controlled one-electron reduction of a relatively stable ferrate(VI) in the neutral to alkaline pH range versus the relatively slow one-electron oxidation of Fe(III) that is only stable in the acid to neutral pH range. Interestingly, the only system in which a well-characterized Fe(IV) is formed as the result of the one-electron reduction of Fe(V) upon reaction with a substrate is when the substrate is cyanide. In all other cases, direct formation of Fe(III) is observed. This suggests that Fe(V) is either a powerful two-electron oxidant or that Fe(IV) is a more reactive species than Fe(V) when aquated.

Acknowledgment

This research was supported by the U.S. Department of Energy Office of Basic Energy Sciences under contract DE-AC02-98CH10086.

References

1. Groves, J.T. *J. Inorg. Biochem.* **2006**, *100*, 434.
2. Shan, X.; Que Jr., L.; *J. Inorg. Biochem.* **2006**, *100*, 421.
3. Decker, A.; Solomon, E.I. *Angew. Chem. Int. Ed.* **2005**, *44*, 2252.
4. Berry, J.F.; Bill, E.; Bothe, E.; George, S.; Mienert, B.; Neese, F.; Wieghardt, K. *Science* **2006**, *312*, 1937.
5. Oliveria, F.T.D.; Chanda, A.; Benerjee, D.; Shan, X.; Mondal, S.; Que, Jr. L.; Bominaar, E.L.; Munck, E.; Collins, T.J. *Science* **2007**, *315*, 835.
6. Goldstein, S.; Meyerstein, D. *Acc. Chem. Res.* **1999**, *32*, 547 and references therein
7. Sharma, V.K.; Mishra, S.K.; Nesnas, N. *Environ. Sci. Technol.,* **2006**, *40*, 7222.
8. Sharma, V.K. Adv. Environ. Res. **2002**, *6,* 143.
9. Sharma, V.K.; Kazama, F.; Hu, J.; Ray, A.K. *J. Water Health* **2005**, *3*, 45
10. Licht, S.; *Science* **1999**, *285*, 1039.
11. Licht, S; TelVered, R. *Chem. Commun.* **2004**, *6*, 628.
12. Waltz, K.A.; Suyana, A.N.; Suyama, W.E.; Sene, J.J.; Zeltner, W.A.; Armacanqui, E.M.; Roszkowski, A.J.; Anderson, M.A. *J. Power Sources.* **2004**, *134(2)*, 318.
13. Sylvester, P., Rutherford, Jr. L.A., Gonzalez-Martin, A. and Kim, *Environ. Sci. Technol.* **2001** *35*, 216.
14. Rush, J.D.; Bielski, B.H.J. *J. Amer. Chem. Soc.* **1986**, *108*, 523.
15. Bielski, B.H. J. *Free Rad. Res. Comms.* **1991**, *12-13*, 469.
16. Rush, J.D.; Bielski, B.H.J. *Inorg. Chem.* **1989**, *28*, 3947.
17. Rush, J.D.; Bielski, B.H.J. *Inorg. Chem.* **1994**, *33*, 5499.
18. Sharma, V.K. *Rad. Phys. Chem.* **2002**, *65*, 349.
19. Sharma, V.K.; Burnett, C.; Yngard, R.; Cabelli, D.E. *Environ. Sci. Technol.* **2005**, *39*, 3849.
20. Sharma, V.K.; O'Connor, D.B.; Cabelli, D.E. *J. Phys. Chem. B* **2001**, *105*, 11529.
21. Sharma, V.K.; Burnett, C.; O'Connor, D.B.; Cabelli, D.E. *Environ. Sci. Technol.* **2002**, *36*, 4182.
22. Sharma, V.K.; O'Connor, D.B.; Cabelli, D.E. *Inorg. Chim Acta* **2004**, *357*, 4587.
23. Sharma, V.K.; Bielski, B.H.J. *Inorg. Chem.* **1991**, *30*, 4306.
24. Bielski, B.H.J.; Sharma, V.K.; Czapski, G. *Rad. Phys. Chem.* **1994**, *44*, 479.
25. Melton, J.D.; Bielski, B.H.J. *Int. J. Rad. Phys. Chem.* **1990**, *36*, 725.

Chapter 11

Identification and Characterization of Aqueous Ferryl(IV) Ion

Oleg Pestovsky* and Andreja Bakac*

Ames Laboratory, 26 Spedding Hall, Iowa State University, Ames, IA 50011

The reaction between ferrous ions and ozone in acidified aqueous solution generates a short-lived species ($t_{½} \approx 7$ sec), which was identified as high-spin pentaaquairon(IV) oxo dication (ferryl) by UV-Vis, Mossbauer, XAS spectroscopies, DFT calculations, ^{18}O isotopic labeling, and conductometric kinetic studies. Kinetic and ^{18}O isotopic labeling studies were used to determine the rate constant for the oxo group exchange between ferryl and solvent water, $k_{ex} = 1400\ s^{-1}$. Oxidation of alcohols, aldehydes, and ethers by ferryl occurs by simultaneous hydrogen atom and hydride transfer mechanisms. Ferryl was also found to be an efficient oxygen atom transfer reagent in the reactions with sulfoxides, a water soluble phosphine, and a thiolato-complex of cobalt(III). A quantitative and fast reaction between ferryl and DMSO ($k_{DMSO} = 1.3 \times 10^5\ M^{-1}\ s^{-1}$) produces methyl sulfone. This and some other findings unambiguously rule out ferryl as a Fenton intermediate.

Nonheme iron(IV) complexes have attracted considerable attention since the discovery that such species serve as reactive intermediates in a number of enzymatic processes (*1-3*). Attempts to unravel and mimic the chemistry of both heme and nonheme iron centers in enzyme active sites have focused on structures, mechanisms, and intermediates (*4-12*). Of particular interest are iron(IV) oxo intermediates, such as that in α-ketoglutarate-dependent taurine dioxygenase (*13*), which are capable of C-H bond activation.

A number of studies directed at preparation of synthetic iron(IV) oxo species resulted in moderately stable complexes with amino-, amido-, and pyridine ligands, which were characterized by X-ray crystallography (*1,14,15*). Similar to their naturally occurring analogs, these complexes were shown to be able to carry out efficient C-H bond hydroxylation (*16*).

Despite all the synthetic and mechanistic effort toward generating nonheme iron(IV) oxo complexes, the chemistry in the absence of stabilizing ligands remains largely unexplored. Among such species, one of the simplest and perhaps most reactive is aqueous ferryl(IV) ion, which was proposed as an alternative to hydroxyl radical in Fenton chemistry (*17-24*), or as an intermediate in some reactions in the atmospheric and environmental chemistry (*25,26*). In acidic and neutral solutions, the existence of a ferryl intermediate, most probably $(H_2O)_5Fe^{IV}{=}O^{2+}$, has not been established independently. Such a complex is, however, believed to be generated in the $Fe(H_2O)_6^{2+}$/ozone reaction (*27,28*), eq 1,which simultaneously yields an equivalent of O_2, which rules out the description of the intermediate as an ozonide complex.

$$Fe_{aq}^{2+} + O_3 \rightarrow Fe_{aq}O^{2+} + O_2 \quad (1)$$

Oxidations of a limited number of organic substrates by aqueous ferryl were reported to take place by a hydrogen atom transfer mechanism (*29*). However, no product analyses were carried out to confirm this assignment, which was based on the kinetics alone. Another report provided kinetic evidence for a 2-electron oxidation of substituted phenols via an Fe(IV)-substrate complex (*30*). In our own work on the chemistry of the closely related aqueous chromyl(IV), it was demonstrated that the opposite trends in reactivity exists, that is, 2-electron oxidation of alcohols and 1-electron oxidation of phenols (*31*). Intuitively, these finding are in contradiction with the thermodynamics of these reactions. The standard reduction potentials for the M_{aq}^{III}/M_{aq}^{II} couples are -0.41 V for chromium and 0.77 V for iron, demonstrating the thermodynamic preference for the formation of Cr_{aq}^{3+} (1-electron product) and Fe_{aq}^{2+} (2-electron product). Therefore, some other, unknown factors must determine the outcome of these reactions, and such knowledge can be used in development of catalytic systems utilizing O_2 and H_2O_2, as the successful catalysis often requires the chemistry to take place in 2-electron steps.

Here we report identification and characterization of aqueous ferryl as a pentaaquairon(IV) oxo dication by a number of spectroscopic and kinetic techniques. The study of reactivity of ferryl in hydrogen atom, hydride, and oxygen atom transfer reactions is also presented. In addition, the unique knowledge of reactivity of genuine aqueous ferryl allows to rule out its participation as an intermediate in Fenton chemistry in aqueous solution.

Preparation and Characterization of Aqueous Ferryl(IV) Ion

The Fe_{aq}^{2+}/O_3 reaction was investigated by conventional and stopped-flow UV-Vis spectrophotometry, as well as a stopped-flow conductometric technique in 0.10 M $HClO_4$ at 25 °C (*32*). With a moderate to large excess of ozone over Fe_{aq}^{2+}, the disappearance of ozone dominated the absorbance changes below 290 nm, and the formation of $Fe_{aq}O^{2+}$ was observed above 290 nm with a maximum absorbance changes at 320nm. Global fitting of the kinetic data to a monoexponential kinetic model afforded a series of pseudo-first-order rate constants, which varied linearly with $[O_3]$ and yielded $k_1 = (8.30 \pm 0.10) \times 10^5$ M^{-1} s^{-1}. These data are in good agreement with an earlier determination (*28*). At longer times, an approximately exponential decay of $Fe_{aq}O^{2+}$ was detected above 300 nm, $k_{2H} = 0.10$ s^{-1} at pH 1.0, eq 2. Solutions of $Fe_{aq}O^{2+}$ in D_2O (7.6 % H) at pD 1.0 were more stable, and decayed with a rate constant $k_{2D} = 0.040$ s^{-1}, yielding a solvent kinetic isotope effect of 2.85.

$$Fe_{aq}O^{2+} + H^{+} \rightarrow Fe_{aq}^{3+} + \frac{1}{4} O_2 + \frac{1}{2} H_2O \qquad (2)$$

In experiments using a large excess (20-30 fold) of Fe_{aq}^{2+} over O_3, the disappearance of ferryl, eq 3, was accompanied by hydrolysis of $Fe_{aq}(OH)_2Fe_{aq}^{4+}$, eq 4 (*33*).

$$Fe_{aq}O^{2+} + Fe_{aq}^{2+} \rightarrow 2\ Fe_{aq}^{3+} / Fe_{aq}(OH)_2Fe_{aq}^{4+} \qquad (3)$$

$$Fe_{aq}(OH)_2Fe_{aq}^{4+} + 2\ H^{+} + \rightarrow 2\ Fe_{aq}^{3+} + 2\ H_2O \qquad (4)$$

A global fit to a biexponential kinetic model afforded $k_3 = (4.33 \pm 0.01) \times 10^4$ M^{-1} s^{-1} and $k_4 = 0.79$ s^{-1}.

A series of stopped-flow conductivity measurements was carried out to determine the number of proton equivalents consumed during $Fe_{aq}O^{2+}$ decay. *Figure 1* shows conductivity changes accompanying the decay of 0.18 mM $Fe_{aq}O^{2+}$. Also shown is a trace obtained by mixing 0.20 mM Fe_{aq}^{2+} and 9.7 mM H_2O_2, eq 5. Nearly identical amplitudes in the two kinetic traces demonstrate that the same number of proton equivalents was consumed in reactions 2 and 5. Since the only persistent charged species in these reactions are H^+, Fe_{aq}^{2+}, Fe_{aq}^{3+}, and ferryl itself, this result establishes a 2+ charge for ferryl.

$$Fe_{aq}^{2+} + \frac{1}{2} H_2O_2 + H^{+} \rightarrow Fe_{aq}^{3+}\ H_2O \qquad (5)$$

Mossbauer and XAS Spectroscopies

Samples of aqueous ferryl(IV) were prepared by stopped-flow mixing of equimolar concentrations of $^{57}Fe(H_2O)_6^{2+}$ and O_3, and freeze-quenching the sample in a rapidly-rotating liquid nitrogen-cooled home-made brass receptacle. The Mössbauer spectra were recorded at 4.2 K in magnetic fields of 0.05 T to 8.0 T. The spectrum in the 0.05 T exhibits a doublet with δ (isomer shift) = 0.38(2) mm/s, and ΔE_q (quadrupole splitting) = 0.33(3) mm/s. The data are consistent with a high-spin (S=2) ferryl species (*34*). In a DFT-optimized structure, the Fe-oxo bond length is 1.63 Å, similar to Fe=O distances in other

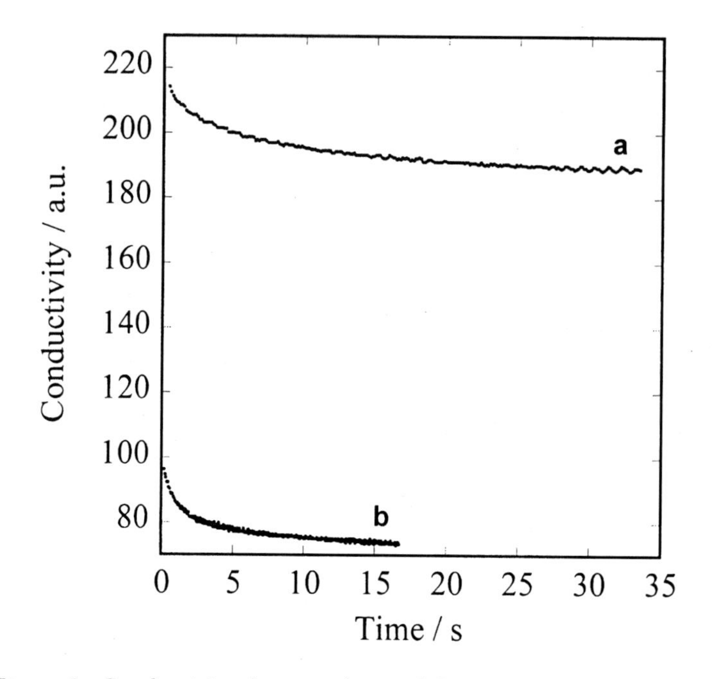

Figure 1. Conductivity changes observed during (a) self-decay of 0.18 mM $Fe_{aq}O^{2+}$ (obtained by mixing 0.20 mM Fe_{aq}^{2+} and 0.25 mM O_3) and (b) in the reaction between 0.20 mM Fe_{aq}^{2+} and 9.7 mM H_2O_2 in 0.10M $HClO_4$ at 25 °C. Adapted from J. Am. Chem. Soc. **2004**, *126*, 13757-13764. *Copyright 2004 American Chemical Society.*

non-heme oxoiron(IV) complexes (*35*), and significantly shorter than the five Fe-OH_2 bond distances (2.04 to 2.09 Å) in the $Fe(H_2O)_5O^{2+}$ cation. In agreement with the Mössbauer spectrum, the DFT calculations yielded a ground state with S=2.

In the X-ray absorption near edge structure (XANES) of the ferryl(IV) ion the edge energy was found at 7126 eV, i.e. between the values found for $Fe(H_2O)_6^{3+}$ (7129 eV) and for nonheme Fe^{IV}=O complexes (7123-7125 eV). This observation is consistent with the strong covalent interaction with the terminal oxo group.

Reactions of Aqueous Ferryl(IV) Ion with Organic Substrates

In its reactions with reducing substrates, $Fe_{aq}O^{2+}$ utilizes one-electron (hydrogen atom transfer, electron transfer) and two electron (hydride transfer, oxygen atom transfer) pathways (*32*). In the reactions with alcohols, the two pathways take place concurrently. This is best illustrated on the example of cyclobutanol, which was oxidized to a mixture of 70% cyclobutanone and 30% (by difference) of ring-opened products formed by rearrangement of the intervening radical. The ^{1}H-NMR spectrum of the product mixture is shown in *Figure 2*.

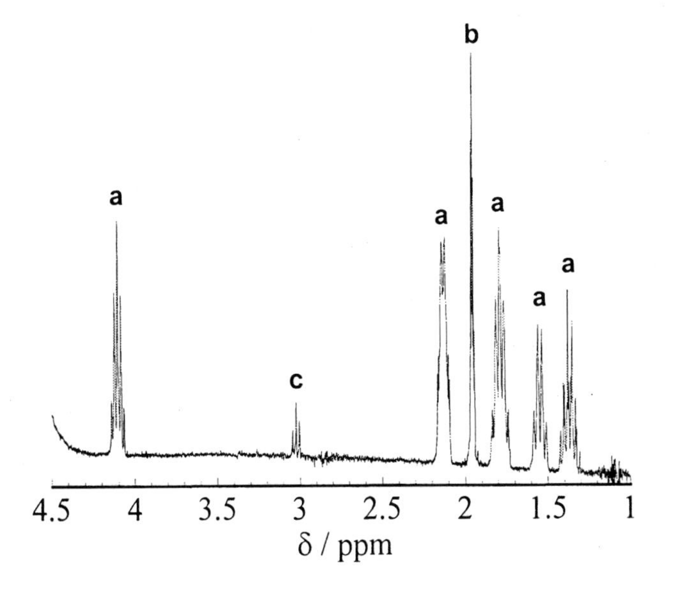

Figure 2. ^{1}H-NMR spectrum of products generated from cyclobutanol (4.16 mM) and 0.18 mM $Fe_{aq}O^{2+}$ in D_2O (0.10 M $DClO_4$, 3.2 % H, 3.2 % CD_3CN), at 25 °C and 0.4 mM O_2. Peak assignment: a) cyclobutanol, b) CD_2HCN, c) cyclobutanone. Adapted from J. Am. Chem. Soc. **2004**, *126*, 13757-13764. *Copyright 2004 American Chemical Society.*

Even though the rate constant for radical ring-opening is not known, all the literature precedents agree that this reaction is too fast for the cyclic radical to be captured by an external substrate and further oxidized to cyclobutanone. Thus the ratio of the two types of products reflects directly the relative contributions of the two pathways, eqs 6 and 7.

$$c-(CH_2)_3CHOH \xrightarrow{1e} c\text{-}(CH_2)_3C^{\bullet}\text{-}OH \xrightarrow{fast} {}^{\bullet}CH_2\text{-}(CH_2)_2\text{-}CHO \quad (6)$$

$$c\text{-}(CH_2)_3CHOH \xrightarrow{2e} c-(CH_2)_3C{=}O \quad (7)$$

The kinetics of disappearance of $Fe_{aq}O^{2+}$ in the reaction with various substrates was determined by stopped-flow, Table I. Unlike in the case of cyclobutanol, the products of oxidation of most of the substrates in Table I do not distinguish between 1-electron and 2-electron mechanisms. Contribution from the individual pathways were obtained by combining the kinetics data for the disappearance of $Fe_{aq}O^{2+}$ with precise reaction stoichiometry, which changes depending on whether Fe_{aq}^{2+} or Fe_{aq}^{3+} is produced in the reduction step. Kinetic simulations, which were essential in both planning of the experiments and in data analysis, afforded the kinetics data for each individual 1-electron or 2-electron pathways, Table II.

Table I. Second-Order Rate Constants for Reactions of Aqueous Ferryl(IV) with Organic Substrates[a]

Substrate	k / M^{-1} s^{-1}	*Substrate*	k / M^{-1} s^{-1}
CD_3OH	1.26×10^2	4-CH_3-Ph-CH_2OH	1.50×10^4
CH_3OH	5.74×10^2	4-CH_3O-Ph-CH_2OH	1.59×10^4
CH_3OD (in D_2O)	5.72×10^2	Cyclobutanol	3.13×10^3
C_2H_5OH	2.51×10^3	CH_2O	7.72×10^2
$(CH_3)_2CHOH$	3.22×10^3	C_2H_5CHO	2.85×10^4
$(CD_3)_2CHOH$	3.07×10^3	PhCHO	2.07×10^4
$(CH_3)_2CDOH$	7.00×10^2	Et_2O	4.74×10^3
$(CD_3)_2CDOH$	6.60×10^2	THF	7.46×10^3
Ph-CH_2OH	1.42×10^4	CH_3COCH_3	3.15×10^1
4-CF_3-Ph-CH_2OH	1.00×10^4	CH_3CN	4.12×10^0
4-Br-Ph-CH_2OH	1.41×10^4		

[a] Adapted from J. Am. Chem. Soc. **2004**, *126*, 13757-13764. Copyright 2004 American Chemical Society.

Table II. Kinetics Data for Reactions of Aqueous Ferryl(IV) with Organic Substrates in Individual Hydrogen Atom Transfer (1e) and Hydride Transfer (2e) Pathways[a]

Substrate	K_{1e} / M^{-1} s^{-1}	k_{2e} / M^{-1} s^{-1}
CH_3OH	5.3×10^2	6.3×10^2
CH_2O	4.0×10^2	1.0×10^3
C_2H_5OH	2.3×10^3	1.8×10^3
$(CH_3)_2CHOH$	1.7×10^3	3.3×10^3
cyclobutanol	2.4×10^3	4.9×10^4
THF	7.5×10^3	2.0×10^3
CH_3COCH_3	3.2×10^1	[b]
CH_3CN	4.1×10^0	[b]

[a] Adapted from J. Am. Chem. Soc. **2004**, *126*, 13757-13764. Copyright 2004 American Chemical Society.

[b] Below detection limit

Two-electron reduction of $Fe_{aq}O^{2+}$ generates Fe_{aq}^{2+} which can be reoxidized with excess ozone to $Fe_{aq}O^{2+}$ to make the reaction catalytic. Obviously, the catalysis will be of limited importance if $Fe_{aq}O^{2+}$ is lost irreversibly in a parallel, 1-electron process, as is the case for all the reactions in Table II. More successful catalytic oxidations were carried out in oxidations taking place by oxygen atom transfer to sulfoxides as described below.

Figure 3 shows kinetic traces for the loss of methyl para-tolyl sulfoxide in a reaction with ozone in the presence and absence of 5 micromolar Fe_{aq}^{2+} (*36*). The interpretation of the large catalytic effect of Fe_{aq}^{2+} appears straightforward

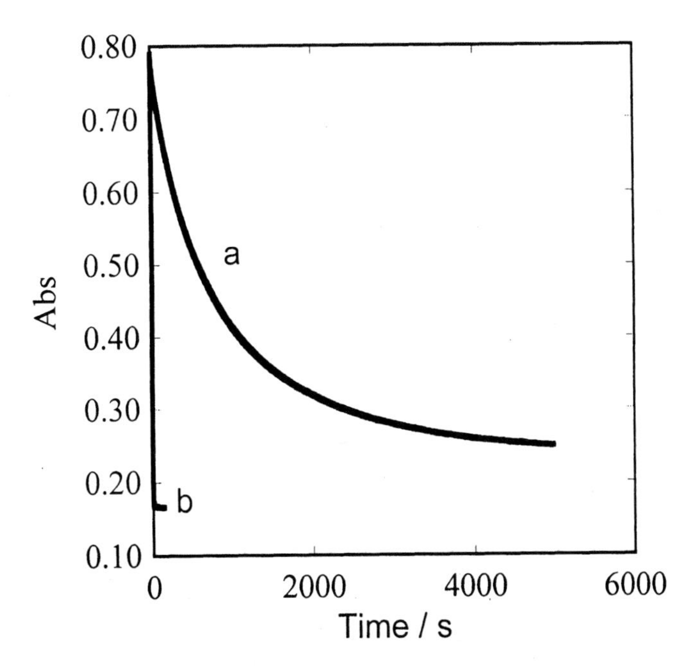

Figure 3. Oxidation of 100 μM methyl para-tolyl sulfoxide with 100 μM ozone catalyzed by Fe_{aq}^{2+} *in 0.10 M aqueous* $HClO_4$*. Kinetic traces at 240 nm at 0 μM* Fe_{aq}^{2+} *(a), and 5 μM* Fe_{aq}^{2+} *(b). Adapted from Inorg. Chem.* **2006**, *126*, 13757-13764. *Copyright 2006 American Chemical Society.*

and suggests that the oxidation of the sulfoxide by $Fe_{aq}O^{2+}$ takes place by oxygen atom transfer , i.e. eq 1 followed by eq 8

$$Fe_{aq}O^{2+} + CH_3S(O)Ar \rightarrow Fe_{aq}^{2+} + CH_3S(O)_2Ar \quad (8)$$

Indeed, the amount of Fe_{aq}^{2+} at the end of the reaction matched exactly the calculated value obtained in simulations under the assumption that only reactions 1, 3 and 8 are involved, i.e. in the complete absence of a 1-e path in the $Fe_{aq}O^{2+}$/sulfoxide reaction.

The kinetics of $Fe_{aq}O^{2+}$/sulfoxide reactions were determined by setting up a competition with excess Fe_{aq}^{2+} and determining the amounts of Fe_{aq}^{3+} at the end of the reaction, according to the following scheme.

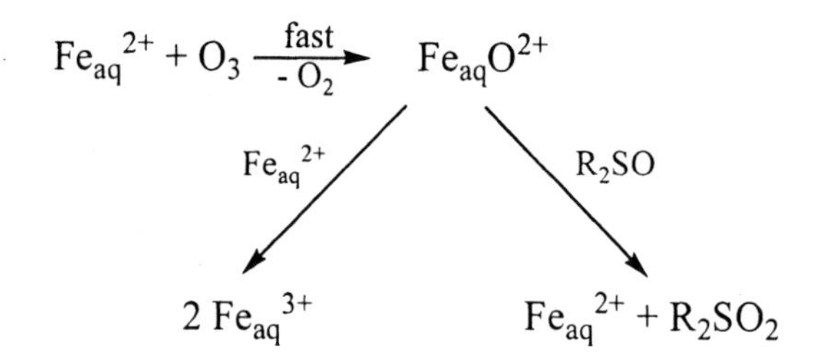

As shown in Table III, the rate constants are similar for all the sulfoxides studied. Moreover, the kinetics are insensitive to the choice of the isotope, H or D, in the alkyl group, as expected for an oxygen atom transfer.

The fast and "clean" oxygen transfer to the sulfoxides provided for an opportunity to determine the rate constant for the oxo oxygen exchange with solvent water. To this goal, the reaction between Fe_{aq}^{2+}, O_3, and dimethylsulfoxide (DMSO) was carried out in $H_2{}^{18}O$, and the isotopic composition of the product sulfone was determined by GCMS. Under the experimental conditions, all the H_2O, including that coordinated to Fe_{aq}^{2+}, contained the ^{18}O label, and O_3 was the only source of ^{16}O. The reaction scheme is shown below.

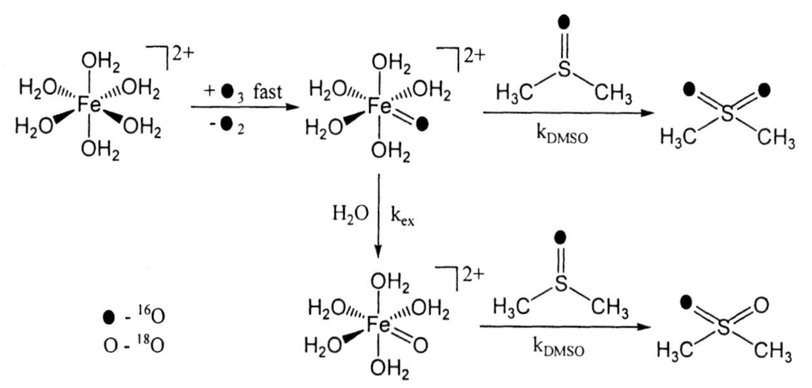

$Fe_{aq}{}^{16}O^{2+}$, generated in the first step, partitions between the reactions with DMSO to generate $DMS(^{16}O)_2$, and the exchange with solvent water, which ultimately gives $DMS(^{16}O)(^{18}O)$. Kinetic analysis of the data yielded the rate constant for the oxo group exchange with solvent water, $k_{ex} = 1.4 \times 10^3\ s^{-1}$.

Once the product of the Fe_{aq}^{2+}/O_3 reaction was identified as $Fe_{aq}O^{2+}$, we were able to compare the chemical behavior of this species with that of the active intermediate in the Fe_{aq}^{2+}/H_2O_2 (Fenton) reaction. This work relied in part

Table III. Second-order Rate Constants for Oxygen Atom Transfer from Aqueous Ferryl(IV)[a]

Substrate	$k / 10^5\ M^{-1}\ s^{-1}$
DMSO	1.26
DMSO-d_6	1.23
4-CH_3-Ph-S(O)-CH_3	1.16
Ph-S(O)-CH_3	1.23
4-Cl-Ph-S(O)-CH_3	0.99
4-CF_3-Ph-S(O)-CH_3	0.79
Ph-CH_2-S(O)-CH_3	1.48
$(4\text{-Cl-Ph})_2SO$	≈ 0.7
TPPMS	≈ 200
$CoSR^{2+}$	≈ 100
$CoS(O)R^{2+}$	≈ 1300

[a] Adapted from Inorg. Chem. **2006**, *126*, 13757-13764. Copyright 2006 American Chemical Society.

on the oxidation of sulfoxides, which, as was shown here, act as oxygen atom acceptors from $Fe_{aq}O^{2+}$. The Fenton intermediate oxidizes DMSO in a one-electron reaction and generates methyl radicals and sulfinic acid. The two are thus clearly different, which rules out $Fe_{aq}O^{2+}$ and supports the other major contender, hydroxyl radical, as the Fenton intermediate.

Acknowledgements

This manuscript has been authored by Iowa State University under Contract No. W-7405-ENG-82 with the US Department of Energy. The United States Government retains and the publisher, by accepting the article for publication, acknowledges that the United States Government retains a nonexclusive, paid-up, irrevocable, world-wide license to publish or reproduce the published form of this manuscript, or allow others to do so, for United States Government purposes.

References

1. Costas, M.; Mehn, M. P.; Jensen, M. P.; Que, L., Jr. *Chem. Rev.* **2004**, *104*, 939-986.
2. Tshuva, E. Y.; Lippard, S. J. *Chem. Rev.* **2004**, *104*, 987-1011.
3. Meunier, B.; Bernadou, J. *Struct. Bonding (Berlin)* **2000**, *97*, 1-35.
4. Karlin, K. D. *Science* **1993**, *261*, 701-708.
5. Merkx, M.; Kopp, D. A.; Sazinsky, M. H.; Blazyk, J. L.; Muller, J.; Lippard, S. J. *Angew. Chem., Int. Ed.* **2001**, *40*, 2782-2807.

6. Ito, M.; Fujisawa, K.; Kitajima, N.; Moro-Oka, Y. *Catalysis by Metal Complexes* **1997**, *19 (Oxygenases and Model Systems)*, 345-376.
7. Que, L., Jr.; Tolman, W. B. *Angew. Chem., Int. Ed.* **2002**, *41*, 1114-1137.
8. Groves, J. T. *Proc. Natl. Acad. Sci. U.S.A.* **2003**, *100*, 3569-3574.
9. Gray, H. B. *Proc. Natl. Acad. Sci. U.S.A.* **2003**, *100*, 3563-3568.
10. De Montellano, P. R. O.; De Voss, J. J. *Nat. Prod. Rep.* **2002**, *19*, 477-493.
11. Green, M. T.; Dawson, J. H.; Gray, H. B. *Science* **2004**, *304*, 1653-1656.
12. Nam, W.; Park, S.-E.; Lim, I. K.; Lim, M. H.; Hong, J.; Kim, J. *J. Am. Chem. Soc.* **2003**, *125*, 14674-14675.
13. Price, J. C.; Barr, E. W.; Tirupati, B.; Bollinger, J. M., Jr.; Krebs, C. *Biochemistry* **2003**, *42*, 7497-7508.
14. Kostka, K. L.; Fox, B. G.; Hendrich, M. P.; Collins, T. J.; Rickard, C. E. F.; Wright, L. J.; Munck, E. *J. Am. Chem. Soc.* **1993**, *115*, 6746-6757.
15. Rohde, J.-U.; In, J.-H.; Lim, M. H.; Brennessel, W. W.; Bukowski, M. R.; Stubna, A.; Muenck, E.; Nam, W.; Que, L., Jr. *Science* **2003**, *299*, 10371039.
16. Kaizer, J.; Klinker, E. J.; Oh, N. Y.; Rohde, J.-U.; Song, W. J.; Stubna, A.; Kim, J.; Muenck, E.; Nam, W.; Que, L., Jr. *J. Am. Chem. Soc.* **2004**, *126*, 472-473.
17. Fenton, H. J. H. *J. Chem. Soc.* **1894**, *65*, 899-910.
18. Haber, F.; Weiss, J. *Proc. Royal Soc. London* **1934**, *147*, 332-351.
19. Bray, W. C.; Gorin, M. H. *J. Am. Chem. Soc.* **1932**, *54*, 2124-2125.
20. Koppenol, W. H. *Free Radical Biol. Med.* **1993**, *15*, 645-651.
21. Masarwa, M.; Cohen, H.; Meyerstein, D.; Hickman, D. L.; Bakac, A.; Espenson, J. H. *J. Am. Chem. Soc.* **1988**, *110*, 4293-4297.
22. Walling, C. *Acc. Chem. Res.* **1998**, *31*, 155-157.
23. Walling, C. *Acc. Chem. Res.* **1975**, *8*, 125-131.
24. Buda, F.; Ensing, B.; Gribnau, M. C. M.; Baerends, E. J. *Chem. Eur. J.* **2003**, *9*, 3436-3444.
25. Hahn, J.; Pienaar, J. J.; Van Eldik, R. *Proceedings of EUROTRAC Symposium '96* **1997**, *1*, 427-431.
26. Jacobsen, F.; Holcman, J.; Sehested, K. *Int. J. Chem. Kinet.* **1998**, *30*, 215221
27. Conocchioli, T. J.; Hamilton, E. J., Jr.; Sutin, N. *J. Am. Chem. Soc.* **1965**, *87*, 926-927.
28. Loegager, T.; Holcman, J.; Sehested, K.; Pedersen, T. *Inorg. Chem.* **1992**, *31*, 3523-3529.
29. Jacobsen, F.; Holcman, J.; Sehested, K. *Int. J. Chem. Kinet.* **1998**, *30*, 215-221.
30. Martire, D. O.; Caregnato, P.; Furlong, J.; Allegretti, P.; Gonzalez, M. C. *Int. J. Chem. Kinet.* **2002**, *34*, 488-494.
31. Scott, S. L.; Bakac, A.; Espenson, J. H. *J. Am. Chem. Soc.* **1992**, *114*, 42054213.
32. Pestovsky, O.; Bakac, A. *J. Am. Chem. Soc.* **2004**, *126*, 13757-13764.
33. Jacobsen, F.; Holcman, J.; Sehested, K. *Int. J. Chem. Kinet.* **1997**, *29*, 17-24.
34. Pestovsky, O.; Stoian, S.; Bominaar, E. L.; Shan, X.; Münck, E.; Que, L., Jr.; Bakac, A. *Angew. Chem., Int. Ed.* **2005**, *44*, 6871-6874.
35. Rohde, J.-U.; In, J.-H.; Lim, M. H.; Brennessel, W. W.; Bukowski, M. R.; Stubna, A.; Münck, E.; Nam, W.; Que, L. Jr. *Science* **2003**, *299*, 1037.
36. Pestovsky, O.; Bakac, A. *Inorg. Chem.* **2006**, *126*, 13757-13764.

Chapter 12

Ferrate(VI) Oxidation of Nitrogenous Compounds

Michael D. Johnson, Brooks J. Hornstein and Jacob Wischnewsky

Department of Chemistry and Biochemistry, New Mexico State University, Las Cruces, NM 88003

The oxidation kinetics of a series of nitrogen containing compounds by ferrate(VI), FeO_4^{2-}, is described. Each of these reactions was studied at 25°C using spectrophotometric techniques. These included stopped-flow, rapid scanning spectrophotometry and convention Diode array spectrophotometry. Mechanistic schemes are proposed for each system studied along with potential intermediates when observed or required by kinetic data.

Introduction

With the investigation of the oxidation reaction mechanisms of metalloproteins, the importance of iron in its high oxidation states has emerged (*1*). For example, in the functioning of catalases and peroxidases, the formation of an iron(IV) intermediate has been postulated as a key step in their enzymatic activity (*2*). In the reactions of cytP450, the generation of an iron(IV) or iron(V) heme complex is crucial in its catalytic cycle (*3*). It has been long recognized that in these enzymes, the iron exists in a porphyrin ring which imparts a unique stabilization of hypervalent iron along with π cation radicals. In contrast, recent studies of *non-heme* enzymes suggest that iron utilizes oxidation states greater than +3 in their catalytic cycles despite the absence of a porphyrin ring. Important representatives of these enzymes include the hydroxylase component of methane monooxygenase(MMO) (*4*), the R2 subunit of ribonucleotide reductase (RNR R2) (*5*), Rieske dioxygenase (*6*) and other less well defined enzymes like squalene epoxidase and tyrosine hydroxylase (*7*). In addition, the involvement of non-heme iron complexes in disease states such as tyrosinemia, phenylketouria and Refusm's disease has been proposed. To date, these non-heme iron systems possess either monomeric or dimeric Fe(II) cores that interact with molecular oxygen to generate species that carry out oxidations. Though model compounds are emerging that mimic these enzymes, difficulties in their preparation and data interpretation exist (*1-8*).

In the past few years, the Que group has undertaken the development of iron(III) and iron(IV) complexes for the understanding of dinuclear non-heme iron enzymes (*1,9*). He has also characterized and developed the chemistry of iron(III)-TPA, and now analogous cyclam complexes (*10*). Recently he has reported on an important new iron(IV) complex as well. His approach has been to start with low oxidation state iron complexes and oxidize them to higher states. This work dovetails well with earlier work by Wieghardt and Eckardt on iron(V) nitrido complexes using cyclam (*11*).

In order to provide a basis for the planning and interpretation of studies on related systems, both enzymes and other model systems, knowledge of the fundamental chemistry of iron in high oxidation states is urgently needed. For example, Valentine has characterized an olefin epoxidation catalyzed by an iron-cyclam complex (*12*). It was suggested that the intermediate responsible for addition across the double bond was a "ferryl" species, (cyclam)Fe=O. Que et al. are now exploring this possibility further. Increased fundamental data on high oxidation state compounds will aid in such studies. Other iron oxidation catalysts have also been suggested to react in a similar fashion and it is interesting to note that the nature of the iron intermediates vary when other oxidants (e.g., iodosylbenzene vs. hydrogen peroxide) are used. These conclusions are based on the nature of products, product yields, Hammett correlations for reactivity, and intra- and intermolecular competitive epoxidation studies. Barton (*13*) and Sawyer (*14*) examined iron catalyzed peroxide oxidations and proposed several active metal intermediates. While limited evidence for these species exists, few high oxidation state *non-heme* iron complexes have been isolated. The elegant work of the Que group has provided the best characterized examples of Fe(IV). While his early work has focused more on the binuclear iron complexes, his Fe(TPA) systems provide an important entry into mononuclear iron(IV) chemistry. Despite his advances, new complexes need to be synthesized, characterized and their reactivity studied in order to understand the broader chemistry involved in this important class of reactions.

Currently, the only "simple" high oxidation state complex of iron that is easily prepared are the ferrates, $[FeO_4]^{2-}$ where iron is in the +6 oxidation state (*15*) In view of the paucity of information on such iron complexes and their importance in a host of reactions, we have studied the chemistry of the FeO_4^{2-} ion to provide a new and unique entry into hypervalent iron chemistry. While ferrate(VI) is a strong oxidant, it appears, from preliminary studies, to be selective in its organic oxidations. We have demonstrated that reaction conditions, such as pH and order of addition can serve to alter the final products.

Experimental

The chemicals used in these studies were reagent grade or higher purity. Potassium ferrate(VI) was synthesized using the method of Schreyer and Ockerman and recrystallized until a >95% purity was achieved. Purities were determined using spectrophotometric determination at 505nm where ferrate has an extinction coefficient of $1175M^{-1}cm^{-1}$.

Reaction rates were monitored using UV-vis detection. Depending on the rates of reaction different instruments were used to carry out the kinetic studies. For reactions occurring in less than 1 minute, rapid scanning spectrophotometry was used as coupled with a stopped-flow as found in the OLIS-RSM1000 system. For slower reactions, an HP8452A Diode Array spectrophotometer was used. In both cases, reaction rate constants were determined using OLIS kinetic packages.

Results and Discussion

Reactions with Hydrazines (N_2H_4 or $CH_3N_2H_3$) (*16*)

We studied the ferrate(VI) oxidation of hydrazines and measured the reaction kinetics in aqueous media. Hydrazine, monomethylhydrazine (MMH) and phenylhydrazine (PH) each produced molecular nitrogen and the latter two produced methanol and phenol respectively. The ferrate was consistently reduced to iron(II). In each case, the following "simple" rate law was observed.

$$\text{Rate} = (k_O + k_H[H^+])\,[FeO_4^{2-}][\text{hydrazine}]$$

Protonation was assumed to occur on the ferrate center since protonated hydrazine is typically oxidized at a slower rate than its deprotonated form.

Checks for a radical mechanism using acrylonitrile showed no evidence for this one-electron pathway for any of the reactions. Support for a two electron pathway for N_2H_4 is supported since unsaturated carboxylic acids were saturated when present in the oxidation mixture. This was assumed to occur via formation of diazine (N_2H_2) which is well known to react with double bonds.

Table 1. Rate Constants of Ferrate(VI) oxidation of hydrazines. Conditions: T = 25°C, I = 1.0M ($NaClO_4$).

Compound	k_O ($M^{-1}s^{-1}$)	k_H ($M^{-1}s^{-1}$)
N_2H_4	5.0×10^3	3.8×10^5
$CH_3N_2H_3$	4.4×10^3	6.2×10^5

Therefore, a series of two electron transfers, presumably inner-sphere, are proposed to account for these observations, Scheme I.

$$HFeO_4^- + N_2H_4 \rightarrow N_2H_2 + Fe(IV) \qquad kO$$
$$HFeO_4^- + N_2H_4 \rightarrow N_2H_2 + Fe(IV) \qquad kH$$
$$Fe(IV) + N_2H_4 \rightarrow N_2H_2 + Fe(II) \qquad rapid$$
$$N_2H_2 + (H)FeO_4^{-(-2)} \rightarrow N_2 + Fe(IV) \qquad rapid$$

Methanol and phenol are produced by oxidation followed by rapid hydrolysis of their corresponding diazene intermediates.

$$RN_2H + HFeO_4^- \rightarrow RN_2^+ + Fe(IV)$$
$$RN_2^+ + H_2O \rightarrow ROH + N_2 + H^+$$

Unfortunately, no other information may be obtained about the existence of the proposed intermediate.

Reactions with Hydroxylamines (RNHOH) (*17*)

Hydroxylamines react rapidly with ferrate to produce a variety of oxidation products and iron(III) as shown in the following reactions.

$$2\ NH_2OH + 4\ H^+ + FeO_4^{2-} \rightarrow N_2O + Fe(II) + 5\ H_2O$$
$$2\ CH_3NHOH + 4\ H^+ + FeO_4^{2-} \rightarrow 2\ CH_3NO + Fe(II) + 4\ H_2O$$
$$2\ PhNHOH + 4\ H^+ + FeO_4^{2-} \rightarrow 2\ PhNO + Fe(II) + 4\ H_2O$$
$$2\ CH_3ONH_2 + 4\ H^+ + FeO_4^{2-} \rightarrow 2\ CH_3OH + N_2O + Fe(II) + 3\ H_2O$$

A general rate law maybe written for each of these reactions as follows,

$$Rate = (k_O + k_H)[H^+])[FeO_4^{2-}][hyroxylamine]$$

where k_O and k_H represent deprotonated and protonated pathways for the oxidation process. These values are shown in Table 2.

Table 2. Rate Constants of Ferrate(VI) oxidation of hydroxylamines Conditions: T = 25°C, I = 1.0M ($NaClO_4$).

Compound	k_O ($M^{-1}s^{-1}$)	k_H ($M^{-1}s^{-1}$)
NH_2OH	4.8×10^3	3.3×10^4
CH_3NHOH	3.5×10^3	1.6×10^5
CH_3ONH_2	1.9	110

For the hydroxylamine or N-substituted hydroxylamines a general reaction mechanism may be written as follows.

(R = H, CH_3 or Ph)

$HFeO_4^- \leftrightarrow FeO_4^{2-} + H^+$	K_{eq}
$FeO_4^{2-} + RNHOH \rightarrow RNO + Fe(IV)$	k_O
$HFeO_4^- + RNHOH \rightarrow N_2H_2 + Fe(IV) + H^+$	k_H
$Fe(IV) + RHNOH \rightarrow RNO + Fe(II)$	rapid

A similar reaction scheme may be written for NH_2OCH_3

$NH_2OCH_3 + FeO_4^{2-} \rightarrow NOR + Fe(IV)$	k_O
$NH_2OCH_3 + HFeO_4^- \rightarrow NOR + Fe(IV) + H^+$	k_H
$NH_2OCH_3 + Fe(IV)^- \rightarrow NOR + Fe(II)$	rapid
$H_2O + 2\,NOR \rightarrow N_2O + 2ROH + Fe(II)$	rapid

The latter reaction is similar to that for the aqueous decomposition of NOH radicals into nitrous oxide and water.

Although no reaction intermediate was observed in any of the system, we propose one where the hydroxylamine is either O-bonded, N-bonded or side-on bonded to the ferrate ion, presumably via expansion of the coordination environment around the iron center. This is "required" since these reactions occur faster than oxygen exchange on the iron(VI) center. Unfortunately, no information regarding the exact coordination environment is known.

Reactions with Anilines (*18*)

Preliminary investigation of the ferrate oxidation of aniline and substituted anilines in aqueous media showed that the ferrate(VI) oxidation proceeds smoothly to produce a single product. *A unique feature of the lower pH process is that the product possesses only a cis-conformation, and represents the only chemical process to form cis-azo complexes.*

With excess aniline at pH 9, ferrate(VI) rapidly (seconds) disappears to form an intermediate that subsequently disappears by a slower reaction (secs-mins) with the committant appearance of azobenzene. This is in contrast to the observations of Eyring et al.[19] In their study, they only monitored the reaction at 505nm. At this wavelength, it is easy to miss the rapid formation of the intermediate and only observe the loss of intermediate. They also did not explore the possible formation of cis-azobenzene.

$$FeO_4^{2-} \xrightarrow{\text{Aniline}} \text{imidoiron(VI) intermediate} \xrightarrow{\text{Aniline}} \text{cis-azobenzene} + Fe(II)$$

milliseconds-seconds *seconds-minutes*

Although the intermediate has not been characterized, a likely prospect is an imidoferrate(VI) species formed by the simple substitution of an oxide (hydroxide) on iron(VI). The presence of an isosbestic point at 480nm shows that this reaction is a clean, single step process with a rate law first order with respect to ferrate(VI), aniline and hydrogen ions.

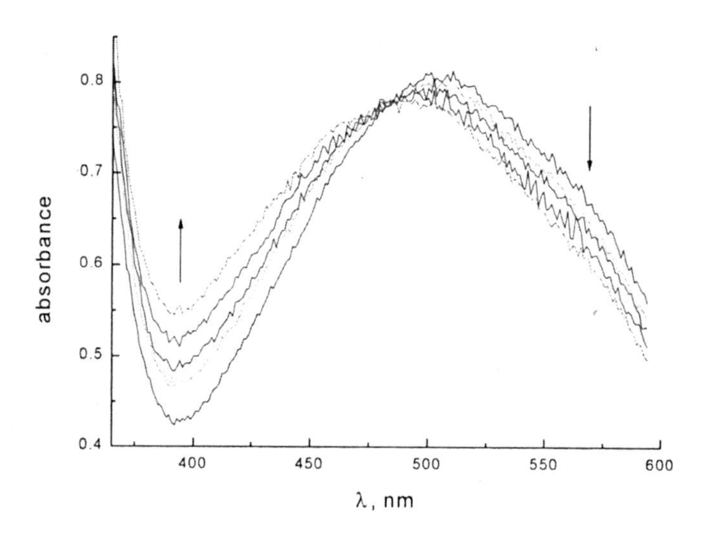

Figure 1. Formation of ferrate(VI) - aniline intermediate. Conditions: $T = 25^{o}C$, pH 9, $[Fe(VI)] \sim 5x10^{-5}M$, $[aniline] = 10^{-3}M$, $I = 1.0M$ ($NaClO_4$).

The kinetic data show a good Hammett correlation with σ^+ to give a ρ equal to a -2 for the formation reaction. This indicates build-up of a high positive charge at the reaction site.

The second step rate law is identical to the first, but first order in intermediate, and well resolved in time, i.e. over ten times slower. See Figure 2. As observed for the first step, the Hammett correlated well with σ^+ to give a ρ equal to a -1.9. We believe this is the addition of a second aniline to the iron center, followed by a series of rapid intramolecular electron transfer steps involving N-N bond formation, to eventually produce *cis*-azobenzene. The iron center apparently acts as a template in this two-step process, see Figure 3.

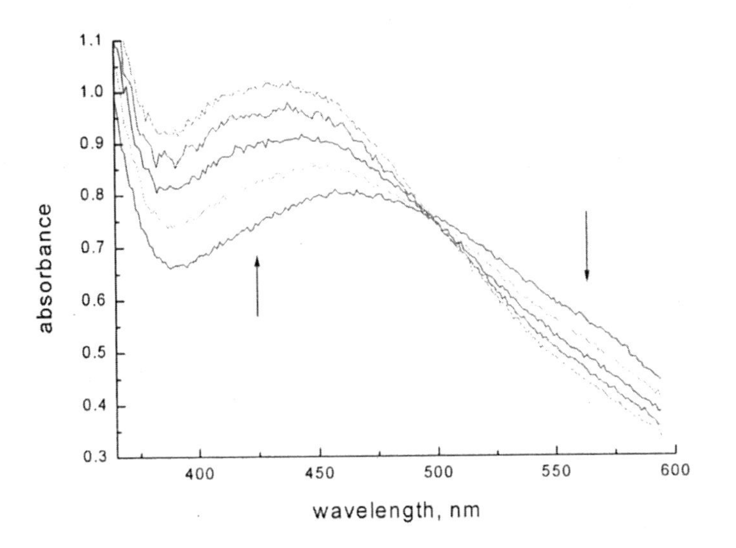

Figure 2. Loss of aniline-ferrate(VI) intermediate. Conditions: $T = 25^{o}C$, pH 9, [intermediate] ~ $5x10^{-5}M$, [aniline] = $10^{-3}M$, $I = 1.0M$ ($NaClO_4$).

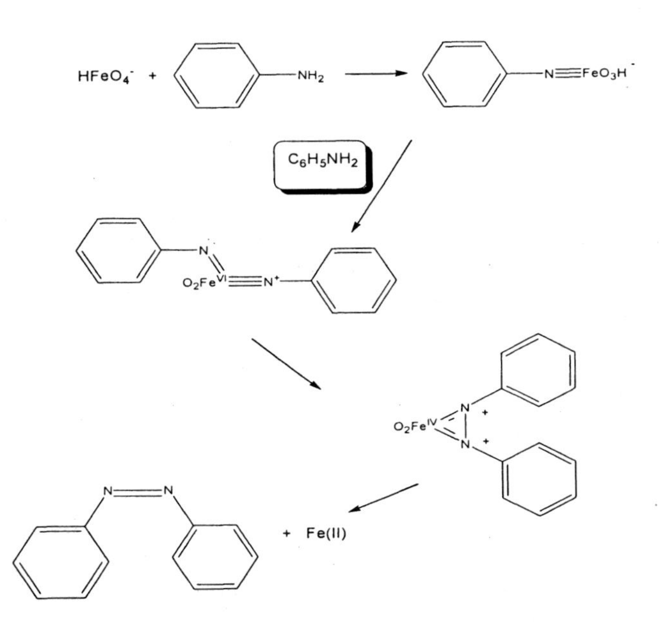

Figure 3. Proposed template reaction for formation of cis-azobenzene.

Several other reactions with substituted anilines were examined in attempts to block either the coupling reaction or intramolecular electron transfer steps to form azo products. Substitution at the para position gave reaction schemes identical to those shown above and azobenzenes were observed. However, when both the ortho and para positions were substituted with methyl groups, we were able to isolate a hydrazine final product. We believe this indicates a change in reaction processes. Unlike the para substituted aniline, these reactions show positive tests for production of radicals. Also, iron(III) instead of iron(II) was produced. This indicates that the reaction is now proceeding via a radical intermediate, similar to the process suggested by Eyring.

Support for this change in mechanism is found by placement of methyl groups on the aniline nitrogen. For example, N, N-dimethylaniline produced the know anilinium type radical, see Figure 4.

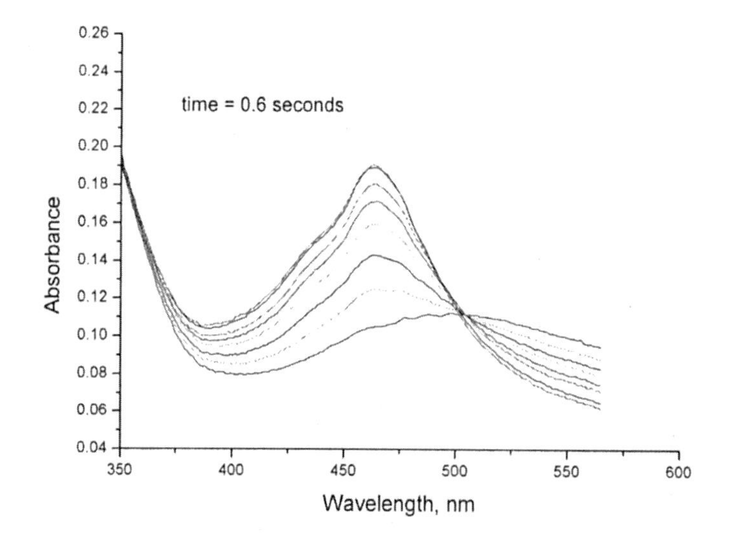

Figure 4. Reaction with N, N-dimethylaniline. Conditions: $T = 25^oC$, pH 9, [intermediate] ~ $5x10^{-5}M$, [N, N-dimethylaniline] = $2x10^{-3}M$, $I = 1.0M$ ($NaClO_4$).

It should be noted that the observed spectrum is very different from that observed with "normal" aniline. Also, the anilinium radical has been observed by Wheeler and Nelson (*19*) to occur at 420nm, vastly different than our observations at ~485nm.

Conclusions

The ferrate(VI) oxidationreactions of nitrogen containing compounds have a vast chemistry that remains to be explored. Preliminary observations may be summarized as follows:

1. Simple hyrazines and hydroxylamines proceed via 2-electron steps.
2. Aniline and para-substituted anilines proceed via 2-electron steps and observation of intermediates is possible.
3. Azobenzene and its para-substituted analogs are all produced as the cis-isomers.
4. Ortho-substituted anlines proceed via 1-electron steps and the final products are hydrazines.
5. Substitution on the nitrogen in aniline causes all the reactions to proceed via radical mechanisms.

Acknowledgements

The authors acknowledge WERC and the ACS/PRF funding agencies for their generous support of this work.

References

1. a) Kumar, D.; Hirao, H.; Que, L., Jr.; Shaik, S., "Theoretical Investigation of C-H Hydroxylation by $(N_4Py)Fe^{IV}{=}O^{2+}$: An Oxidant More Powerful than P450?" *J. Am. Chem. Soc.* **2005**; *127*, 8026. Bassan, Arianna; Blomberg, Margareta R. A.; Siegbahn, Per E. M.; Que, Lawrence, Jr. "A density functional study on a biomimetic non-heme iron catalyst: insights into alkane hydroxylation by a formally HO-Fe^{V}=O oxidant." *Chemistry-A European Journal* **2005**, *11*, 692.

b) Sastri, C. V.; Park, M. J.; Ohta, T.; Jackson, T. A.; Stubna, A.; Seo, M. S.; Lee, J.; Kim, J.; Kitagawa, T.; Munck, E.; Que, L., Jr.; Nam, W., "Axial Ligand Substituted Nonheme Fe^{IV}=O Complexes: Observation of Near-UV LMCT Bands and Fe=O Raman Vibrations", *J. Am. Chem. Soc.* **2005**, *127*, 12494. Quinonero, David; Morokuma, Keiji; Musaev, Djamaladdin G.; Mas-Balleste, Ruben; Que, Lawrence, Jr., "Metal-Peroxo versus Metal-Oxo Oxidants in Non-Heme Iron-Catalyzed Olefin Oxidations: Computational and Experimental Studies on the Effect of Water" *J. Am. Chem. Soc* .**2005**, *127*, 6548. Rohde, J.; Lim, M.; Brennessel, W.; Bukowsky, M.; Stubna, A.; Munck, E. Nam, W.; Que, L. "Crystallographic and Spectroscopic Characterization of a Nonheme Fe(IV)=O complex." *Science* **2003**, *299*, 1037.

2. Kovacs, J. "Synthetic Analogues of Cysteinate-Ligated Non-Heme Iron and Non-Corrinoid Cobalt Enzymes." *Chem. Rev.* **2004**, *104*, 825. Kovacs, J. "Biochemistry: how iron activates O_2." *Science* **2003**, *299*, 1035.

3. Wasser, I.M.; Constantinus, F. M.; Verani, C.N.; Rentschler, E.; Huang, H.; Loccoz, P.M.; Zakharov, L.N.; Rheingold, A.L.; Karlin, K.D. "Synthesis and Spectroscopy of Oxo (O^{2-})-Bridged Heme/Non-heme Diiron Complexes: Models for the Active Site of Nitric Oxide Reductase." *Inorg. Chem.* **2004**, *43*, 651. Wasser, I.M.; de Vries, S.; Moenne-Loccoz, P.; Schroeder, I.; Karlin, K.D. "Nitric Oxide in Biological Denitrification: Fe/Cu Metalloenzyme and Metal Complex NOx Redox Chemistry." *Chem. Rev.* **2002**, *102*, 1201.

4. Solomon, E..; Brunold, Thomas C.; Davis, Mindy I.; Kemsley, Jyllian N.; Lee, Sang-Kyu; Lehnert, Nicolai; Neese, Frank; Skulan, Andrew J.; Yang, Yi-Shan; Zhou, Jing. "Geometric and Electronic Structure/Function Correlations in Non-Heme Iron Enzymes." *Chem Revs.* **2000**, *100*, 235. Decker, Andrea; Solomon, Edward I. "Comparison of Fe^{IV}=O heme and non-heme species: Electronic structures, bonding, and reactivities." *Angew. Chem., Intl. Ed.* **2005**, *44*, 2252. Neidig, Michael L.; Decker, Andrea; Kavana, Michael; Moran, Graham R.; Solomon, Edward I. "Spectroscopic and computational studies of NTBC bound to the non-heme iron enzyme (4-hydroxyphenyl)pyruvate dioxygenase: Active site contributions to drug inhibition." *Biochem. Biophys. Res.Commun.* **2005** *338*, 206.

5. Lippard, S.J.; Berg, J.M. "Principles of Bioinorganic Chemistry" University Science Books, NY **1994**.

6. Feig, Andrew L.; Becker, Michael; Schindler, Siegfried; van Eldik, Rudi; Lippard, Stephen J.. "Mechanistic Studies of the Formation and Decay of Diiron(III) Peroxo Complexes in the Reaction of Diiron(II) Precursors with Dioxygen." *Inorg. Chem.* **1996**, *35*, 2590.

7. Ortiz de Montellano, P.R. "Cytochrome P450: Structure, Mechanism, and Biochemistry" Plenum, NY **1985**.

8. Rosenzweig, A.C.; Frederick, C.A.; Lippard, S.J.; Nordlund, P. "Crystal structure of a bacterial non-heme iron hydroxylase that catalyzes the biological oxidation of methane." *Nature* **1993**, *366*, 537. Fontecave M; Nordlund P; Eklund H; Reichard P "The redox centers of ribonucleotide reductase of Escherichia coli." *Adv. Enzymol.* **1992**, *675*, 147. Lipscomb J. D. "Biochemistry of the soluble methane monooxygenase." *Ann. Rev. Microbiol.* **1994**, *48*, 371.

9. Dong, Y.; Fujii, H.; Hendrick, M.P.; Leising, R.A.; Pan, G.; Randall, C.R.; Wilkinson, E.; Zang, Y. Que, L.; Fox, B.G.; Kauffman, K.; Munck, E. "A High-Valent Nonheme Iron Intermediate. Structure and Properties of $[Fe_2(\mu\text{-}O)_2(5\text{-}Me\text{-}TPA)_2](ClO_4)_3$." *J. Am. Chem. Soc.* **1995**, *117*, 2778. Leising, R.A.; Brennan, B.A.; Que, L.; Fox, B.G.; Munck, E. "Models for non-heme iron oxygenases: a high-valent iron-oxo intermediate." *J. Am. Chem. Soc* **1991**, *113*, 3988. Leising, Randolph A.; Kojima, Takahiko; Que, Lawrence, Jr. "Alkane functionalization at nonheme iron centers: mechanistic insights." *Act. Dioxygen Homogen. Catal. Oxid., [Proc. Int. Symp.], 5th* **1993**, 321.

10. Kauppi, B.; Lee, K.; Carredano, E.; Parales, R.E.; Gibson, D.T.; Eklund, H.; Ramaswamy, S. "Purification and crystallization of the oxygenase component of naphthalene dioxygenase in native and selenomethionine-derivatized forms." *Structure* **1998**, *6*, 571.

11. Que, L.; True, A.E. Progress in Inorganic Chemistry: Bioinorganic Chemistry: S.J. Lippard, ed., Wiley, NY **1990**. Nam, W.; Valentine, J.S. "Zinc(II) complexes and aluminum(III) porphyrin complexes catalyze the epoxidation of olefins by iodosylbenzene." *J. Am. Chem. Soc* **1990**, *112*, 4987.

12. Holz, R.C.; Elgren, T.E.; Pearce, L.L.; Zhang, J.H.; O'Connor, C.J.; Que, L., "Spectroscopic and electrochemical properties of (μ-oxo)diiron(III) complexes related to diiron-oxo proteins. Structure of $[Fe_2O(TPA)_2(MoO_4)](ClO_4)_2$." *Inorg. Chem.* **1993**, *32*, 5844. Chen, K.; Que, L. "cis-Dihydroxylation of olefins by a non-heme iron catalyst: a functional model for Rieske dioxygenases." *Angew. Chem.* **1999**, *38*, 2227.

14. Zang, Y.; Kim, J.; Dong, Y.H.; Wilkinson, E.C.; Appleman, E.H.; Que, L. " Models for Nonheme Iron Intermediates: Structural Basis for Tuning the Spin States of Fe(TPA) Complexes." *J. Am. Chem. Soc* **1997**, *119*, 4197.

15. Meyer, K., E. Bill, B. Mienert, T. Weyhermüller and Wieghardt, K. "Photolysis of cis- and trans-$[Fe^{III}(cyclam)(N_3)_2]+$ Complexes: Spectroscopic Characterization of a Nitridoiron(V) Species." *J. Am. Chem. Soc.* **1999**, *121*, 4859. Grapperhaus, C. A., B. Mienert, E. B., Weyhermüller, T. and Wieghardt, K. "Mononuclear (Nitrido)iron(V) and (Oxo)iron(IV) Complexes via Photolysis of $[(cyclam\text{-}acetato)FeIII(N_3)]+$ and Ozonolysis of $[(cyclam\text{-}acetato)Fe^{III}(O_3SCF_3)]+$ in Water/Acetone Mixtures." *Inorg. Chem.* **2000**, *39*, 5306.

16. Johnson, M. D.; Hornstein, B. J. "Kinetics and Mechanism of the Ferrate Oxidation of Hydrazine and Monomethylhydrazine" *Inorg. Chim. Acta* **1994**, *225*, 145.

17. Johnson, M. D.; Hornstein, B. J. "Kinetics and Mechanism of the Ferrate(VI) Oxidation of Hydroxylamines" *Inorg. Chem.* **2003**, *42*, 6923.

18. Johnson, M. D.; Hornstein, B. J. "Unexpected selectivity on the oxidation of arylamines with ferrate-preliminary mechanistic observations" *J. Chem. Soc., Chem. Commun.* **1996**, 965.

19. Huang, H.; Sommerfeld, D.; Dunn, B.C.; Lloyd, C.R.; Eyring, E.M. "Ferrate(VI) oxidation of aniline." *J. Chem. Soc. Dalton Trans.* **2001**, 1301.

20. Wheeler, J.; Nelson, R. "Spectrophotometric observation of anilinium radicals"*J. Phys. Chem.* **1973**, *77*, 2490.

Chapter 13

Kinetics and Product Identification of Oxidation by Ferrate(VI) of Water and Aqueous Nitrogen Containing Solutes

James D. Carr

Department of Chemistry, University of Nebraska-Lincoln, Lincoln NE 68588-0304

Water is shown to be oxidized by ferrate(VI) via a pathway dominated by protonated ferrate and leading to molecular oxygen via hydrogen peroxide. Nitrogen-containing solutes are oxidized in competition with water oxidation. Products of oxidation are determined by a variety of analytical methods. Iron(II) is shown to be an intermediate state of iron which originated as Fe(VI) but Fe(III) is the final product in the absence of a trapping reagent for Fe(II). Nitroxyl is shown to be an important intermediate species in the oxidation of azide and hydroxylamine.

Kinetics of Water Oxidation

The oxidation kinetics of water by ferrate(VI) was studied by monitoring the absorbance of ferrate(VI) at 505 nm by both conventional and stopped-flow spectrophotometry over the pH range of 2-10 buffered with phosphate, borate, or pyrophosphate, all at or near 20°C. Equation (1) shows the rate law which shows terms which are first-order and second-order in iron(VI).[1]

$$\text{rate} = -d[\text{Fe(VI)}]/dt = k_1[\text{Fe(VI)}] + k_2[\text{Fe(VI)}]^2 \quad (1)$$

The reaction is much faster at low pH and the rate constant k_1 was resolved into specific terms for the three differently protonated forms of ferrate.

$$k_1[Fe(VI)] = k_{H2FeO4}[H_2FeO_4] + k_{HFeO4}[HFeO_4^-] + k_{FeO4}[FeO_4^{2-}] \qquad (2)$$

The value of k_{HFeO4} is general acid catalyzed and depends on the concentration of dihydrogenphosphate ion. The resolved rate constant for monoprotonated ferrate therefore is further resolved as follows, where k^c_{HFeO4} is the rate constant for the general acid catalyzed term.

$$k_{HFeO4}[HFeO_4^-] = k_{HFeO4}[HFeO_4^-] + k^c_{HFeO4}[HFeO_4^-][H_2PO_4^-] \qquad (3)$$

Resolved values of the rate constants are

$k_{H2FeO4} = 20\ sec^{-1}$
$k^c_{HFeO4} = 1.74$ L/mol sec
$k_{HFeO4} = < 5 \times 10^{-2}\ sec^{-1}$
$k_{FeO4} = < 1 \times 10^{-6}\ sec^{-1}$

Figure 1. Shows the level of agreement of observed values of k_1 with the predicted values obtained from using the resolved values shown here.

Absorbance at time zero was plotted vs pH (Figure 2) and used to estimate the acidity constants of H_2FeO_4 and molar absorptivities of H_2FeO_4 , $HFeO_4^-$, and FeO_4^{2-}. Our best values for these quantities are $pK_{a1} = 3.2$, $pK_{a2} = 7.8$,$_{,.H2FeO4}$ = 50, $_{.HFeO4}$ = 267 and $_{.FeO4}$ = 1170 L/mol cm.

The term k_2 in the water oxidation rate law is interpreted to be the reaction of diferrate ion, assumed to be in rapid equilibrium with ferrate according to equation (4)

$$2\ FeO_4^{2-} + 2\ H^+ \rightarrow Fe_2O_7^{2-} + H_2O \qquad (4)$$

The equilibrium constant for reaction 4 is termed

$$K_D = [Fe_2O_7^{2-}] / [FeO_4^{2-}]^2[H^+]^2$$

The value of K_D is not known but is estimated based on the similar reaction of chromate to form dichromate, whose equilibrium constant = 1×10^{14}.

Therefore, $k_2[FeO_4^{2-}]^2 = k_D[Fe_2O_7^{2-}] = k_DK_D[FeO_4^{2-}][H^+]^2$

Finally, $k_2 = k_DK_D[H^+]^2$

Over a considerable pH range, the value of k_2 is, indeed, second order in $[H^+]$ (Figure 3) and a value of $k_D = 2.1 \times 10^3$ is calculated if $K_D = 1 \times 10^{14}$.

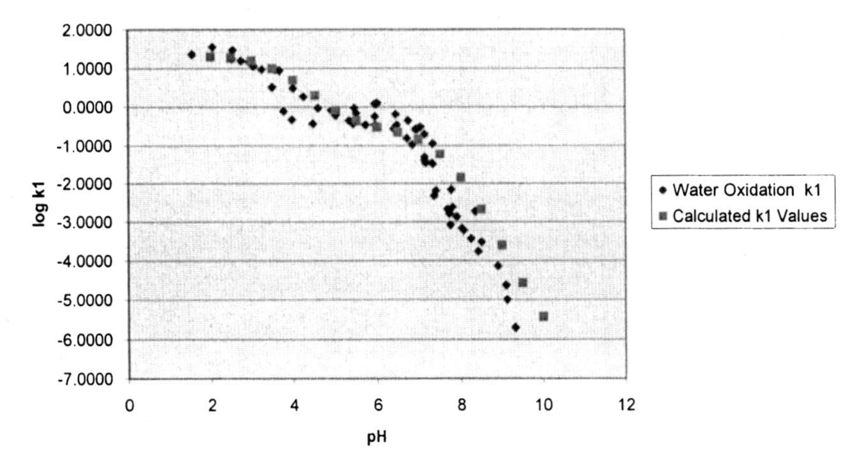

Figure 1. Water Oxidation by Ferrate

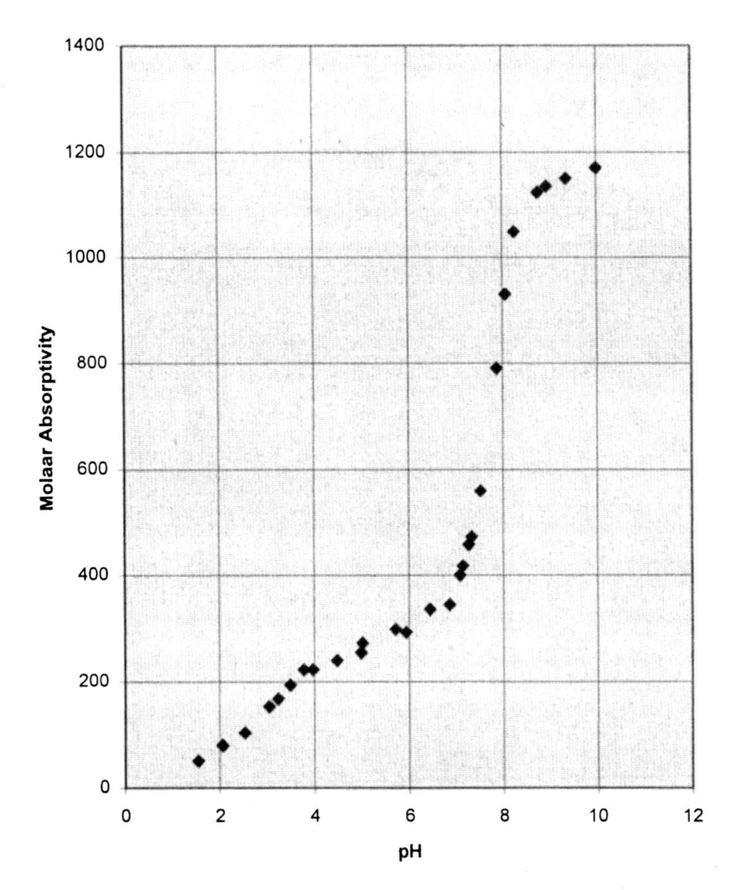

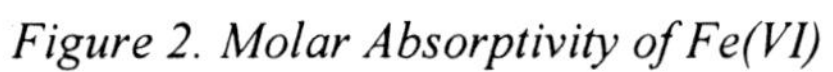
Figure 2. Molar Absorptivity of Fe(VI)

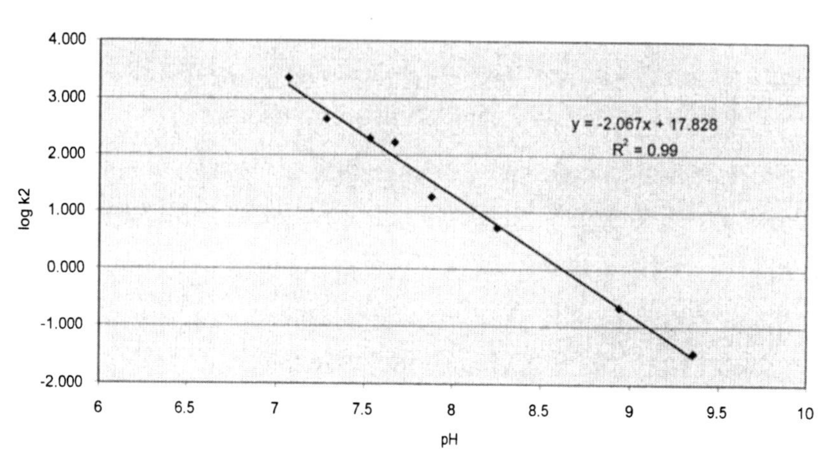

Figure 3. Second-Order Water Oxidation

Product Analysis of Water Oxidation

Product analysis of water oxidation was carried out by monitoring dissolved oxygen (Clark=s electrode voltammetry) and evolved gaseous oxygen (manometry and gas chromatography) as well as measuring hydrogen peroxide by pulse voltammetry. When gaseous and dissolved oxygen as well as hydrogen peroxide were considered, product formation matched expected values based on iron going from +6 oxidation state to +3. When these measurements were made of reactions carried out in the presence of 1,10-phenanthroline to chelate the suspected intermediate, Fe(II), the yield of oxygen increased to match the predicted values based on the amount of $Fe(phen)_3^{2+}$ as final product.

Oxidation Kinetics of Nitrogen-Containing Solutes

The oxidations of aqueous solutions of azide, hydrazine, cyanide, hydroxylamine, ammonia, nitrite, and methyl amine were measured by similar methods and, in general, were found to follow a rate law with an additional term k_s which describes the rate at which the solute, S, reacts. In most cases, solute was in >10-fold excess over ferrate and the k_s term was so large as to cause the water oxidation terms to be insignificant. This was not the case for low concentrations of solute and for solutes that were oxidized only slowly. This competition between solute and solvent oxidation will limit the usefulness of ferrate in water treatment procedures.

$$\text{rate} = -d[\text{Fe(VI)}]/dt = k_1[\text{Fe(VI)}] + k_2[\text{Fe(VI)}]^2 + k_S[\text{Fe(VI)}][\text{S}] \quad (5)$$

The values of k_s, the oxidation rate constant of the solute, varied considerably among these solutes (Figures 4a and 4b). Reactions of hydroxylamine and cyanide were very fast, those of nitrite and azide not quite that fast, and of ammonia quite slow. Interestingly, even though very little ammonia was oxidized, the presence of ammonia speeded the formation of molecular oxygen via, apparently catalyzed, oxidation of water. Trimethylamine reacts much more slowly than either mono- or dimethylamine.

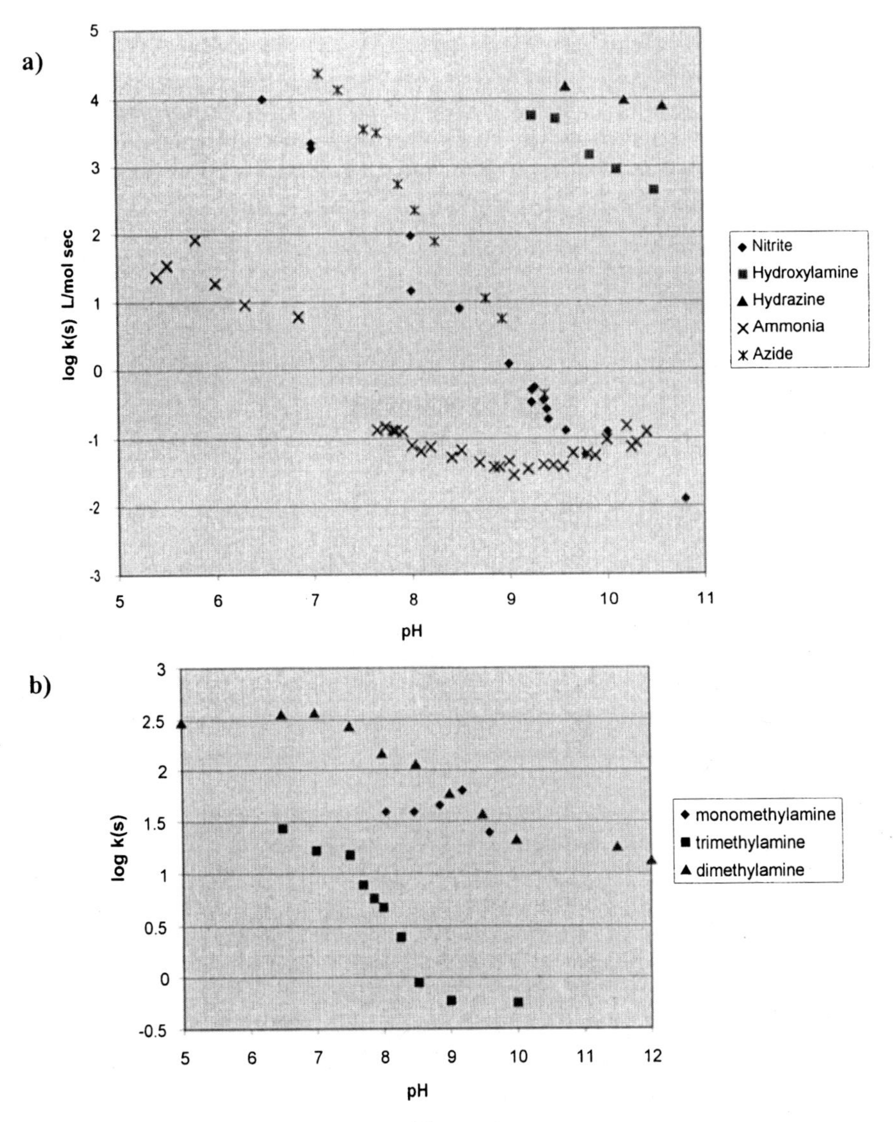

Figure 4.

In all cases, the reactions are faster in acid solution, signifying that protonated ferrate reacts more rapidly than the ferrate dianion. The observed, pH dependent values of k_s are resolved into values corresponding to the rates of reaction of the solute with diprotonated, monoprotonated, and unprotonated ferrate. For instance, rate constants for the reaction of nitrite with H_2FeO_4, $HFeO_4^-$, and FeO_4^{2-} respectively are 3.2×10^7, 33, and $< 1 \times 10^{-2}$ L/mol-s respectively.

Product Analysis of Nitrogenous Solutes

Methods of product analysis of these oxidations were various. Gaseous nitrogen and nitrous oxide were measured (along with gaseous oxygen) by gas chromatography. Total gaseous products were measured by manometry. Nitrate and nitrite along with unreacted azide were measured by ion chromatography. Colorimetric methods were also used to determine nitrate and nitrite. Selective potentiometric electrodes were used to measure nitrate and ammonia.

In the oxidation of azide ion, nitroxyl (HNO or its conjugate base NO^-) is a key intermediate, and leads to a distribution of products including nitrous oxide, molecular nitrogen, nitrite ion and nitrate ion.

$$FeO_4^{2-} + N_3^- \rightarrow N_2 + NO^- + FeO_3^{2-}$$

$$FeO_4^{2-} + NO^- \rightarrow NO_2^- + FeO_3^{2-}$$

$$FeO_4^{2-} + NO_2^- \rightarrow NO_3^- + FeO_3^{2-}$$

$$2\ HNO \rightarrow N_2O + H_2O$$

Mass balance calculations show excellent agreement with predictions based on a $3e^-$ reduction of Fe(VI) to Fe(III) unless 1,10-phenanthroline is present in which case a $4e^-$ reduction occurs to the extent that $Fe(phen)_3^{2+}$ is formed. Oxidation of azide in ^{18}O enriched water showed that the oxygen of N_2O did not come from the solvent, but rather came from oxygen atom transfer from ferrate. Oxidation of azide in which a terminal nitrogen was enriched in ^{15}N showed that the N_2 formed in the first step involves the central nitrogen atom and one of the terminal atoms of the azide ion. The nitrogen atom of nitroxyl and therefore of nitrous oxide originated at one end of the azide.

Ferrate oxidation of hydroxylamine also shows evidence of nitroxyl being a key intermediate, presumably via the following reaction sequence. Again, when carried out in ^{18}O enriched water, the label does not go to the nitrous oxide. It is suspected that the oxygen in HNO and therefore in N_2O was originally part of the hydroxylamine molecule.

$$FeO_4^{2-} + NH_2OH \rightarrow HNO + H_2O + FeO_3^{2-}$$

$$FeO_4^{2-} + NO^- \rightarrow NO_2^- + FeO_3^{2-}$$

$$FeO_4^{2-} + NO_2^- \rightarrow NO_3^- + FeO_3^{2-}$$

$$HNO + NH_2OH \rightarrow N_2 + 2\,H_2O$$

$$2\,HNO \rightarrow N_2O + H_2O$$

In contrast to earlier reports, we find that ammonia is not readily oxidized by ferrate to molecular nitrogen. Rather, the product is principally molecular oxygen. Only at very high ammonia concentrations (> 0.10M) at high pH (pH > 10) is as much nitrogen formed as oxygen.

Cyanide is very rapidly oxidized by ferrate to form cyanate, (OCN^-) and its isomer, fulminate (CNO^-). Cyanate is unreactive toward ferrate but slowly decomposes in water to form ammonia and bicarbonate. Fulminate is further oxidized to form nitrate and bicarbonate. Formaldehyde, formate, and carbon monoxide were not detected upon oxidation of cyanide.

In the oxidation of hydrazine by ferrate, the only nitrogen containing product is molecular nitrogen, presumably by two sequential reactions as shown.

$$FeO_4^{2-} + N_2H_4 \text{ c> } N_2H_2 + H_2O + FeO_3^{2-}$$

$$FeO_4^{2-} + N_2H_2 \text{ c> } N_2 + H_2O + FeO_3^{2-}$$

Ferrate oxidation of formamide leads to the single product, cyanate.

$$FeO_4^{2-} + HCONH_2 \text{ c> } HOCN + H_2O + FeO_3^{2-}$$

Methylformamide does not react at an appreciable rate with ferrate.

Ferrate oxidation of methylamine leads to $HCO3^-$, NCO^-, and N_2. A proposed sequence is the formation of an imine followed by formamide and cyanate.

$$FeO_4^{2-} + CH_3NH_2 \text{ c> } CH_2{=}NH + H_2O + FeO_3^{2-}$$

$$CH_2{=}NH + FeO_4^{2-} \text{ c> } HCO\text{-}NH_2 + FeO_3^{2-}$$

$$HCO\text{-}NH_2 + FeO_4^{2-} \text{ c> } HOCN + H_2O + FeO_3^{2-}$$

Dimethylamine and trimethylamine do not follow a similar path, but rather form formic acid presumably from the intermediate formation of formaldehyde.

Molecular nitrogen is the major nitrogen containing compound formed when the carbon-nitrogen bond is broken. Perhaps the N_2 is formed from intermediate diimide (HN=NH). Ammonium, nitrite, and nitrate are specifically not found as products of the oxidation of any of the methylamines.

All products which are described here are the result of oxygen atom transfer from high oxidation state iron to the solute molecule.

The final iron-containing product is Fe(III), the exact species depending on pH and buffer components. This is the case unless 1,10-phenanthroline has been added to the buffered solute mixture before adding ferrate. In that case, some of the iron is trapped as $Fe(phen)_3^{2+}$ with its characteristic orange color. This implies that Fe(II) is an intermediate species, probably formed by reduction of Fe(IV), the initial reduction product, FeO_3^{2-}, which is also a good oxidant. If phenanthroline is added after the oxidation reaction, no $Fe(phen)_3^{2+}$ is formed. A sequence like the following is suspected, although other oxidants than Fe(IV) are possibly more important in converting Fe(II) into Fe(III).

Fe(VI) + S B> Fe(IV) + SO

Fe(IV) + S B> Fe(II) + SO

Fe(II) + Fe(IV) B> 2 Fe(III)

Acknowledgements

Graduate students John Erickson, Michael Humphrey, David Splichal, Ted Ericson, Paul Kelter, Charles Buller, Mark Cherwin, and William McLaughlin.

References

1. Carr, J.D. et al. *Properties of Ferrate(VI) in Aqueous Solution: An Alternate Oxidant in Wastewater Treatment*: Water Chlorination Chemistry, Environmental Impact and Health Effects. Proceedings of the Fifth Conference on Water Chlorination, June 1984, **1985**, 1285-1298

Chapter 14

Recent Advances in Fe(VI) Charge Storage and Super-Iron Batteries

Stuart Licht and Xingwen Yu

Department of Chemistry, University of Massachusetts at Boston, 100 Morrissay Boulevard, Boston, MA 02125–3393

An overview of Super-Iron batteries, introduced and demonstrated in 1999, is presented. The batteries are based on an unusual Fe(VI) redox couple and multiple electron cathodic charge storage. Such Fe(VI) cathodes can be "green", cost effective and store considerably higher charge than conventional cathode materials. For example, the Fe(VI) salt K_2FeO_4 holds 406 mA/g, is readily prepared, and the ferric oxide discharge product is environmentally benign. Fe(VI) salts that have been synthesized and demonstrated as Super-Iron cathodes include compounds with a 3-electron cathodic charge capacity, such as Li_2FeO_4, Na_2FeO_4, K_2FeO_4, Rb_2FeO_4, Cs_2FeO_4 (alkali Fe(VI) salts), as well as alkali earth Fe(VI) salts $BaFeO_4$, $SrFeO_4$, and also a transition Fe(VI) salt Ag_2FeO_4 which exhibits a 5-electron cathodic charge storage. Four classes of Super-Iron battery have been demonstrated: Super-Iron Primary Alkaline Batteries, Super-Iron Primary Lithium Batteries, Super-Iron Rechargeable Alkaline Batteries and Super-Iron Rechargeable Nonaqueous batteries which exhibit higher energy storage capacity than conventional batteries. Configuration optimization, enhancement and mediation of Fe(VI) alkaline and nonaqueous cathode charge transfer are summarized. Composite Fe(VI)/Mn(IV or VII), Fe(VI)/Ag(II) and zirconia coating stabilized Fe(VI)/Ag(II) cathode alkaline batteries are also illustrated. Thin Fe(VI/III) film cathodes are reversible in both alkaline and nonaqueous (containing Li salt) electrolyte media for rechargeable Fe(VI)/metal hydride and Fe(VI)/Li batteries.

Fe(VI) species have been known for over a century, although its chemistry remains relatively unexplored (*1-3*). Lower valence, Fe(III) compounds had been studied both as cathode (*4*) and anode (*5*) materials in electrochemical storage cells. However, higher valence, greater charge capacity iron salts had not been previously considered. The fundamental solubility and stability constraints on high valence iron chemistry were not well established. Indeed, the perception that Fe(VI) species were intrinsically unstable was incorrect (*6*). Recently, we introduced cathodes incorporating hexavalent, Fe(VI), which sustain facile, energetic, cathodic charge transfer (*6-32*). Resources to prepare Fe(VI) salts are plentiful and clean. Iron is the second most abundant metal in the earth's core, and the Fe(VI) reduction product is non-toxic ferric oxide. Fe(VI) salts can exhibit substantially higher than conventional cathodic storage capacities (*6-32*). Due to their highly oxidized iron basis, multiple electron transfer, and high intrinsic energy, we have defined high oxidation state iron compounds as 'super-iron's and the new electrochemical storage cells containing them as 'Super-Iron' batteries (*6*).

Capacity, power, cost, and safety factors have led to the annual global use of approximately $6x10^{10}$ primary (single discharge) batteries, and society's use of secondary (rechargeable) energy storage remains dominated by lead acid battery technologies. Rechargeable lithium and metal hydride anodes have increased the energy capacity of batteries, but further advances are limited by the low energy capacity of their cathodes. New higher capacity, environmentally benign and cost effect cathode materials are needed. International interest in Fe(VI) electrochemistry is growing, including research efforts in China, Canada, Europe, and Japan (*33-36*). Favorable battery cathodes characteristics are low solubility, stability, facile charge transfer, and high charge capacity, and a high oxidative electrochemical potential. K_2FeO_4, prepared as described in our ref. 11, is particularly robust. We have also probed the controlled synthesis of a range of Fe(VI) salts (*9, 11-13, 18, 25*), and shown a direct route for their electrochemical synthesis from iron metal (*23*).

Insoluble cathodes avoid solution phase decomposition, prevent cathode diffusion, chemical reaction with the anode, and self-discharge. Fe(VI) salts are insoluble in a wide variety of nonaqueous solvents (*7, 9*), including electrolytes conducive to studies of Li electrochemistry such as acetonitrile, propylene and ethylene carbonate, γ-butyrolactone, and dimethoxyethane. Nonaqueous, lithium and Li-ion compatible Super-Iron cathodes containing, either a Li_2FeO_4, K_2FeO_4, $SrFeO_4$ or $BaFeO_4$ cathode, have been studied and exhibit a high cathodic discharge capacities respectively approaching 600, 400, 380 and 310 mAh/g (*9, 21*). We found that the nonaqueous discharge of Fe(VI) also incorporates a $3e^-$ Fe(VI) reduction, together with a reaction of Li ion (*32*).

The most recent development in Super-Iron cathode chemistry is that it is reversible, which has led to the demonstration of rechargeable Super-Iron batteries. In accord with our recent studies (*22, 24, 30, 32*), the principal limitation to Fe(VI) reversibility has been passivation of the charge transfer due to resistive buildup of low-conductivity ferric (Fe(III) salts. Recently, we have shown that ultrathin Fe(VI) layers are reversible cathodes; that is they are rechargeable in electrolytes respectively conducive to both lithium-ion (*32*), *or* metal hydride (*24, 30*), anode batteries, and that thicker films, established on an appropriate conductive matrix, are also reversible (*30*).

The synthesis and use of Fe(VI) chemical oxidants is presented elsewhere in this volume. Topics presented in this overview include:

Introduction to
Recent Advances in Fe(VI) Charge Storage & Super-Iron Batteries

1. Fundamentals of Primary Alkaline Super-Iron Batteries

2. Alkaline Fe(VI) K_2FeO_4 & $BaFeO_4$ Charge Transfer
    2i. Zirconia coating stabilization of the alkaline K_2FeO_4 cathode
    (Zirconia's hydroxide mediation of Fe(VI/III) charge transfer)
    2ii. Chemical (manganese) mediation of Fe(VI/III) charge transfer
    (Composite Mn/Fe Super-Iron Batteries)
    2iii. Electronic (silver) mediation of Fe(VI/III) charge transfer
    (Highest energy and power alkaline Super-Iron Batteries)
    2iv. Titanate, Ba(II), Mn(IV) & Co(III), Fe(VI) energy modifiers

3. Alkaline Fe(VI) $Na(K)FeO_4$, $Rb(K)FeO_4$, Cs_2FeO_4,
$SrFeO_4$ or Ag_2FeO_4 Charge Transfer

4. Conductors for Alkaline Super-Iron cathodes

5. Alkaline reversible Fe(VI) charge transfer
for Rechargeable Super-Iron Metal Hydride Batteries.

6. Nonaqueous primary Fe(VI) charge transfer
for Super-Iron/Lithium Batteries

7. Nonaqueous reversible Fe(VI) charge transfer
for Rechargeable Super-Iron Lithium Batteries.

1. Fundamentals of Primary Alkaline Super-Iron Batteries

In 1999, we introduced a class of batteries, referred to as Super-Iron batteries, containing a cathode utilizing a common material (iron) in an unusual +6 or "super-oxidized" valence state (*6*). The cathode is based on abundant starting materials, and is compatible with an alkaline electrolyte and common zinc or metal hydride anodes. Batteries utilizing a zinc anode and manganese dioxide (MnO_2) cathode have remained the dominant primary (single discharge) battery on the world market for over a half century due to its performance and low cost. The storage capacity of the aqueous MnO_2/Zn battery is severely limited, constrained by the charge capacity of its cathode storage material, which can retain up to a maximum of 308 mAh/g based on the one electron oxidation of MnO_2, which is low compared to that of its anode storage material Zn which can retain up to 820 mAh/g based on the 2e$^-$ (two electron) oxidation of zinc. Replacement of the MnO_2 cathodes in these cells with a more energetic cathode such as the Super-Iron cathode utilizing Fe(VI) compounds can substantially increase the energy storage capacity of these cells. For example, using the same zinc anode and electrolyte, Fe(VI) cathode batteries were shown to provide 50% more energy capacity than conventional alkaline batteries at low discharge rates, and several fold higher energy capacity at high rates of discharge (*6*).

The use of Fe(VI) salts as alkaline cathodic charge storage materials is based on the energetic and high-capacity 3e$^-$ reduction of Fe(VI) to a ferric oxide or hydroxide product. In a manner analogous to the alkaline oxidation product of zinc, whose zincate product varies with the discharge and the composition of the electrolyte, the degree of hydration and any associated cation of Fe(VI)'s ferric product, will depend on the extent of reduction and the composition of the hydroxide electrolyte. The 3e$^-$ cathodic charge storage of Fe(VI) is presented in eq 1.1 or eq 1.2 via the reduction of the alkaline Fe(VI) species, FeO_4^{2-}, respectively to the ferric hydroxide or anhydrous oxide product) (*6*).

$$FeO_4^{2-} + 3H_2O + 3e^- \rightarrow FeOOH + 5OH^- \qquad (1.1)$$

$$FeO_4^{2-} + 5/2H_2O + 3e^- \rightarrow 1/2Fe_2O_3 + 5OH^- \quad E = 0.5\text{-}0.65 \text{ V vs. SHE} \qquad (1.2)$$

Various Fe(VI) salts synthesized in our lab, exhibit the typical Super-Iron alkaline cathodic discharge. These include the high purity alkali Fe(VI) salts $K_xNa_{(2-x)}FeO_4$, K_2FeO_4, Rb_2FeO_4, Cs_2FeO_4, and as well as alkali earth Fe(VI) salts $SrFeO_4$, $BaFeO_4$, and also Ag_2FeO_4. The theoretical 3e$^-$ charge capacity of the Fe(VI) salts are determined as: $3F \times MW^{-1}$, from the salt molecular weight, MW(g/mol) and the Faraday constant (F = 96485 coulomb/mol = 26801 mA·hours/mol). The theoretical capacities of various alkali and alkali earth Fe(VI) salts are listed in Table 1.1.

A primary alkaline super iron battery contains an Fe(VI) cathode, and can utilize the zinc anode and alkaline electrolyte from a conventional alkaline batteries. In a zinc alkaline battery, the zinc anode generates a distribution of zinc oxide and zincate products, and similarly the final Fe(VI) product will depend on the depth of discharge. The general discharge of alkaline electrolyte cells utilizing a Zn anode and Fe(VI) cathodes is expressed as:

$$MFeO_4 + 3/2Zn \rightarrow 1/2Fe_2O_3 + 1/2ZnO + MZnO_2$$
$$M = Li_2, Na_2, K_2, Ru_2, Cs_2, Sr, Ba \qquad (1.3)$$

Ag_2FeO_4 is of interest, this Fe(VI) salt has an intrinsic cathodic capacity that includes not only the $3e^-$ Fe(VI) reduction, but also the single electron reduction of each of two Ag(I), for at total 5 Faraday per mole or 399.3 mAh/g intrinsic capacity, eq 1.4 (*26*).

$$Ag_2FeO_4 + 5/2H_2O + 5e^- \rightarrow 2Ag + 1/2Fe_2O_3 + 5OH^- \quad (1.4)$$

Table 1.1. Theoretical 3-Electron Charge Capacities of Various Fe(VI) Salts

Fe(VI) Salts	*Li_2FeO_4*	*Na_2FeO_4*	*K_2FeO_4*	*Rb_2FeO_4*	*Cs_2FeO_4*	*$SrFeO_4$*	*$BaFeO_4$*	*Ag_2FeO_4*
Charge storage	$3e^-$	$3e^-$	$3e^-$	$3e^-$	$3e^-$	$3e^-$	$3e^-$	$5e^-$
Intrinsic Capacity mAh/g	601	485	406	276	209	388	313	399

In accordance with eq 1.4, the discharge of alkaline Ag_2FeO_4 cathode, Zn anode Super-Iron cell, will be expressed as:

$$Ag_2FeO_4 + 5/2Zn \rightarrow 1/2Fe_2O_3 + 2Ag + 5/2ZnO \quad (1.5)$$

Alkaline super iron batteries, are readily studied in either a 'coin cell' or common cylindrical cell, such as the 'AAA cell' configuration, with the cathode composite formed by mixing a specified mass of Fe(VI) salt with an indicated weight percent of various carbon (carbon black or graphite) as the conductive matrix, or other additives. Super-Iron coin cells with a zinc anode were prepared using a conventional 1.1 cm diameter 'button' battery cell. The cells were opened, the anode and separator retained, and the cathode replaced with the new Fe(VI) cathode in the cell. In Super-Iron AAA experiments, components were removed from standard commercial alkaline cells (a cylindrical cell configuration with diameter 10.1mm, and a 42 mm cathode current collector case height), and the outer cathode MnO_2 mix, replaced with a pressed Fe(VI) mix, followed by reinsertion of the separator, Zn anode mix, gasket, and anode collector and resealing of the cell. Cells were discharged with a constant current, constant load or at a constant power, and the variation in time of cell's discharge potential was measured via LabView Data Acquisition. Cumulative discharge, as ampere hours or as watt hours, was determined by subsequent integration of the discharge current. The theoretical charge capacity is calculated from the measured mass of the Fe(VI) cathode salt using the (3 F mol^{-1}, F = Faraday, converted to ampere hours) values summarized in Table 1.1. The $3e^-$ Fe(VI) faradaic efficiency is determined by comparison of the measured cumulative ampere hours of discharge to the theoretical charge capacity.

High purity K_2FeO_4 and $BaFeO_4$ are readily synthesized. Coupled with a conventional Zn anode in a KOH electrolyte, the open-circuit potential of $BaFeO_4$ alkaline Super-Iron battery is approximately 1.85 V, and 0.1 V higher than that the 1.75 V potential of the K_2FeO_4 alkaline battery. On the basis of open-circuit potential and the mass of reactants in eq 1.3, the K_2FeO_4/Zn and $BaFeO_4$/Zn batteries have a respective maximum energy capacity of 475 Wh/kg, and 419 Wh/kg, both higher than the theoretical of 323 Wh/kg for the pervasive 1.55 V MnO_2/Zn alkaline battery.

Practical energy capacities are considerably less than the theoretical capacity. This discharge capacity further decrease at higher discharge rates. For example, consistent with Figure 1.1a, the observed experimental energy capacity of the conventional alkaline MnO_2/Zn alkaline battery is ~150 Wh/kg which falls to less than half that value at high rate (smaller ohmic load) conditions. The measured energy capacities of K_2FeO_4, $BaFeO_4$ and conventional MnO_2 cathode, alkaline primary batteries with a Zn anode are compared in Figure 1.1a.

In both the low- (6000 Ω, current density J = ~0.25 mA/cm^2) and high- (500 Ω, J = ~3 mA/cm^2) discharge domain, the K_2FeO_4 cell generates significantly higher capacity than does the MnO_2 cell. Of the three cells examined, the $BaFeO_4$ cathode cell exhibits the highest coulombic efficiency at high discharge rates (J > 10 mA/cm^2). Despite the lower intrinsic charge capacity of $BaFeO_4$ compared to K_2FeO_4, better $BaFeO_4$ charge transfer, results in the observed higher energy capacity. The benefit of the facile charge-transfer capabilities of the conductive $BaFeO_4$ salt is evident in a cylindrical cell configuration (Figure 1.1b). Discharged to 1 V a high constant power of 0.7 W, the $BaFeO_4$ cell provides 200% higher energy, compared to an advanced MnO_2 alkaline cylindrical cell (*6*).

2. Alkaline Fe(VI) K_2FeO_4 & $BaFeO_4$ Charge Transfer

2i. Zirconia coating stabilization of the alkaline K_2FeO_4 cathode

Zirconia's hydroxide mediation of Fe(VI/III) charge transfer

K_2FeO_4 exhibits higher solid state stability (< 0.1% decomposition per year) and higher intrinsic 3e$^-$ capacity than pure $BaFeO_4$, but the rate of charge transfer is higher in the latter. Charge transfer is effectively enhanced several fold in K_2FeO_4 by small additions of AgO, and AgO/K_2FeO_4 composite cathode exhibits higher capacity than pure $BaFeO_4$ and MnO_2 cathodes. However during cell storage, with or without silver additives, although the bulk Fe(VI) can remain active, the Fe(VI) in contact with the electrolyte forms a ferric overlayer over the bulk material. This overlayer is resistive, and tends to passivate the alkaline cathode towards further discharge.

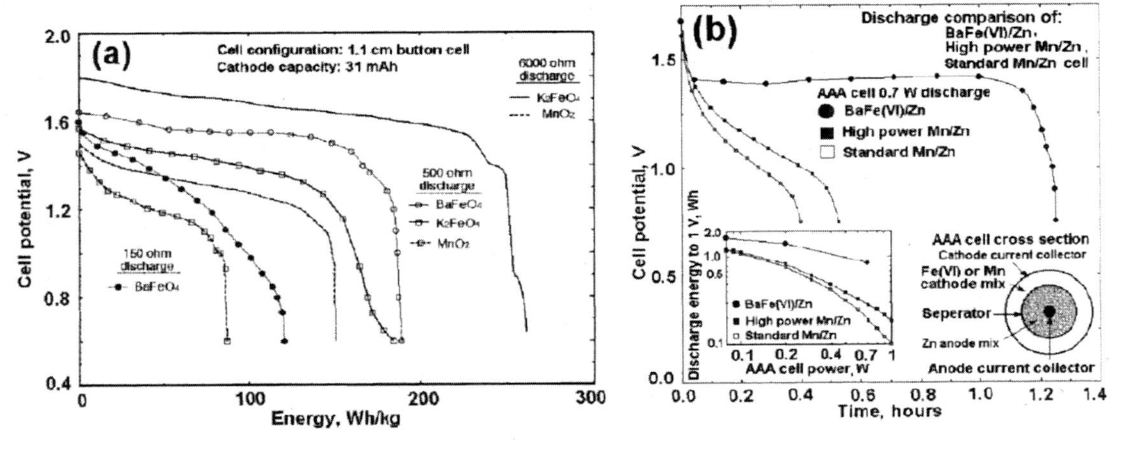

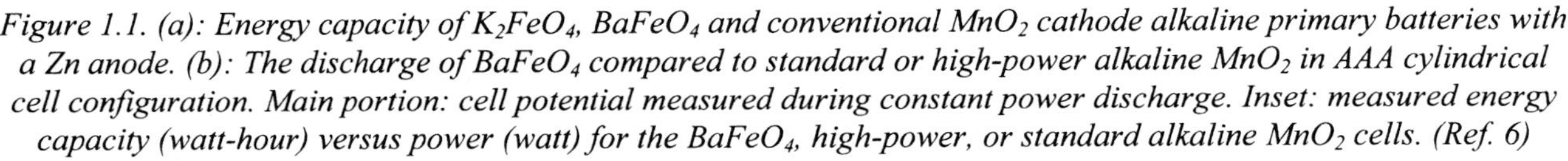

Figure 1.1. (a): Energy capacity of K_2FeO_4, $BaFeO_4$ and conventional MnO_2 cathode alkaline primary batteries with a Zn anode. (b): The discharge of $BaFeO_4$ compared to standard or high-power alkaline MnO_2 in AAA cylindrical cell configuration. Main portion: cell potential measured during constant power discharge. Inset: measured energy capacity (watt-hour) versus power (watt) for the $BaFeO_4$, high-power, or standard alkaline MnO_2 cells. (Ref. 6)

Due to its extreme stability over a wide temperature and environmental range, zirconia has been used as a protective coating for a variety of materials (*41, 42*). It has been explored to a lesser extent to protect in aqueous alkaline media, as typical zirconia deposition methods such as spray pyrolysis, plasma deposition, and colloidal deposition tend to deactivate or only partially cover electroactive surfaces (*41-43*). In aqueous alkaline media, zirconia is practically insoluble (Ksp = 8×10^{-52}) and stable (*44*). We developed a novel zirconia coating, derived from an organic soluble zirconium salt. The coating method and the formation/protection mechanism for zirconia coated alkaline cathodes are detailed in the reference 31.

The passivation of K_2FeO_4 cathode without zirconia coating is seen in Figure 2.1, in which the fresh pure K_2FeO_4 alkaline cathode discharges when a large fraction (25 wt%) of graphite is added as a supporting conductive matrix. However, the capacity decreases by an order of magnitude after 7 days of storage. A 1% zirconia coating dramatically improves the capacity after storage, which is further improved with a 5% KOH additive. As will be discussed in latter sections, a low level AgO additive to the cathode, not only facilitates charge transfer, sustaining an effective discharge with a smaller conducting support (10%, rather than 25% graphite), but as seen in the figure yields an even greater discharge capacity than the uncoated, fresh K_2FeO_4.

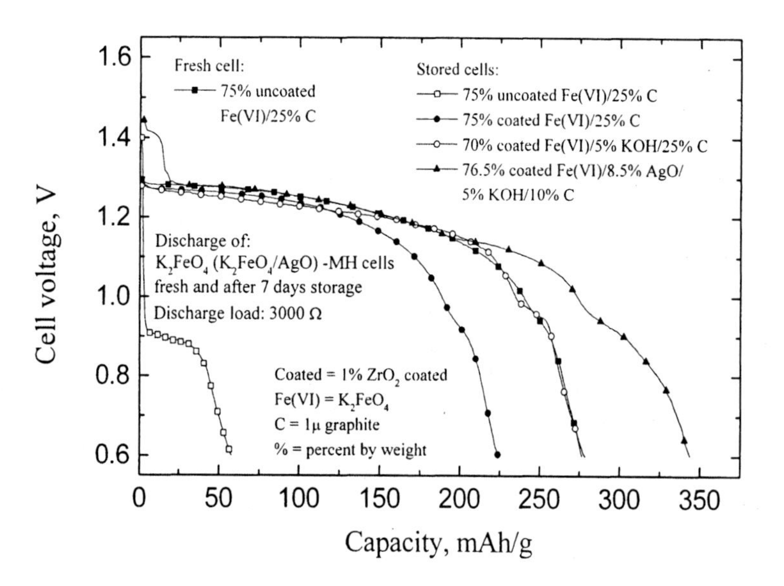

Figure 2.1. Discharge capacity of K_2FeO_4 (uncoated or coated, and K_2FeO_4 composite) –MH button cells fresh and after 7 days storage. (Ref. 31)

2ii. Chemical (manganese) mediation of Fe(VI) charge transfer

Inhibition of Fe(VI) Charge Transfer

The alkaline galvanostatic reduction of solution phase Fe(VI) on Pt generates an observable solid Fe(III) overlayer and sustains cathodic current densities of less than 100 $\mu A/cm^2$. As seen in comparison in Figure 2.2, solid Fe(VI) cathodes can sustain 2 orders of magnitude higher current density, highest when mixed with several percent of graphite.

As represented in the scheme included within the figure, Fe(VI) charge transfer inhibition, when occurring, appears directly related to buildup of a low conductivity Fe(III) reduction product at the Fe(VI)/cathode current collector interface. We have used FTIR, ICP, and X-ray powder diffraction to study the Fe(III) products (*12*). From the FTIR, these Fe(VI) discharge products contain hydroxide and Fe(III) salts, but the amorphous nature of the observed spectra does not lead to identification of the specific $M_aFe(III)O_x(OH)_y(H_2O)_z$ product ($M = K_2$ or Ba). It is reasonable to assume that the product will vary with pH, the extent of hydration, and the degree of Fe(VI) discharge; coexisting product stoichiometries may include:

$$Fe_2O_3 + xH_2O \Leftrightarrow 2FeOOH \cdot (x-1)/2H_2O \quad (2.1)$$

$$Fe_2O_3 + xBa(OH)_2 \Leftrightarrow Ba_xFe_2O_{(3+x)} \cdot yH_2O + (x-y)H_2O \quad (2.2)$$

$$Fe_2O_3 + 2xKOH \Leftrightarrow K_{2x}Fe_2O_{(3+x)} \cdot yH_2O + (x-y)H_2O \quad (2.3)$$

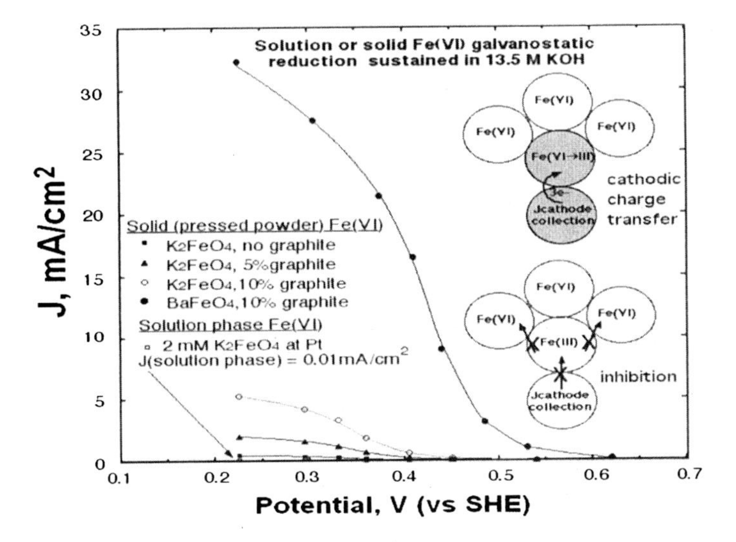

Figure 2.2. Galvanostatic reduction on Pt for 2 mM K_2FeO_4 in 13.5 M KOH or the indicated pressed Fe(VI) powders on Pt in 13.5 M KOH. (Ref. 20)

It is evident from the observed higher relative coulombic efficiencies of $BaFeO_4$ cathodes in Figure 1.1a, and from the high current densities sustained in

Figure 2.2, the barium product of a fresh $BaFeO_4$ reduction does not inhibit charge transfer to the degree of inhibition of the potassium product of K_2FeO_4 reduction.

Chemical Mediation of Fe(VI) Charge Transfer

We have probed the Mn(VI), Mn(VII) improvement of (both potassium and barium) Fe(VI) charge transfer (*8, 14*). A mechanism consistent with Mn(VII) facilitated charge transfer is proposed as Figure 2.3.

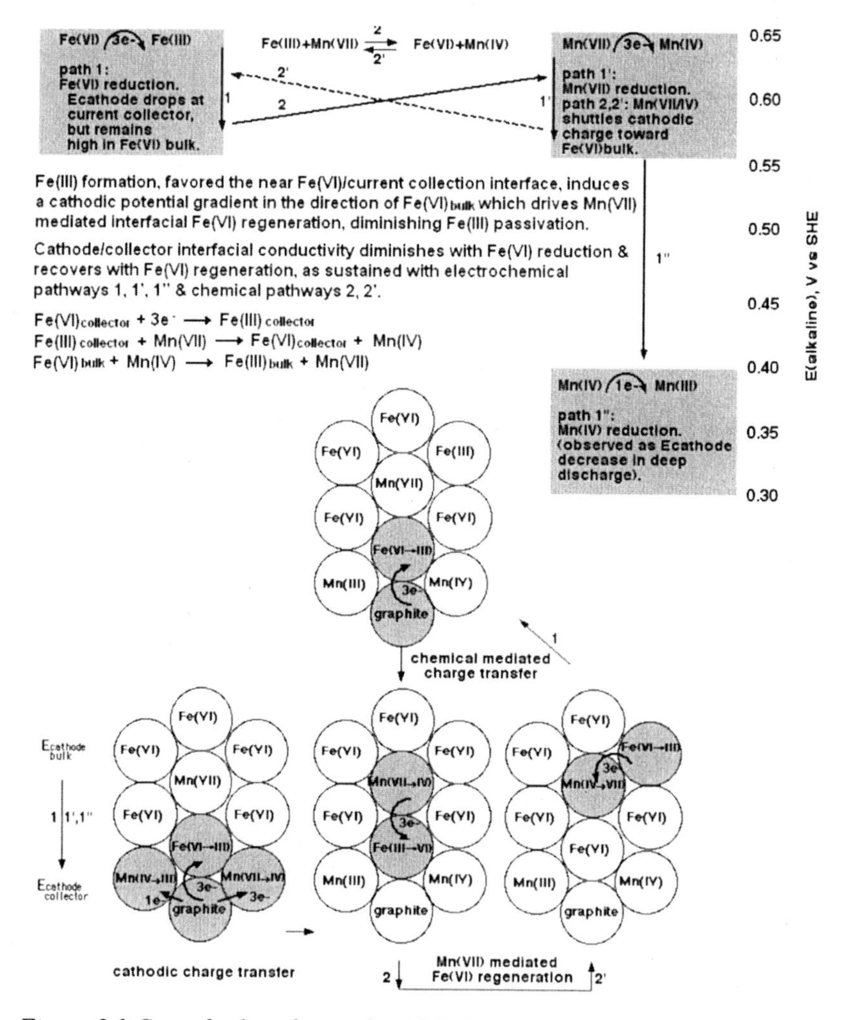

Figure 2.3 Co-cathode redox mediated Fe(VI) charge transfer, exemplified by Mn(VII) addition. Composites with near lying redox processes provide multiple pathways to facilitate Fe(VI) charge transfer; illustrated here by energy, mechanism, and chemical schematic representation. (Ref. 20)

The process utilizes the overlapping energetics of Mn(VII) and Fe(VI) redox chemistry to provide alternate pathways to minimize the effect of Fe(III) charged transfer inhibition and promoting Fe(VI) regeneration. Both Mn(VII) (shown in the figure and Mn(VI) (not shown in the figure) have a similar alkaline potential, and an analogous description for a manganate, Mn(VI), facilitated Fe(VI) process will be evident from eq 2.5, but for clarity is not included in the Figure 2.3 description. The driving force for the Fe(VI) regeneration, is the chemical and potential gradient that will spontaneously arise as the cathode discharges, and which as described in Figure 2.3, creates an anodic shift in more highly reduced portions of the cathode. Fe(III) at these sites is spontaneously regenerated to nonpassivating Fe(VI) by Mn(VII), as expressed by:

$$2MnO_4^- + Fe_2O_3 + 2OH^- \leftrightarrow 2MnO_2 + 2FeO_4^{2-} + H_2O \qquad (2.4)$$

or via manganate as expressed by:

$$3MnO_4^{2-} + H_2O + Fe_2O_3 \leftrightarrow 3MnO_2 + 2OH^- + 2FeO_4^{2-} \qquad (2.5)$$

This same potential gradient will drive Mn(IV) regeneration to Mn(VII) via interior (bulk) Fe(VI) and the reverse reaction in eq 2.4. This provides a charge shuttle to access bulk Fe(VI), which is only possible due to the near lying redox potentials of the Fe(VI/III) and Mn(VII/IV) half reactions. As detailed in Figure 2.3, the process may be summarized by the co-cathode chemical mediation of the Fe(VI/III) redox reaction to prevent Fe(VI) depletion near the cathode current collector and provide facile charge-transfer according to:

$$Fe(VI)_{collector} + 3e^- \rightarrow Fe(III)_{collector} \qquad (2.6)$$

$$Fe(III)_{collector} + Mn(VII) \rightarrow Fe(VI)_{collector} + Mn(IV) \qquad (2.7)$$

$$Fe(VI)_{bulk} + Mn(IV) \rightarrow Fe(III)_{bulk} + Mn(VII) \qquad (2.8)$$

K_2FeO_4/Mn(VII or VI) Composite Super-Iron Batteries

The synergistic improvement of a K_2FeO_4 cathode with either $KMnO_4$ or $BaMnO_4$ is presented in the top 2 section of Figure 2.4 and further detailed in Table 2.1. As seen in Table 2.1, besides $KMnO_4$ and $BaMnO_4$, The K_2FeO_4 cathode can also be improved by inclusion of other activators individually or together with Mn(VII or VI). As detailed in the table, added salts, such as LiOH or NaOH do not increase a K_2FeO_4 cathode discharge. Also included in the table are relatively small, but significant, improvements of the K_2FeO_4 discharge energy with 10% addition of CsOH and $Ba(OH)_2$.

A composite K_2FeO_4/$BaMnO_4$ cathode yields a significantly higher energy capacity than pure K_2FeO_4 cathode. As indicated in the top section of Figure 2.4 as the open or solid small circles, added barium manganate can significantly enhance the discharge energy of the K_2FeO_4 cathode. This is a synergistic effect, increasing the energy of either pure cathode alone. For the K_2FeO_4/$BaMnO_4$ composites, a maximum 2.8 Ω discharge energy of 0.78 Wh is measured for the cell containing 45 wt% K_2FeO_4 and 55 wt% $BaMnO_4$, which is more than double that seen for the pure K_2FeO_4 cathode. As detailed in Table 2.1 at the low rate (constant 75 Ω load), the $K_2FeO_4/BaMnO_4$ composite cathode exhibits a nearly constant maximum energy capacity of 1.2 Wh over a wide composition range varying from 33%:67% to 67%:33%. KOH or Al_2O_3 impairs the discharge effectiveness of $BaMnO_4/K_2FeO_4$ composite cathode. But $Ba(OH)_2$ modestly increases the discharge energy of the composite $BaMnO_4/K_2FeO_4$ cathode due to the improvements of $Ba(OH)_2$ on the pure K_2FeO_4 electrode (*20*).

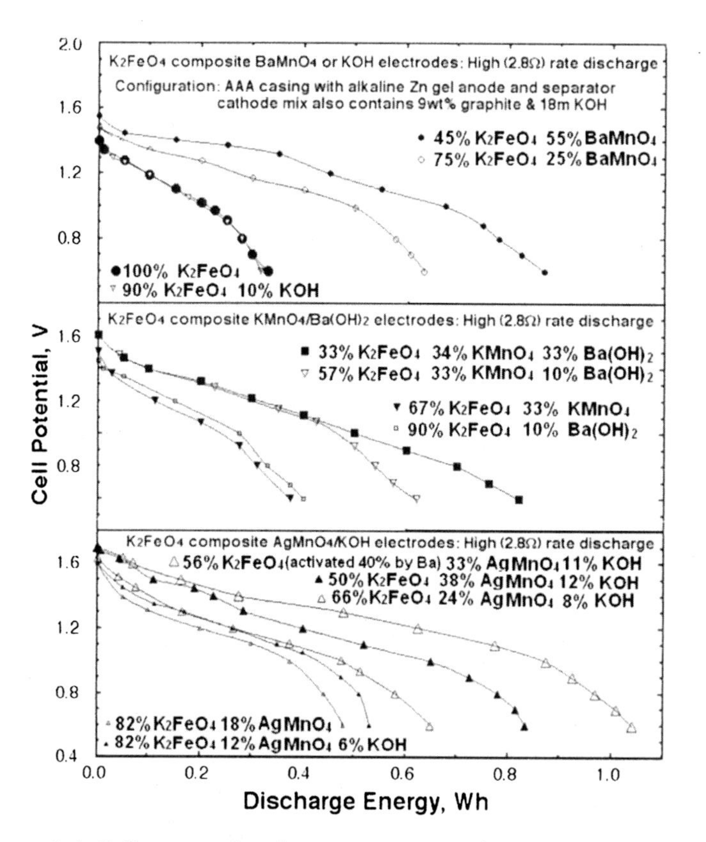

Figure 2.4. Cell potential and energy capacity of alkaline cells with K_2FeO_4 composite cathodes containing various relative amounts of $BaMnO_4$, $KMnO_4$, $AgMnO_4$, $Ba(OH)_2$, or KOH, compared to K_2FeO_4 in the cathode mix, during discharge at a high constant load rate of 2.8 Ω. In the composite cells, the combined mass of the K_2FeO_4 and other salts is intermediate to the mass of the pure salts (a pure cathode contains 3.5g K_2FeO_4, 4.2g $BaFeO_4$, 4.1g $BaMnO_4$, 3.5g $KMnO_4$, or 4.6g $AgMnO_4$). Cells use an alkaline AAA configuration including 9 wt% graphite and 18 M KOH electrolyte. (Ref. 20)

Table 2.1. Comparison of the Discharge Behavior in Alkaline AAA Cell of a Cathode Composite Containing $K_2FeO_4/BaMnO_4$ or $K_2FeO_4/KMnO_4$ (dry cathode composition, by mass discharge to 0.8 V, at constant load) (*Ref. 20*)

						2.8Ω		*75Ω*	
Fe Salt	*Wt %*	*Mn Salt*	*Wt %*	*Salt*	*Wt %*	*E* (Wh)	V_{av} (V)	*E* (Wh)	V_{av} (V)
K_2FeO_4	100					0.28	1.17	0.68	1.36
K_2FeO_4	90			$Ba(OH)_2$	10	0.35	1.15	0.79	1.44
K_2FeO_4	90			LiOH	10	0.30	1.09	0.63	1.31
K_2FeO_4	90			NaOH	10	0.24	1.05	0.60	1.35
K_2FeO_4	90			CsOH	10	0.34	1.17	0.83	1.44
K_2FeO_4	0	$BaMnO_4$	100			0.37	1.16	0.96	1.19
K_2FeO_4	5	$BaMnO_4$	95			0.44	1.16	1.03	1.26
K_2FeO_4	10	$BaMnO_4$	90			0.54	1.17	1.14	1.31
K_2FeO_4	25	$BaMnO_4$	75			0.59	1.17	1.16	1.39
K_2FeO_4	33	$BaMnO_4$	67			0.65	1.19	1.20	1.40
K_2FeO_4	45	$BaMnO_4$	55			0.78	1.20	1.20	1.41
K_2FeO_4	50	$BaMnO_4$	50			0.67	1.20	1.20	1.41
K_2FeO_4	67	$BaMnO_4$	33			0.66	1.20	1.19	1.44
K_2FeO_4	75	$BaMnO_4$	25			0.57	1.22	1.12	1.45
K_2FeO_4	90	$BaMnO_4$	10			0.38	1.21	0.98	1.45
K_2FeO_4	95	$BaMnO_4$	5			0.38	1.19	0.71	1.43
K_2FeO_4	33	$BaMnO_4$	57	Al_2O_3	10	0.62	1.19	1.10	1.37
K_2FeO_4	57	$BaMnO_4$	33	Al_2O_3	10	0.56	1.17	1.13	1.43
K_2FeO_4	10	$BaMnO_4$	57	$Ba(OH)_2$	33	0.44	1.26	0.68	1.28
K_2FeO_4	33	$BaMnO_4$	34	$Ba(OH)_2$	33	0.60	1.28	0.92	1.51
K_2FeO_4	33	$BaMnO_4$	57	$Ba(OH)_2$	10	0.81	1.27	1.17	1.41
K_2FeO_4	50	$BaMnO_4$	25	$Ba(OH)_2$	25	0.78	1.28	1.15	1.51
K_2FeO_4	57	$BaMnO_4$	33	$Ba(OH)_2$	10	0.81	1.29	1.21	1.46
K_2FeO_4	57	$BaMnO_4$	10	$Ba(OH)_2$	33	0.76	1.29	1.27	1.61
K_2FeO_4	57	$KMnO_4$	10	$Ba(OH)_2$	33	0.63	1.26	1.04	1.16
K_2FeO_4	50	$KMnO_4$	50			0.47		1.41	

The discharge of $K_2FeO_4/KMnO_4$ composite cathodes was optimized with various additives. Here, the $Ba(OH)_2$ was the most effective additive, and the high rate discharge is summarized in the midsection of Figure 2.4. In the presence of both Mn(VII) salts and Fe(VI), competing alkali earth hydroxide effects are complex. For example, in the reaction with $Ba(OH)_2$, the $KMnO_4$ reaction product $Ba(MnO_4)_2$ is highly soluble. However, the alternative reaction K_2FeO_4 with $Ba(OH)_2$ forms $BaFeO_4$, which is insoluble in water. As seen in the midsection portion of Figure 2.4, the addition of $Ba(OH)_2$ to the $K_2FeO_4/KMnO_4$ cathode results in a significant increase in discharge energy, and at an average discharge potential greater than that observed for the $K_2FeO_4/KMnO_4$ composite without $Ba(OH)_2$. At both high and low rate, a maximum discharge energy is observed with the 33:57:10 wt.% K_2FeO_4:$KMnO_4$:$Ba(OH)_2$ composition which provides 0.73 and 1.62 Wh respectively over either 2.8 Ω or 75 Ω load discharges (*20*).

$AgMnO_4$ provides an unusual salt in that the Ag valence acts in a manner intermediate to Ag(I) and Ag(II), that is as for Ag(I + x)Mn(VII - x)O_4, where $0 < x < 1$ (*16, 39*). Of the permanganate and manganate salts explored to date, $AgMnO_4$ promotes one of the larger increases in the K_2FeO_4 alkaline cathode discharge, a phenomenon consistent with the observed Ag activation of Fe(VI) (will be presented in next section), but the $AgMnO_4$ activation phenomenon is only substantial in the presence of KOH (added as a solid salt to the mix) (*16*). This is observed in the lowest section of Figure 2.4. In the absence of KOH, the 2.8 Ω discharge of the K_2FeO_4 cathode increases from to 0.3 Wh to ~0.4 Wh with addition of 18% $AgMnO_4$, but is enhanced to 0.5 Wh using only 12 wt% $AgMnO_4$ with KOH (6 wt%). This increases to ~0.8 Wh with inclusion of 38 wt% $AgMnO_4$ and 12 wt% KOH. Finally, as also seen in the figure, a K_2FeO_4, partially converted to the barium salt with a $Ba(OH)_2$ wash and mixed with $AgMnO_4$ and KOH, provides a cathode with a high rate discharge similar to the desired capacity of the $BaFeO_4$ cathode, exhibiting a higher energy capacity, but lower average discharge potential. Aspects of the interesting KOH activation of the pure $AgMnO_4$ alkaline cathode (without K_2FeO_4) are explored in reference 16, and the $AgMnO_4$ activation of K_2FeO_4 is still lower than that observed for the AgO mediation of K_2FeO_4 charge transfer in a later section.

$BaFeO_4$/Mn(VII) Composite Super-Iron Batteries

The 75 Ω energy capacity of pure $BaFeO_4$ is 1.75 Wh with a discharge profile highly similar to that of the 95% $BaFeO_4$/5% $KMnO_4$ composite cathode (capacity of 1.8 Wh to a 0.8V discharge) detailed in the top section of Figure 2.5. As seen in the figure, cells containing high fractions of both $BaFeO_4$ and $KMnO_4$ discharge to a higher capacity. Hence, a cathode with either 50:50 or 75:25 relative weight percent of $BaFeO_4$ to $KMnO_4$ yields respective discharge capacities of 1.83 and 1.95 Wh. Utilization of a CsOH, rather than KOH electrolyte further enhanced the discharge energy to 2.1-2.2 Wh. The resultant volumetric capacity of 600 Wh/cc is high compared a maximum 400 Wh/cc for a high performance MnO_2 cathode alkaline AAA cell.

$BaFeO_4/MnO_2$ Composite Super-Iron Batteries

Due to restrictions on the use of barium salts by the US EPA (*40*), it is of interest to probe the $MnO_2/BaFeO_4$ composite cathodes with reduced barium levels. As seen in Figure 2.6, due to the lower average voltage discharge for the

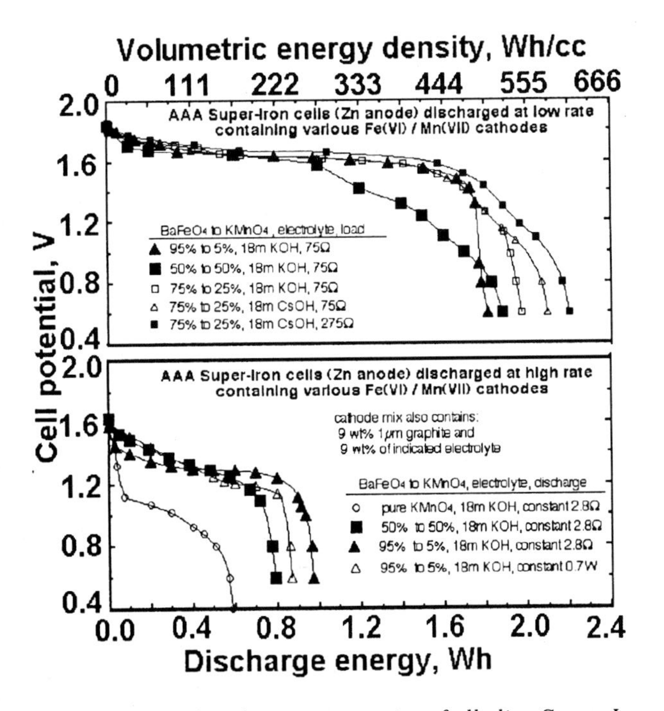

Figure 2.5. Cell potential and energy capacity of alkaline Super-Iron AAA cells. Cells contain various indicated weight fractions of Fe(VI), Mn(VII), and Mn(VI) salts, in the cathode mix, and use either a KOH or CsOH electrolyte. (Ref. 14)

pure MnO_2 cathode, its 0.7 W discharge is diminished compared to the 2.8 Ω. In comparison, high capacities are shown for both a constant load (2.8 Ω) discharge and for a constant power (0.7 W) discharge, for the cell containing a cathode composed primarily of $BaFeO_4$. As also presented in the figure, $BaFeO_4/MnO_2$ composites, containing less barium, exhibit significantly higher discharge energy compared to the MnO_2 cathode in conventional alkaline cells, and intermediate to the values observed for the MnO_2 free, $BaFeO_4$ cathode. Under the same 2.8 Ω load, high rate conditions, the 3:1 composite $MnO_2/BaFeO_4$ cathode cells yields 0.78 Wh to an 0.8 V discharge cut-off, providing ~30% additional capacity compared to the 2.8 Ω pure MnO_2 cell discharge, and more than double the capacity of the pure MnO_2 0.7 W discharge. The 1:1 composite $MnO_2/BaFeO_4$ cathode cells yield 0.85 Wh to an 0.8 V discharge cut-off, providing ~40% additional capacity compared to the 2.8 Ω pure MnO_2 cell discharge, and triple the capacity of the pure MnO_2 0.7 W discharge. Note: a 3:1 "MnO_2" to "Ba" composite cathode was prepared with 75% by mass of the dry MnO_2 cathode mix and 25% of the dry $BaFeO_4$ cathode mix, for a total dry mass of 4.9 g; a 1:1 "MnO_2" to "Ba" composite cathode was prepared with 50% by mass of the dry MnO_2 cathode mix and 50% of the dry $BaFeO_4$ cathode mix, for a total dry mass of 4.8 g. A 0.4 g of saturated KOH is added to each of the various cathode mixes (*19*).

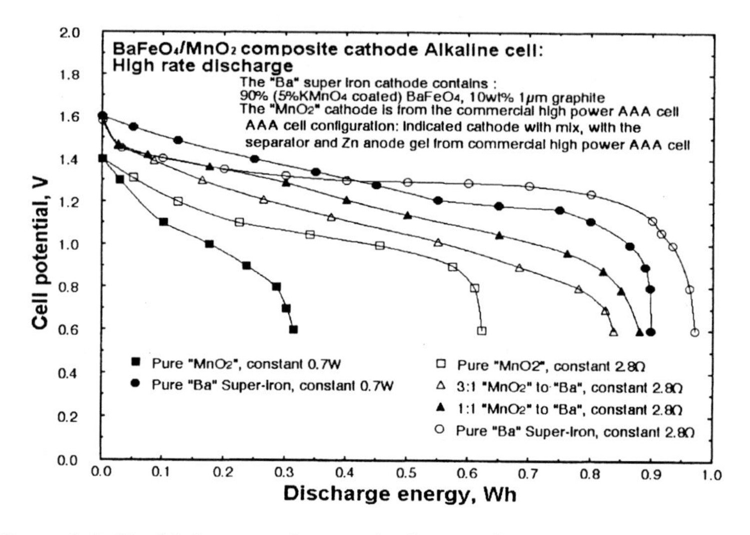

Figure 2.6. The high power domain discharge of Super-Iron ($BaFeO_4$ and $BaFeO_4/MnO_2$ composites) compared to the discharge of MnO_2 cathode cells, each in a cylindrical (AAA) cell configuration. (Ref. 19)

2iii. Electronic (silver) mediation of Fe(VI) Charge Transfer Highest energy and power alkaline Super-Iron Batteries

Simultaneous Chemical & Electronic Mediation of Fe(VI) Charge Transfer

As previously discussed, high valence manganese centers can chemically mediate Fe(VI) cathode charge transfer. High valence silver centers can better this process, providing simultaneous pathways for both the chemical and electronic mediation, and further improvement, of the $3e^-$ alkaline reduction of Fe(VI). For example, it will be shown that cathodic current densities sustained by an alkaline K_2FeO_4 cathode, can approach the very high rates previously only observed for $BaFeO_4$, when Ag(II) oxides are added. Figure 2.7 is a model of the process for co- electronic and redox mediated Fe(VI) charge transfer in which conductive composites, with energetically near lying redox processes, provide multiple pathways to facilitate Fe(VI) charge transfer (*20*). In this figure, the process is exemplified by addition of an Ag(II) salt, in which the cathode/collector interfacial conductivity (a) increases with Ag(I) and Ag(0) formation, (b) diminishes with Fe(VI) reduction, and (c) recovers with Fe(VI) regeneration. An Ag(II) salt, such as AgO, as an Fe(VI) co-cathode is analogous to Mn(VII) and Mn(VI) as it exhibits intrinsic 2 separate alkaline cathodic redox couples in the same potential domain as the single $3e^-$ Fe(VI) redox couple.

$$AgO + 1/2H_2O + e^- \rightarrow 1/2Ag_2O + OH^- \quad E = 0.6V \quad VS\ SHE \qquad (2.9)$$
$$1/2Ag_2O + 1/2H_2O + e^- \rightarrow Ag + OH^- \quad E = {\sim}0.35V \quad VS\ SHE \qquad (2.10)$$

In comparison to the chemical mediation model of Figure 2.3, in Figure 2.7 the additional pathway for Fe(VI) charge transfer is evident in the mid-left chemical schematic labeled electronic mediated charge transfer. Concurrent with increasing cathode consumption is an increasing buildup of conductive Ag, providing electronic access to bulk Fe(VI). Silver is a superlative metallic conductor; as the AgO discharges, the concentration of reduced silver grows and provides a growing conductive matrix to increasingly facilitate the Fe(VI) reduction (*20*).

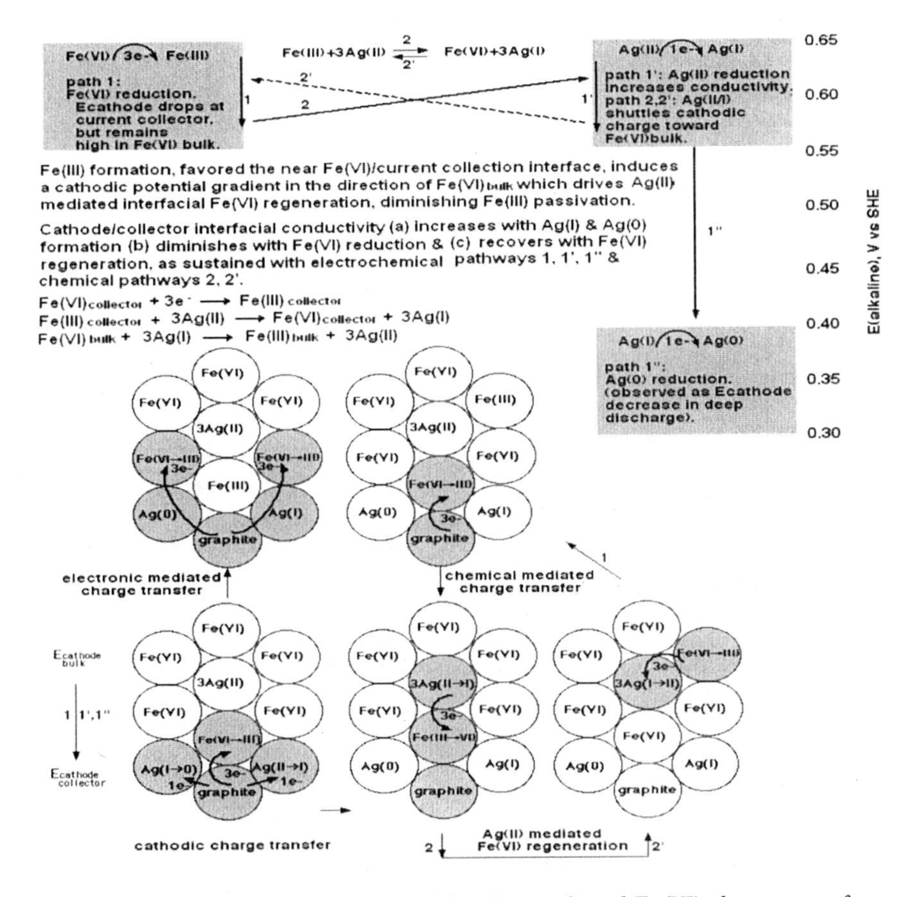

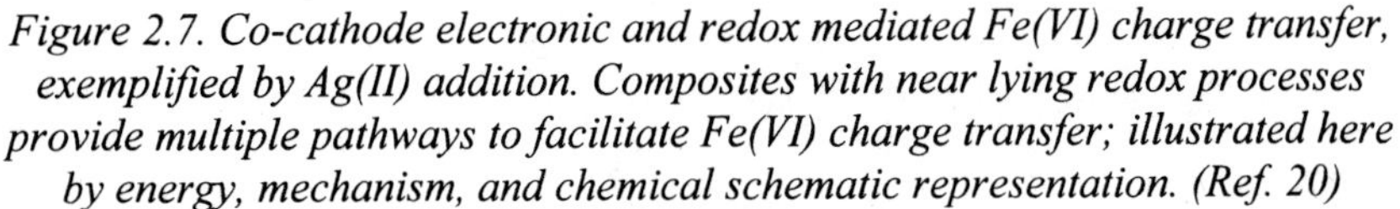
Figure 2.7. Co-cathode electronic and redox mediated Fe(VI) charge transfer, exemplified by Ag(II) addition. Composites with near lying redox processes provide multiple pathways to facilitate Fe(VI) charge transfer; illustrated here by energy, mechanism, and chemical schematic representation. (Ref. 20)

$MFeO_4$ (M=K_2 or Ba)/AgO Composite Cathode Super-Iron Batteries

The activation of $BaFeO_4$ and K_2FeO_4 by AgO in Figure 2.8 is substantial compared to that by $KMnO_4$ or $BaMnO_4$ observed in either Figure 2.4, Figure 2.5 or Table 2.1. As seen in Figure 2.8b, as little as 7 wt% AgO composite with K_2FeO_4 yields a discharge energy comparable or larger than the 50 wt% the $KMnO_4/K_2FeO_4$ composite cathode. At larger AgO fractions, high rate discharge energies as great as 1.5 Wh are observed. These discharge energies are substantially higher than conventional alkaline MnO_2, and are also higher than AgO (or $BaFeO_4$) alone.

The AgO has a synergistic activation on $BaFeO_4$ or K_2FeO_4 in which the combined discharge capacity of the composite Ag(II)/Fe(VI) cathode is larger than that of either cathode alone. As with K_2FeO_4, an unusually high energy discharge also occurs for an AgO cathode composite with the $BaFeO_4$ Fe(VI) salt. In Figure 2.8a $BaFeO_4$/AgO cathode mix maintains the unusual high-power characteristic known for the $BaFeO_4$ cathode without AgO. Hence, the $BaFeO_4$ cathode both with, and without, AgO generates a power of at least 0.7 W over a constant 2.8 Ω load. However, in addition, the Fe(VI) composite cathode unexpectedly discharges for ~170 min and generates 1.5 Wh, whereas under the same conditions, the $BaFeO_4$ cathode without AgO discharges for ~80 to 90 min and generates 0.9 Wh. The $BaFeO_4$/AgO composite cathode exhibits a maximum discharge energy higher than either component alone. The discharge capacity is ~5-fold higher than the equivalent constant power discharge of the conventional alkaline MnO_2 cell, or ~3- fold higher than a constant resistive load discharge (*20*).

Constant Power Comparison of MnO_2, $BaFeO_4$, and AgO/K_2FeO_4 Cathodes

The unusually high specific energy/specific power of various Fe(VI) alkaline batteries is summarized in the inset of Figure 2.8a. Of relevance to both practical electronics and as a fundamental energy comparison, a constant power density, rather than constant load or constant current density, is a more stringent comparison of cathode capabilities. In this discharge the lower average cathode potential of the MnO_2 cathode compared to Fe(VI) must be compensated by a higher average current density, and this will further impair the MnO_2 charge transfer. As in Figure 1.1, under conditions of constant, rapid 0.7 W discharge in an AAA cell configuration, the MnO_2 discharges to a maximum of 0.52 h (0.36 Wh), whereas a 5% $KMnO_4$/95% $BaFeO_4$ cathode (containing (4.0g $BaFeO_4$) discharges for 1.26 h to 0.88 Wh (*12*). Under the same conditions, for the composite AgO/K_2FeO_4 cathodes, a 8 wt% (0.3 g) AgO/92 wt% K_2FeO_4 cell discharges for 1.28 h to 0.90 Wh, a 20 wt% (0.7 g) AgO/80 wt% K_2FeO_4 cell discharges for 1.58 h to 1.11 Wh, and a 39 wt% (1.5 g) AgO/61 wt% K_2FeO_4 cell discharges for 2.13 h to 1.49 Wh (*20*).

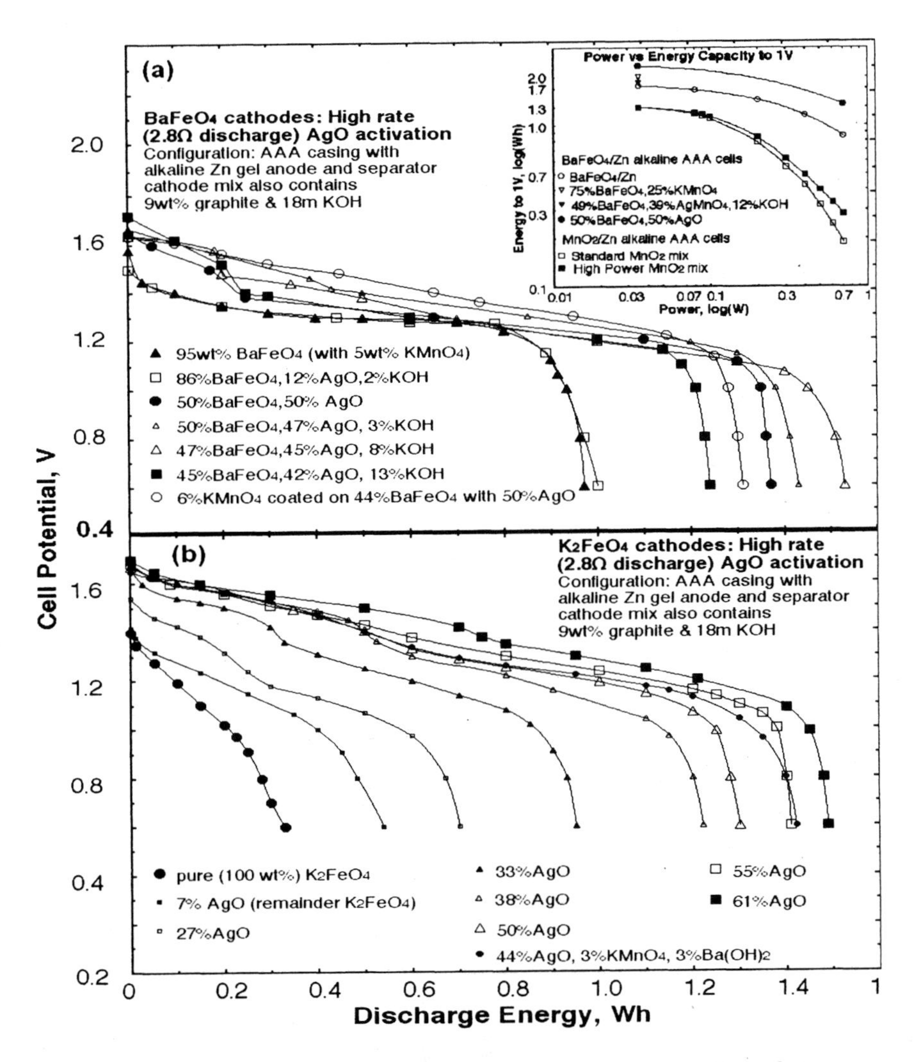

Figure 2.8. Cell potential and energy capacity of alkaline cells with (a): $BaFeO_4$, (b): K_2FeO_4 cathode composites containing various weight fractions of AgO during discharge at a high constant load rate of 2.8 Ω. Cells use an alkaline AAA configuration including in the cathode 9 wt% graphite and 18 M KOH electrolyte. Inset of (a): Energy at various powers during discharge of these and comparative cells. (Ref. 20)

2iv. Titanate, Ba(II), Mn(IV) and Co(III) Modifiers

Discharge Efficiency of K_2FeO_4, $BaFeO_4$ Alkaline Super-Iron Batteries with $SrTiO_3$, $Ba(OH)_2$ Additives

At high load (current densities above 1 mA/cm^2) the faradaic efficiency of Fe(VI) reduction is significantly higher for a pure $BaFeO_4$ cathode compared to a pure K_2FeO_4 cathode, as shown in Figure 1.1. However, additives can considerably alter (facilitate) the charge transfer behavior of the Fe(VI) cathodes. In KOH electrolytes, titanates (Ti(IV) salts) have been utilized to improve the high current density utilization of the one-electron reduction of MnO_2 (*37*). A similar phenomenon is also observed for the K_2FeO_4 faradaic efficiency. The improvement in K_2FeO_4 charge transfer was observed for a $SrTiO_3$ additive, which as presented in Figure 2.9 under the indicated 500 Ω discharge load (equivalent to a current density of ~3 mA/cm^2), improves the three-electron utilization from ~68 to 77% (as measured by the fraction of the available 31 mAh discharged to a 0.8 V cutoff).

The fundamental solubility and stability of K_2FeO_4 and $BaFeO_4$ in alkaline electrolyte have been probed (*7*). K_2FeO_4 solubility may be controlled over several orders of magnitude through judicious choice of alkali solution, and that very low (sub millimolar) K_2FeO_4 solubility can be achieved in a KOH electrolyte also containing $Ba(OH)_2$. In highly concentrated hydroxide, both KOH and CsOH electrolytes suppress K_2FeO_4 solubility, $BaFeO_4$ is insoluble in water, and has a solubility of less than 2×10^{-4} M in 5M KOH containing $Ba(OH)_2$. The resultant very low Fe(VI) solubility is advantageous, diminishing the possibility for self discharge in the Super-Iron battery (*7*). Electrolyte choices for alkaline batteries is based on the low solubility of Fe(VI) salts. Similarly, a small amounts of $Ba(OH)_2$ added to the Fe(VI) cathode can constructively enhance the Fe(VI) insolubility in alkaline electrolyte. To a K_2FeO_4 cathode, added $Ba(OH)_2$ can enhance charge transfer of the battery due to the partial conversion of K_2FeO_4 to $BaFeO_4$ in the alkaline electrolyte (*7, 10*).

$$K_2FeO_4 + Ba(OH)_2 \rightarrow BaFeO_4 + 2KOH \qquad (2.11)$$

Mn(IV), Co(III) Modifiers and Their Effects on Super-Iron Batteries

The Fe(VI) battery potential can be shifted and controlled by specific modifiers. Small modifications (on the order of a 0.1 to 10% by weight) of the Fe(VI) cathode can be used to either substantially enhance or diminish the discharge potential. Specifically, as shown in Figure 2.10a, the average discharge potential of the $BaFeO_4$ cell is increased by an average of ~150 mV when a few percent of Co_2O_3, is mixed with the $BaFeO_4$. Alternately, the average discharge potential is decreased by an average of ~200 mV when 1% of MnO_2 is used to modify the $BaFeO_4$.

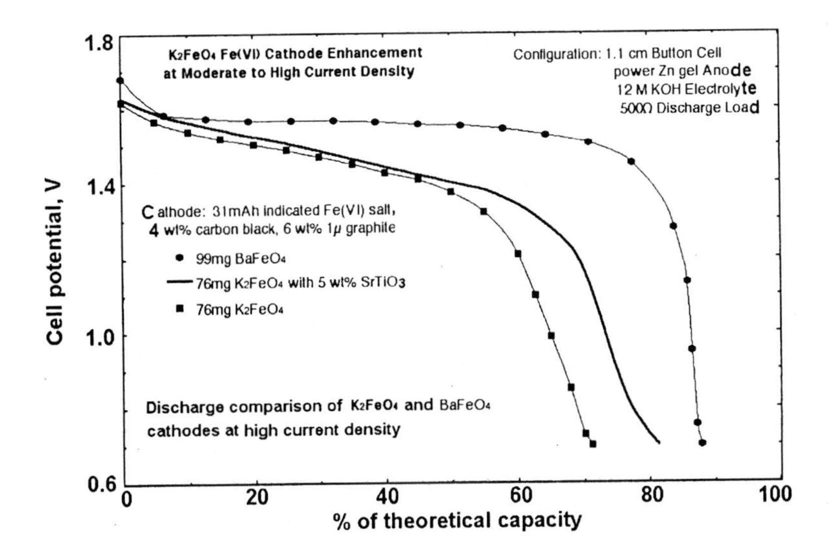

Figure 2.9. The improved faradaic efficiency of a K_2FeO_4 cathode cell with a titanate additive. The percent of theoretical capacity is determined by the measured cumulative ampere hours, compared to the calculated ampere hours in the 3 F mol^{-1} measured mass of the Fe(VI) salt. (Ref. 10)

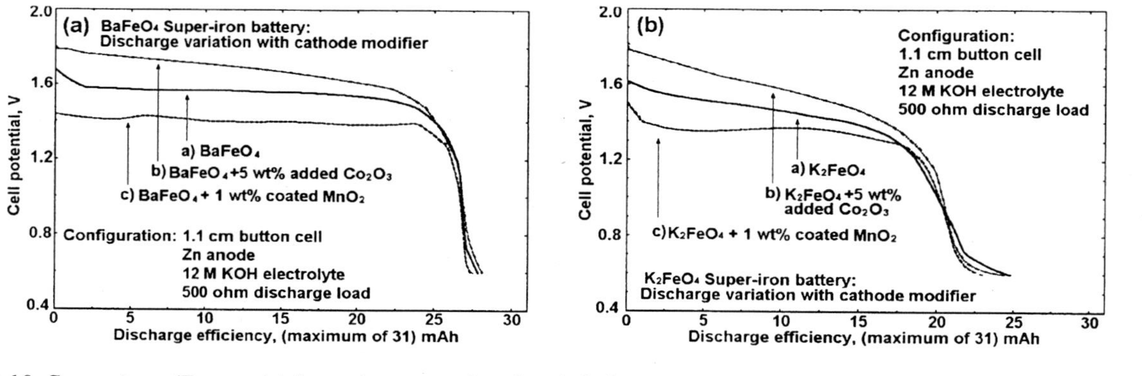

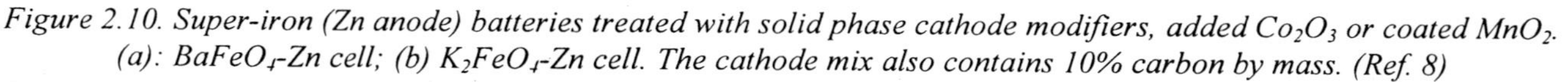
Figure 2.10. Super-iron (Zn anode) batteries treated with solid phase cathode modifiers, added Co_2O_3 or coated MnO_2. (a): $BaFeO_4$-Zn cell; (b) K_2FeO_4-Zn cell. The cathode mix also contains 10% carbon by mass. (Ref. 8)

Interestingly, these modifiers do not significantly affect the open circuit cathode potential of 1.84 to 1.89 V measured versus the Zn anode (*8*). Nor do these modifications significantly affect the high faradaic efficiency of the three-electron reduction. Such effects are useful to provide control of the super iron battery discharge potential. These effects are also generalized to K_2FeO_4 cathodes (Figure 2.10b).

Chemical mediation of Fe(VI) charge transfer with permanganate and manganate were presented in other sections. Here, We take advantage of Mn(VII), Mn(VI) remediation of Fe(VI) to coat the Fe(VI) particles with a small overlayer of MnO_2, which can have a dramatic effect on the discharge potential of the Super-Iron battery. The coating procedure is detailed in reference 8. Normally, if K_2FeO_4 powder is coated with an excess of manganate or permanganate, the cell potential is not significantly affected. However, when the K_2FeO_4 powder is coated to a low level, as seen in Figure 2.10b with the 1% $KMnO_4$-coated K_2FeO_4 powder, the cell exhibits a discharge potential diminished by 150 mV.

The following proposed model appears to be consistent with the observed decrease in potential of this Mn-coated super-iron: the decreased Fe(VI) potential is due to the intermediate reduction of Mn(IV) surface states which are electrocatalytically active in accord with eq 2.12:

$$3MnO_2 + 3/2H_2O + 3e^- \rightarrow 3/2Mn_2O_3 + 3OH^- \quad E = 0.3 \text{ V vs. SHE} \quad (2.12)$$

Following reduction, discharged manganese sites are reactivated by the remaining Fe(VI) in the chemical step:

$$3/2Mn_2O_3 + FeO_4^{2-} + H_2O \rightarrow 3MnO_2 + 1/2Fe_2O_3 + 2OH^- \quad (2.13)$$

The renewed MnO_2 can then continue to catalyze the Fe(VI) reduction, in accord with the electrochemical and chemical steps summarized by a repetitions of eq 2.12 and eq 2.13. In accord with this mechanism, Fe(VI) reduction will occur at a potential shifted towards the MnO_2 reduction potential, nevertheless accessing the intrinsic faradaic capacity of the Fe(VI) cathode.

Understanding of the observed Co(III) effect of a significant potential increase shown in Figure 2.2 is facilitated by review of Co alkaline electrochemistry. Of significance is the emphasis by Gohr (*38*), indicating a Co(IV) oxide may be reduced in alkaline solutions at a potential which is 0.1 to 0.2 V higher than the Fe(VI) potential, and therefore is consistent with our observed increase in the Fe(VI) cathode discharge potential:

$$CoO_2 + 2H_2O + e^- \rightarrow Co(OH)_3 + OH^- \quad E = 0.7 \text{ V vs. SHE} \quad (2.14)$$

Eq 2.14 indicates that the Fe(VI) potential is insufficient for Co(III) oxidation to Co(IV). Nevertheless the potentials are close, and one hypothesis is that a Nernst shift of potentials, sufficient for Fe(VI) to oxidize Co(III), may occur, for example, in an excess of Fe(VI), and at the electrode | concentrated hydroxide interface. This hypothesis suggests a mechanism for the observed Co(III) enhancement of the super-iron potential via an electrochemically active Co(IV) intermediate state:

$$3Co(OH)_3 + FeO_4^{2-} \rightarrow 3CoO_2 + 1/2Fe_2O_3 + 7/2H_2O + 2OH^- \qquad (2.15)$$

The electrochemical discharge of the Co(IV) intermediate follows eq 2.14 and thereby releases Co(III) for continued catalysis of the Fe(VI) reduction in accord with eqs 2.14 and 2.15 (*8*).

3. Alkaline Fe(VI) $Na(K)FeO_4$, $Rb(K)FeO_4$, Cs_2FeO_4, $SrFeO_4$ or Ag_2FeO_4 Charge Transfer

To expand the library of Super-Iron cathode salts beyond K_2FeO_4 and $BaFeO_4$, other Fe(VI) salts including the alkali salts $Na(K)FeO_4$, $Rb(K)FeO_4$, Cs_2FeO_4, the alkali earth salt $SrFeO_4$, and a transition metal Fe(VI) salt Ag_2FeO_4 have been successfully synthesized in our lab. This section focuses the cathodic charge transfer of these Fe(VI) salts in alkaline battery systems.

Figure 3.1a compares the constant load discharge of Super-Iron batteries containing the Na, K, Rb, Cs and Ba Super-Iron cathodes discharged at the same constant load conditions (*25*). As observed in the figure, the alternate Cs, Rb and Na mix cathodes discharge to a significant fraction of their respective intrinsic capacities of 209 mAh/g for Cs_2FeO_4, 290 mAh/g for $Rb_{1.7}K_{0.3}FeO_4$, and 445 mAh/g for $Na_{1.1}K_{0.9}FeO_4$. Each of the Fe(VI) cathodes is similar in discharge potential, but does not generate quite as a high coulombic efficiency as the $BaFeO_4$ cathode (*25*).

Figure 3.1b compares the discharge of various pure (K_2FeO_4, Cs_2FeO_4, and $BaFeO_4$), not mixed ($Na_{1.1}K_{0.9}FeO_4$ or $Rb_{1.7}K_{0.3}FeO_4$) alkaline Fe(VI) cathodes in a configuration utilizing as a constraint a fixed volume of cathode, as compared to the configuration which utilized a fixed mass (300 mg) of cathode. The fixed cathode volume constraint of this cylindrical AAA cell configuration favors a higher mass packing of the cesium (4.9 g Cs_2FeO_4) and barium (4.4g $BaFeO_4$) compared to potassium (3.2g K_2FeO_4), which is due to the lower density of the latter. However, this can be compensated for by the substantially higher intrinsic $3e^-$ 406 mAh/g capacity for the reduction of the potassium, compared to the cesium, or barium Fe(VI) salts. As seen in Figure 3.1b, the barium Fe(VI) cathode alkaline cell generates both at high and low load, a substantially higher discharge energy than the equivalent cesium cathode cell. This is also the case for the potassium, compared to the barium, discharge in the figure. However, as included in the figure, silver activation of the potassium salt can lead to higher discharge capacities for a K_2FeO_4/AgO composite cathode, compared to the $BaFeO_4$ cathode cell (*25*).

Figure 3.2a compares the $SrFeO_4$, K_2FeO_4 and $BaFeO_4$ Super-Iron batteries under the same constant load discharge condition. Under these conditions, the strontium Fe(VI) cathode discharges to ~1.5 Wh, a significantly higher energy than generated by the K_2FeO_4 cathode, and approaching that of the $BaFeO_4$ cathode cell. The strontium cathode cells typically exhibit ~20 to 40mV higher open circuit voltage than the equivalent barium cell, and as seen in the figure, a higher potential average is exhibited when the strontium cell is discharged under low (75Ω) load. Also of significance in Figure 3.2a, is the discharge of the cell under conditions of high power (small load). At a constant 2.8Ω, the strontium cell discharges to 0.6 Wh. In addition (not shown here), a range of Fe(VI) salts were synthesized with a variety of strontium to barium ratios ($Sr_xBa_{(1-x)}FeO_4$; x =0.05 to 0.95) and it is of interest that in the high power (2.8Ω) domain, the cell discharges to ~0.9 Wh at high potential, and this discharge potential is indistinguishable from that observed in the pure barium cathode when either a $Sr_{0.25}Ba_{0.75}FeO_4$ or $Sr_{0.05}Ba_{0.95}FeO_4$ was utilized (*13*).

The Ag_2FeO_4-Zn Super-Iron battery has an observed open circuit voltage of 1.86(±0.04) V. The potential under load is significantly lower (Figure 3.2b), due to Fe(III) polarization losses. As seen in Figure 3.2b, for the Ag_2FeO_4 & 10wt% graphite discharge curve, the measured coulombic efficiency of the Ag_2FeO_4 salt is substantially less than that observed under the same conditions for either the K_2FeO_4 or AgO/K_2FeO_4 composite cathodes. As previously discussed for other Fe(VI) salts, these losses appear to be related to an Fe(III) blocking layer near the cathode/conductor interface. Specially, for Ag_2FeO_4, these losses are expected to be particularly acute due the relatively large, 13% ferric oxide/hydroxide impurity present as an artifact of the Ag_2FeO_4 synthesis (*26*). This will intersperse insulating Fe(III) sites throughout the cathode, and inhibit charge transfer. This inhibition can be minimized by improving the conducting collector, which is in contact with the cathode salt. Hence, improvement of the cathode conductive matrix is effectuated by incorporating 30%, rather than 10%, graphite in the Ag_2FeO_4 salt cathode mix, resulting in an observed substantial increase in coulombic efficiency in Figure 3.2b; accessing more than 80% of the theoretical $5e^-$ capacity of the Ag_2FeO_4. A two step potential is evident during the discharge process. Consistent with the expectation that three, of the five, electrons, are accessed in the process of Fe(VI) reduction, approximately 60% of the discharge occurs at observed higher potential. Consistent with two, of the five, electrons accessed in the process of Ag(I) reduction, approximately 40% of the discharge is observed to occur in a second step, at a lower potential. To improve Ag_2FeO_4 charge transfer, a variety of alternative syntheses, to decrease the observed 13% Fe(III) impurity in Ag_2FeO_4 salt were conducted (*26*). These syntheses included variation of $Ag(NO_3)_3$ concentrations, drying temperature, and filtration conditions. Another alternative synthesis included drying under O_2, rather than vacuum, to attempt to minimize wet salt decomposition losses. It is this latter synthesized salt whose cathode product is included as the "alternate synthesis" discharge in Figure 3.2b (*26*).

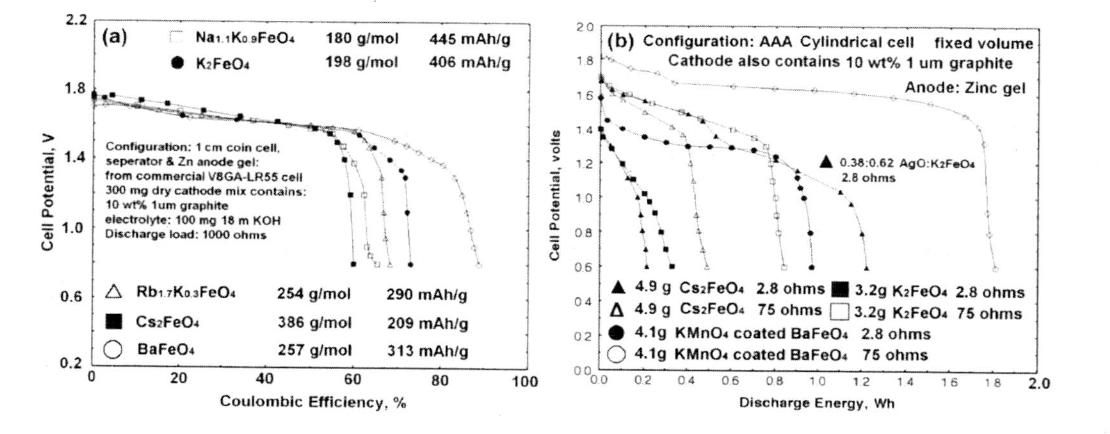

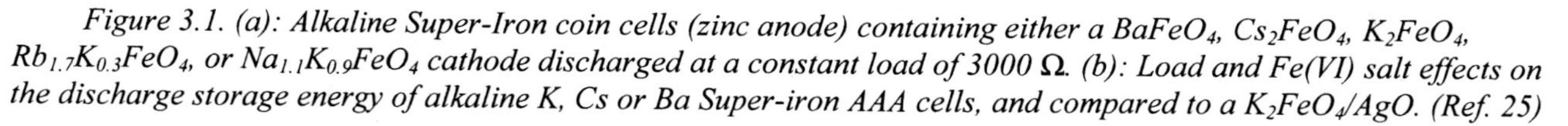

Figure 3.1. (a): Alkaline Super-Iron coin cells (zinc anode) containing either a $BaFeO_4$, Cs_2FeO_4, K_2FeO_4, $Rb_{1.7}K_{0.3}FeO_4$, or $Na_{1.1}K_{0.9}FeO_4$ cathode discharged at a constant load of 3000 Ω. (b): Load and Fe(VI) salt effects on the discharge storage energy of alkaline K, Cs or Ba Super-iron AAA cells, and compared to a K_2FeO_4/AgO. (Ref. 25)

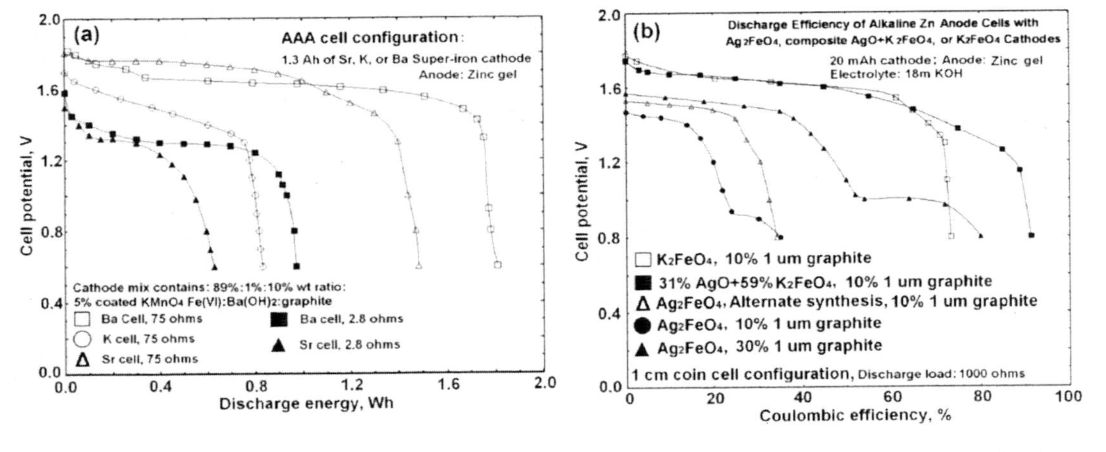

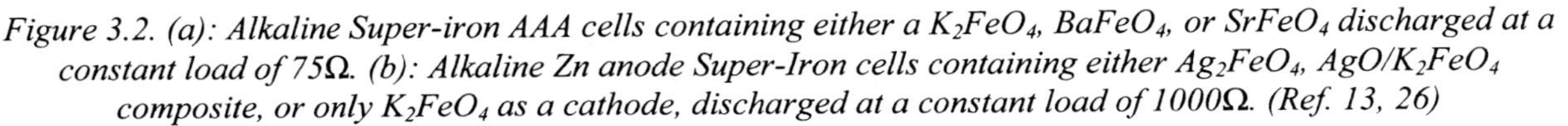

Figure 3.2. (a): Alkaline Super-iron AAA cells containing either a K_2FeO_4, $BaFeO_4$, or $SrFeO_4$ discharged at a constant load of 75Ω. (b): Alkaline Zn anode Super-Iron cells containing either Ag_2FeO_4, AgO/K_2FeO_4 composite, or only K_2FeO_4 as a cathode, discharged at a constant load of 1000Ω. (Ref. 13, 26)

4. Conductors for Alkaline Super-Iron Batteries

A variety of carbon materials were probed, added as a conductive matrix to support Fe(VI) reduction. Of these, 1 μm graphite and carbon black are the two best conductive matrix to support Fe(VI) cathodes discharge. As seen in Figure 5.1, other carbons yield a low Fe(VI) discharge capacity. These observed low discharge capacities may be due to several factors including (i) high intrinsic resistance of the added 'conductor', (ii) poor electron transfer between the clean conductor | Fe(VI) interface, and (iii) reaction and passivation of the conductor | Fe(VI) interface (*10*).

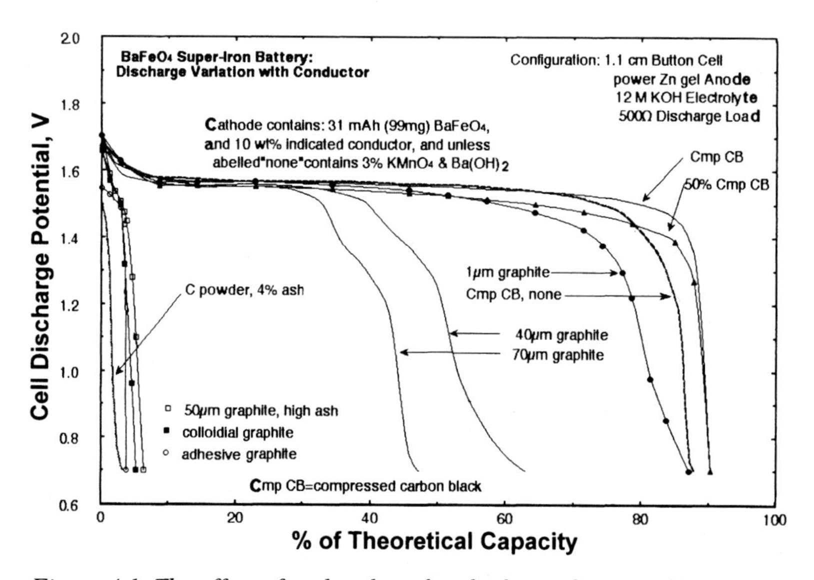

Figure 4.1. The effect of carbon based cathode conductive additive on the discharge of Super-Iron ($BaFeO_4$-Zn) batteries. (Ref. 10)

The carbon based materials added to the Fe(VI) cathode are observed to be described by 3 categories in Figure 4.1, which consist of materials supporting either efficient, intermediate, or inefficient Fe(VI) charge transfer. The inefficient category includes carbon powder with 4% ash, high ash, colloidal, and adhesive graphites. The intermediate category includes the 40 and 70 μm particle size graphites. Finally, the efficient Fe(VI) charge transfer category comprises carbon black (either 50% or fully compressed) and 1 μm graphite. As seen in the figure, the fully compressed carbon black provides improved Super-Iron battery discharge characteristics compared to the 50% compressed carbon black cathode conductor. As also noted in the figure, a small (3%) addition to the cathode of $KMnO_4$ and $Ba(OH)_2$ can further improve the discharge characteristics (*10*).

Fluorinated Polymer Graphite Conductor and Its Effects on Super-Iron Batteries

Fluorinated polymer graphite has an interesting effect on the discharge of super iron-battery (*10, 15*), as presented in Figure 4.2.

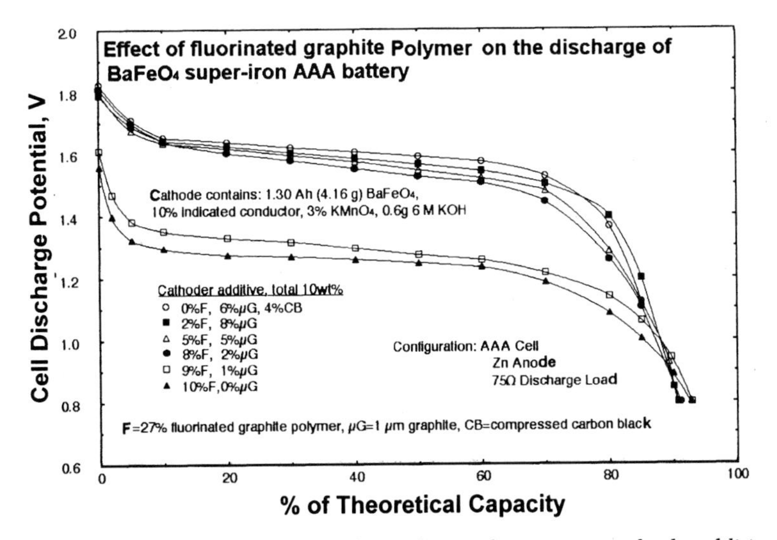

Figure 4.2. The effect of fluorinated graphite polymer as a cathode additive on the discharge of $BaFeO_4$-Zn AAA cylindrical batteries. (Ref. 10)

As seen in the top curve of Figure 4.2, and consistent with the results of Figure 2.1, a mix of 1 μm graphite and carbon black sustains a high discharge potential throughout an efficient three-electron reduction of the Fe(VI). The fluorinated polymer graphite can be used to control effectively the cell discharge potential, while maintaining a high discharge efficiency. The lowest curve in Figure 4.2 was obtained with a Fe(VI) cathode mix containing only the 27% fluorinated polymer graphite (FG(27%)). A lower, but significant discharge voltage of 1.2 to 1.3 V is sustained. Despite these significant polarization losses, this conductor supports a somewhat higher Fe(VI) faradaic efficiency than that observed with the graphite/carbon black conductor. As seen in the figure, replacement of 1% of the FG(27%) with the 1 μm graphite has little effect on the discharge potential. The observed faradaic efficiency of approximately 92% is yet higher. Replacement of higher amounts of the FG(27%) with the 1 μm graphite provides control of the discharge voltage. The discharge voltage jumps by 300 mV when 2% of 1 μm graphite is utilized and, as seen in the figure, the discharge potential continues to increase when 5% of 1 μm graphite is utilized. Finally, as seen comparing the top two curves of the figure, the Fe(VI) mix, containing 2% FG(27%) and 8% 1 μm graphite, provides a marginally higher faradaic efficiency, albeit at marginally lower discharge potential, compared to the cathode mix containing 1 μm graphite and carbon black cathode.

Further presented in Figure 4.3a, a cell with a 10 wt% of FG(27%) conductor, does not exhibit an evident greater storage capacity than the graphite/carbon black cell. However, cells using 10 wt% of FG(58%), not only exhibit a higher storage capacity, but as seen in the figure, this nominal storage capacity is over 100% of the theoretical capacity of $BaFeO_4$. This effect was further explored with the additional set of cathode mixes containing a higher weight fraction of the added conductor, as shown in Figure 4.3b.

As shown in Figure 1.1, at high load, the faradaic efficiency of K_2FeO_4 cathode is low. But at lower current densities and at higher fractions of graphite, K_2FeO_4 is also expected to approach discharging to the theoretical $3e^-$/Fe(VI). As seen in Figure 4.3b, the discharge of 10 wt% K_2FeO_4 in 90 wt% 1 μm graphite, leads to the expected near 100% of the theoretical $3e^-$ storage capacity (406 mAh/g K_2FeO_4). Unexpectedly, replacement of this high fraction of graphite with an equal mass (90 wt%) of FG(27%), leads to nearly 200% of the theoretical storage capacity. Combining with $KMnO_4$, a cell containing both 5 wt% K_2FeO_4 and 5 wt% $KMnO_4$ with 90 wt% of this FG exhibits over 175% of the theoretical three-electron storage capacity. A cell containing both 5 wt% $BaFeO_4$ and 5 wt% $KMnO_4$ with 90 wt% of this FG(27%) exhibits nearly 200% of the theoretical three-electron storage capacity (*15*).

The unusual effects of FG conductors on the discharge of Super-Iron batteries are related to the intrinsic alkaline cathode capacity in FGs which does not occur in conventional graphites or carbon blacks (*15*). The FG polymers are observed to simultaneously maintain two roles in the cathode composite; functioning both as a conductive matrix, and also adding intrinsic capacity to the cathodes (*15*).

5. Rechargeable Alkaline Fe(VI/III) Thin Film Cathodes

While primary charge transfer of Super-Iron batteries has been extensively demonstrated, reversible charge transfer of Fe(VI) cathode had been problematic. The poor reversibility of Fe(VI) cathode is attribute to the conductivity constraints imposed by the ferric oxide product of reduced Fe(VI), as illustrated in section 3 (Figure 3.1 inset). However, a sufficiently thin film Fe(III/VI) cathode prepared on a conductive matrix (such as Pt or extended conductive matrix) should facilitate sufficient electronic communication to sustain cycled charge storage (*24*). In this section this reversibility is demonstrated, using increasingly thick Fe(VI/III) thin films, ranging from nanometers to 100's of nanometers in thickness. The thin Fe(VI/III) film half cell charge/discharge behavior was investigated using a three-electrode, sandwiched electrochemical cell with a cation selective membrane (Nafion-350) to isolate the cathodic and anodic cell compartments (isolate the Fe(VI/III) film electrode with the counter and reference electrode). A Saturated Calomel Electrode (SCE) was used as the reference electrode, and a nickel sheet was used as the counter electrode. In full cell studies, the counter electrode and reference electrodes were replaced with a metal hydride anode removed from a Powerstream® 40 mAh Ni-NiOOH coin cells.

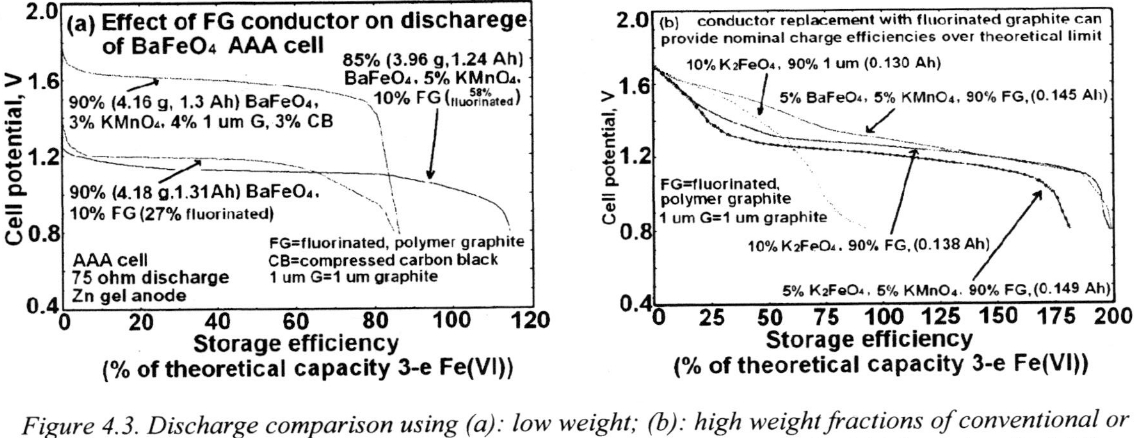

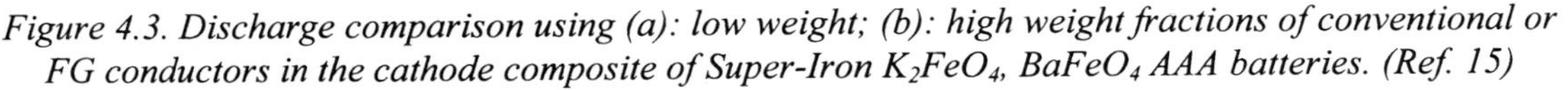
Figure 4.3. Discharge comparison using (a): low weight; (b): high weight fractions of conventional or FG conductors in the cathode composite of Super-Iron K_2FeO_4, $BaFeO_4$ *AAA batteries. (Ref. 15)*

Reversible Fe(VI/III) Ultra-Thin Film on Smooth Pt

In 2004, we reported that rechargeable ferric films can be generated, as formed by electrodeposition onto conductive substrates from solution phase alkaline Fe(VI) electrolytes (*24*). A thin film, conducive to Fe(VI) charge cycling, is generated on a smooth Pt foil electrode from micro-pipette controlled, microliter volumes of dissolved Fe(VI) in alkaline solution. We observe that a nanofilm (for example, a 3 nm thickness Fe(VI) film), formed by electrodeposition is highly reversible. The preparation of this thin film is detailed in the reference 24. The film is rigorous, and when used as a storage cathode, exhibits charging and discharging potentials characteristic of the Fe(VI) redox couple, and extended, substantial reversibility. Figure 5.1 presents a three Faraday capacity, reversible Fe(III/VI) cathode. As seen in the figure, a full 80% DOD (depth of discharge) of the 485 mAh/g capacity Na_2FeO_4 film is readily evident after 100 galvanostatic cycles (*24*).

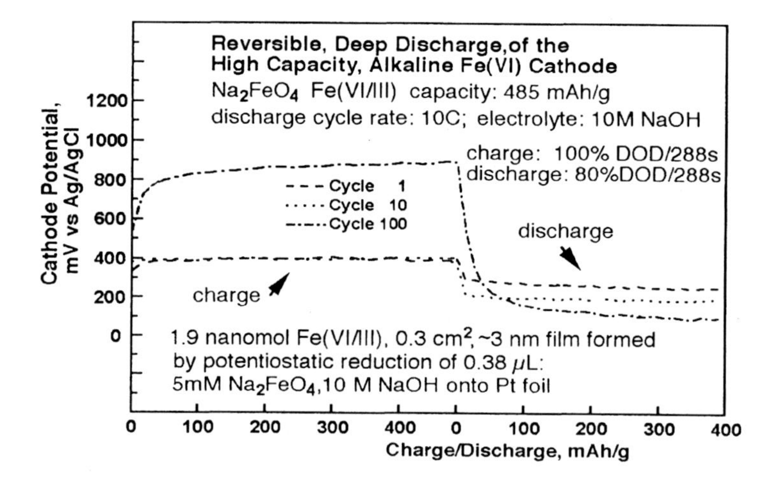

Figure 5.1. Reversible charge storage of a Fe(VI) of a 485 mAh/g capacity Na_2FeO_4 nanofilm. Each galvanostatic storage cycle is 100% DOD charge followed by 80% DOD discharge, at the 10C discharge rate. (Ref. 24)

Passivation of Thick Fe(III/VI) Charge Transfer on Smooth Platinum

As shown in Figure 5.1, ultrathin (3 nm thick) super-iron films can sustain over 100 charge-discharge cycles. However, thicker films were not rechargeable due to the irreversible buildup of passivating (resistive) Fe(III) oxide, formed during film reduction, as illustrated in Figure 5.2.

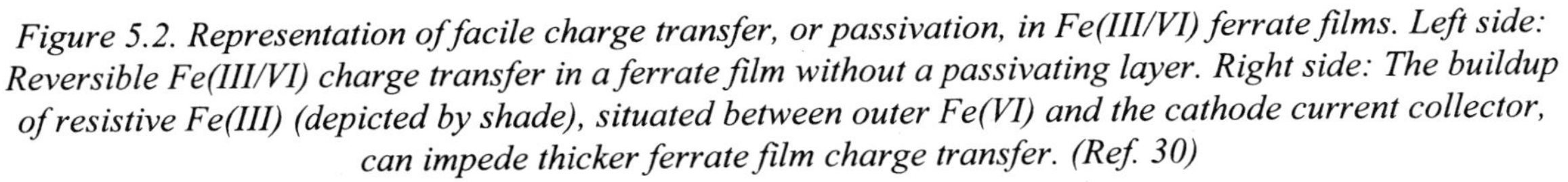

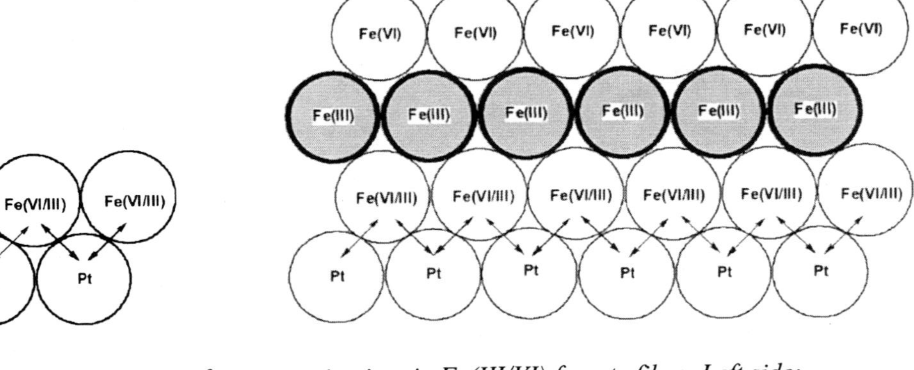

Figure 5.2. Representation of facile charge transfer, or passivation, in Fe(III/VI) ferrate films. Left side: Reversible Fe(III/VI) charge transfer in a ferrate film without a passivating layer. Right side: The buildup of resistive Fe(III) (depicted by shade), situated between outer Fe(VI) and the cathode current collector, can impede thicker ferrate film charge transfer. (Ref. 30)

3nm and 19 nm Fe(III/VI) films on Pt are compared in Figure 5.3 (Left side of Figure 5.3 is the expanded results from Figure 5.1). As evident, a 3 nm Fe(VI) film exhibits reversible behavior throughout 20 galvanostatic charge/discharge cycles, whereas a 19 nm film rapidly passivates. Specifically, each film is repeatedly subject to a 0.5 mA/cm^2 galvanostatic charge, followed by a deep 0.005 mA/cm^2 galvanostatic discharge (to 80% of the Fe(III/VI) film theoretical, intrinsic $3e^-$ capacity). The 3 nm film, smoothly approaches a plateau discharge potential between 0.2 to 0.3 V (Figure 5.3, left B). This contrasts with the discharge behavior of a 19 nm super-iron film (Figure 5.3, right B), in which this discharge plateau is only sustained during the initial portion of the discharge, after which the potential begins to sharply decline. A stable plateau between 0.38 V and 0.2 V and a rapid polarization decay to -0.2 V was observed on the discharge of the 19 nm film. The plateau discharge potential was sustained to approximately 75% of the full capacity during the first cycle, and diminished to only 25% of this capacity within 20 cycles. Unlike the 19 nm film, the 3 nm ferrate film will also evidence sustained charge transfer reversibility at higher cathodic currents. Hence, as presented in Figure 5.3 left C, at an order of magnitude higher discharge rate of 0.05 mA/cm^2, the 3 nm Fe(III/VI) film also sustains repeated, deep discharge. Compared to the 0.005 mA/cm^2 cathodic cycling, by the 25th cycle, the polarization losses are higher at this higher current density. For example in this last cycle, at 80% depth of discharge, the potential diminishes to 0.13 V vs SCE at 0.05 mA/cm^2, whereas the discharge potential is 90 mV higher at 0.005 mA/cm^2.

Conductive Matrix Facilitated Fe(III/VI) Charge Transfer

The conductive matrix to support Fe(VI/III) charge transfer, can be extended from a 2 dimensional structure to a high surface area 3 dimensional structure, by platinization of Pt and Ti substrates, or co-deposition of platinum and gold onto a Ti substrate, as detailed in reference 30. The effective surface area of the platinized substrate increases with the degree of platinization. On a Pt substrate, up to a ~1000-fold increase in electroactivity was obtained with 10.4 mg cm^{-2} of Pt deposit. On a Ti substrate, a normalized electroactivity ~326 was obtained when 4.5 mg cm^{-2} of Pt loaded. By co-depositing Pt with Au on the Ti substrate, a stable surface containing up to 14 mg cm^{-2} of Pt with 15 mg cm^{-2} of Au deposit were achieved. The relationship between the normalized electroactivity and the amount of platinization on Pt or Ti substrate, as well as the preparation of Fe(VI/III) on these extended conductive matrix, were detailed in reference 30.

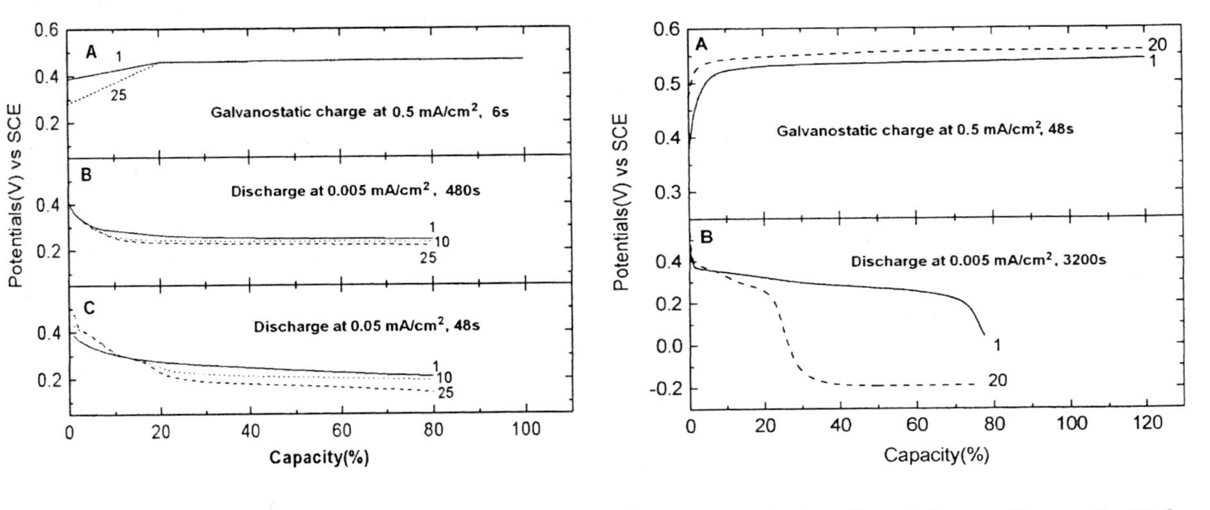

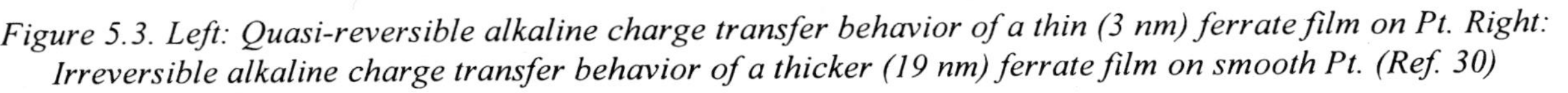

Figure 5.3. Left: Quasi-reversible alkaline charge transfer behavior of a thin (3 nm) ferrate film on Pt. Right: Irreversible alkaline charge transfer behavior of a thicker (19 nm) ferrate film on smooth Pt. (Ref. 30)

Platinum substrates for Fe(III/VI) charge transfer

A substantial improvement to sustain thick film charge transfer is observed when an extended conductive matrix was utilized as the film substrate. The left side of Figure 5.4 A, summarizes the discharge behavior during repeated charge/discharge cycles of a 19 nm super-iron film on platinized platinum, which contained 2.6 mg cm^{-2} of Pt deposits. Compared to the film on smooth platinum presented in Figure 5.3 right, the film was able to sustain substantially higher discharge current (0.05 compared to 0.005 mA cm^{-2}), as well as significantly higher reversibility (over 500 discharge cycles) without onset of significant passivation. As shown in Figure 5.4 left B, a 150 nm super-iron film on a platinized platinum containing 10 mg cm^{-2} of Pt deposits, can sustain over 50 charge/discharge cycles (*30*).

Figure 5.4 right summarizes the charging (the other half of the discharge/charge cycle) of the 19 nm Fe(III/VI) film on the extended, conductive matrix. Figure 5.4 right compares the measured charging potentials over a range of applied galvanostatic oxidation current densities. As seen in the figure, charging potentials during initial stages do not exhibit simple trends, and were observed to be significantly effected by cycle number, thickness and current density. In latter stages, a consistent increase of overpotential is observed with increasing charging current density. An order of magnitude increase in the relative charging current, to 0.25 mA/cm^2, generates only approximately ~40 mV charging overpotential. The next order of magnitude increase, to 2.5mA/cm^2, incurs a substantially larger over potential (~200 mV). For thicker super-iron films, a minimum current density of 0.5 mA cm^{-2} was useful to sufficiently regenerate at least 80% of the Fe(VI) discharge. To investigate both thin and thick super-iron films,- a consistent charging current density of 0.5 mA cm^{-2} was utilized to facilitate oxidation to Fe(VI), while sustaining relatively low charging overpotentials. The alkaline thermodynamic potential for oxygen evolution occurs at 0.16 V vs SCE, in accord with:

$$4OH^- \rightarrow O_2 + 2H_2O + 4e^-; \quad E = 0.40V \text{ vs SHE} \qquad (5.1)$$

FeO_2^- species in NaOH have been observed to diminish the oxygen evolution potential by 0.1 V (*45, 46*). However, at room temperature, Pt exhibits a high overpotential to oxygen evolution, and below 0.5 to 0.6 V vs SCE, the rate of oxygen evolution is not significant (*47*). Similarly no observable oxygen evolution was evident for charging the Fe(III/VI) films at these potentials. As can be seen in Figure 5.4 right A, at charging currents $\leq$ 0.5 mA cm^{-2}, the charging potential is generally less than 0.5 V vs SCE. The super-iron films remained intact even after 500-cycles. In Figure 5.4 right B the limiting value of the charging potentials gradually increased with the increasing number of cycles and reaching a maximum of 0.95 V only at the end of 500 cycles (*30*).

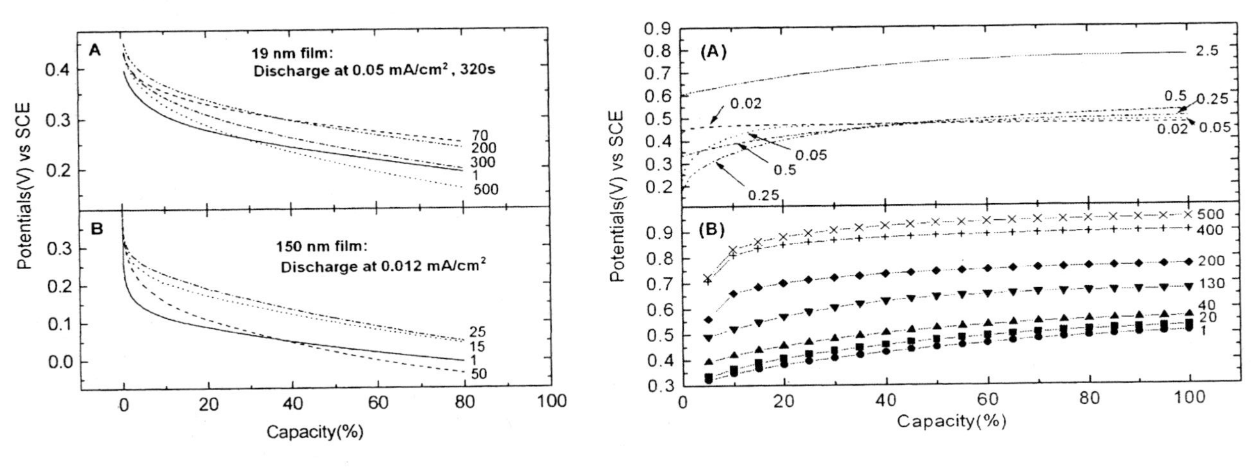

Figure 5.4. Left: Cathode discharge potential during cycling of (A) 19 nm and (B) 150 nm ferrate film on a platinized Pt substrate (2.6 and 10 mg/cm² of Pt load respectively) in a galvanostatic configuration in 10 M NaOH. The cycling consists of galvanostatic charge at 0.5 mA/cm² followed by galvanostatic discharge at 0.05 mA/cm² for 320 seconds (19 nm) and at 0.012 mA/cm² (150 nm). Right: Cathode charging potential of a 19 nm film on a platinized (2.6 mg Pt/cm²) Pt substrate: (A) for 20 mC/cm² of charge, during the 1st cycle, at various current densities, (B) for various cycles at a fixed galvanostatic charge of 0.5 mA/cm² for 40 seconds. (Ref. 30)

Both with, and without, the extended conductive matrix, thicker ferrate films exhibited greater overpotential (lower cathode potential) during discharge. This is observed by comparing A and B section of Figure 5.4 left. As seen in the figure, the final discharge potential was 0.0 V vs SCE for the thicker film (150 nm), but 0.15 V higher for the thinner film (19 nm) (*30*).

The importance of an effective, enhanced conductive matrix substrate was evident in attempts to reversible cycle thicker (greater than 20 nm) super-iron films, as summarized in the Table 5.1. When a 50 nm super-iron film was formed on a normalized electroactivity =550 platinized Pt substrate, the film passivated when discharged at a current density of 0.025 mA cm^{-2}, but discharged effectively at a current density of 0.012 mA cm^{-2}. In the latter case, the 50 nm film reversibly cycled 100 times before the onset of passivation. When a 50 nm super-iron film was deposited onto platinized Pt with normalized electroactivity of 725, the film could sustain 100 cycles, at a higher discharge current density of 0.025 mA cm^{-2}. Further enhancement in the conductive matrix also resulted in longer cycle life, and 160 reversible cycles were sustained when the film was deposited onto a surface with normalized electroactivity of 990 (discharged at a current density of 0.025 mA cm^{-2}). Greater reversibility was again observed, when discharging at a lower current density, and the film sustained 250 cycles at a discharge current density of 0.012 mA cm^{-2} (*30*).

Table 5.1. The Influence of Normalized Electroactivity and Discharge Current Density on Cycle Life of a 50 nm Film

Normalized electroactivity	*Current Density (mA cm^{-2})*	*DOD (%); DOD= depth of $3e^-$ discharge*	*Cycles to polarization deactivation*
550	0.012	75	100
725	0.025	80	100
725	0.012	80	250
990	0.025	80	160

The formation of passivating, irreversible Fe(III) centers is more likely in the case of thicker films. Passivation is evident in increased charge and discharge overpotentials, and in the inability to discharge to a significant fraction of the intrinsic charge capacity. The facilitated super-iron charge transfer, upon platinization, as a result of the expanded conductive matrix to facilitate charge transfer is represented in Figure 5.5. Without direct contact with the substrate, the shaded Fe(III) centers in the figure had posed an impediment to charge transfer. This is partially (Figure 5.5 left side) and fully alleviated (right side), by intimate contact with the enhanced conductive matrix which maintains extended direct contact with the substrate (*30*).

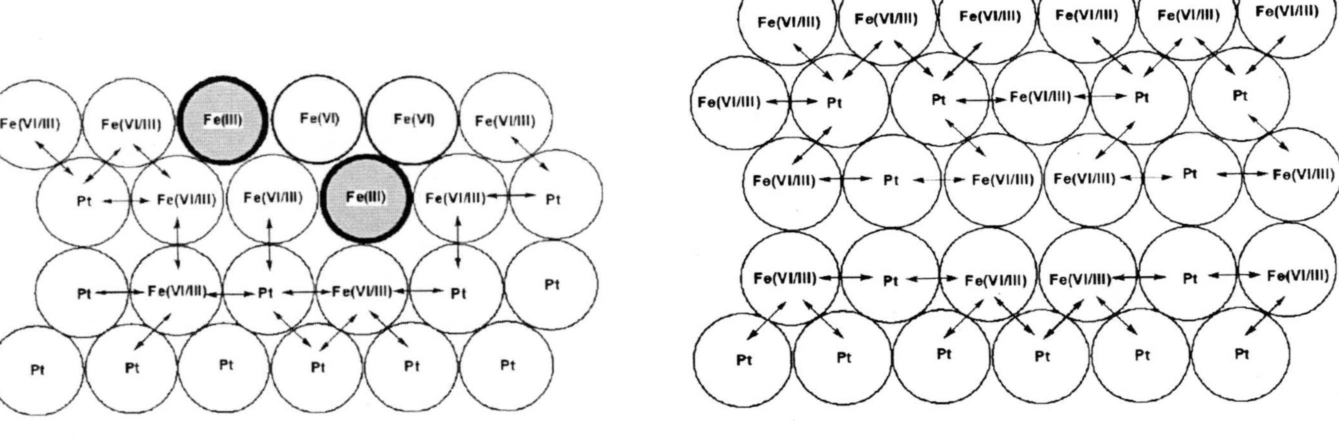

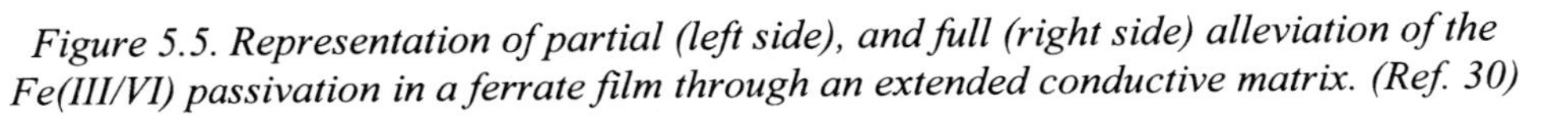
Figure 5.5. Representation of partial (left side), and full (right side) alleviation of the Fe(III/VI) passivation in a ferrate film through an extended conductive matrix. (Ref. 30)

Platinized titanium substrates for Fe(III/VI) charge transfer

Platinized Ti substrates also can be used as effectively as platinized platinum. To effectively utilize super-iron films on the platinized Ti substrate, it was observed that during the initial phases of the cycling (for cycles 2 through 10) an extended charging time, equivalent to 150% charge of the intrinsic capacity, was required otherwise the full discharge could not be accessed. Following this, the conventional charge (equivalent to 120% charge of the intrinsic capacity) was sufficient to sustain extended charge/discharge cycling. Initial morphological changes can occur in a high surface areas, stressed, and crystallographically disordered platinized Pt surface (*48, 49*). The electronic changes associated with the formation of Ti-oxides may initiate this process (*46*). Hence, this latter effect could arise from improved contact, of the underlying Pt layer with cycling.

The charge/discharge curves of a 50 nm film on 7.5 mg Pt cm^{-2} platinized Ti are given in Figure 5.6 left. The one drawback encountered with the platinized Ti substrate is the difficulty to obtain larger Pt deposits, more than 7.5 mg cm^{-2} which resulted in unstable powdery surface, and due to this it was difficult to deposit more than a 70 nm super-iron film on platinized Ti (*30*). However, a substantial improvement in the stability and upper limit of the thickness of the film was observed when Pt-Au co-deposited Ti surface was used as the substrate. A 300 nm super-iron film displayed a moderate cycle life of 20. Charge-discharge profiles are presented in Figure 5.6 right.

Metal-Hydride Anode Compatibility with the Rechargeable Super-Iron cathode Films

Facilitated charge transfer, thicker, Fe(III/VI) films compatibility with metal-hydride anodes was explored in conjunction with full cell charge storage (*30*). Generally in alkaline media, the metal hydride anode charge storage has been utilized in conjunction with the nickel oxyhydroxy cathode (*50*). The half, and full, cell reactions in the charge storage process of the super-iron/metal hydride full cell can be described in an analogous manner to the NiOOH/metal hydride cell. During discharge FeO_4^{2-} is reduced in the Na_2FeO_4 cathode and the metal hydride (MH_x) is oxidized to the metal (M) in accordance with:

$$MH_x + xOH^- \rightarrow M + xH_2O + xe^-; \qquad E = -0.82 \text{ V vs SHE} \qquad (5.2)$$

The process is reversed during the charge. Combined with the super-iron cathode reaction:

$$Na_2FeO_4 + 2H_2O + 3e^- \rightarrow 1/2Na_2Fe_2O_4 + NaOH + 3OH^- \qquad (5.3)$$

the overall cell discharge reaction will be:

$$Na_2FeO_4 + 3/xMH_x \rightarrow 1/2Na_2Fe_2O_4 + 3/xM + NaOH + H_2O; \qquad E = 1.42 \text{ V} \quad (5.4)$$

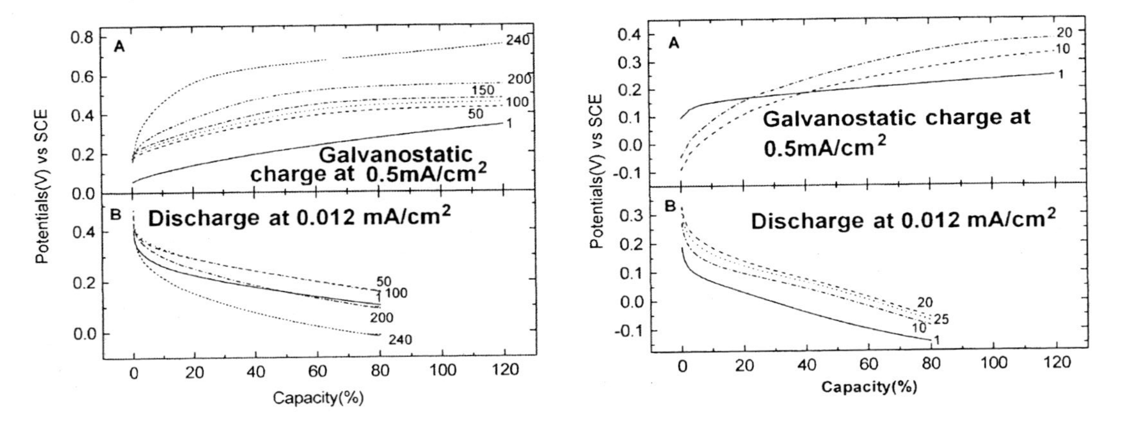

Figure 5.6. Left: Quasi-reversible alkaline charge transfer behavior of a 50 nm ferrate film on a platinized Ti (7.5 mg/cm² of Pt load). A) galvanostatic charge at 0.5 mA cm⁻². B) galvanostatic discharge at 0.012 mA cm⁻². Right: Quasi-reversible alkaline charge transfer behavior of a 300 nm ferrate film on a Gold-Platinum co-deposited Ti substrate (14 mg cm⁻² of Pt and 15 mg cm⁻² of Au loads). A) galvanostatic charge at 0.5 mA cm⁻². B) galvanostatic discharge at 0.012 mA cm⁻². (Ref. 30)

Interestingly, super-iron films deposited on the extended, conductive matrix substrate were characterized by a significantly longer cycle life in the full, metal hydride storage cell, compared to those measured in the half cell configuration (*30*). In the cell, a 25 nm film Fe(III/VI) cathode, deposited on 10 mg cm^{-2} Pt platinized substrate, displayed high cell voltage. A discharge voltage of 1.2 V is sustained through the end of 300 cycles in the full cell, and within 0.22 V of the eq. 5.4 thermodynamic cell rest potential (*30*).

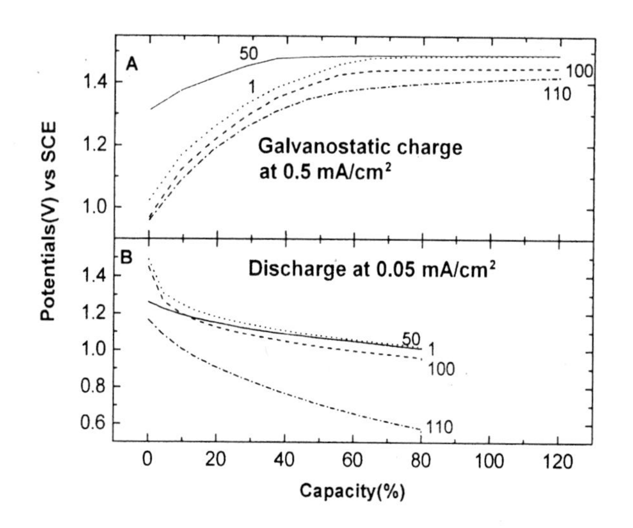

Figure 5.7. Two electrode rechargeable behavior of 150 nm ferrate cathode (deposited on 10 mg cm^{-2} Pt platinized substrate) with a MH anode. A) Full cell potential during a galvanostatic charge at 0.5 mA cm^{-2}. B) Electrode film potential during a galvanostatic discharge at 0.05 mA cm^{-2}. (Ref. 30)

Although the cycling of the cathode in the full cell is consistently higher than that of the similar half cell measurements. In both cases, as the thickness of the film increased, the cycle life decreased. The observed charge/discharge profiles of a 150 nm Fe(III/VI) film, in a super-iron metal hydride cell are presented in Figure 5.7. Consistent with the prior half reaction measurements, the film was charged to 120% of its theoretical capacity at a constant current density of 0.5 mA cm^{-2}, and subsequently discharged to 80% at a constant current density of 0.05 mA cm^{-2}. As seen in the lower portion of the figure, discharge commenced at ~1.3 V and decayed to 1.0 V at the end of 80% capacity discharge. Subsequent to this initial cycle (through charge/discharge cycle 100), discharge commenced at a higher voltage of ~1.5 V. The rate of voltage decay during the discharge increased after 100 discharge cycles, and as seen in the figure, by cycle 110 a diminished voltage of 0.6 V was observed at the end of 80% capacity discharge. A 250 nm film, capable of storing 264 mC of intrinsic capacity per cm^2, sustained stable, reversible charge storage only for

40-cycles. A 50% longer cycle life was achieved with MH anode in comparisons to cycle life of the half reaction, compared to the three electrode cell which had utilized a nickel counter electrode. This is indicative of greater compatibility of the Fe(III/VI) film cathode in this metal hydride anode cell (*30*).

6. Nonaqueous Primary Fe(VI) Charge Transfer for Super-Iron Lithium Batteries

In this section, the cathodic chemistry of Fe(VI) compounds is considered in nonaqueous electrolytes for use as a lithium, or lithium-ion, anode electrochemical storage system. Primary nonaqueous Super-Iron/Lithium battery is presented, and the initial results demonstrating reversibility of Fe(VI) cathodes in nonaqueous electrolyte system are also illustrated.

Primary Nonaqueous Super-Iron Battery

We introduced the utilization of Fe(VI) salts as a nonaqueous cathode, in conjunction with a lithium anode in nonaqueous media, in 2000 (*9*). The Fe(VI) salts, such as $BaFeO_4$ and K_2FeO_4, are insoluble in a variety of organic solvents, including solvents conducive to studies of lithium electrochemistry such as acetonitrile (ACN), propylene carbonate (PC), ethylene carbonate (EC), g-butyrolactone (BLA), and dimethoxyethane (DME). In addition, these Fe(VI) salts remain insoluble when necessary lithium salts are dissolved to increase the electrolytic conductivity. In the longest duration stability studies available, there is no Fe(VI) dissolution or reaction of either $BaFeO_4$ or K_2FeO_4 powders immersed in dry ACN, PC, DME, or BLA either with or without the dissolved lithium perchlorate. Insolubility and stability of Fe(VI) salts in nonaqueous electrolyte are attractive physical chemical properties for a nonaqueous Super-Iron battery.

Super-iron primary cells with a lithium anode were prepared in a conventional CR1216 coin cell case. The cathode composite was formed by mixing the indicated mass of Fe(VI) salt (either K_2FeO_4, $BaFeO_4$, $SrFeO_4$, or Li_2FeO_4) with the indicated weight percent of conductors (graphite or carbon black) or other additives. In each situation, the button cell was opened, the anode retained, an appropriate electrolyte added, and the new cathode placed in the cell. Cells were discharged with a constant load. Cell potential variation over time was measured via LabView Data Acquisition on a PC, and cumulative discharge current, as millampere-hours, determined by subsequent integration.

In the alkaline Super-Iron battery, Fe(VI) cathodes discharge through a 3-electron electrochemical reduction reaction as eq 1.1 or eq 1.2. The cathodic discharge mechanism of Fe(VI) in nonaqueous Super-Iron battery is complex. In 2000, we noted that the Fe(VI) discharge in nonaqueous media was consistent with the storage of three electrons per Fe center, and consistent with other nonaqueous cathodes, suggested a mechanism of charge transfer via the insertion of Li^+ (our more recent, alternate mechanism will be presented later in this section) (*9*) :

$$MFeO_4 + xLi^+ + xe^- \rightarrow Li_xMFeO_4 \quad M = K_2, Ba, Sr, Li_2, etc \qquad (6.1)$$

Nonaqueous primary Super-Iron batteries were initially investigated with Fe(VI) cathodes comprised of either Li_2FeO_4, K_2FeO_4, $SrFeO_4$, or $BaFeO_4$. The 601 mAh/g theoretical, intrinsic cathodic, charge capacity of Li_2FeO_4 based on the insertion of 3 Li^+ per Fe, is substantially higher than the 406 mAh/g for K_2FeO_4, 388 mAh/g for $SrFeO_4$, or 313 mA/g for $BaFeO_4$. Figure 6.1 compares the discharge of Li_2FeO_4, K_2FeO_4, $SrFeO_4$, and $BaFeO_4$ cathode Super-Iron lithium batteries with a LiTFB PC:DME electrolyte. Under equivalent conditions, the $BaFeO_4$ cathode exhibits lower polarization losses, and a higher faradaic efficiency (based on equivalents of Li^+ insertion per equivalent of Fe(VI)), than the K_2FeO_4 cathode. However, due to its lighter mass the observed specific capacity of the K_2FeO_4 is marginally higher. The lower discharge voltage of the lithium, compared to barium, strontium or potassium Super-Iron cathodes is likely due to the lower purity and resultant higher polarization of this cathode. Note that unlike K_2FeO_4, $SrFeO_4$, or $BaFeO_4$, the measured purity of the Li_2FeO_4, is very low (~20%), requiring 7.5 mg of salt to provide the indicated 1.5 mg of Li_2FeO_4 in the cathode, and the mass normalized measured capacity is based on this small active component of this impure material.

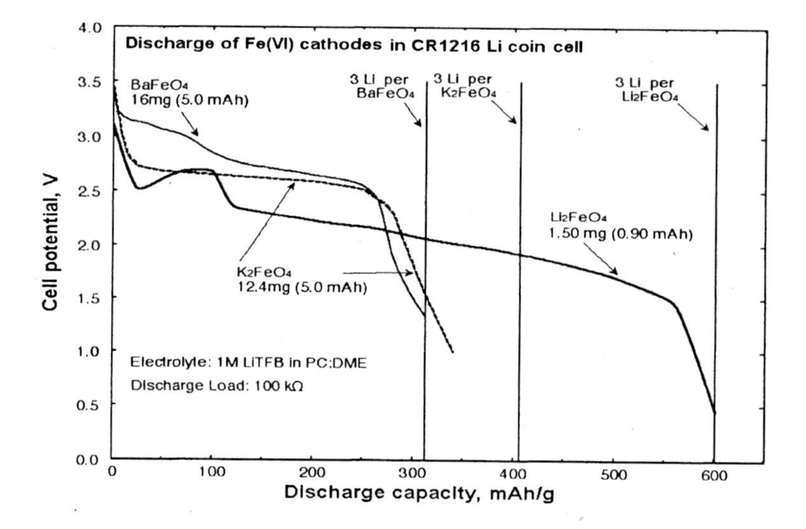

Figure 6.1. Constant load discharge of cells containing a Li_2FeO_4, K_2FeO_4, $SrFeO_4$, or $BaFeO_4$ Fe(VI) cathode and a lithium anode. Electrolyte is 1 M LiTFB PC:DME. (Ref. 9)

The configuration of the primary Super-Iron/Li batteries was studied to enhance nonaqueous Fe(VI) cathodic charge transfer for both K_2FeO_4 and $BaFeO_4$ cathodes. Table 6.1 presents solvent and electrolyte effects on the cathodic charge capacity of a $BaFeO_4$ cathode (*21*).

As summarized, the specific capacity of $BaFeO_4$ increased from 28, 84, 113, to 241 mAh/g when the solvent for 1 M $LiClO_4$ electrolyte was changed from acetonitrile, 1:1 (v/v) EC-DEC, γ-BLA to 1:1 (v/v) PC-DME, respectively. In addition to the solvent, the variation of the electrolyte supporting salt also has a substantial effect on Fe(VI) cathodic capacity. As summarized in the table, in a range of solvents, dissolved LiTFMS did not effectuate charge transfer as efficiently as $LiClO_4$, which in turn did not promote the high degree of cathode capacity observed in Fe(VI) cells with the $LiBF_4$ and $LiPF_6$. As also evident from the table, an electrolyte prepared with a PC-DME co-solvent yields higher $BaFeO_4$ capacity (260 mAh/g in 1 M $LiPF_6$) than the other studied solvents, although the γ-BLA solvent was also effective (206 mAh/g with 1 M $LiPF_6$).

Figure 6.2a presents in a PC-DME electrolyte prepared with a range of lithiated and nonlithiated supporting salts, the nonaqueous discharge of cells containing a $BaFeO_4$/carbon black cathode. Nonlithium electrolytes cells may exhibit significantly impaired cathodic charge transfer. In these electrolytes, the only lithium ion source in the electrolyte is the buildup that occurs during mass transport from the discharging anode. This impaired cathodic charge transport is clearly evident in Figure 6.2a, and seen as the low cathodic specific capacity measured during discharge of the lithium free TEA-PF_6 and TEA-TFB PC-DME electrolytes. For the lithiated salt PC-DME electrolytes, the highest cathodic capacity was measured for the 1 M $LiPF_6$ electrolyte (*21*).

While $LiPF_6$ supported higher cathodic capacities in Fe(VI) cells, $LiBF_4$ facilitated charge transfer. This is indicated by the lower $LiPF_6$ plateau discharge potential in Figure 6.2a. This facilitated charge transfer is probed in the inset to Figure 6.2a, and which indicates in the process of Fe(VI) cathodic storage, a lower polarization in the 1 M $LiBF_4$ compared to the steep polarization losses in the 1 M $LiPF_6$ electrolyte (polarization of 3.0 mV μA^{-1} cm^2 *vs.* 5.7 mV μA^{-1} cm^2 in $LiPF_6$). These polarizations represent charge transfer rates which are low compared to the equivalent aqueous Fe(VI) systems. For example polarization losses are several orders of magnitude smaller for the three electron faradaic reduction of aqueous alkaline $BaFeO_4$ cathodes, and in this aqueous media, Fe(VI) has been observed to sustain levels of cathodic charge transfer at ampere, rather than milliampere, currents. Interestingly, as also seen in the figure inset, a lithium iodide electrolyte considerably decreases polarization losses to 0.6 mV μA^{-1} cm^2, albeit at a lower onset potential of 3.0 V *vs.* 3.5 V observed for the $LiBF_4$ electrolyte (*21*).

Polarization toward nonaqueous cathodic charge transfer can be expected to be effected by modifications of the cathode's supporting conductive matrix. Also, polarization losses should decrease for smaller cathode particle size, and with an increase in the discharge temperature. In Figure 6.2b, with variation of

Table 6.1. Variation of the $BaFeO_4$ Cathodic Capacity with Electrolyte Supporting Salt or with Electrolyte Solvent. (*Ref. 21*)

Solvent	*Supporting salt*	*Average potential (V)*	*$BaFeO_4$ Capacity (mAh/g)*
Acetonitrile	$LiClO_4$	2.29	28
Acetonitrile	LiTFMS	2.50	30
Acetonitrile	$LiBF_4$	2.49	38
Acetonitrile	$LiPF_6$	2.15	56
EC-DEC	$LiClO_4$	2.50	84
EC-DEC	LiTFMS	2.23	26
EC-DEC	$LiBF_4$; not 1 M soluble		
EC-DEC	$LiPF_6$	2.13	48
γ-BLA	$LiClO_4$	2.64	113
γ -BLA	LiTFMS	2.26	59
γ -BLA	$LiBF_4$	2.18	78
γ -BLA	$LiPF_6$	2.52	206
PC-DME	$LiClO_4$	2.65	241
PC-DME	LiTFMS	2.41	129
PC-DME	$LiBF_4$	2.67	247
PC-DME	$LiPF_6$	2.57	260

NOTE: Cathodes are composed of 80 wt% $BaFeO_4$ and 20 wt% carbon black, with capacity determined as $BaFeO_4$ mass normalized mAh of discharge measured to 2 V. The cell is discharged over a 67 kΩ load at 25°C. The tabulated potential is measured as the average discharge potential through a cutoff voltage of 2 V.

the constant load during 25°C discharge, the need for improved cathodic charge transfer is evident in the 1 M $LiBF_4$ PC-DME electrolyte 80% $BaFeO_4$ (<150 μm particle), 20% carbon black cathode cell. In the figure, compared to the 67 kΩ load, the lighter 100 kΩ load discharge substantially increased the average cell potential and increased specific capacity by over 10%. The cathode capacity fell substantially for discharge at high rates, such as during the 33 kΩ discharge, and this cathode/electrolyte configuration did not support even higher rate discharge. As seen in the figure, the cell potential did not recover during cell discharge at 15 kΩ (*21*).

Figure 6.2c explores the effects of variations of the cathode conductive additive in nonaqueous media. For the 1 M $LiBF_4$ PC-DME $BaFeO_4$ cell, discharge was particularly ineffective for the noncarbon conductor (indium tin/oxide, ITO). Replacement of the 20 wt% carbon black, by an equal mass of 1 μm graphite in the $BaFeO_4$ cathode mix, lowered capacity from 247 to 196 mAh/g. Of the Fe(VI) 20 wt% additives investigated, carbon black consistently supported the highest cathodic capacity, both in fluorinated and nonfluorinated electrolytes, or compared to fluorinated graphites (*21*).

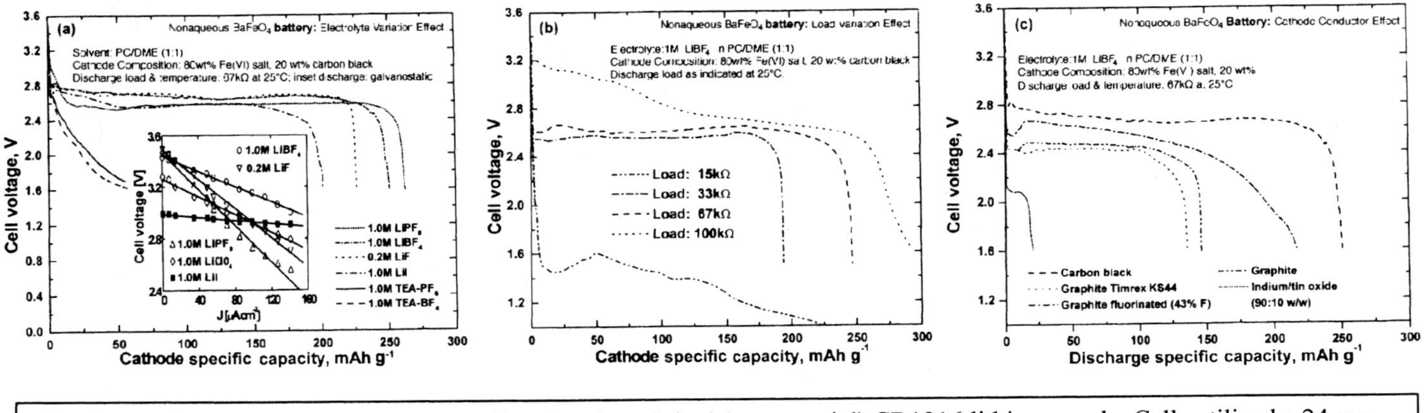

Cell configuration: 1 cm diam button cells using the original (commercial) CR1216 lithium anode. Cells utilized a 24 mg cathode containing 80 wt% $BaFeO_4$ (<150 μm), 20 wt% indicated conductor, and the indicated electrolyte

Figure 6.2. (a): Electrolyte effect on the $BaFeO_4$ cathode capacity in Li anode cell discharged with PC/DME electrolytes containing a variety of different salts. Inset: Polarization measured as a function of potential measured with constant current at 25°C. Conductor: carbon black; Discharge load: 67 kΩ. (b): Constant load variation effect on $BaFeO_4$ cathode capacity in Li anode cell with 1 M $LiBF_4$ PC/DME electrolyte. Conductor: carbon black. (c): Cathode conductive matrix effect on $BaFeO_4$ cathode capacity in Li anode cells with 1 M $LiBF_4$ PC/DME electrolyte. Discharge load: 67 kΩ. (Ref. 21)

Decrease of the cathode salt particle size can substantially increase the interfacial surface area, and in principal improve electrolytic mass diffusion and diminish the solid state diffusion path length. As summarized in Table 6.2, we observed that decrease of Fe(VI) salt particle size improved the Fe(VI) cathode specific storage capacity. In the 1 M $LiPF_6$ PC-DME electrolyte, $BaFeO_4$ particle size decreased from 111 to 54 μm, or <35 μm increased cathode capacity respectively from 241, to 252, or 290 mAh/g (*21*).

Table 6.2. Variation of the $BaFeO_4$ Cathodic Capacity with Either Temperature or $BaFeO_4$ Particle Size. (*Ref. 21*)

Temperature (°C)	*$BaFeO_4$ Particle Size (μm)*	*Average potential*	*$BaFeO_4$ capacity (mAh/g)*
25	< 150	2.57	260
25	73-149	2.68	241
25	35-73	2.62	252
25	< 35	2.62	290
0	< 35	2.56	15
25	< 35	2.62	290
40	< 35	2.73	295
50	< 35	2.63	298

NOTE: Cathodes with the indicated particle size $BaFeO_4$ are composed of 80 wt% $BaFeO_4$ and 20 wt% carbon black. The cells use 1 M $LiPF_6$ in (1:1 v:v) PC/DME, electrolyte, and are discharged over a 67 kΩ load at the indicated temperature.

It is likely that the Fe(VI) cathode will exhibit further improvements in cathodic charge transfer with further optimization of the Fe(IV) particle morphology, and a further enhancement of cathodic charge transfer can occur with discharge temperature. This activation was observed for the smallest particle range explored <35 μm $BaFeO_4$; as also included in Table 6.2, this exhibited increased cathodic capacity to 298 mAh/g with increase of discharge temperature from 25 to 50°C (*21*).

The top portion of Figure 6.3 left compares the specific cathode capacity of small particle (<35 μm) K_2FeO_4 or $BaFeO_4$ cathodes. The combined benefits of the smaller particle size, change of supporting salt and increased discharge temperature also result in a significant increase in specific capacity for the K_2FeO_4 cathode, ~370 mAh/g (compared with 290 mAh/g shown in Figure 6.1).

As presented in section 3, in aqueous system, MnO_2 substantially enhance Fe(VI) charge transfer resulting in a synergistic increase in charge storage for a Fe/Mn co-cathode salt. As seen in Figure 6.3 right, a nonaqueous cocathode consisting of K_2FeO_4 and lithiated MnO_2 also provided an effective cathode. As observed, the co-cathode cell discharged to the same capacity but utilizing only 100 mg of cathode, compared to 153 mg of cathode in the original lithiated

MnO_2 cell, and hence the co-cathode evidences significantly higher capacity than the MnO_2 cathode (without K_2FeO_4). Inset of Figure 6.3 right, are the relative capacities compared for Fe(VI) or Fe(VI) co-cathode cells. In this determination, discharge capacity is measured to 2 V, and relative capacity is calculated normalized to the capacity measured in discharge of the 100 kΩ load cell. The co-cathode is as described above, while the Fe(VI) cathode cell contained only K_2FeO_4 and carbon black (without lithiated MnO_2). In particular, inclusion of MnO_2 as a co-cathode facilitates the cathode discharge. It is evident from the figure inset, that under equivalent heavy loads, the co-cathode discharged to a significantly higher storage capacity than the Fe(VI) cathode alone (*21*).

7. Nonaqueous Reversible Fe(VI) Charge Transfer for Rechargeable Super-Iron Lithium Batteries

Electrochemical Behavior Fe(VI) in Nonaqueous Electrolyte & Reversible Super-Iron/Li Battery

As presented at the beginning of this section, the Fe(VI) storage mechanism for cell discharges in Li containing nonaqueous media had been considered as the intercalation and insertion of Li^+. However, the more recent evidence suggests that the iron centers undergo a partially reversible faradaic reduction in the compounds from 6^+ to 3^+ states, central to this interpretation is direct Mössbauer measurement of the iron valence state of the cathode during charge storage (*32*).

Figure 7.1 shows typical cyclic voltammograms (CVs, experimental details are in reference 32) of composite electrodes comprising K_2FeO_4 or $BaFeO_4$, carbon black and PVdF, in PC-$LiClO_4$ solution.

These voltammograms show that both compounds, K_2FeO_4 and $BaFeO_4$, are electrochemically active. A reduction process is clearly seen in the 1.5–2.5 V (Li/Li^+) range for K_2FeO_4 and in the 2–3 V range for $BaFeO_4$, with corresponding oxidation processes at around 3–4 and 3.5–4 V, respectively. The broad peaks with the large hysteresis between the cathodic and anodic peaks (1.5 V) indicate that the electrochemical process is complex and can suffer significant kinetic limitations. $BaFeO_4$, however, shows a slightly smaller hysteresis than K_2FeO_4 during lithiation and delithiation processes, although the cathodic and anodic peaks of the CVs of $BaFeO_4$ electrodes are also broad. Adding larger amounts of carbon additive to the composite electrodes leads in both cases (K_2FeO_4 and $BaFeO_4$ electrodes) to a decrease in the hysteresis between the cathodic and anodic peaks. 20%, rather than 10%, carbon black in a $BaFeO_4$ electrode results in a twofold increase of the observed electrode capacity (100 instead of 50 mAh/g) (*32*).

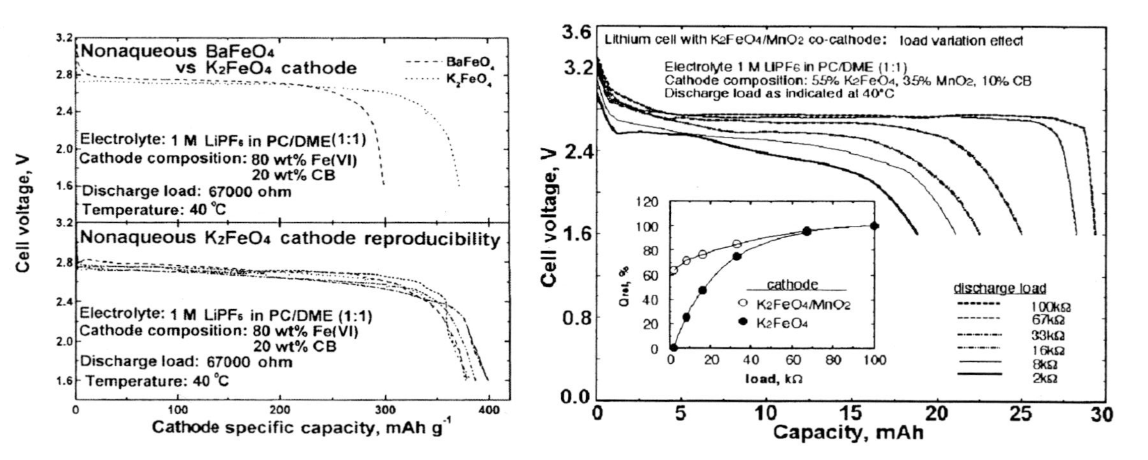

Figure 6.3. Left: Fe(VI) cathode/Li anode cells with 1 M $LiPF_6$ PC/DME electrolyte. Top: The specific cathode capacity of 24 mg cathodes containing 80 wt% K_2FeO_4 or $BaFeO_4$ (small particle < 35 μm diam salt) and 20 wt% carbon black. Bottom: K_2FeO_4 cathode discharge reproducibility. Right: Discharge of Fe(VI)/MnO_2 co-cathode/Li anode cells with PC/DME electrolyte. The cathode comprises 100 mg dry mass of cathode (100 mg total mass of K_2FeO_4, MnO_2 and CB) to balance the available anode capacity of 29 mAh. Cells are discharged at 40°C utilizing the indicated constant load. (Ref. 21)

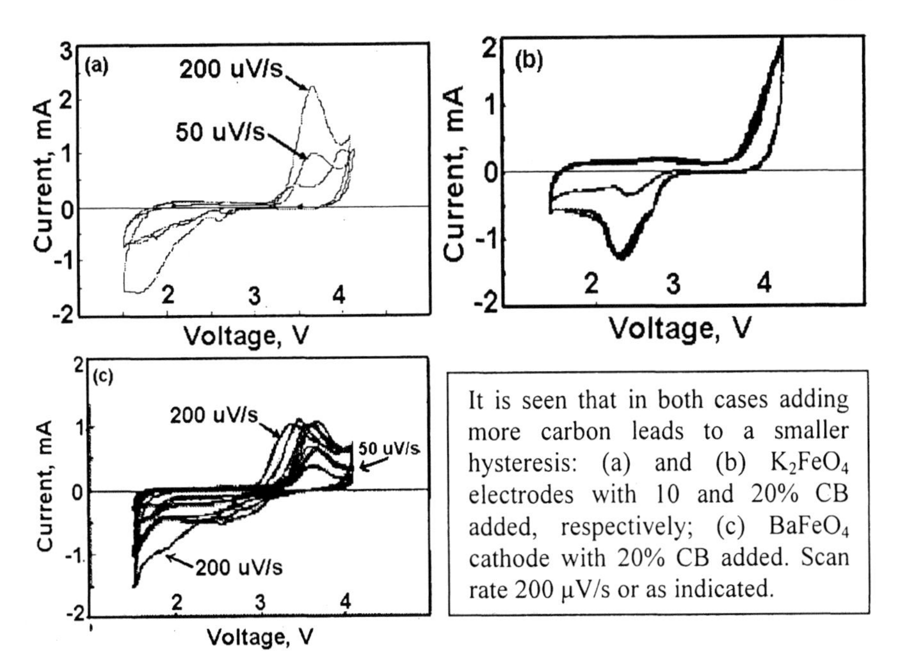

Figure 7.1. CV of K_2FeO_4 and $BaFeO_4$ cathodes with a different amount of added carbon. (Ref. 32)

The electrochemical response of the nonaqueous Fe(VI) cathodes reflects several kinetic challenges related to both the intrinsic properties of the active mass, and to the electrical contact among the particles. For example, there is a positive shift of the anodic peak in the CVs of the composite K_2FeO_4 electrodes when the amount of the carbon black (CB) is increased from 10 to 20% and the active mass is lowered accordingly to 70 from 80% (Figure 7.1). In parallel, the specific capacity of the active mass is higher for the electrode comprising 70% active mass, probably due to the higher amount of CB which improves the interparticle electrical contact. These electrodes (70% active mass) reach their maximal capacity during cycling faster than electrodes comprising higher active mass but lower carbon content. The electrochemical activity of these electrodes does not appear to be affected by solution reactions, and the charge storage reactions occur within the known electrochemical window of the electrolytic solutions. The highest capacity for K_2FeO_4 composite electrodes was observed with 10% CB. For further investigations, 20 and 10% CB for $BaFeO_4$ and K_2FeO_4 electrodes, respectively, were chosen as optimal (*32*).

Figure 7.2a and b shows typical chronopotentiograms of K_2FeO_4 and $BaFeO_4$ composite electrodes, respectively, in repeated galvanostatic cathodic and anodic polarization in EC-PC 1:1/$LiClO_4$ 1 M and $LiAsF_6$ 1 M THF solutions, as indicated. Results from different solutions are also presented to

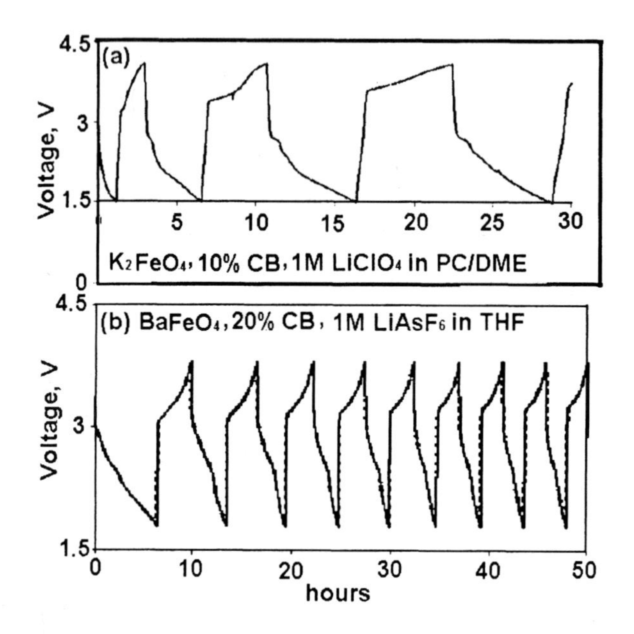

Figure 7.2. (a): A typical chronopotentiogram of a K_2FeO_4 composite electrode, 10% CB, 10% PVdF, cycled in a coin-type cell in a $LiClO_4$ 1 M PC/DME solution. (b): A typical chronopotentiogram of a $BaFeO_4$ composite electrode, 10% CB, 10% PVdF, cycled in a $LiAsF_6$ 1 M THF solution. Galvanostatic cycling at C/10 rates. (Ref. 32)

demonstrate that the role of the electrolyte solution used is minor. The sloping potential profiles between 3 and 1.5 V during the cathodic processes and the sloping potential profiles in the range of 2.5–4 V during the anodic processes correspond to the broad peaks that characterize the voltammetric response presented in Figure 7.1. Elemental analysis of cycled electrodes using atomic absorption (experimental details are in reference 32) shows that lithium is inserted into the active mass in amounts corresponding to the charge measured. Hence, the electrochemical processes presented in Figure 7.1 and Figure 7.2 are consistent with the lithiation and delithiation of the electrodes studied (*32*). Anodic polarization does not remove all the lithium from the active mass, as can be seen by a comparison of the charge involved in the first cathodic polarization and that related to the first anodic process and subsequent cycles and the corresponding element analyses.

Figure 7.3a shows curves of capacity vs cycle number (galvanostatic charge-discharge cycling) of K_2FeO_4 electrodes containing different amounts of carbon, as indicated. As seen in this figure, it takes the electrodes a few cycles to reach a maximal capacity close to the theoretical value (~310 mAh/g), depending on the electrode's composition. The capacity then fades from cycle to cycle and stabilizes at a low value (approximately 1/3 of the observed

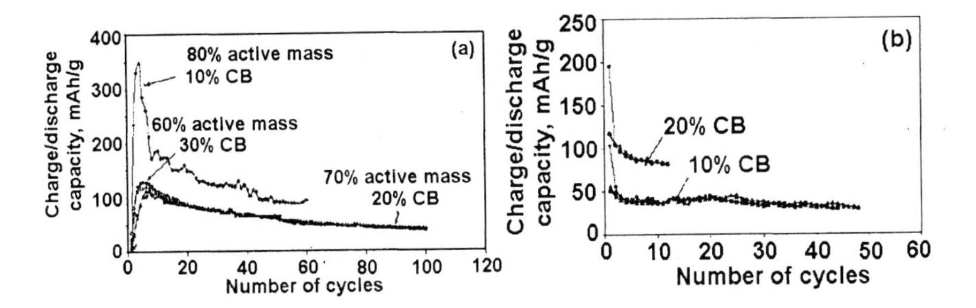

Figure 7.3. (a): Charge and discharge capacity vs cycle number of composite $BaFeO_4$ electrodes with 10, 20, and 30% of CB. (b): Charge and discharge capacity vs cycle number of composite K_2FeO_4 electrodes with 10, and 20% of CB. (Ref. 32)

irreversible capacity). These electrodes could be cycled reversibly hundreds of times in the alkyl carbonate-based solutions, showing low (100 mAh/g) but stable capacity. Similar capacity curves (Figure 7.2b) for $BaFeO_4$ electrodes start from the highest values (above 200 mAh/g) in the first cycle, and then the capacity decreases slightly, but monotonously, until stabilization at around 80–100 mAh/g. Note that during the first cycle of $BaFeO_4$ electrodes, the charge that relates to the first lithiation process is higher than that of the following delithiation process (*32*).

Table 7.1 summarizes a series of electrochemical measurements carried out with different $MFeO_4$ composite electrodes (cycled vs Li counter electrodes in coin-type cells).

Table 7.2 summarizes the elemental analysis of composite $BaFeO_4$ composite electrodes by atomic absorption (AA). The relative amounts of lithium and iron were analyzed in order to measure the stoichiometry of the lithiated compounds at different stages of their lithiation–delithiation process. The relevant capacities were measured electrochemically and are also listed. As seen in Table 7.2, the lithium content of electrodes that underwent cathodic polarization as a single step, measured by AA, was always lower than what would be expected from the charge measurements. This is consistent with either the occurrence of side reactions, which do not involve lithiation (e.g., solvent reduction that forms solution species), or a partial self-discharge of the lithiated electrodes, as they are disconnected from the potentiostat at the low potential. Oxidation/delithiation of these electrodes leaves an appreciable amount of lithium trapped in the electrodes, as measured by both electrochemistry and elemental analysis. For electrodes that underwent a lithiated–delithiated cycle, there is a good match between the AA data and the charge measurements. This supports the assumption that the lithiated electrodes show lower lithium content than the corresponding charge measured, due to some delithiation that occurs during treatment (*32*).

Table 7.1. Summary of Galvanostatic Measurements of Various $MFeO_4$ Salts. (*Ref. 32*)

Type of electrode	*Electrolytic solutions*	*General electrochemical behavior*	*Capacity and capacity fading*
$BaFeO_4$ 10%, 20% CB 10%, 20% KS6 10% KS6-10% CB	$LiClO_4$-EC/PC, $LiAsF_6$-THF, $LiAsF_6$-2Me-THF, $LiAsF_6$-DMC	Good cyclibility at prolonged charge-discharge processes. Pronounced hysteresis between the cathodic and anodic processes.	Peak value, about 200 mAh/g in the first lithiation, may be cycled at a steady capacity of around 90 mAh/g at hundreds of cycles.
K_2FeO_4 10%, 20% CB	$LiClO_4$-EC/PC, $LiAsF_6$-THF, $LiAsF_6$-2M-THF, $LiAsF_6$-DMC	Good cyclibility at prolonged charge-discharge processes. The capacity increases to a peak value and then fades to low values and stabilizes.	Peak value, about 310 mAh/g, may be cycled with capacities of 85 mAh/g during hundreds of cycles.
Sr_2FeO_4 10%, 20% CB	$LiClO_4$-EC/PC	Poor cyclibility, low capacity	Less than 20 mAh/g

There is evidence from elemental analysis for dissolution of K^+ ions from K_2FeO_4 electrodes into $LiClO_4$ 1 M EC/DME solutions. No dissolution of Fe ions could be detected. Hence, it is suggested that some K^+ ions are substituted by Li^+ ions in the active mass (*32*).

Unlike thicker films, thin Fe(VI/III) films exhibit extensive, high capacity nonaqueous rechargeablity (over 300 mAh/g capacity for >> 50 charge/discharge cycles). Reversibility of the electrodes was also probed using thin films of K_2FeO_4 on Pt foils in a PC/DME 2:1 1 M $LiClO_4$ solution (*32*). The electrodes are galvanostatic cycled at C rates (i.e., the entire capacity is discharged and charged during less than 1 h per process). In these experiments, the complications related to the composite structure of the usual electrodes were eliminated, and hence, the intrinsic behavior of the active mass could be better examined. As seen in Figure 7.4a, K_2FeO_4 thin-film electrodes could be cycled reversibly at a capacity above 300 mAh/g with a little capacity fading during prolonged cycling. The fact that the discharge capacity is slightly higher than the charge capacity can be explained by the reduction of PC at the lowest potentials that the electrodes reach in these experiments (1.5 V vs Li/Li^+) (*51*). As seen

Table 7.2. Correlation Between Charge Measured (Chronopotentiometry) and Li Content (AA) of $BaFeO_4$ Electrodes During First Lithiation and First Lithiation-Delithiation Cycle. (Ref. 32)

Electrochemical process	*Capacity measured during first lithiation (mAh/g)*	*Capacity measured during first delithiation (mAh/g)*	*Capacity measured from AA (mAh/g)*
Slow polarization to 1.5 V at C/45 rates	236.5	Not measured	177.37
Slow polarization to 1.5 V at C/45 rates	180	Not measured	154.41
Slow polarization to 1.5V at C/45 rates	207	Not measured	140.85
Slow polarization to 1.5 V at C/45 rates	207	Not measured	125.2
Slow polarization to 1.5 V and then to 4 V at C/45 rates	222.6	104.33	112.47
Slow polarization to 1.5 V and then to 4 V at C/45 rates	236.4	111.28	121.46

from the potential profiles in Figure 7.4b (galvanostatic cycling in the $LiClO_4$ 1 M PC/DME solution), there are two processes that take place during the film electrode's oxidation. It is possible that the Fe^{6+} compound is reformed by a two-step process via a Fe^{4+} intermediate process (*32*).

Mössbauer spectroscopy confirmed that cathodic polarization (lithiation) of K_2FeO_4 changes the oxidation state of the iron from 6^+ to 3^+, and that the reverse anodic process (delithiation) of the reduced material regenerates an iron 6^+ compound. The Mössbauer spectroscopic measurements provided direct evidence that the corresponding cathodic and anodic processes of these compounds, as described above (Figure 7.1-7.4), involve a reversible change in the oxidation state of the central iron atoms of the active materials. The Mössbauer spectroscopic measurements showed similar changes in the oxidation state of iron during galvanostatic cycling of $BaFeO_4$ composite electrodes. The data related to the Mössbauer spectroscopic measurements of composite $BaFeO_4$ electrodes is summarized in Table 7.3. In general, there is a correlation between the charge measured electrochemically and the quantitative analysis by the Mössbauer spectroscopy. Cathodic polarization transforms Fe^{6+} to Fe^{3+}, while a consecutive polarization reforms the Fe^{6+} compound. It can be clearly seen that about 17% of the active electrode's Fe^{6+} compounds decomposed or underwent irreversible changes during prolonged cycling in several experiments (*32*).

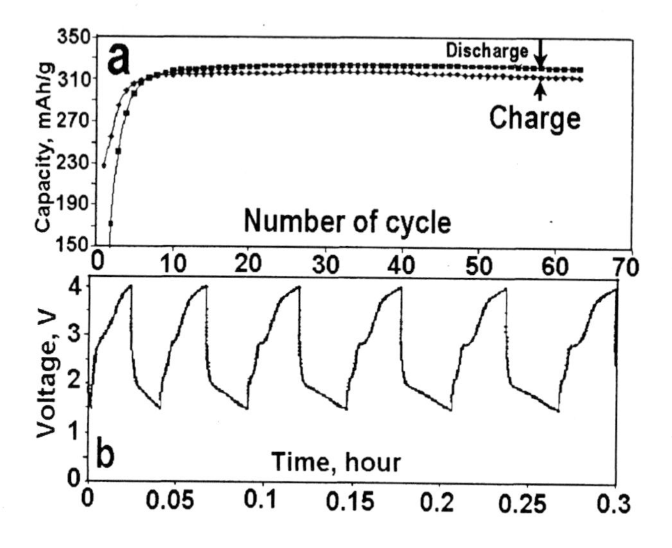

Figure 7.4. The electrochemical behavior of a thin-film K_2FeO_4 on a Pt electrode, stabilized in NaOH 10 M, and then cycled in a $LiClO_4$ 1 M PC/DME 2:1 solution: (a) charge and discharge capacity vs cycle number (galvanostatic cycling at C/10 rates) and (b) typical potential profile of these electrodes upon galvanostatic cycling. (Ref. 32)

Table 7.3. Summary of Mössbauer Spectroscopy Analysis of $BaFeO_4$ Electrodes. (*Ref. 32*)

Type of process $LiClO_4$ 1 M PC-DME 2:1 solution	*Measured capacity during electrochemical reduction and oxidation processes (mAh/g)*	*Percentage of Fe^{6+} (%)*	*Percentage of Fe^{3+} (%)*
Polarization to 1.8 V and then to 4 V at C/45 rates.	Reduction 68 Oxidation 30	89	11
Ten complete cycles in the 1.8-4 V range and removed at 1.8 V, C/45 rates.	Average of charge-discharge capacity 35	83	17
One reduction/oxidation cycle, 1.5-4 V range, at C/45 and removed at 3 V.	Charge and discharge capacity 35	83	17
Slow polarization to 1.8 V at a C/45 rate.	Reduction 85	79	21
20 complete cycles in the 1.8-4 V potential range and removed at 3·V.	Average of charge-discharge 37	83	17
$BaFeO_4$ pristine electrode		92	8

The results above show that the Fe(VI) compounds can be electrochemically reduced, and re-oxidized in nonaqueous Li salt solutions. The charge/discharge process of Fe(VI)/Fe(III) in nonaqueous medium involves the lithiation/delithiation of the active mass and the reduction/oxidation of the iron (from 6^+ to 3^+ or from 3^+ to 6^+ states). (confirmed by AA, Mössbauer spectroscopy, and charge measurements in the electrochemical processes) (*32*).

Conclusions

We recently introduced a novel battery type based, "Super-Iron" batteries, on a class of cathodes incorporating Fe(VI), and sustaining facile, energetic, cathodic charge transfer. Due to their highly oxidized iron basis, multiple electron transfer, and high intrinsic energy, we have defined Fe(VI) compounds as 'super-iron's, and such compounds can provide useful cathode salts for Super-Iron batteries.

Fe(VI) salts are capable of efficient three-electron reduction and sustains higher electrochemical storage capacity than conventional cathode materials. Discharge products of Fe(VI) are environmental benign. A series of Fe(VI) compounds have been synthesized and explored in Super-Iron batteries in our lab, include the synthesized Fe(VI) salts with 3-electron cathodic charge capacity Li_2FeO_4, Na_2FeO_4, K_2FeO_4, Rb_2FeO_4, Cs_2FeO_4 (alkali Fe(VI) salts), as well as alkali earth Fe(VI) salts $BaFeO_4$, $SrFeO_4$, and also a transition Fe(VI) salt Ag_2FeO_4 which exhibits a 5-electron cathodic charge storage. Four battery types, with primary and rechargeable Fe(VI) cathodes, in aqueous (alkaline) and nonaqueous electrolyte systems have been developed.

Solid Fe(VI) salts, such as K_2FeO_4 are stable and Fe(VI) salts exhibit low solubility in concentrated KOH electrolytes, such as commonly used in alkaline batteries. A easily deposited zirconia overlayer on the Fe(VI) salt protects (stabilizes) the salt when in contact with an alkaline solution. Primary alkaline Super-Iron battery chemistry was established with Fe(VI) cathodes and zinc anodes, and sustain several fold higher capacity than conventional alkaline batteries. Configuration optimization, enhancement and mediation of Fe(VI) cathode charge transfer of primary Fe(VI) alkaline batteries have been detailed. Small particle (1 μm) graphite and compressed carbon black are shown to provide a superior conductive matrix for the alkaline Super-Iron cathodes. Fluorinated polymer graphites provide an unusual additive to Fe(VI) cathode, and are observed to simultaneously maintain two roles in the cathode; not only acting as a conductive matrix, but also adding intrinsic capacity to the cathode. Several inorganic additives (such as $SrTiO_3$) can also improve the faradaic efficiency of Fe(VI) reduction. The alkaline Fe(VI) cathode potential can be shifted and controlled by the solid phase modifiers MnO_2 (decreasing ~200 mV) and Co_2O_3 (increasing ~150 mV). High capacity Fe(VI) salts (such as K_2FeO_4) tend to be passivated in alkaline media and hence the charge transfer is inhibited. 2 effective mediations and a coating were demonstrated. The chemical mediation of Fe(VI), with a Mn(VII) or Mn(VI) additive, improved cathodic discharge efficiency of K_2FeO_4. The combined chemical and electronic Ag(II) mediation of Fe(VI) redox chemistry, by Ag(II), significantly improves alkaline Fe(VI) cathodic charge transfer. Composite Fe(VI)/Mn(VI or VII), Fe(VI)/MnO_2, as well as Fe(VI)/Ag(II) cathodes provide much higher power energy capacity than the single cathodes. A novel Zirconia coating derived from an organic soluble zirconium salt ($ZrCl_4$) through an organic media, significantly stabilized Fe(VI) cathodes and extended the storage life of Super-Iron batteries.

Reversible Fe(VI) chemistry was developed with Fe(VI/III) thin film cathodes. The films were deposited, by electrochemical reduction of Na_2FeO_4, with an intrinsic high capacity 3e^- cathodic storage of 485 mAh/g, on either smooth or on extended conductive matrixes composed of high-surface-area Pt, Ti and Au. Whereas ultra-thin (nano-scale) Fe(VI/III) film exhibited high degree of reversibility, thicker films had been increasingly passive toward the Fe(VI) charge transfer. The extended conductive matrix facilitated a 2 orders of magnitude enhancement in charge storage for reversible Fe(III/VI) Super-Iron thin films. A 100 nm Fe(VI) cathode, on the extended conductive matrixes, sustained 100-200 reversible three-electrode charge/discharge cycles, and a 19 nm thin film cathode sustained 500 such cycles. With a metal hydride anode, in a full cell, a 250 nm super-iron film cathode film sustained 40 charge/discharge cycles, and a 25 nm film was reversible throughout 300 cycles.

Super-iron cathodes can also be discharged in conjunction with Li anode in nonaqueous media. Through selection of electrolytes (including solvent and supporting Li-contained salt), use of carbon black as the conductor, reduction of cathode particle size, and moderate increase of temperature (from room temperature to 40°C), the cathodic charge transfer of Fe(VI) in nonaqueous electrolyte was enhanced. Co-cathodes comprised of Fe(VI)/MnO_2 exhibited high storage capacity.

Fe(VI) compounds can be electrochemically reduced in nonaqueous Li salt solutions and Fe(VI) cathode also exhibited reversibility in nonaqueous battery system. The charge/discharge process of Fe(VI)/Fe(III) in nonaqueous medium involves the lithiation/delithiation of the active mass and the reduction/oxidation of the iron (from 6^+ to 3^+ or from 3^+ to 6^+ states). However, similar to that in the aqueous system, only the thin-film Fe(VI) electrodes can sustain high reversibility involve the full theoretical capacity in nonaqueous battery system.

References

1. Mellor J. W. *A Comprehensive Treatise on Inorganic and Theoretical Chemistry;* Longmans, Green and Co.: London, **1934**; Vol. XIII, p 930.
2. BeMiller J.; Kumari G.; Darling S. *Tetrahedron Lett.* **1972**, *40*, 41.
3. Schink T.; Waite T. D. *Water Research* **1980**, *14*, 1705.
4. Sakurai Y.; Arai H.; Okada S.; Yamaki J. *J. Power Sources*, **1997**, *68*, 711.

5. Hua S.; Cao G.; Cui Y. *J. Power Sources*, **1998**, *76*, 112.
6. Licht S.; Wang B.; Ghosh S. *Science* **1999**, *285*, 1039.
7. Licht S.; Wang B.; Ghosh S.; Li J.; Naschitz V. *Electrochem. Comm.* **1999**, *1*, 522.
8. Licht S.; Wang B.; Xu G.; Li J.; Naschitz V. *Electrochem. Comm.* **1999**, *1*, 527.
9. Licht S.; Wang B. *Electrochem. Solid- State Lett.* **2000**, *3*, 209.
10. Licht S.; Wang B.; Li J.; Ghosh S.; Tel-Vered R. *Electrochem. Comm.* **2000**, *2*, 535.
11. Licht S.; Naschitz V.; Ghosh S.; Liu B.; Halperine N.; Halperin L.; Rozen D. *J. Power Sources* **2001**, *99*, 7.
12. Licht S.; Naschitz V.; Lin L.; Chen J.; Ghosh S.; Lui B. *J. Power Sources* **2001**, *101*, 167.
13. Licht S.; Naschitz V.; Ghosh S.; Lin L.; Lui B. *Electrochem. Comm.* **2001**, *3*, 340.
14. Licht S.; Ghosh S.; Naschitz V.; Halperin N.; Halperin L. *J. Phys. Chem., B* **2001**, *105*, 11933.
15. Licht S.; Ghosh S.; Dong Q. *J. Electrochem. Soc.* **2001**, *148*, A1072.
16. Licht S.; Naschitz V.; Ghosh S. *Electrochem. Solid-State Lett.* **2001**, *4*, A209.
17. Licht S.; Tel-Vered R.; Halperin L. *Electrochem. Comm.* **2002**, *4*, 933.
18. Licht S.; Naschitz V.; Wang B. *J. Power Sources*, **2002**, *109*, 67.
19. Licht S.; Ghosh S. *J. Power Sources* **2002**, *109/2*, 465.
20. Licht S.; Naschitz V.; Ghosh S. *J. Phys. Chem. B* **2002**, *106*, 5947.
21. Tel-Vered R.; Rozen D.; Licht S. *J. Electrochem. Soc.* **2003**, *150*, A1671.
22. Ghosh S.; Wen W.; Urian R. C.; Heath C.; Srinivasamurthi V.; Reiff W. M.; Mukerjee S.; Naschitz V.; Licht S. *Electrochem. Solid-State Lett.* **2003**, *6*, A260.
23. Licht S.; Tel-Vered R.; Halperin L. *J. Electrochem. Soc.* **2004**, *151*, A31.
24. Licht S.; Tel-Vered R. *Chem. Comm.* **2004**, 628.
25. Licht S.; Naschitz V.; Rozen D.; Halperin N. *J. Electrochem. Soc.* **2004**, *151*, A1147.
26. Licht S.; Yang L.; Wang B. *Electrochem. Comm.* **2005**, *7*, 931.
27. Licht S.; Yu X. *Environ. Sci. Technol.* **2005**, *39*, 8071.
28. Nowik I.; Herber R. H.; Koltypin M.; Aurbach D.; Licht S. *J. Phys. Chem. Solids* **2005**, *66*, 1307.
29. Koltypin M.; Licht S.; Tel-Vered R.; Naschitz V.; Aurbach D. *J. Power Sources* **2005**, *146*, 723.
30. Licht S.; DeAlwis C. *J. Phys. Chem. B* **2006**, *110*, 12394.
31. Licht S.; Yu X.; Zheng D. *Chem. Comm.* **2006**, 4341.
32. Koltypin M.; Licht S.; Nowik I.; levi E.; Gofer Y.; Aurbach D. *J. Electrochem. Soc.* **2006**, *153*, A32.
33. Yang W.; Wang J.; Pan T.; Xu J.; Zhang J.; Cao C. *Electrochem. Comm.* **2002**, *4*, 710.
34. Bouzek K.; Schmidt M. J.; Wragg A. *Electrochem. Comm.* **1999**, *1*, 370.

35. Lee J.; Tryk D.; Fujishima A.; Park S. *Chem. Comm.* **2002**, 486.
36. De Koninck M.; Brousse T.; Belanger D. *Electrochim. Acta* **2003**, *48*, 1425.
37. Nardi J. C.; Swierbut W.M. US Patent 5895734. **1999**.
38. Gohr H. *Electochim. Acta* **1966**, *11*, 827.
39. Mehne, L. F.; Wayland, B. B. *J. Inorg. Nucl. Chem.* **1975**, *37*, 1371.
40. *US Federal Register;* **1997**; Vol. 62, p 367.
41. Martinz H.P.; Nigg B.; Matej J.; Sulik M.; Larcher H. Hoffmann A. *Int. J. Refract. Met. Hard Mater.* **2006**, *24*, 283.
42. Ibanez R.; Martin F.; Ramos-Barrado J. R.; Leinen D. *Surf. Coat. Technol.* **2006**, *200*, 6368.
43. Thackeray M. M.; Johnson C. S.; Kim J.-S.; Lauzze K. C.; Vaughey J. T.; Dietz N.; Abraham D.; Hackney S. A.; Zeltner W.; Anderson M. A. *Electrochem. Comm.* **2003**, *5*, 752.
44. Linke W. F. *Solubilities of Inorganic and Metal-Organic Compounds;* Van Nostrand: Princeton, N. J., **1958**; 4th ed., p 1695.
45. Bouzek, K.; Roušar, I.; Bergmann, H.; Hertwig, K. *J. Electroanal. Chem.* **1997**, *425*, 125.
46. Kamnev; A. A., Ezhov; B. B. *Electrochim. Acta* **1992**, *37*, 607.
47. Licht, S.; Halperin, L.; Kalina, M.; Zidman, M.; Halperin, N. *Chem. Comm.* **2003**, 3006.
48. Mentus; S. V. *Electrochim. Acta* **2005**, *50*, 3609.
49. Hu; C-C., Liu; K-Y. *Electrochim. Acta* **1999**, *44*, 2727.
50. Ovshinsky S. R.; Fetcenko M.; Ross J. *Science* **1993**, *260*, 176.
51. Aurbach D.; Daroux M. L.; Faguy P.; Yeager E. *J. Electroanal. Chem. Interfacial Electrochem.* **1991**, *297*, 225.

Chapter 15

Transformation of Iron(VI) into Iron(III) in the Presence of Chelating Agents: A Frozen Solution Mössbauer Study

Mössbauer Investigation of the Reaction between Iron(VI) and Chelating Agents in Alkaline Medium

Zoltán Homonnay[1,3], Nadine N. Noorhasan[2], Virender K. Sharma[2], Petra Á. Szilágyi[1], and Ernő Kuzmann[3]

[1]Laboratory of Nuclear Chemistry, Institute of Chemistry, Eötvös Loránd University, Pázmány P. s. 1/A, Budapest, Hungary
[2]Department of Chemistry, Florida Institute of Technology, 150 West University Boulevard, Melbourne, FL 32901
[3]Laboratory of Nuclear Chemistry of the Chemical Research Center of the Hungarian Academy of Sciences, Pázmány P. s. 1/A, Budapest, Hungary

Mössbauer spectroscopy is a good tool to look at the transformations of iron containing species in various reactions using the frozen solution technique. This study presents the fate of iron(VI) as being reduced to iron(III) in the presence of complexing agents, glycine and ethylenediaminetetraacetic acid (EDTA). The intermediate oxidation states, Fe^{V} and Fe^{IV} of the reaction were not seen within the time scale of sample freezing (≥ 5 s). The characterization of ferric species formed from Fe^{VI} reactions with glycine and EDTA was carried out.

Introduction

Although Fe^{VI} has been known for many years, its aqueous chemistry has not been well established. Fe^{VI} has been of great interest because of its role as an oxidant and a hydroxylating agent in remediation and industrial processes (*1-5*). Other high oxidation states of iron, Fe^{V} and Fe^{IV}, are postulated as intermediates in the decay kinetics of Fe^{VI} as well as in the oxidation of inorganic and organic substrates (1,6-10). No direct experimental evidence has been provided to confirm intermediate oxidation states of iron, Fe^{V} and Fe^{IV}, in the oxidation processes carried out by Fe^{VI}. This paper seeks to observe intermediates, Fe^{V} and Fe^{IV} in oxidation of glycine (NH_2CH_2COOH) and ethylenediaminetetraacetate, EDTA ($(CH_2)_2N_2(CH_2COOH)_4$), by Fe^{VI} in alkaline medium.

Glycine serves as a model for the more complex amino acids (*11,12*). It may also be used as a model for amino polycarboxylic acids such as EDTA. EDTA is a representative of complexing agents in inorganic chemistry and as being used in decontamination technologies, may be present in low-level nuclear waste (*13*). In natural waters, the origin of organic nitrogen compounds is not only from humic substances but also from total amino acids that can be found in several micrograms per liter in surface waters (*14,15*). Within this category, glycine represents one of the major amino acids found in natural waters. Determination of intermediates in oxidation of glycine and EDTA by Fe^{VI} species is thus relevant to both biological and environmental processes and is not fully understood. Interestingly, the interaction between transition metals and EDTA has been shown to promote the formation and stabilization of higher oxidation states of the corresponding metals (*16*).

This paper presents the use of Mössbauer spectroscopy in conjunction with rapid-freeze solution technique to observe Fe^{V} and Fe^{IV} in oxidation of glycine and EDTA by Fe^{VI} in alkaline medium. Similar experiments were also conducted in the reaction of Fe^{III}EDTA with Fe^{VI} in basic pH solution. Fe^{V} and Fe^{IV} are relatively more stable at basic pH; thus, reactions were carried out in alkaline medium to make sure that intermediate iron species may be stabilized and become observable in Mössbauer measurements. The reaction of ferrate promotes the coagulation process, and subsequently forms a ferric ion that greatly enhances the aggregation and settling process (*17*). Fe^{VI}, as a coagulant, has thus been proven effective for removing organic matter, nutrients, and some metals, at low levels (*2,4,17,18*). However, the nature of ferric oxide/hydroxide formed from Fe^{VI} has not been studied. This paper also presents the detail of Mössbauer experiments on the final product(s) of reactions of Fe^{VI} with glycine and EDTA and of a reaction of Fe^{II}-EDTA system with aerial oxygen in order to fully characterize the Fe^{III} species formed from in the reactions.

Experimental

In the frozen solution Mössbauer spectroscopy experiments, the reactants were allowed to react for 5-15 seconds and the reaction mixture was immediately frozen on an aluminum slab soaked in liquid nitrogen, with small holes drilled in its surface. This freezing technique provides a cooling rate which is safely over the critical cooling rate necessary to preserve the original structure of the solution and the iron species in the liquid state (*19*). The frozen beads obtained from liquid droplets were collected from these holes and were transferred into the Mössbauer sample holder. The sample was then inserted into a bath type cryostat in order to keep the temperature of solid sample at 83±2 K.

The Mössbauer spectra were recorded in transmission geometry with a conventional Mössbauer spectrometer (Wissel). The spectrometer was operated in constant acceleration mode and a ^{57}Co source of 400 MBq provided the gamma rays. Isomer shifts are given relative to α-Fe at room temperature. All measurements were carried out at 83±2 K. The Mössbauer spectra were analyzed by the least squares fitting using Lorentzian lines with the help of the MOSSWINN code (*20*).

Results and Discussion

Reaction Between Ferrate(VI) and Glycine

In this experiment, solid potassium ferrate(VI) was dissolved in 5M NaOH to obtain a 0.1 M Fe^{VI} solution. Sufficient amounts of solid glycine were added and the final ratios of glycine to Fe were 0.5 : 1.0 and 1.5 : 1.0 in the mixed solution. The deep purple color of Fe^{VI} promptly disappeared and the solution became yellow and opalescent. The Mössbauer spectra of the frozen solutions (see *b* and *c* in Figure 1) showed a single component, a doublet with parameters typical for Fe^{III} (isomer shift, δ=0.43(2) mm/s, quadrupole splitting, Δ=0.72(2) mm/s). This indicates the formation of "amorphous" $Fe(OH)_3$ gel (ferrihydrite) (*26, 21*) as a final product of Fe^{VI} reaction with glycine. The features of the Mössbauer spectrum did not change appreciably with the concentration of glycine. The small asymmetry of doublet *c* in Fig. 1 is most probably due to two (or more) slightly different Fe-species; suggesting that the degradation products of glycine may be coordinating to iron(III) in the ferrihydrite giving rise to another chemical environment for iron. The intermediate oxidation states, Fe^{V} and Fe^{IV} of the reaction were thus not seen. There is a possibility that Fe^{V} and Fe^{IV} species formed from the reaction are either further reacting with glycine (*22,23*) or self-decomposing to give Fe^{III} (*6,24,25*) within the time scale of sample freezing (≥ 5 s).

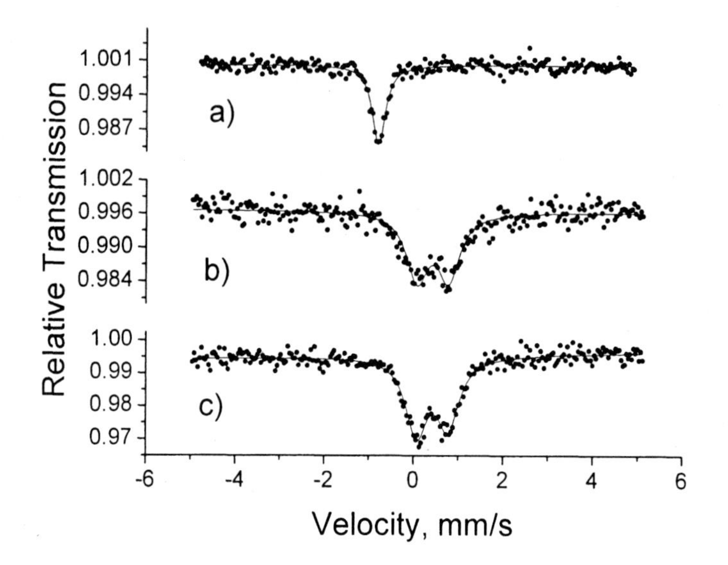

Figure 1. Mössbauer spectra of the frozen solutions obtained after reacting ferrate(VI) in 5 M NaOH with glycine. Molar ratios of glycine to Fe: 1.5 (b) and 0.5 (c). a) represents the frozen solution spectrum of the initial ferrate.

Reaction Between Ferrate(VI) and EDTA

In this experimental set-up, solid EDTA as a disodium salt was added into 0.1 M Fe^{VI} solution in 5M NaOH and ratio of EDTA to Fe was 1 : 0.5. Similar to the Fe^{VI} reaction with glycine, the deep purple color of Fe^{VI} in the mixed solution immediately disappeared in presence of EDTA. This solution was also yellow and opalescent. The Mössbauer spectrum of the frozen solution (Figure 2a) showed a single component, a doublet representing ferrihydrite. The Fe^{V} and Fe^{IV} species as intermediates of the reaction were also not observed.

Next, the amount of EDTA added was tripled to have the presence of excess EDTA in the mixed solution ([EDTA]/[Fe]=1.5). It was expected that if half a mole of EDTA was able to reduce Fe^{VI} to Fe^{III}, the extra moles of EDTA would be enough to form a chelate and keep Fe soluble. However, the Mössbauer spectrum of the reaction mixture showed exactly the same doublet (Figure 2 b). Thus, EDTA was not able to chelate Fe^{III}. Kinetic studies from our laboratory have shown that the oxidation of EDTA by Fe^{VI} is very slow (*27*). Therefore, observed insoluble Fe^{III} in the reaction can possible be due to some catalytic decomposition of ferrate. The high concentration of OH^- did not allow the formation of Fe^{III}-EDTA complex. It though seems to be that EDTA reduced Fe^{VI} to Fe^{III} and did not form a possible soluble chelate, the μ-oxo-bridged complex $O[Fe(EDTA)]_2$ expected at moderately high pH (*28*).

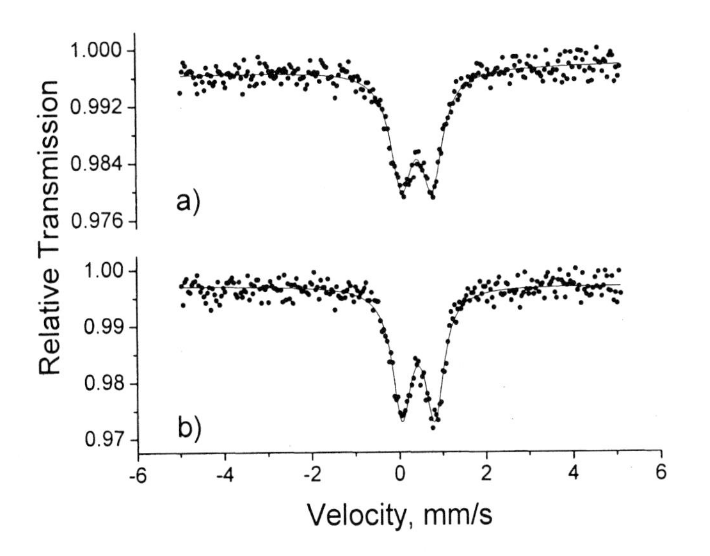

Figure 2. Mössbauer spectrum of the frozen solution obtained after reacting ferrate(VI) in 5 M NaOH with solid EDTA. Molar ratios of EDTA to Fe: 1.5 (a) and 0.5 (b).

Fe^{VI} was also found to be stable in a 0.001 M borate and 0.005 M phosphate buffer (*1*). This buffer can solubilize Fe^{III} and prevents precipitation of Fe^{III} as oxide/hydroxide. The 0.1 M Fe^{VI} solution was thus prepared in borate/phosphate buffer rather than 5 M NaOH. The pH of the final solution was approximately 10.0. It should be pointed out that the ferrate concentration (0.1 M) is much larger than that used in spectrophotometry and the borate/phosphate buffer may not be considered as a real buffer. This low concentration though avoided the possibility of the presence of extra species in the Mössbauer spectra due to the formation of phosphato or borato complexes. A 0.15 M EDTA was used to react with 0.1 M ferrate(VI). The reaction mixture indicated precipitation, however, the precipitate was a deep purple colored and not light brown as expected for regular amorphous $Fe(OH)_3$. The reaction mixture was frozen without separating the solution and the precipitate.

Surprisingly, the Mössbauer spectrum (Figure 3) indicated one single component, evidently the doublet of the μ-oxo dimer species $O[Fe(EDTA)]_2$ (δ= 0.46 mm/s, Δ=1.63 mm/s). This shows that (i) ferrate did not destroy all the EDTA (ii) coordination of OH^- to Fe^{3+} could not compete with the coordination of EDTA in the presence of borate and phosphate apparently acting as a catalyst. The precipitation seems to be due to low solubility of $K_4O[Fe(EDTA)]_2$ (or a mixed Na-K-salt).

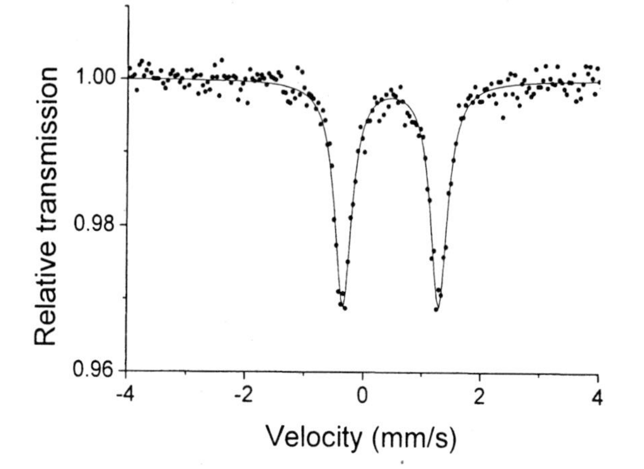

Figure 3. Mössbauer spectrum of the frozen solution obtained after reacting ferrate(VI) with solid EDTA in a borate- phosphate "buffer" at pH≥10.

Reaction Between Ferrate(VI) and Fe^{III}EDTA

It is interesting to study what happens when ferrate(VI) is reacted with EDTA already bound to Fe^{III}. A 0.5 M ferrate(VI) solution in 5M NaOH was reacted with $NaFe^{III}$EDTA (purchased from Aldrich) at the same concentration. The purple color of reaction mixture did not disappear before freezing. The resultant Mössbauer spectrum is shown in Figure 4.

It is clear that the original ferrate(VI) is present (the major singlet on the left side with δ=-0.79 mm/s). The rest of the spectrum indicates ferrihydrite with the same Mössbauer parameters as those observed in the reactions of Fe^{VI} and EDTA in 5M NaOH. From the relative intensity of ferrate(VI) and ferrihydrite being 53:47, it is obvious that the ferrate did not decompose appreciably before freezing the reaction mixture. The deviation from 50:50 can be due to experimental error and different Mössbauer-Lamb factors.

From this experiment, it is not clear whether the formation of ferrihydrite was influenced by the ferrate(VI) or $NaFe^{III}$EDTA just decomposed in the highly alkaline medium. To focus on the fate of iron formed from Fe^{III}EDTA, another experiment was performed using Fe^{III}EDTA prepared from Fe^{III}-nitrate enriched in ^{57}Fe (95%) and stoichiometric EDTA. Due to the low natural abundance of ^{57}Fe (2.17%), in this experiment, the products formed from the EDTA complex can only be seen. The Fe-species that are formed from ferrate(VI) are overshadowed. The resultant Mössbauer spectrum can be seen in Figure 5a.

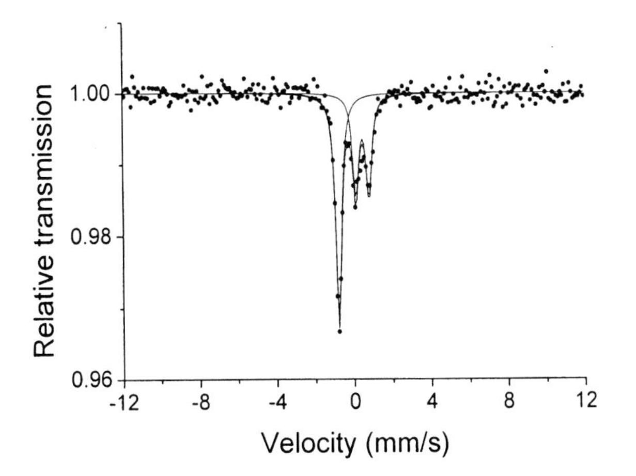

Figure 4. Frozen solution Mössbauer spectrum of the products of the reaction between ferrate(VI) and Fe^{III}EDTA.

The spectrum contains the doublet of ferrihydrite in addition to a sextet with Mössbauer parameters not identical to any known iron hydroxides, oxyhydroxides or oxides (δ= 0.48 mm/s, 2ε=-0.23 mm/s, B=48.8 T). In order to see the role of ferrate(VI), the same experiment was performed but without adding ferrate(VI). The result was the same (Figure 5b). It was concluded from these results that ferrate(VI) may react slowly with EDTA already bound in an Fe-complex.

In Figure 5a, however, the singlet of ferrate(VI) is not present. We assume that the $^{57}Fe^{III}$EDTA prepared from Fe^{III}-nitrate and EDTA probably contained some free EDTA due to lower pH than that of the simple aqueous solution of NaFeEDTA. This accelerated the decomposition of the ferrate and the product (max. ~2.5%) showed the same doublet of ferrihydrite as that formed from $^{57}Fe^{III}$EDTA.

The possible iron oxides and oxyhydroxides which can form at high pH in oxidative environment and are magnetic at around 90 K are hematite, goethite and akaganéite (*21*). Akaganéite may be excluded because of the absence of chloride (which is preferred for the formation of this phase), and the Mössbauer parameters are closest to those of goethite (α-FeOOH). It is remarkable that goethite (a goethite-like phase) forms so rapidly in this reaction. Normally, goethite can be produced from ferrihydrite by slow transformation at 60-90°C. In addition, it is also known that hematite can be synthesized from metal chelates where the formation of goethite is a strong competitive reaction mostly at low temperature. It is very likely that the role of EDTA was crucial in the formation

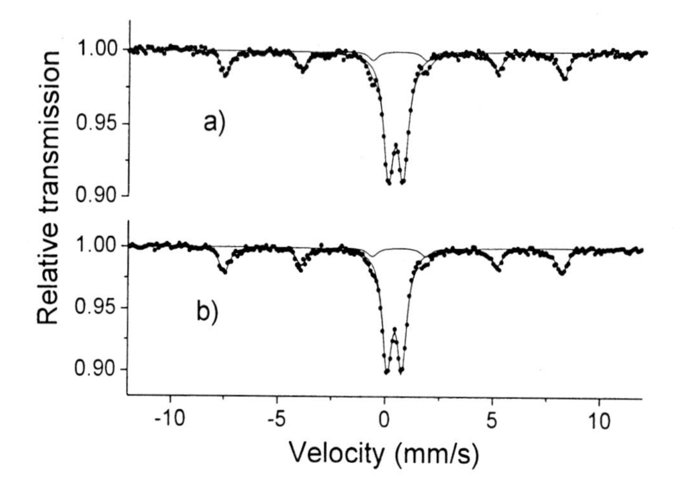

Figure 5. Mössbauer spectra of the species formed from Fe^{III}EDTA after reaction with 5M aqueous NaOH solution, a): containing ferrate(VI), b): not containing ferrate(VI).

of this goethite-like phase. Since the solution(s) became opalescent after the reaction, the colloidal particles formed may have a size distribution that results in superparamagnetism and therefore the doublet (or at least a fraction of it) and the sextet can be assigned to the same phase.

In the other experiment (Figure 4), where enriched iron was not used, the magnetic sextet might have remained hidden in the statistics of the baseline of the Mössbauer spectrum.

Summarizing the results of these experiments, it can be seen that in 5M NaOH, the Fe^{III} product of the decomposition of ferrate(VI) could not be stabilized as a μ-oxo-bis(ethylenediaminetetraacetato-ferrate(III)) dimer complex and precipitated as ferrihidrite. This may be explained by using kinetic and thermodynamic arguments. It is known that ferrihidrite can readily transform into different oxyhydroxides and finally to oxides at high pH (*21*), which can react with EDTA at a very low rate (*29*). This transformation to Fe^{III}-compounds insoluble to EDTA appears to be much faster than the chelation to EDTA reaction, which may give the dimer complex mentioned above. Although this reasoning is certainly valid at about neutral pH, one has also take it into account that due to the solubility product of $Fe(OH)_3$ being $\sim 10^{-39}$ (*30*) and the stability constant of Fe^{III}(EDTA) being 2.0×10^{-26} (*31*), the expected concentration of the chelate is in the nanomolar range. (These data should be considered as rough estimates because (i) the stoichiometry of ferrihydrite is uncertain and (ii) the cited stability constant refers to the monomeric Fe-EDTA complex.)

Nevertheless, it is a fact that at moderately high pH (around 10), when a borate-phosphate buffer was used, ferrate(VI) gave pure μ-oxo-bis(FeEDTA) dimer in its reaction with EDTA. This can be explained by the lower pH and also by a possible catalytic effect of the phosphate and tetraborate ions. Note that under similar conditions but in the absence of borate and phosphate, the regular ferrihydrite precipitation was observed. Selective tracing of iron in $Fe^{III}EDTA^-$ with the help of ^{57}Fe "Mössbauer tracer" while reacting with ferrate(VI) in 5 M NaOH revealed that the iron content of the chelate is transformed partially to ferrihydrite and partially to a goethite-like compound without the contribution of the ferrate. Ferrate(VI) was found to be stable and unaffected by EDTA already bound to iron.

Acknowledgement

V.K. Sharma and Z. Homonnay acknowledge the support U.S.-Hungary Partnership Program supported by the National Science Foundation, Hungarian Academy of Sciences and National Fundamental Research Programs, Hungary (85-46138). Helpful discussions with Prof. Attila Vértes are gratefully acknowledged.

References

1. Carr, J.D.; Kelter, P.B.; Tabatabai, A.; Spichal, D.; Erickson, L.; McLaughlin, C.W. *Proceedings of the Conference on Water Chlorination and Chemical Environment Impact Health Effects*, **1985**, 1285-1298.
2. Waite, T.D.; Gilbert, M.J. *Water Poll. Contr. Fed.* **1978**, 543-551.
3. Johnson, M.D., Hornstein, B.J. *Chem. Commun.* **1996**, 965-966.
4. Sharma, V.K. *Adv. Environ. Res.* **2002**, *6*, 143-156.
5. Sharma, V.K.; Kazama, F.; Jiangyong, H.; Ray, A.K. *J. Water Health*, **2005**, *3*, 45-58.
6. Sharma, V.K.; Bielski, B.H.J. *Inorg. Chem.* **1991**, *30*, 4306-4311.
7. Sharma, V.K.; Burnett, C.R.; Yngard, R.; Cabelli, D. *Environ. Sci. Technol.* **2005**, *39*, 3849-3855.
8. Johnson, M.D.; Hornstein, B.J. *Inorg. Chem.* **2003**, *42*, 6923-6928.
9. Johnson, M.D.; Read, J.F. *Inorg. Chem.* **1996**, *35*, 6795-6799.
10. Huang, H.; Sommerfield, D.; Dunn, B.C.; Lloyd, C.R.; Erying, E.M. *J. Chem. Soc. Dalton Trans.* **2001**, 1301-1305.
11. Hug, G.L.; Carmichael, I.; Fessenden, R.W. *J. Chem. Soc. Perkin Trans 2*, **2000**, 907-908.

12. Armstrong, D.A.; Asmus, K-D.; Bonifacic, M. *J. Phys. Chem. A* **2004**, *108*, 2238-2246.
13. Meisel, D.; Camaioni, D.; Orlando, T. In *ACS Symposium Series;* American Chemical Society: Washington, DC, **2001**, *778*, 342-361.
14. Berger, P.; Leitner, N.K.V.; Dore, M.; Legube, B. *Water Res.* **1999**, *33*, 433-441.
15. Leitner, N.K.V.; Berger, P.; Legube, B. *Environ. Sci. Technol.* **2002**, *36*, 3083-3089.
16. Aust, S.D.; Morehouse, L.A.; Thomas, C.E. Role of metals in oxygen radical reactions. *J Free Radicals Biol. Med.* **1985**, *1*, 3-25.
17. Jiang, J.Q.; Wang, S.; Panagoulopoulos, A. *Chemosphere* **2006**, *63*, 212.
18. Jiang, J.Q.; Lloyd, B. *Wat. Res.* **2002**, *36*, 1397-1408.
19. Vertes, A.; Nagy, D.L. In *Mössbauer spectroscopy of frozen solutions*, Akademiai Kiado, Budapest, **1990**.
20. Klencsar, Z.; Kuzman, E.; Vertes, A.; *J. Radioanal. Nucl. Chem.* **1996**, *210*, 105-112.
21. Scwertmann, U.; Cornell, R. M. *Iron oxides in the laboratory*, VCH: Weinheim, New York, Basel, Cambridge, **1991**.
22. Rush, J.D.; Bielski, B.H.J. *Free Rad. Res.* **1995**, *22*, 571-579.
23. Bielski, B.H.J.; Sharma, V.K.; Czapski, G. *Radiat. Phys. Chem.* **1994**, *44*, 479.
24. Rush, J.D.; Bielski, B.H.J. *Inorg. Chem.* **1994**, *33*, 5499.
25. Menton, J.D.; Bielski, B.H.J. *Radiat. Phys. Chem.* **1990**, *36*, 725.
26. Bowen, L.H. *Mössbauer Effect Reference and Data Journal* **1979**, *2*, 76.
27. Sharma, V.K.; Noorhasan-Smith, N.; Mishra, S.K.; Nesnas, N. Preprints of Extended Abstracts presented at the ACS National Meeting, American Chemical Society, Division of Environmental Chemistry **2006**, *46*, 611-615.
28. Schugar, R.L.; Hubbard, A.T.; Anson, F.C.; Martell, A.E. *J. Amer. Chem. Soc.* **1969**, *67*, 576-582.
29. Ballesteros, M. C.; Rueda, E. H.; Blesa, M. A. *J. Coll. Interface Sci.* **1998**, *201*, 13.
30. Liu, X.; Millero, F.J. Mar. Chem. **2002**, *77*, 43.
31. Furia, T.E., in "CRC handbook of Food Additives", 2nd ed. 1972

Applications

Chapter 16

Electrochemical Fe(VI) Water Purification and Remediation

Stuart Licht, Xingwen Yu, and Deyang Qu

Department of Chemistry, University of Massachusetts at Boston, Boston, MA 02125

A novel on-line electrochemical Fe(VI) water purification methodology is introduced, which can quickly oxidize and remove a wide range of both inorganic and organic water contaminants. Fe(VI) is an unusual and strongly oxidizing form of iron, which provides a potentially less hazardous water purifying agent than chlorine. However, a means of Fe(VI) addition to the effluent had been a barrier to its effective use in water remediation, as solid Fe(VI) salts require complex (costly) syntheses steps, and solutions of Fe(VI) decompose. On-line electrochemical Fe(VI) water purification avoids these limitations, in which Fe(VI) is directly prepared in solution from an iron anode as the FeO_4^{2-} ion, and is added to the contaminant stream. Added FeO_4^{2-} decomposes, by oxidizing water contaminants. Demonstration of this methodology is performed with inorganic contaminants sulfides, cyanides and arsenic; and water soluble organic contaminants including phenol, aniline and hydrazine. In addition, removal of the oxidation products by following an activated carbon filter at the downstream of the on-line configuration is also presented.

Chlorination has been a mainstay of tertiary water treatment and drinking water science. As a stepping stone to the future, water utilities have sought alternative, less hazardous, broadly applicable and more cost effective oxidative methodologies to chlorination. Fe(VI) is an unusual and strongly oxidizing form of iron. As a strong oxidant, Fe(VI) has been investigated as a less hazardous water purifying agent for several decades as a safer alternative to the chlorination purification of water (*1-3*). The Fe(VI) oxidant for water treatment is present as the soluble aqueous FeO_4^{2-} species, which is then reduced to environmentally benign ferric oxide products. Our generalized oxidation mechanism for Fe(VI) treatment of a contaminant is presented in eq 1, with the Fe(III) iron product ferric oxide in various states of hydration ($1/2Fe_2O_3$, FeOOH, etc.):

$$aFe(VI) + bContaminant \rightarrow cFe(III) + dDe\text{-}toxified(reduced)Contaminant \quad (1)$$

The field of Fe(VI) compounds for charge storage was introduced in 1999, and at that time the term super-iron was coined to refer to the class of materials which contain "super-oxidized" iron in the unusual hexavalent state *(4)*. The synthesis of alkali and alkali earth super-irons has been presented and the charge transfer chemistry of the super-irons in both aqueous and non-aqueous media has been investigated *(4-32)*.

In previous water purification studies, the Fe(VI) salt, K_2FeO_4, has been used to oxidize and remove arsenic (*3*, *33*), and for the removal from water of ammonia, cyanide, thiocyanate and sulfide (*34*), and been used for the effective treatment of biosolids (*35*). Viral inactivation by K_2FeO_4 has been demonstrated (*36*), as well as algae removal (*37*). Fe(VI) effectively oxidizes a wide variety of organics including phenol (*38*), alcohols (*39-43*), toluene and cycloalkanes (*39*), ketones and hydroquinones (*42*), carbohydrates (*44*), and aminobenzene (*45*).

Super-iron, FeO_4^{2-}, provides an environmentally friendly, potentially cost effective oxidant, in lieu of chlorine, but delivery challenges of this oxidant is a principal limitation to its implementation by water utilities (*46*). In the new on-line electrochemical water purification methodology, Fe(VI) is now directly prepared in an alkaline solution from the Fe metal anode as the FeO_4^{2-} ion, and is added to the contaminant stream for water treatment on an "as need" basis (*3*). Prior barriers for Fe(VI) addition to an effluent for use in water remediation, were due to synthesis and solution phase stability challenges. The chemical synthesis of solid super-iron salts, principally by hypochlorite reaction with ferric salts, can be complex and costly, and only highly purified K_2FeO_4 maintains long-term retention of the hexavalent state. Solutions of dissolved super-iron salts will oxidize water, and hence cannot be effectively stored or shipped as an oxidant for water treatment. On-line delivery system for super-iron water purification will avoids these limitations (*3*). In addition to chemical syntheses, we have recently explored solid super-iron salt syntheses utilizing Fe(VI) electrochemically generated from a simple iron metal (*3-29*). On-line electrochemical super-iron, rapid, effective, treatment of a range of contaminants will be presented and summarized in this paper.

Electrochemical Fe(VI) Water Remediation Configurations

On-line electrochemical Fe(VI) water remediation configuration.

The on-line electrochemical Fe(VI) water remediation configuration is shown as Scheme 1 (*3*). This configuration generally includes: 1) a Fe(VI) generation section; 2) an effluent oxidation section. The Fe(VI) generation section is a 2-electrode electrochemical flow liquid cell. An alkaline solution is used as the electrolyte for Fe(VI) generation. The anode may consist of an iron sheet, but a higher surface area iron anode (*15,21,32*), increases the rate per volume of Fe(VI) formed in solution. The cathode in this on-line Fe(VI) remediation is limited to materials which are stable when immersed in an alkaline, reductive environment. Cathodes used have included nickel and nickel oxide, platinum, gold, graphite, carbon black, iridium oxide or ruthenium oxide. The anode and the cathode electrode are controlled by an external power supply. As seen in Scheme 1, the Fe(VI) formation compartment may have an optional separator between the cathode and anode. When the anode is downstream of the cathode, the separator is generally not required. If the anode were in proximity of the cathode, then a portion of the anodically formed FeO_4^{2-} could be lost at the cathode. When used, a separator between the iron anode and the cathode can be either a non conductive separator configured with open channels, grids or pores, such as a ceramic frit, or a membrane to impede FeO_4^{2-} transfer, such as a cation selective membrane (eg. Nafion 350, Dupont®). The water to be purified, and the solution containing electrochemically synthesized FeO_4^{2-} are brought together by means of a pump to the oxidation section.

In our study, all Fe(VI) formation electrolytes are prepared with NaOH (analytical grade), and triply deionized water. The concentration of NaOH can

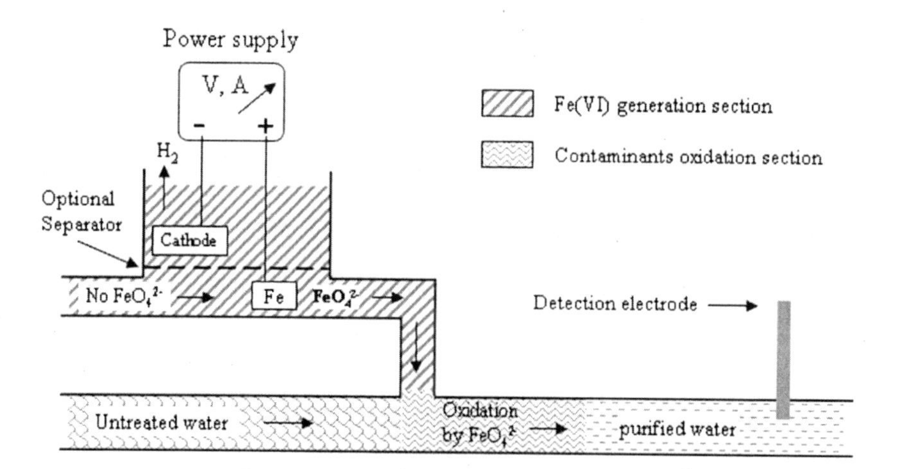

Scheme 1. On-line electrochemical Fe(VI) water purification.

range from 1 M to saturation. A concentrated alkaline solution provides higher current efficiency, and higher rates of Fe(VI) formation at lower overpotential (*21*). However, the solubility of Fe(VI) will decrease with increasing NaOH concentration (*4,5*), and therefore 10 M NaOH is observed as an optimal concentration. Small diameter, high surface area, coiled iron wire is used as the anode, prepared from Fluka (0.1% Cu, 0.1% Ni and 0.7% Mn) iron wire, d=0.2 mm diameter, L = 128 m coiled length, and has a surface area, $A = L\pi d = 800$ cm^2. Anode pretreatment consists of 3 minutes sonication in 1:3 HCl (concentrated HCl:water), followed by washing with triply deionized water (Nanopure water system of 18.2 megaohm resistivity) to pH=7. 80 cm^2 planar nickel sheet is used as the cathode. A 2 mA/cm^2 current density is observed to provide the highest current efficiency for Fe(VI) yield, therefore, constant current of $2 \times 800 = 1600$ mA (1.6 A) is applied for Fe(VI) formation. The flow cell is operated at room temperature (22 ^{0}C).

In the oxidation section, the contaminated effluent steam to be purified is brought together with the Fe(VI) stream [solution phase Fe(VI), as FeO_4^{2-}] generated in the Fe(VI) generation section. The flow rate of the effluent to be treated, as well as the flow rate of the generated Fe(VI) solution, are separately managed with two Control Company variable flow mini-pumps, Model number 3385, using 0.5-4.8mm (I.D.) tubing. Upon mix of the two streams, Fe(VI) contacts the contaminants and the contaminants are oxidatively decomposed by Fe(VI), which reduces to Fe(III). The length of the oxidation section can be modified, to control the time for the reaction between Fe(VI) and contaminants. For a relatively slower reaction rate, a longer oxidation section is required to provide sufficient time for the mixed streams to interact. Effectiveness of the contaminants removal is probed downstream of the configuration, by measuring the contaminants' concentration.

In-line electrochemical Fe(VI) water purification configuration

An in-line electrochemical Fe(VI) water purification system, as represented in Scheme 2, is an alternate configuration (*3*) for effluent treatment. This configuration, utilizes the water to be treated as an electrolyte for the electrochemical formation of Fe(VI) species, and the water to be purified is in contact with, and flows over, the FeO_4^{2-} generating anode. However with this in-line configuration, the iron electrode is exposed to the untreated water and vulnerable to fouling.

On-line electrochemical Fe(VI) – activated carbon water purification configuration.

Based on the on-line electrochemical Fe(VI) water purification configuration introduced above, an on-line electrochemical Fe(VI) – activated

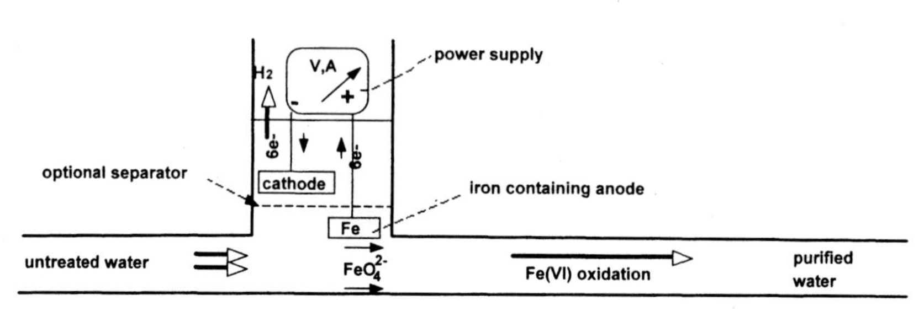

Scheme 2. In-line electrochemical Fe(VI) water purification.

carbon water purification configuration is further developed by extending downstream an activated carbon filter, and probed for the rapid removal of organic toxins. This expanded configuration is shown as Scheme 3. In addition to the 1) Fe(VI) generation section, and 2) effluent oxidation section, this exapnded configuration also includes 3) a products removal section.

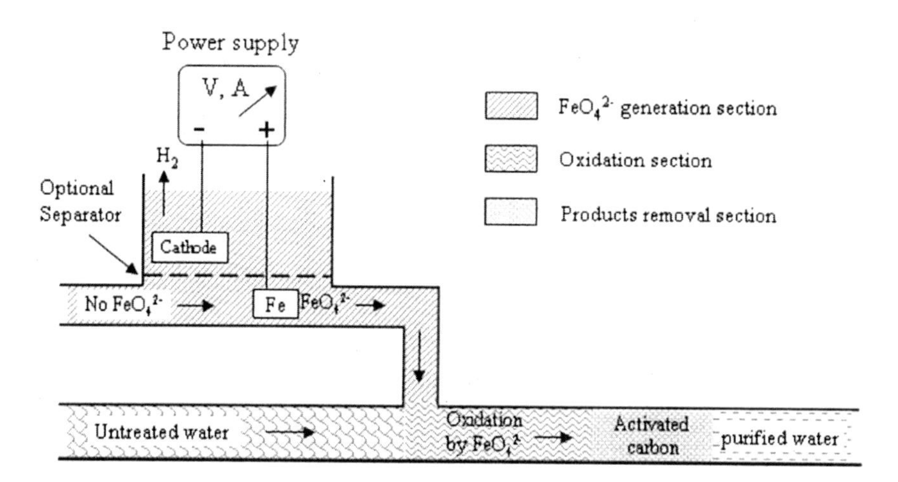

Scheme 3. on-line electrochemical Fe(VI) – activated carbon water purification.

The Fe(VI) generation section and the effluent oxidation section in the Scheme 2 on-line electrochemical Fe(VI) – activated carbon water purification configuration, are the same as Scheme 1. The final products removal section in this configuration is prepared as a 5 cm long activated carbon filter. It is used to absorb the insoluble oxidation products of the treated water, such as suspensions of minute particles, or those present as emulsions of water insoluble liquids. If a large amount of insoluble oxidation products are produced from the effluent that is treated, it would be necessary to separate the bulk insoluble oxidation products prior to the stream flowing through the activated carbon filter; however, this was not observed as necessary, to treat the relatively small volumes of contaminants used in this study. Hence, the stream is directly introduced into the product removal section after the oxidation section.

In this paper, the on-line electrochemical Fe(VI) water remediation configuration shown as Scheme 1 will be used for the demonstration of inorganic contaminated water purification. While, organic contaminated water remediation will be demonstrated with on-line electrochemical Fe(VI) – activated carbon water purification shown as Scheme 3.

Electrochemical Fe(VI) Generation and Analysis

Fe(VI) Formation Fundamentals

Fe(VI) can be prepared from alkaline hypochlorite oxidation of Fe(III) salts (*4, 10, 12*). The electrochemical Fe(VI) water purification methodology is based on the novel on-line electrochemical formation and addition of Fe(VI) to oxidize and treat contaminants. Fe(VI) is added in the form of FeO_4^{2-} ion. Electrochemical formation of the FeO_4^{2-} is accomplished at a positively biased iron anode in contact with an alkaline aqueous solution. High pH acts to enhance the rate of formation of the Fe(VI) species, favoring utilization of concentrated hydroxide electrolytes, similarly high surface area anodes, such as wound iron wire, favor rapid FeO_4^{2-} formation (*21*). At sufficiently anodic potentials in alkaline media, an iron anode is directly oxidized to the Fe(VI) species, FeO_4^{2-}, in accord with the oxidation reaction:

$$Fe + 8OH^- \rightarrow FeO_4^{2-} + 4H_2O + 6e^- \quad (2)$$

The process occurs at potentials > 0.6 V vs the standard hydrogen electrode, SHE, consistent with the alkaline rest potentials related to the formation of Fe(II), Fe(III) and Fe(VI) from Fe(0):

$$Fe + 2OH^- \rightarrow Fe(OH)_2 + 2e^-;\ E° = -0.8\ V\ vs\ SHE \quad (3)$$

$$Fe(OH)_2 + OH^- \rightarrow FeOOH + H_2O + e^-;\ E° = -0.7\ V\ vs\ SHE \quad (4)$$

$$FeOOH + 5OH^- \rightarrow FeO_4^{2-} + 3H_2O + 3e^-;\ E° = +0.6\ V\ vs\ SHE \quad (5)$$

Isolation of Fe(VI) away from the cathode minimizes losses due to the cathodic back-reduction of the synthesized Fe(VI), eq 6, and instead favoring the cathodic evolution of hydrogen, eq 7:

$$FeO_4^{2-} + 3H_2O + 3e^- \rightarrow FeOOH + 5OH^- \quad (6)$$

$$6H_2O + 6e^- \rightarrow 6OH^- + 3H_2 \qquad (7)$$

With 6 Faraday of charge transfer through the anode and cathode per equivalent of Fe, eq 2 and 7 combined yield the net Fe(VI) solution phase alkaline synthesis reaction:

$$Fe + 2OH^- + 2H_2O \rightarrow FeO_4^{2-} + 3H_2 \qquad (8)$$

Fe(VI) Analyses

Fe(VI), dissolved as FeO_4^{2-}, is evident as a purple color in aqueous solution, and has a distinctive UV/Vis spectrum. The visible absorption of Fe(VI) in highly alkaline solution is presented in the inset of Figure 1 (*13*). This shows a maximum absorption at 505 nm, an absorption shoulder at 570 nm. The molar absorptivity measured at 505 nm is 1070±30 $M^{-1}cm^{-1}$. The molar absorptivity remained constant up to 200 mM. Similarly, at a fixed ferrate concentration, the measured absorbance was highly independent of the alkali hydroxide cation or hydroxide concentration in solution (*4, 13*). Figure 1 shows the linear relationship between the 505 nm absorbance and FeO_4^{2-} concentrations, and with the slope about 1070 M^{-1}. In our study, FeO_4^{2-} concentration is determined with a Hewlett Packard 8453 UV/Vis photodiode array spectrophotometer by the 1070 M^{-1} cm^{-1} molar absorptivity at λ=505 nm with 385 nm baseline correction, and with dilution as required to reach the analyzable FeO_4^{2-} concentration range (*13*).

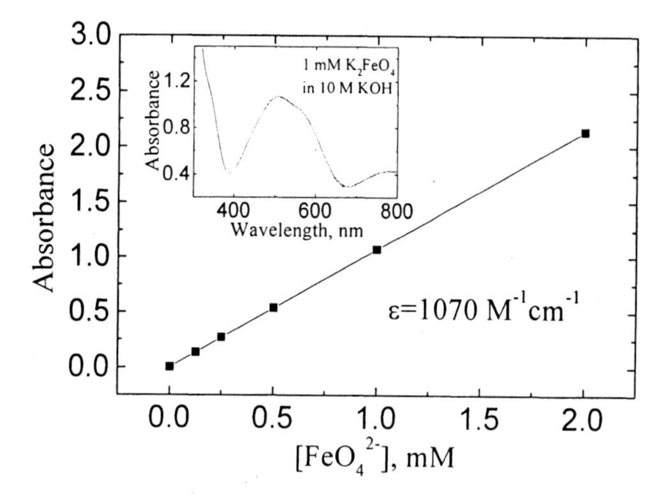

Figure 1. The 505 nm absorbance of ferrate (VI) solutions of various concentrations at 505 nm. Inset: Vis. absorption spectra of FeO_4^{2-} in alkaline aqueous solution.

Fe(VI) generation in the on-line electrochemical flow cell

Study of the parameters which influence Fe(VI) generation, such as electrolyte (NaOH) concentration, iron anode surface area and current density, and temperature, have been studied in details elsewhere in order to optimize yield for the electrochemical synthesis of Fe(VI) salts (*15, 21*). In this water purification study, the following conditions provide effective dissemination of the Fe(VI) for remediation: surface area of Fe wire anode: 800 cm^2 in a 40ml anode chamber of the flow cell; NaOH electrolyte concentration: 10 M; current density applied to the Fe anode: $2mA/cm^{2-}$ (total current applied: 800 $cm^2 \times 2$ mA/cm = 1.6 A), operated at room temperature (22^0C). Under these conditions, the cell reached a steady state Fe(VI) concentration in approximately half an hour as shown in Figure 2. Specifically, anodic oxidation of 10, 20, 30 or 60 minutes respectively, generated concentrations of 21, 33, 38 and 39 mM FeO_4^{2-}, respectively (measured by 505 nm absorption spectroscopy as discussed above).

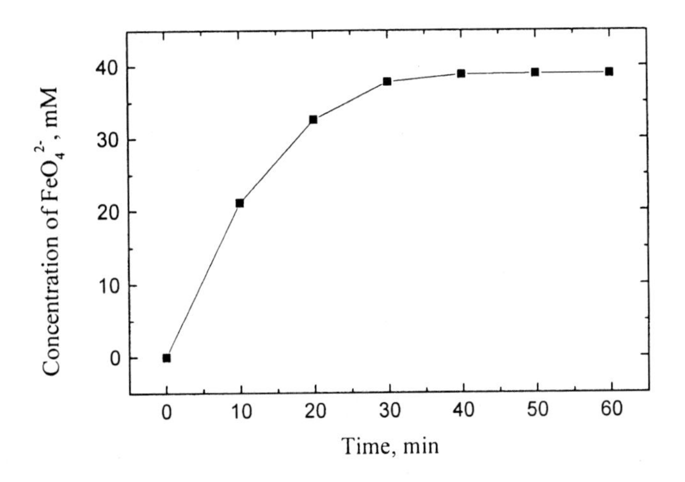

Figure 2. FeO_4^{2-} concentration increase with the time of application of a constant anodic current in the flow cell.

Inorganic Contaminated Water Purification

Inorganic contaminants remediation for water purification will be demonstrated with the <u>on</u>-line electrochemical Fe(VI) water purification shown as Scheme 1. Contaminated water samples are prepared with sulfide, cyanide and arsenite.

Analyses of inorganic contaminants sulfide, cyanide and arsenite

Inorganic contaminants water samples are prepared with sulfide, cyanide and arsenite. Sulfide concentration is analyzed with a model 27504-28 Cole-Parmer® Silver/Sulfide electrode and cyanide with a model number 27504-12 Cole-Parmer® Cyanide electrode, both using an Orion Research expandable ionAnalyzer EA 920 measured as a function of time using LabView data acquisition. Unlike sulfide, and cyanide, ion selective analysis was not available for arsenic determination. Instead arsenic concentration is determined by 3 electrode cyclic voltammetry on an 0.5 mm^2 platinum electrode scanned at 50 mV/s, with a Pt counter electrode and silver/silver chloride reference electrode; accomplished by monitoring the relative decrease in the observed As(III) peak current on Pt at +0.13V, which generates a linear response of peak current over the 0.5 to 10 mM concentration $NaAsO_2$ in 1 M NaOH, as shown in Figure 3.

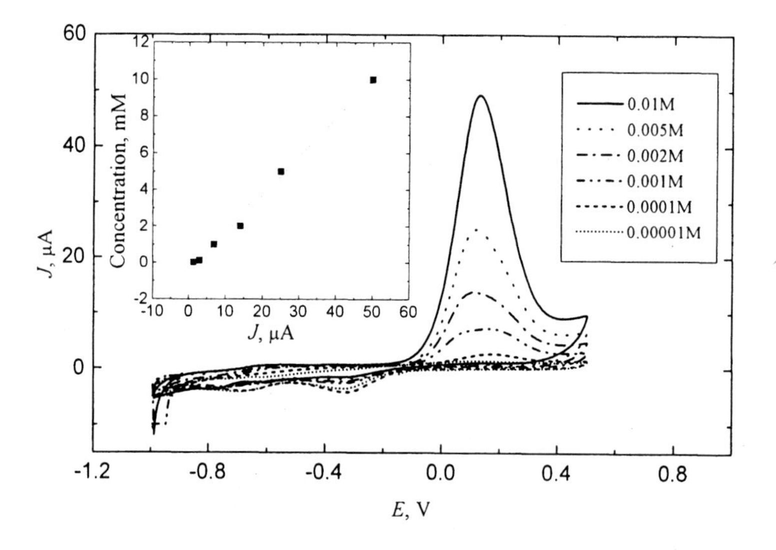

Figure 3. cyclic voltammetry of the inert Pt electrode in different concentration arsenite solution. Inset: linear relationship between peak current and arsenite concentrations.

Sulfide, cyanide and arsenite contaminated water treatment with on-line electrochemical Fe(VI) water purification configuration

Fe(VI) was readily formed on-line as the FeO_4^{2-} species in the flowing 10 M NaOH solution, and was added to an effluent to be treated in accord with the Scheme 1. The effect of Fe(VI) on a contaminant was probed by cycling at various controlled time intervals, a constant Fe(VI) formation current. Fe(VI) was added in this stepped manner to a purposely prepared, sample contaminated effluent, by switching on and off, the anodic (Fe metal electrode) oxidation current, as represented in Figure 4. The contaminant's concentration were measured down-line of the Fe(VI) addition, after the combined effluent and Fe(VI) streams had intermingled for 3 minutes.

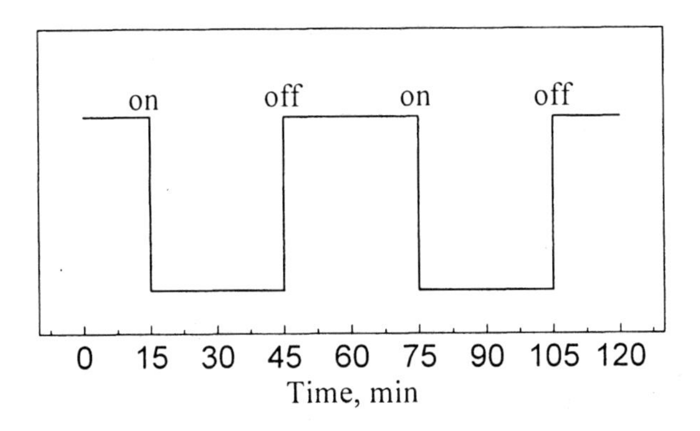

Figure 4. Sketch of stepped addition of Fe(VI) to effluent and time control arrangement. On: switch on the constant anodic current (lower horizontal line); off: switch off the anodic current (upper horizontal line).

Sulfide was the first contaminant studied. Aqueous sulfide and sulfur (polysulfide) chemistry is complex. In all but the most alkaline solutions, sulfide solutions predominantly contain the (hydrosulfide) HS^-, rather than the S^{2-}, species, and partial oxidation of the sulfide solution generates a mix of polysulfides, S_x^{2-}, and thiosulfate, $S_2O_3^{2-}$, whose speciation varies with sulfide concentration and pH (*47-52*). H_2S is fully oxidized by excess ferrate to sulfate, SO_4^{2-}, whereas, at high pH sulfite and thiosulfate products can also occur (*2*), and the latter can predominate (*47*). Theoretically, oxidative removal of sulfide to thiosulfate, $S_2O_3^{2-}$, requires a 4/3 molar ratio consistent with eq. 1, when treated by Fe(VI) as:

$$4FeO_4^{2-} + 3S^{2-} + 23/2H_2O \rightarrow 4Fe(OH)_3 + 3/2S_2O_3^{2-} + 11OH^- \quad (9)$$

We observe that the oxidation product is thiosulfate (i) via thiosulfate's characteristic FTIR absorption peak at 996 cm^{-1}, and (ii) the stoichiometric ratio of Fe(VI) required to fully remove sulfide, as measured by the sulfide ion selective electrode. By this latter technique, a molar ratio $\geq$ 1.4: 1 of FeO_4^{2-} to Na_2S, is observed to remove all sulfide from 0.01 M Na_2S in 1 M NaOH (while a 1.3:1 molar left trace sulfide). In accord with the stoichiometric ratio of eq. 9, the current controlled concentration of added FeO_4^{2-} was 4/3 that of the sulfide concentration. The variation of sulfide concentration in time of on-line electrochemical Fe(VI) treated 0.01 M Na_2S in 1 M NaOH was measured using sulfide selective electrode analysis. This was calibrated using the same experimental configuration, but periodically interrupting the sulfide solution flow, and instead of the Fe(VI) solution, flowing a sulfide free (and Fe(VI) free) solution. The sulfide calibration is presented in the bottom section of Figure 5. Figure 5 also presents the on-line electrochemical treatment of sulfide by Fe(VI), with, or without anode controlled FeO_4^{2-} addition. As seen in the top section of the figure, the sulfide is completely removed by the Fe(VI).

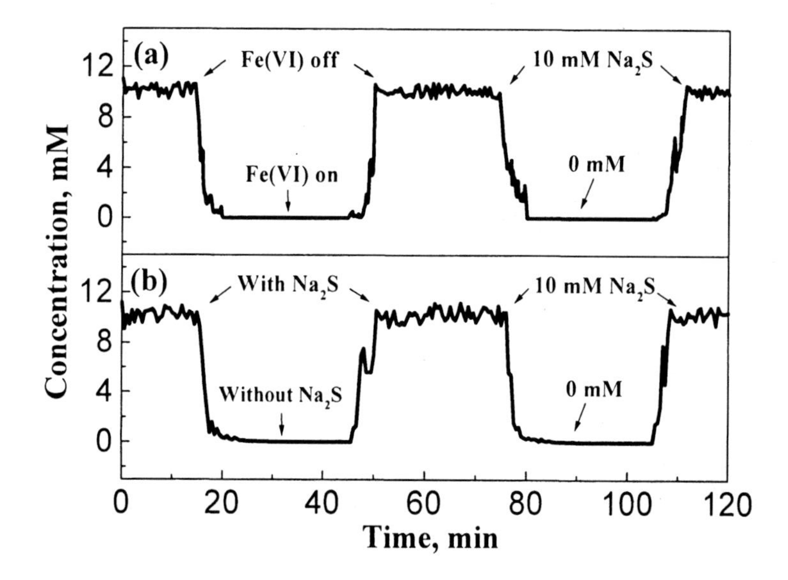

Figure 5. Top (a): Remediation of sulfide by on-line Fe(VI) treatment. The variation of sulfide concentration upon FeO_4^{2-} addition to sulfide effluent flow. The iron anode is alternately cycled from 0 current to constant oxidation current mode, as described in Figure 4. Bottom (b): Sulfide concentration measurement and calibration, exposing the downstream sulfide ion selective electrode to the 1 M NaOH flowing solution alternately with, and without, 0.01 M Na_2S.

Generally in this study, the effluent studied was 1 M NaOH, at approximately pH 14 (*53*), containing different contaminants. However, the FeO_4^{2-} is also effective for oxidation remediation of water at lower pH. Figure 6, summarizes sulfide removal at lower pH by Fe(VI), in both pH 12 and 13 solutions, and again the complete removal of sulfide is observed. Water to be

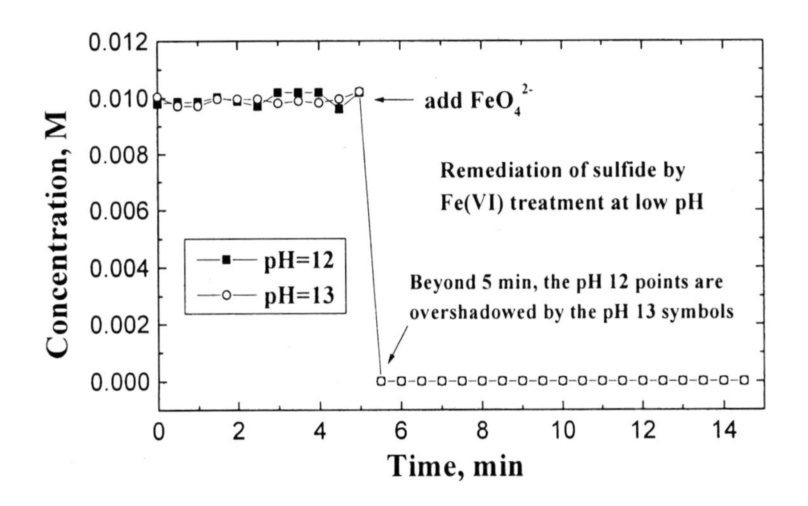

Figure 6. Remediation of sulfide by Fe(VI) treatment at lower pH. The variation of sulfide concentration upon FeO_4^{2-} addition. pH values of 12 and 13 controlled by NaOH addition to 0.01 M Na_2S with FeO_4^{2-} added as K_2FeO_4.

treated may also be at neutral pH, although the treated stream will be raised to higher pH by addition of the alkali media used to generate the Fe(VI).

Aqueous alkaline oxidation of cyanide generates cyanate (*54, 55*), which will form upon Fe(VI) addition, in accord with:

$$2FeO_4^{2-} + 3CN^- + 5H_2O \rightarrow 2Fe(OH)_3 + 3CNO^- + 4OH^- \qquad (10)$$

Aqueous cyanide may be treated by Fe(VI) through oxidative decomposition as described by eq 10. Calibration of the cyanide concentration, as measured by ion selective electrode analysis, for a 1 M NaOH solution with, or without, 0.01 M cyanide is presented in the bottom section of Figure 7. Removal of cyanide is in accord with the stoichiometric ratio of 2 Fe(VI) per 3 cyanide in eq 10. However, the rate of oxidation is slower than that observed for sulfide. As seen in Figure 8, whereas the Fe(VI) rate of sulfide oxidation occurs in less than 1 minute, the rate of cyanide oxidation is longer. It is seen in the figure that only 90% of the initial cyanide concentration is removed in the first 3 minutes. This also occurs for on-line Fe(VI) remediation of cyanide, when the effluent and Fe(VI) streams are permitted to intemix for only 3 minutes prior to downstream analysis. As seen in the top section of Figure 7, approximately 90% of the cyanide, to 1mM, is removed by the Fe(VI) addition during this time. In accord with Figure 8, the fraction of cyanide removed increases with increasing mixing time of the cyanide effluent and Fe(VI) streams.

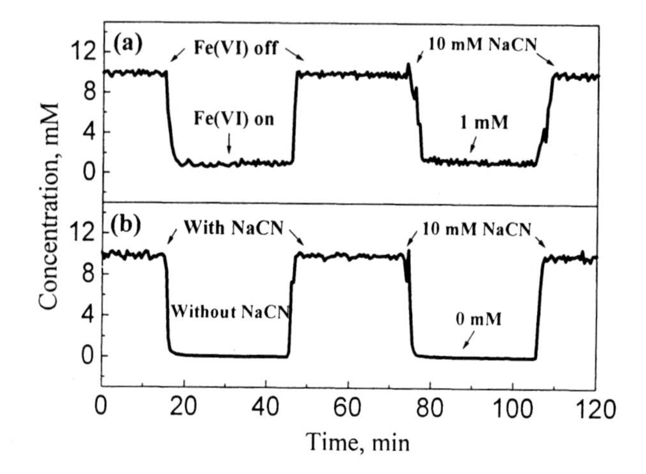

Figure 7. Top (a): Remediation of cyanide by on-line Fe(VI) treatment. The variation of cyanide concentration upon FeO_4^{2-} addition to the 0.01 M cyanide effluent flow. The iron anode is alternately cycled from 0 current to constant oxidation current mode, as described in Figure 4. Bottom (b): Cyanide concentration measurement and calibration, exposing the downstream cyanide ion selective electrode to the 1 M NaOH flowing solution alternately with, and without, 0.01 M NaCN.

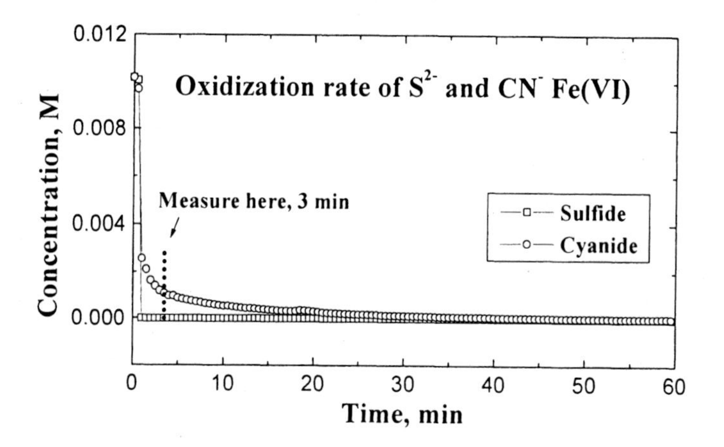

Figure 8. Comparison of the rate of 0.01 M S^{2-} or 0.01 M CN^- oxidization by FeO_4^{2-}. Respective cyanide and sulfide concentrations are measured by ion selective electrode.

For the arsenic contaminant, arsenite, As(III) (for example as added $NaAsO_2$), the oxidation product upon Fe(VI) addition is arsenate, As(V), with the overall reaction given by:

$$FeO_4^{2-} + 3/2AsO_2^- + OH^- + H_2O \rightarrow Fe(OH)_3 + 3/2AsO_4^{3-} \qquad (11)$$

Hence, the FeO_4^{2-} required to oxidize 1 mole of AsO_3^{3-} is 2/3 mole. Figure 9 presents the result of on-line electrochemical treatment of arsenic by Fe(VI). As seen in figure, in a manner analogous to the sulfide treatment, the Fe(VI) removal of arsenic, as As(III) in the 0.01 $NaAsO_2$ M effluent, is complete.

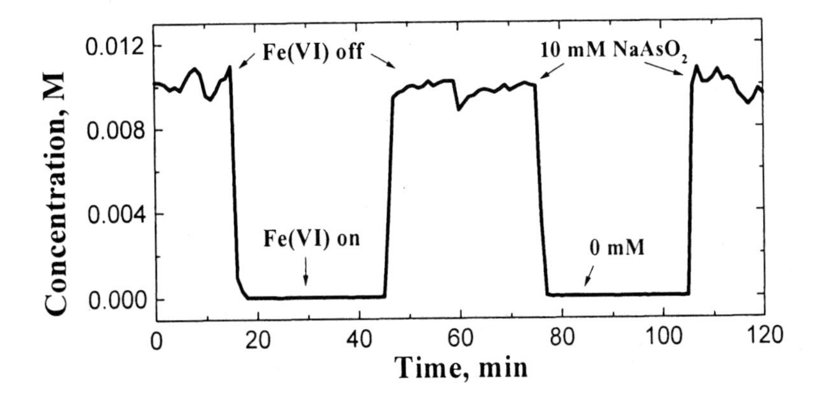

Figure 9. Remediation of arsenic by on-line Fe(VI) treatment. The variation of arsenic concentration upon FeO_4^{2-} addition to the 0.01 M $NaAsO_2$ in 1 M NaOH effluent flow. The iron anode is alternately cycled from 0 current to constant oxidation current mode, as described in Figure 4.

Simultaneous treatment of multiple contaminants in an effluent flow by Fe(VI) on-line remediation was also found to be effective. Figure 10, presents the treatment, by on-line electrochemically generated Fe(VI), of a solution containing both 0.01 M sulfide and cyanide. The Fe(VI) concentration is anodically controlled in accord with the sum of eqs 9 and 10. The concentration of sulfide and cyanide in the treated effluent, is observed to be consistent with the treatment of the individual contaminant components. Specifically, as observed in Figure 10 the sulfide removal is complete, and 90% of the initially cyanide is oxidized as limited by the (3 minute) intermix time, prior to detection, of the effluent Fe(VI) and contaminant streams.

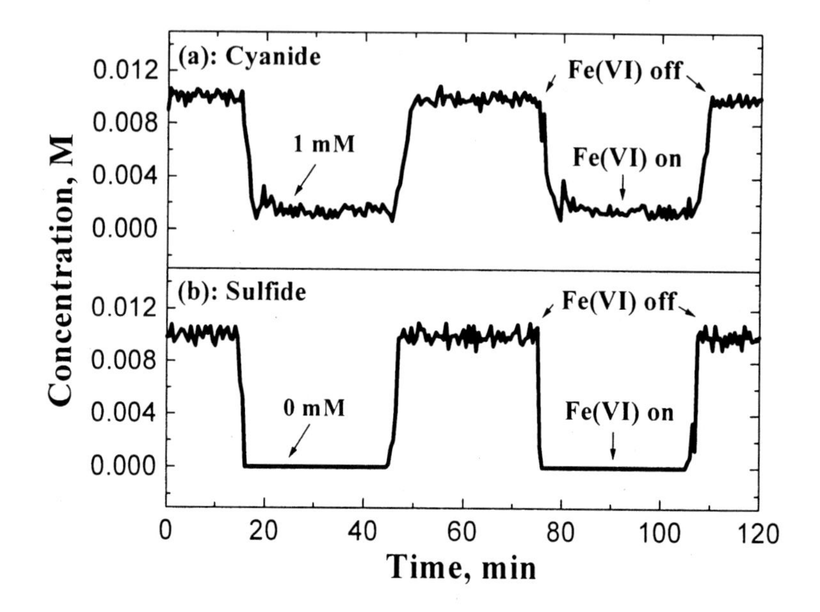

Figure 10. Simultaneous remediation of cyanide and sulfide by on-line Fe(VI) treatment. The effluent contains 0.01 M sulfide and cyanide. The time variation of the concentration of each [Top (a): cyanide and Bottom (b): sulfide], upon FeO_4^{2-} addition to the effluent flow, is shown. The iron anode is alternately cycled from 0 current to constant oxidation current mode, as described in Figure 4.

Organic Contaminated Water Purification

Organic contaminants water purification was demonstrated with the on-line electrochemical Fe(VI) – activated carbon water purification shown as Scheme 3. Organic contaminated water samples were prepared with aniline, phenol and hydrazine.

Measurement of aniline, phenol and hydrazine.

Both aniline and phenol are quantitatively analyzed with UV/Vis spectroscopy using a series of standard concentrations of each. As seen in Figure 11a, two characteristic peaks of aniline at 230 nm and 280 nm are candidates for the quantitative analysis of aniline in water. However in the UV region, the absorbance by hydroxide overlaps and interferes with analine's 230 nm absorbance. Hence, although aniline's absorbance is larger at 230 nm than at 280 nm, in a medium in which the hydroxide concentration can vary, the 280 nm absorption provides a more quantitative measure for aniline analysis. In a similar manner, it is evident in Figure 11b, that phenol absorption at 270 nm is more useful for quantitative phenol analysis, due to the close proximity of the UV hydroxide absorbance to the alternate 211 nm phenol absorption. The linear variation of absorbance, A, with either mM concentration of aniline, C_a, or phenol, C_p, is presented in the insets of Figure 11a and Figure 11b respectively, and yields the linear relationships for respective aniline and phenol UV/Vis analysis:

$$\text{For aniline: } A_a(280\text{ nm}) = 1.288\, C_a \quad (12)$$

$$\text{For phenol: } A_p(270\text{ nm}) = 1.457\, C_p \quad (13)$$

The hydrazine concentration is analyzed with PDMAB method (ASTM D 1385-01), using a 0-500 ppm VACUettes Kit®.

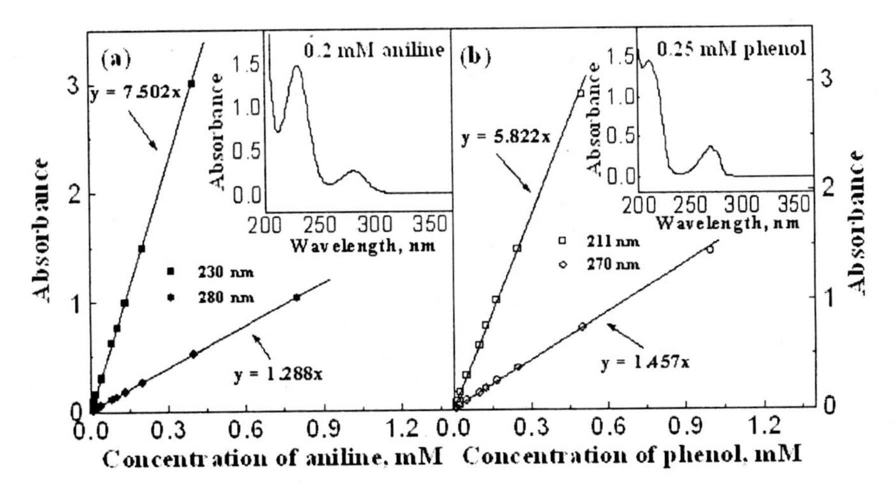

Figure 11. Quantitative analyses of aniline and phenol with UV/Vis spectroscopy. (a): The 230 nm and 280 nm absorbance of aniline solutions at various concentrations; (b): The 211 nm and 270 nm absorbance of phenol solutions at various concentrations. Inset: Representative UV/Vis spectra obtained at the indicated concentration of aniline and phenol solutions.

Oxidative treatment and removal of aniline, phenol and hydrazine.

Hydrophilic groups (amido and hydroxyl) in aniline, phenol and hydrazine facilitate a high degree of water solubility, and each is known as a hazard contaminant in industrial wastewater (*56-58*). Each is in a relatively reduced state, and in principal should be susceptible to oxidation in aqueous medium. Furthermore, the oxidation products of these organics are either insoluble or gas products which under the appropriate conditions can facilitate product removal. The oxidation product of aniline by Fe(VI) is azobenzene (*39*). The oxidation of phenol by Fe(VI) is more complex, with the principal oxidation product determined as p-benzoquinone (*38*). The respective oxidative reaction of aniline or phenol by FeO_4^{2-} may be generalized as:

$$3C_6H_7N + 2FeO_4^{2-} + 2H_2O \rightarrow 3/2C_6H_5N{=}NH_5C_6 + 2Fe(OH)_3 + 4OH^- \quad (14)$$

$$3C_6H_5OH + 4FeO_4^{2-} + 7H_2O \rightarrow 3C_6H_4O_2 + 4Fe(OH)_3 + 8OH^- \quad (15)$$

Based on eq 14 and eq 15, the oxidative removal of aniline or phenol to azobenzene and benzoquinone respectively requires either a 2/3, or a 4/3, molar ratio of FeO_4^{2-}. 8 mM aniline and 5 mM phenol solution are prepared as sample contaminated water effluents, and treated with various flow rates of the Fe(VI) stream. In accord with Figure 2, a constant concentration of 39±1 mM Fe(VI) is obtained in half an hour after initiation of oxidative current in the flow cell. The contaminant water effluent (including aniline or phenol) and Fe(VI) stream are then brought together into the oxidation section of Scheme 3 with the flow ratio of Fe(VI) : contaminated water, varying over a twenty fold range (from= 1:1 to 1:20). This provides Fe(VI) to contaminant molar ratios varying from ~5:1 up to 1:4 for Fe(VI) to aniline; and from ~8:1 to 1:2.5 for phenol. Upon 30 minutes treatment, the residual aniline or phenol in the treated water is analyzed before the mixed streams flow through the activated carbon. Figure 12a and Figure 12b show that a flow rate ratio of 1:4 for Fe(VI) stream to contaminated water stream is necessary to sufficiently oxidize aniline or phenol contaminants. Molar ratios of Fe(VI) required for completely oxidizing aniline or phenol derived from Figure 12a and Figure 12b are 1.2:1 for Fe(VI) to aniline, and 2:1 for Fe(VI) to phenol. Both are higher than theoretical calculation based on the stoichiometric ratios in eq 14 and eq 15 (molar ratio > 2/3 for Fe(VI) to aniline; and > 4/3 for Fe(VI) to phenol).

Contaminated water sample prepared with 8 mM aniline and 5 mM phenol are then fully treated with the on-line electrochemical Fe(VI) – activated carbon water treatment configuration shown as Scheme 3, with a flow rate ratio of Fe(VI) : effluent contaminant equal to 1 : 4. After the two streams of Fe(VI) and contaminated water (containing aniline or phenol) are brought together, the mixed streams flow through the oxidation section for either 5 min, 15 min or 30

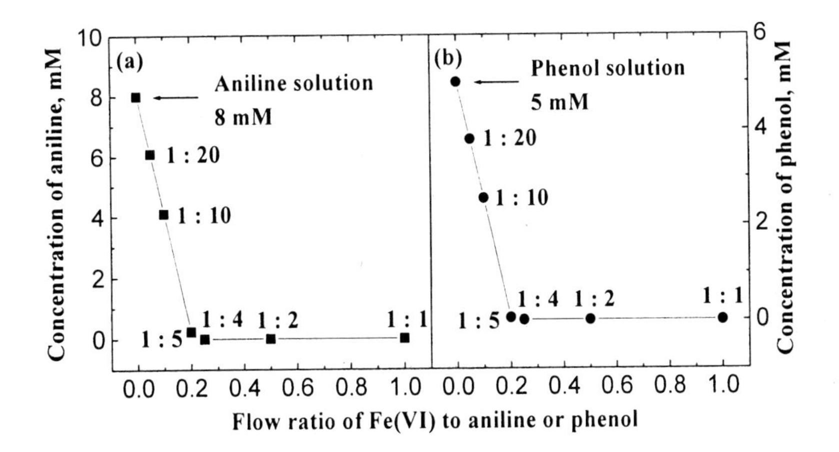

Figure 12. The oxidation of aniline and phenol by Fe(VI) at different flow ratio of FeO_4^{2-} (using anodically generated 39±1 mM FeO_4^{2-}) to (a): 8.0 mM aniline and (b): 5.0 mM phenol solution.

min, and then flow through the activated carbon filter. After this treatment time, the samples are diluted tenfold (to the within the concentration range of the analysis) and analyzed with UV/Vis spectroscopy. As in the solid-circle curves in Figure 13a and Figure 13b, the characteristic peaks for aniline at 280 nm and phenol at 270 nm fully disappear after 5 minutes of Fe(VI)/effluent mix. Spectra measured after 15 or 30 minutes (not shown) are virtually identical to those measured. Hence, 5 minutes provides an upper limit to the requisite reaction time for complete effluent remediation. Also notable in the solid-circle curves is the efficacy of the activated carbon, evident in the absence of any product peak in the 300-320 nm range. Whereas, these product peaks are evident in the figure, when either effluent is treated and removed by Fe(VI) in the absence of an activated carbon filter. For comparison, the pure azobenzene (the oxidation product of aniline, evident at 320 nm) and p-benzoquinone (the oxidation product of phenol, evident at 300 nm) are also included in Figure 13.

In principal, oxidative removal of hydrazine to nitrogen requires a 4/3 molar ratio of FeO_4^{2-} to N_2H_4 in accordance with (*55*):

$$3N_2H_4 + 4FeO_4^{2-} + 4H_2O \rightarrow 3N_2 + 4Fe(OH)_3 + 8OH^- \quad (16)$$

Hydrazine oxidation is performed in a simplified Scheme 3 configuration, without inclusion of the activated carbon filter to remove the oxidation product, as the oxidation product of hydrazine is nitrogen (N_2) (*59*). The water contaminant effluent sample is prepared as 5.6 mmol/L hydrazine. Rapid reaction is readily observed between Fe(VI) and hydrazine; seen as a

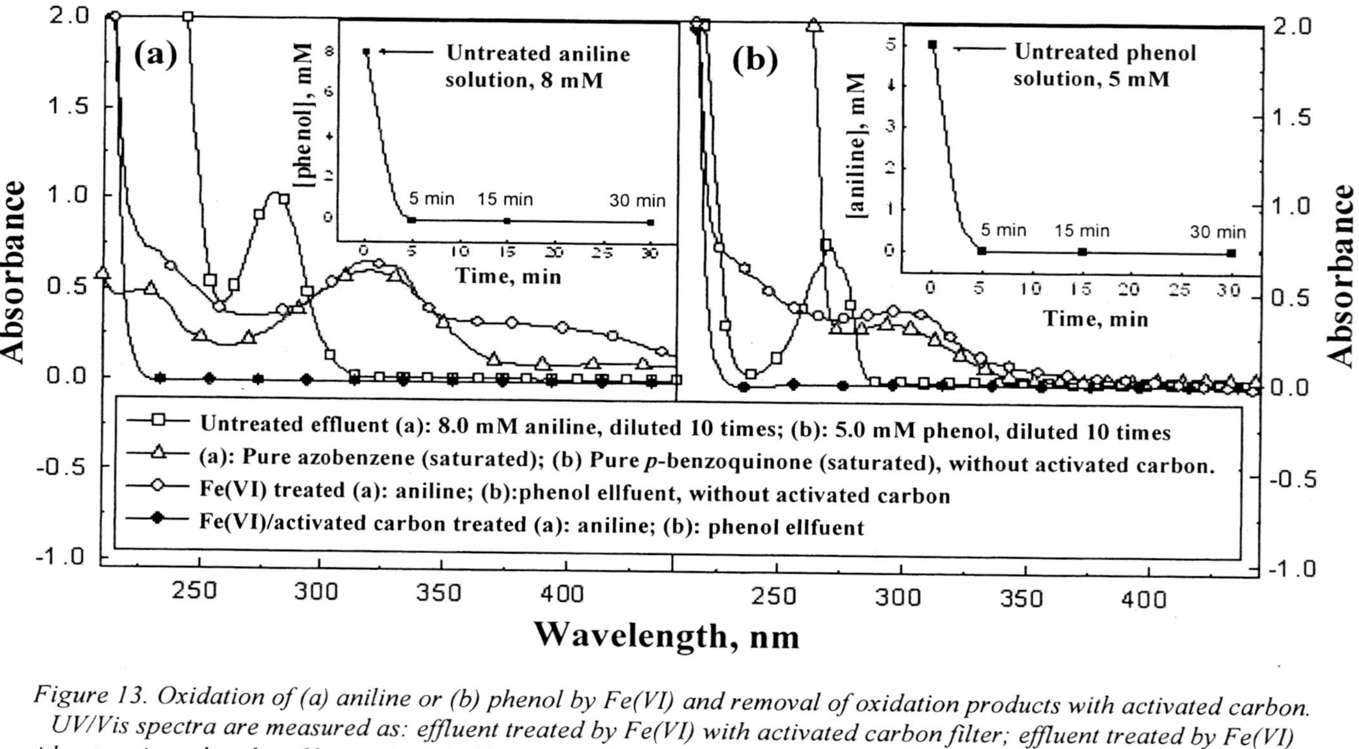

Figure 13. Oxidation of (a) aniline or (b) phenol by Fe(VI) and removal of oxidation products with activated carbon. UV/Vis spectra are measured as: effluent treated by Fe(VI) with activated carbon filter; effluent treated by Fe(VI) without activated carbon filter; original effluent sample (8 mM aniline or 5 mM phenol after 10-fold dilution); and pure azobenzene and p-benzoquinone. Inset: Concentration change of (a): aniline or (b): phenol before & after treatment.

disappearance of the purple color FeO_4^{2-}, after the two streams of contaminated water sample and Fe(VI) meet within the oxidation section. Figure 14 shows the hydrazine oxidation at different flow ratios between the Fe(VI) and hydrazine solutions, ranging from Fe(VI) : hydrazine = 1:1 (as a molar ratio of ~7.0 : 1), up to 1: 20 (as a molar ratio of 1: ~2.9). At the flow ratio of Fe(VI) : hydrazine = 1: 4 (molar ratio is 1.7~1.8), full remediation of the effluent is observed, hydrazine is fully oxidized and removed.

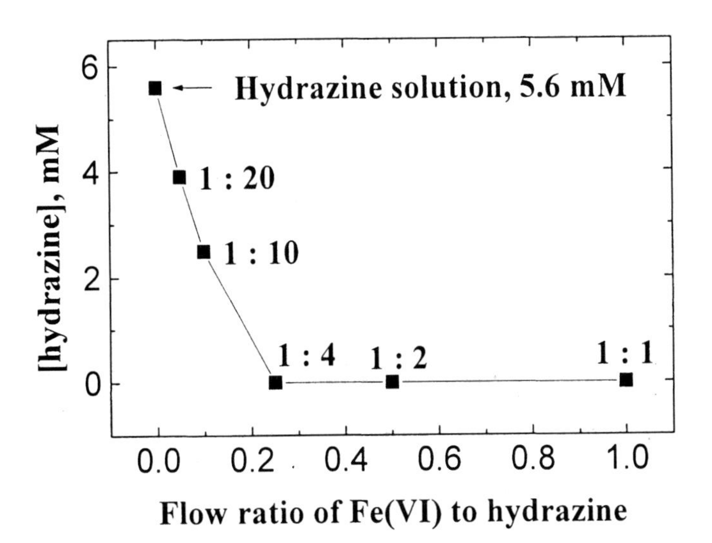

Figure 14. Hydrazine oxidation by Fe(VI) at different flow ratio of anodically generated 39±1 mM Fe(VI) to 5.6 mM hydrazine. The concentration of hydrazine is measured by the PDMAB method (ASTM D 1385-01) by means of VACUettes Kit®.

The simultaneous treatment of multiple contaminants in an effluent flow by Fe(VI) on-line remediation is also effective. Figure 15, presents this simultaneous treatment, in which on-line electrochemically generated Fe(VI), reacts with a multiple component effluent containing 8 mM aniline, 5 mM phenol and 5.6 mM hydrazine. A 1 : 1 flow rate ratio of Fe(VI) to the effluent is maintained. With the Fe(VI) treatment, all contaminants are fully eliminated, as observed in the figure with the complete disappearance of the initial aniline, phenol and hydrazine absorptions. Also evident in the figure, and again consistent with the treatment of the individual contaminant components in Figures 12-14, is the activated carbon efficacy, which additionally removes the products of the remediation.

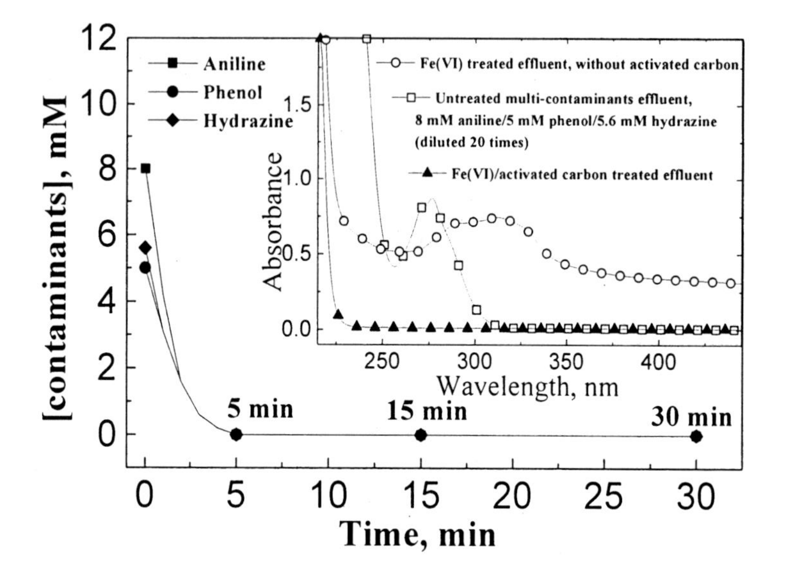

Figure 15. Simultaneous oxidation of aniline, phenol and hydrazine by on-line Fe(VI) treatment including oxidation products removal with activated carbon. Concentration changes of aniline, phenol and hydrazine before and after treatment. Inset: UV/Vis spectra are measured as: original effluent sample (8 mM aniline, 5 mM phenol and 5.6 mM hydrazine); effluent treated by Fe(VI) either with or without the activated carbon filter. Concentration of hydrazine is measured with PDMAB method (ASTM D 1385-01) by means of VACUettes Kit®.

Conclusions

On-line electrochemical Fe(VI) water treatment methodology can be an effective and straight forward process for water purification. Fe(VI) is an unusual and strongly oxidizing form of iron, which provides a potentially less hazardous water purifying agent than chlorine. The mechanism for Fe(VI) addition had been a barrier to its effective use in water remediation, as solid Fe(VI) salts require complex (costly) syntheses steps, and solutions of Fe(VI) decompose. On-line electrochemical Fe(VI) water purification avoids these limitations, as Fe(VI) is directly prepared in solution from an iron anode as the solution phase FeO_4^{2-} species, for immediate addition to the contaminant stream. The added FeO_4^{2-} decomposes, by oxidation, a wide range of water contaminants including inorganics (demonstrated in this paper are sulfides, cyanides and arsenic), organics (aniline, phenol and hydrazine are demonstrated), algae and viruses.

Downstream of the treated effluent, an activated carbon filter is shown to facilitate effective removal of the remediation oxidation products. Aniline, phenol and hydrazine are industrial organic water contaminants, and are water soluble due to containing amido and hydroxyl hydrophilic groups. They are individually readily oxidized by aqueous FeO_4^{2-}, to aqueous insoluble (solid or gas phase) products. These organic pollutants are removed with the on-line generated Fe(VI), and the remediation products are effectively removed with activated carbon. Simultaneous remediation of multiple contaminants is also demonstrated with a mix of either inorganic (cyanide and sulfide) or organic (aniline, phenol and hydrazine contaminants) contaminants. The latter (organic) mix was treated by the expanded system of on-line electrochemical Fe(VI) – activated carbon water purification.

Future efforts for this studies include modification of the iron electrode and the electrolyte configuration for Fe(VI) formation. Important issues to be addressed are the treatment of lower pH wastewater, and also the subsequent chemistry of the products of the redox reaction. Future studies should include specific comparisons of this Fe(VI) on-line remediation process to conventional chlorination remediation, as well as kinetic studies, not only of the Fe(VI) reaction rates, but also of the lifetime and fate of the Fe(III) products, and the fate of the oxidized contaminants in the presence of the Fe(III) product.

References

1. Carr J. D.; Kelter P. B.; Tabatabai A.; Splichal D.; Erickson J.; Mclaughlin C. W. *Proceedings of the 5th Conference, Water Chlorination;* Jolly R. L., Eds.; Lewis: Chelsea, MI, 1985; p. 1285.
2. Sharma V. K.; Smith J. O.; Millero F. J. *Environ. Sci. Tech.* **1997**, *31*, 2486.
3. Licht S.; Yu X. *Environ. Sci. Technol.* **2005**, *39*, 8071.
4. Licht S.; Wang B.; Ghosh S. *Science* **1999**, *285*, 1039.
5. Licht S.; Wang B.; Ghosh S.; Li J.; Naschitz V. *Electrochem. Comm.* **1999**, *1*, 522.
6. Licht S.; Wang B.; Xu G.; Li J.; Naschitz V. *Electrochem. Comm.,* **1999**, *1*, 527.
7. Licht S.; Wang B.; Li J.; Ghosh S.; Tel-Vered R. *Electrochem. Comm.* **2000**, *2*, 535.
8. Licht S.; Wang B. *Electrochem. Solid- State Lett.* **2000**, *3*, 209.
9. Licht S.; Ghosh S.; Dong Q. *J. Electrochem. Soc.* **2001**, *148*, A1072.
10. Licht S.; Naschitz V.; Ghosh S.; Lin L.; Lui B. *Electrochem. Com.* **2001**, *3,* 340.
11. Licht S.; Naschitz V.; Ghosh S. *Electrochem. Solid-State Lett.* **2001**, *4*, A209.
12. Licht S.; Naschitz V.; Ghosh S.; Liu B.; Halperine N.; Halperin L.; Rozen D. *J. Power Sources* **2001**, *99*, 7.
13. Licht S.; Naschitz V.; Lin L.; Chen J.; Ghosh S.; Lui B. *J. Power Sources* **2001**, *101,* 167.

14. Licht S.; Ghosh S.; Naschitz V.; Halperin N.; Halperin L. *J. Phys. Chem., B* **2001**, *105*, 11933.
15. Licht S.; Tel-Vered R.; Halperin L. *Electrochem. Com.* **2002**, *4,* 933.
16. Licht S.; Naschitz V.; Wang B. *J. Power Sources*, **2002**, *109,* 67.
17. Licht S.; Ghosh S. *J. Power Sources* **2002**, *109/2*, 465.
18. Licht S.; Naschitz V.; Ghosh S. *J. Phys. Chem. B* **2002,** *106*, 5947.
19. Ghosh S.; Wen W.; Urian R. C.; Heath C.; Srinivasamurthi V.; Reiff W. M.; Mukerjee S.; Naschitz V.; Licht S. *Electrochem. Solid-State Lett.* **2003**, *6*, A260.
20. Licht S.; Rozen D.; Tel-Vered R.; Halperin L. *J. Electrochem. Soc.* **2003**, *150*, A1671.
21. Licht S.; Tel-Vered R.; Halperin L. *J. Electrochem. Soc.* **2004**, *151*, A31.
22. Licht S.; Naschitz V.; Rozen D.; Halperin N. *J. Electrochem. Soc.* **2004**, *151*, A1147.
23. Licht S.; Tel-Vered R. *Chem. Comm.* **2004**, 628.
24. Nowik I.; Herber R. H.; Koltypin M.; Aurbach D.; Licht S. *J. Phys. Chem. Solids* **2005**, *66*, 1307.
25. Koltypin M.; Licht S.; Tel-Vered R.; Naschitz V.; Aurbach D. *J. Power Sources* **2005**, *146*, 723.
26. Licht S.; Yang L.; Wang B. *Electrochem. Comm.* **2005**, *7*, 931.
27. Koltypin M.; Licht S.; Nowik I.; levi E.; Gofer Y.; Aurbach D. *J. Electrochem. Soc.* **2006**, *153*, A32.
28. Licht S.; DeAlwis C. *J. Phys. Chem. B* **2006**, *110*, 12394.
29. Licht S.; Yu X.; Zheng D. *Chem. Comm.* **2006**, 4341.
30. Lee J.; Tryk D.; Fujishima A.; Park S. *Chem. Comm.* **2002**, 486.
31. Yang W.; Wang J.; Oan T.; Xu J.; Zhang J.; Cao C.. *Electrochem. Com.* **2002**, *4,* 710.
32. De Konnick M.; Belanger D. *Electrochim. Acta* **2003**, *48*, 1435.
33. Lee Y.; Um I. H.; Yoon J. *Environ. Sci. Technol.* **2003**, *37*, 5750.
34. Sharma V. K. *Adv. Environ.* Res. **2002**, *6*, 143.
35. Luca S. J.; Idle C. N.; Chao A. C. *Wat. Sci. Tech.* **1996**, *33*, 119.
36. Kazama F. *Wat. Sci. Tech.* **1995**, *31*, 165.
37. Ma J.; Liu W. *Wat. Res.* **2002**, *36*, 871.
38. Huang H.; Sommerfeld D.; Dunn B.; Eyring E.; Lloyd C. *J. Phys. Chem. A* **2001**, *105*, 3536.
39. Delaude L.; Laszlo P. *J. Org. Chem.* **1996**, *61*, 6360.
40. Ohata T; Kamachi T.; Shiota Y.; Yoshizawa K. *J. Org. Chem.* **2001**, *66*, 4122.
41. Norcross B.; Lewis W.; Gai H.; Noureldin N.; Lee D. *Canad. J. Chem.* **1997**, *75*, 129.
42. Firouzabdi H.; Mohajer D.; Entezari-Moghadam M. *Bull. Chem. Soc. Jpn.* **1988**, *61*, 2185.
43. Audette R.; Quail J.; Smith P. *J. Chem. Soc., Chem. Comm.* **1972**, 38.
44. BeMiller J.; Kumari G.; Darling S. *Tetrahedron Lett.* **1972**, *40*, 41.

45. Huang H.; Sommerfeld D.; Dunn B.; Lloyd C.; Eyring E. *J. Chem. Soc., Dalton Trans.* **2001**, 1301.
46. Jiang J.-Q.; Lloyd B. *Water Res.* **2002**, 36, 1397.
47. Licht S.; Davis J. *J. Phys. Chem.* **1997**, *101*, 2540.
48. Peramunage D.; Forouzan F.; Licht S. *Anal. Chem.* **1994**, *66*, 378.
49. Licht S.; Longo K.; Peramunage D.; Farouzan F. *J. of Electroanal. & Interfac. Electrochem.* **1991**, *318*, 111.
50. Licht S.; Fardad D.; Forouzan F.; Longo K. *Anal. Chem.* **1990**, *62*, 1356.
51. Licht S. *J. Electrochem. Soc.* **1988**, *135*, 2971.
52. Licht S.; Manassen J.; Hodes, G. *Inorgan. Chem.* **1985**, *25*, 2486.
53. Licht S. *Anal. Chem.* **1985**, *57*, 514.
54. Licht S.; Peramunage D. *J. Electrochem. Soc.* **1992**, *139*, L23.
55. Licht S.; Peramunage D. *Nature* **1991**, *354*, 440.
56. Choudhary G.; Hansen H. *Chemosphere.* **1998**, *37*, 801.
57. Laszlo K. *Colloids and Surfaces A: Physicochemical and Engineering Aspects.* **2005**, *265*, 32.
58. Niu J.; Conway B. E. *J. Electroanalytical Chem.* **2002**, *536*, 83.
59. Johnson M.; Hornstein B. *Inorg. Chim. Acta* **1994**, *225*, 145.

Chapter 17

Evaluating the Coagulation Performance of Ferrate: A Preliminary Study

Khoi Tran Tien[1], Nigel Graham[1,*], and Jia-Qian Jiang[2]

[1]Department of Civil and Environmental Engineering, Imperial College London, South Kensington Campus, London SW7 2AZ, United Kingdom
[2]Centre for Environmental Health Engineering, School of Engineering, C5, University of Surrey, Guildford GU2 7XH, United Kingdom
[*]Corresponding author: n.graham@imperial.ac.uk; fax: 44 207 5946124

Ferrate is cited as having a dual role in water treatment, both as oxidant and coagulant. Few studies have considered the coagulation effect in detail, mainly because of the difficulty of separating the oxidation and coagulation effects. This paper summarises some preliminary results from an ongoing laboratory-based project that is investigating the coagulation reaction, dynamically via a PDA instrument, between ferrate and a suspension of kaolin powder at different doses and pH values, and comparing the observations with the use of ferric chloride. The PDA output gives a comparative measure of the rate of floc growth and the magnitude of floc formation. The results of the tests show some similarities and significant differences in the pattern of behaviour between ferrate and ferric chloride. This paper presents and discusses these observations and provides some comparative information on the strength of flocs formed.

Introduction

Ferrate (Fe[VI]) is widely cited as having a dual role in water treatment, both as a powerful oxidant and as a coagulant, the latter as a consequence of its chemical reduction via Fe[V] to Fe[III]. Among the many studies of ferrate as water treatment chemical, Jiang *et al* (*1*) found that the maximum turbidity removal (almost 100%) was achieved at pH 7.5 for ferrate dosages from 2 to 12 mg/L as Fe and ferrate performed better than ferric sulphate in treating upland coloured water at low doses. In addition, ferrate showed a better performance in removing UV_{254} absorbance and dissolved organic carbon for waters containing humic and fulvic acids in comparison with ferric sulphate (*2*). Ma and Liu (*3;4*) demonstrated that pretreatment with ferrate clearly enhanced the removal of surface water turbidity and algae by coagulation with alum.

Whilst these and other previous reports have referred to the coagulation effect with ferrate, there appear to be very few studies that have considered the effect in detail; it is assumed that this is mainly because of the difficulty of separating the oxidation and coagulation phenomena when ferrate is applied. This paper summarizes some early results from an ongoing laboratory-based project that is investigating the coagulation effect. The study has employed a inert-type particle suspension (kaolin) and organic-free water in order to simplify the nature of the ferrate reaction by minimizing the influence of oxidation effects. The results are compared with those from identical tests using a conventional iron salt (ferric chloride) to observe any differences in behaviour.

Materials and Methods

Chemicals

All water used was deionised RO purified waster from a Purite – Neptune water purification system. Solid potassium ferrate was produced in the laboratory by the wet oxidation method of Li *et al.* (*5*) in which ferric nitrate was oxidized with potassium hypochlorite, and solid phase ferrate was precipitated in strong alkaline conditions. The purity of the ferrate, estimated to be ≥ 90%, and the concentration of ferrate in solution, were determined by visible light absorbance spectrophotometry; both by the conventional method (at 510 nm, pH 9) and by a novel, indirect spectrophotometric method involving ABTS and an absorption λ of 415 nm (*6*). Ferric chloride stock solution (0.01 M) was made by dissolving 2.71 g $FeCl_3 \cdot 6H_2O$ (from BDH, UK) in 1L deionised water before the commencement of the tests. Kaolin stock suspensions of 5 g/L and 10 g/L were prepared by adding kaolin (light) powder (from Fisher, UK) into 1L deionised distilled water and mixing thoroughly. Bicarbonate buffer solution (0.1 M) was made by dissolving 8.41 g $NaHCO_3$ (from Fisher, UK) in 1L of deionised water. Boric acid/NaOH buffer solution (pH 9.0) was prepared from 50 ml 0.1 M boric acid/0.1 M KCl solution, and 20.8 ml 0.1 M NaOH, then diluted to 100 ml with deionised water.

Apparatus

Conventional jar test methods, as used in coagulation tests, are limited in terms of their sensitivity and practical convenience, although they do sometimes provide a useful visual and semi-quantitative simulation of full-scale performance. In our tests the coagulation process was followed by use of a photometric dispersion analyzer (PDA 2000, Rank Brothers, Cambridge, UK) in a modified jar test procedure. This is a relatively new approach to evaluating coagulation performance which has been found to provide a sensitive and rapid response, although the output of the PDA itself is qualitative in nature. The PDA is an instrument for observing rapidly changing particle suspensions via an optical technique that analyses the light transmitted through a flowing suspension (Figure 1); the instrument measures the average transmitted light intensity (dc value) and the root-mean-square (RMS) value of the fluctuating component. The output, either the RMS or RMS/dc ratio, serves as a relative measure of the change in particle size and density distributions (*7*). The output is referred to as the Flocculation Index (FI).

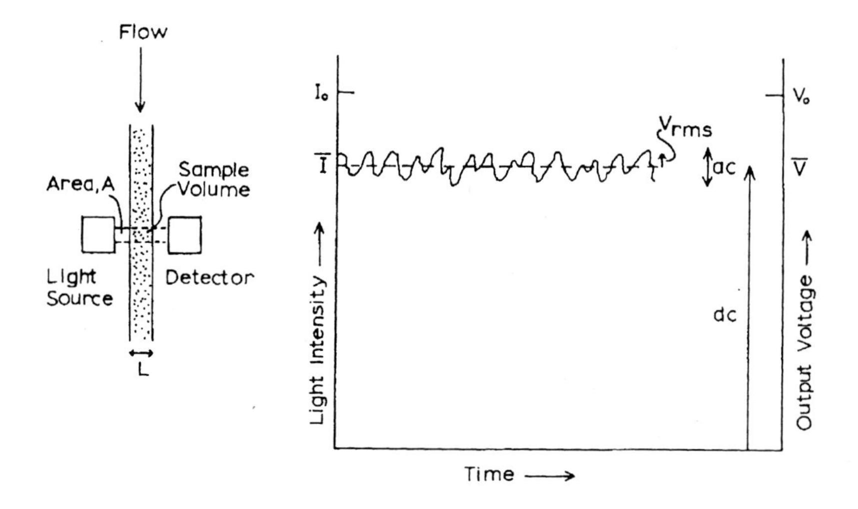

Figure 1. Principle of photometric dispersion analysis (Reproduced from J. Colloid Interface Sci. ***1985***, 105(2). *Copyright 1985.*

Tests were carried out by connecting the PDA to a stirred reactor where the ferrate or ferric chloride is added to the kaolin suspension under defined mixing conditions (Figure 2). The reactor was calibrated so that the mean velocity gradient (G, s^{-1}) of solutions undergoing stirring was directly determined from the rotational speed of the paddle. In the tests a constant flow from the stirred reactor to the PDA optical sensor was maintained by the use of tubing and a peristaltic pump (Matson 505S, UK) operating at about 25 ml/min; the flow passes through the optical sensor where it is illuminated by a narrow light beam (850nm wavelength). The optical data were recorded every second and the results were logged by computer for subsequent spreadsheet analysis. Although inherently qualitative, it is believed that the FI value is correlated with floc size and always increases as flocs grow larger.

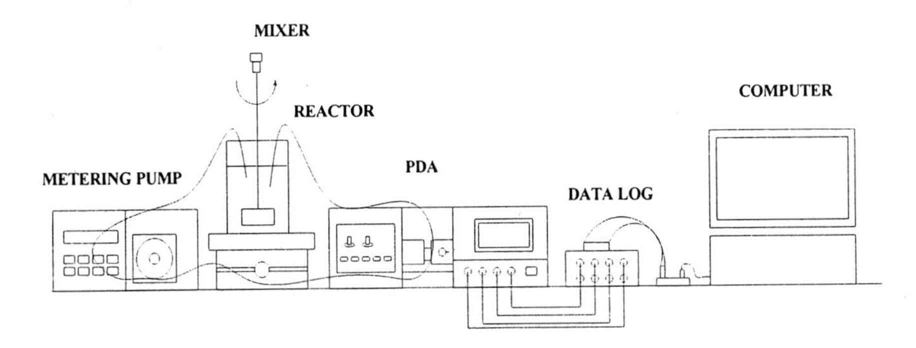

Figure 2. The experimental apparatus.

Experimental Procedures

Coagulation Tests

Our studies involved observation of the reactions between ferrate and an aqueous suspension of kaolin powder (50 mg/L) at different ferrate doses (5 to 200 μM) and pH values (4 to 8), and comparing the observations with the use of ferric chloride (a conventional chemical coagulant) under identical conditions. Kaolin is assumed to be inert to the oxidation effects of ferrate, so the observations of particle (floc) growth are believed to be the result of predominantly coagulation effects by Fe-hydrolysis species. Set pH values were controlled using bicarbonate and boric acid buffers..

Floc Strength Tests

The strength of flocs formed during coagulation depends on many factors, such as the type and amount of coagulant, the nature and quantity of interacting contaminants and the hydrodynamic conditions prevailing during floc formation. An individual floc will break if the stress applied at its surface is larger than the bonding strength within the floc. Gregory (*8*) has observed that when comparing different flocs formed by the same coagulant, the size of the floc (or indirectly the flocculation index, FI) for a given shear rate indicates floc strength. Using the 2L stirred reactor, an indication of relative floc strength can be obtained by applying a sudden increase in shear rate to the formed aggregates and relating velocity gradient applied to the maximum floc size resulting.

In these tests, the samples were prepared identically to those corresponding to the optimum performance for the coagulation tests at pH 7, where the optimum Fe dose was 200 μM. The kaolin suspension after the addition of the ferrate or $FeCl_3$ was initially mixed at 200 rpm ($G = 350\ s^{-1}$) for 60 seconds, and then the stirring was reduced and maintained at 50 rpm ($G = 50\ s^{-1}$) for an appropriate period of time so as to obtain the maximum flocculation (about 400 seconds). A sudden rise in the shear rate was then applied by increasing the stirring speed to 400 rpm (600 s^{-1}) for 120 seconds, followed by a reduction/return in the speed to 50 rpm for the remainder of overall test period. The flocculation index was observed and recorded continuously during the test so that semi-quantitative values, corresponding to a *strength factor* and *recovery factor* could be determined and used for comparison. These terms are defined later.

Results and Discussion

The output from the PDA, although qualitative in nature, gives a comparative measure of the rate of floc growth and the magnitude of floc formation. Two examples of the output of the PDA are given in Figures 3 and 4; in these figures the PDA response, the Floc Index (RMS), is an optical index quantified on the y-axis, with time from the beginning of the test on the x-axis. The particular results shown in Figures 3 and 4 summarise the observed variation in PDA response with pH, at a given dose (200 μM) of ferric chloride and potassium ferrate, respectively. Additional results have been obtained with different doses of the chemicals.

Interpretation of the overall results is complicated, partly by the well-established dependence of coagulation performance on both chemical dose and pH, and partly by the qualitative nature of the PDA response. With the latter, to provide a quantitative basis for comparison, two values can be extracted from

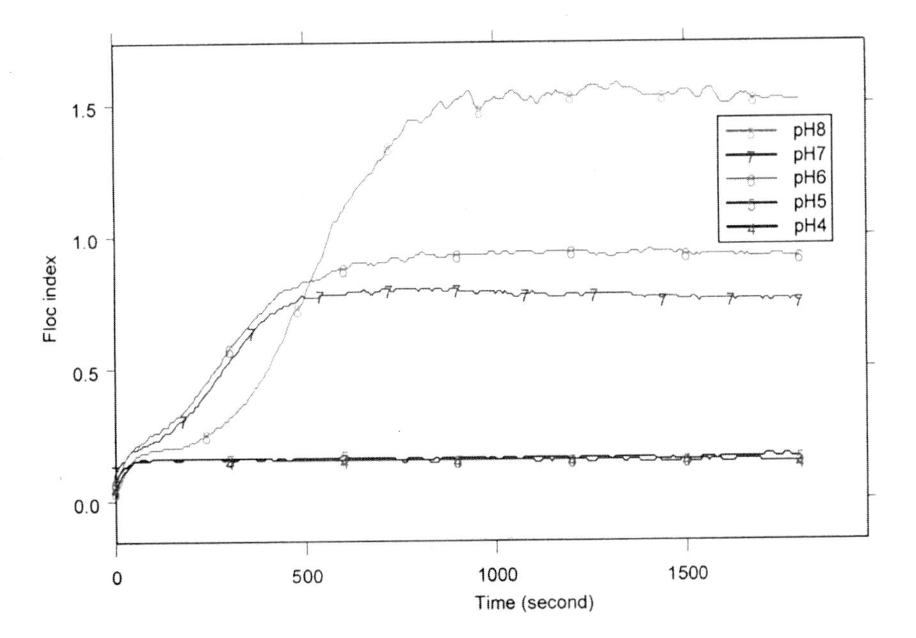

Figure 3. Coagulation of 50mg/L kaolin with 200μM $FeCl_3$ (1mM $NaHCO_3$ buffer solution).

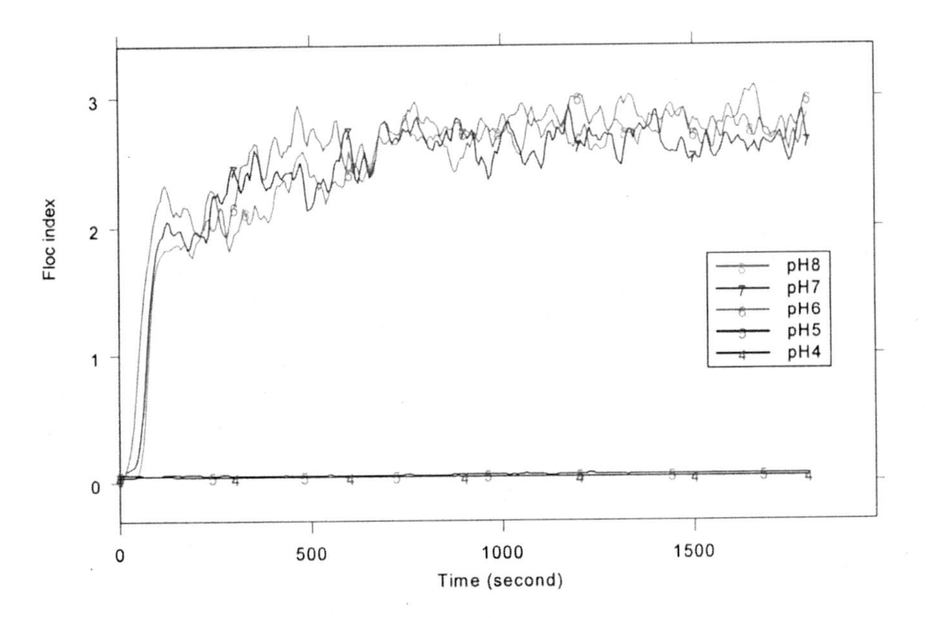

Figure 4. Coagulation of 50mg/L kaolin with 200μM Ferrate (1mM $NaHCO_3$ /Boric acid buffer solution).

each PDA response; these are the T_{50} and FI_{max} (RMS_{max}). The T_{50} is the time elapsed since the start of the test for the RMS value to reach 50% of its maximum value (RMS_{max}). Thus, good coagulation is demonstrated by a low T_{50} (i.e. rapid destabilization and floc growth) and a high FI_{max} (RMS_{max}) (ie. large, voluminous floc particles).

The results of the tests with ferrate and ferric chloride show some similarities and significant differences in the coagulation behaviour. In general, substantial floc formation was observed with both chemicals at the highest Fe concentrations (50, 100 and 200 μM), at pH 6, 7 and 8 (see Figures 3 and 4). This is consistent with the so-called 'sweep flocculation' mechanism whereby particles (kaolin) are incorporated within amorphous ferric hydroxide precipitates.

The solubility of amorphous ferric hydroxide has been reported to be a minimum at pH 8, and with only a minor increase in the range of 6 to 8. With ferric chloride the lowest T_{50} value was 66 s, corresponding to 200 μM Fe at pH 6; whilst for ferrate the lowest T_{50} value was 258 s, corresponding to 200 μM Fe at pH 7. However, the FI_{max} values for the two cases were quite different (viz. 3.09 ferric chloride, 0.95 ferrate), indicating significant differences in the nature of the flocs, and the conclusion that overall coagulation performance by ferrate was inferior to that with $FeCl_3$.

As can be seen from Figure 5 and 6, coagulation effects with both ferrate and ferric chloride were observed at doses from 5 to 200 μM at pH 6. Under these conditions it is believed that coagulation occurs by a combination of 'sweep flocculation' via ferric hydroxide precipitation *and* charge interaction between the kaolin and soluble cationic iron species. Good coagulation occurred under conditions of higher pH (6-8) and high dose (50 to 200 μM), which corresponds to the sweep floc region, and relatively poor coagulation occurred at low pH (4-5) and low dose (5, 10 and 15 μM). The performance increased proportionally with Fe dose. Overall, the coagulation performance of ferrate was not as good as ferric chloride, with generally higher values of T_{50} and lower values of FI_{max} observed with ferrate.

At lower pH values, 4 and 5, and [Fe]>15 μM, no or very little floc formation was evident with both chemicals. This is consistent with previous studies (*9*) which suggest that under these conditions the coagulant exists primarily as charged Fe(III) hydrolysis species (e.g. $Fe(OH)^{2+}$, $Fe(OH)_2^+$) that re-stabilize the kaolin suspension. However, at lower Fe doses (5, 10 and 15 μM), the balance of charges is such that some degree of destabilization occurs, leading to coagulation. At pH 5, and with 10 μM Fe dose, the ferrate coagulation performance was superior to the ferric chloride (see Figure 7). Some tests at pH 4 and 5 were repeated with a higher concentration of kaolin suspension and evidence of a dose stoichiometry (Fe:kaolin ratio) was found, with both ferrate and ferric chloride, supporting the assumption that charge neutralization is the predominant coagulation mechcanism under these pH—dose conditions.

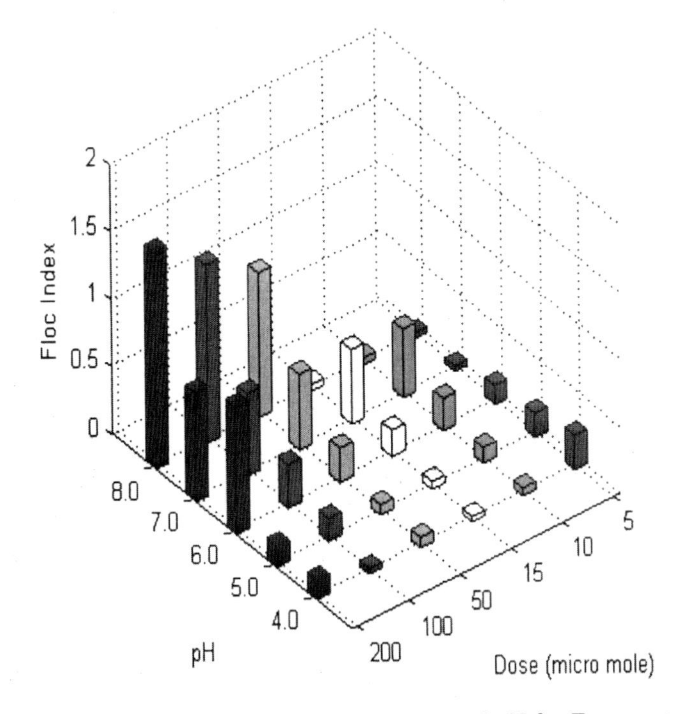

Figure 5. Variation of FI_{max} with dose and pH for Ferrate.

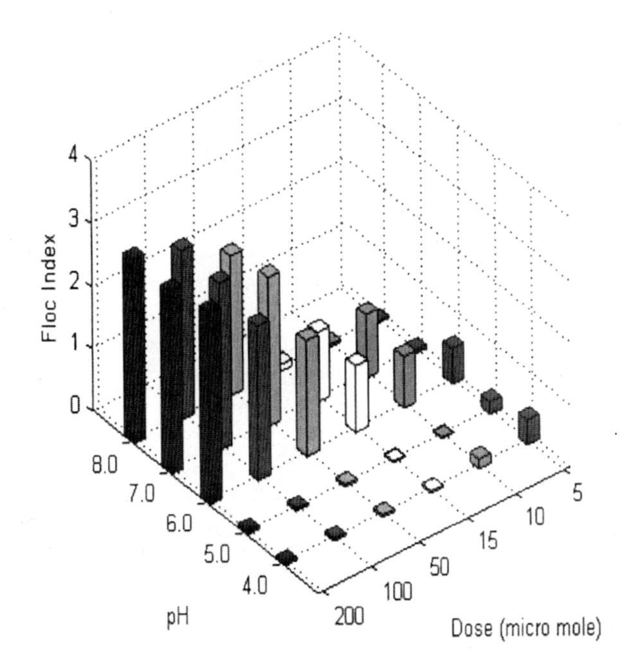

Figure 6. Variation of FI_{max} with dose and pH for Ferric Chloride.

Figure 8 presents the results of the floc strength tests with ferrate and ferric chloride under common optimal coagulation conditions (ie. 200 μM at pH 7). From Figure 8, it can be seen that by increasing the stirring rate from 50 rpm (50 s^{-1}) to the higher rate of 400 rpm (600 s^{-1}), there is an immediate and rapid decrease in FI, corresponding to a rapid breakage of flocs. When the stirring speed was reduced back, after 120 s, to 50 rpm (50 s^{-1}) there was no real evidence of a recovery or re-growth of flocs; in other coagulation systems limited re-growth of flocs is often observed. Thus, in these limited tests the results have indicated complete and irreversible floc breakage with both chemicals.

In general, semi-quantitative values, or indices, can be determined from the FI response curves in order to compare alternative coagulation conditions. Representative Flocculation Index values for the initial (FI_1), broken (FI_2) and reformed (FI_3) flocs can be used as surrogates for floc size in order to determine indices for the floc strength. These are as follows, as defined by Gregory and Yukselen (*10*) as:

$$\textit{Strength factor} = (FI_2 / FI_1) \cdot 100$$

$$\textit{Recovery factor} = [(FI_3 - FI_2) / (FI_1 - FI_2)] \cdot 100$$

It is believed that the higher the value of the strength factor, the stronger the flocs, since they are less sensitive to breakage as a result of the increased shear rate. The recovery factor is a measure of the capability of the floc to re-form and is of particular relevance to coagulation performance in practice, where floc disturbance due to flow irregularities is typical.

The results shown in Figure 8 indicate similar floc strengths for the two coagulants (strength factors: 0.3 for ferrate and 0.35 for ferric chloride), with ferrate flocs being slightly weaker than those from ferric chloride. The generally similar response suggests that the flocs could have similar physical structures arising from the aggregation of the insoluble Fe hydrolysis products and kaolin particles; this aspect will be investigated further. With both coagulants it is apparent that an irreversible breakage (recovery factor = 0) occurred when the shear rate increased. This is consistent with the well known effect in practice that flocs formed by hydrolysing coagulants tend to be weak and not fully reversible when formed in the sweep coagulation range (*7*); the zero recovery factor may be a consequence of the use of organic-free model water.

In some studies concerning the oxidation performance of ferrate the complicating influence of coagulation effects are minimized by the use of a phosphate buffer and high pH (e.g. ref *11*). It is assumed that Fe(III) forms soluble complexes with phosphate, and this phenomenon was briefly studied here. Figure 9 shows the effect of 5mM phosphate on the ferrate interaction with the kaolin. The results showed that the effectiveness of the phosphate in preventing coagulation effects at all pH values, with the exception of pH 4 where

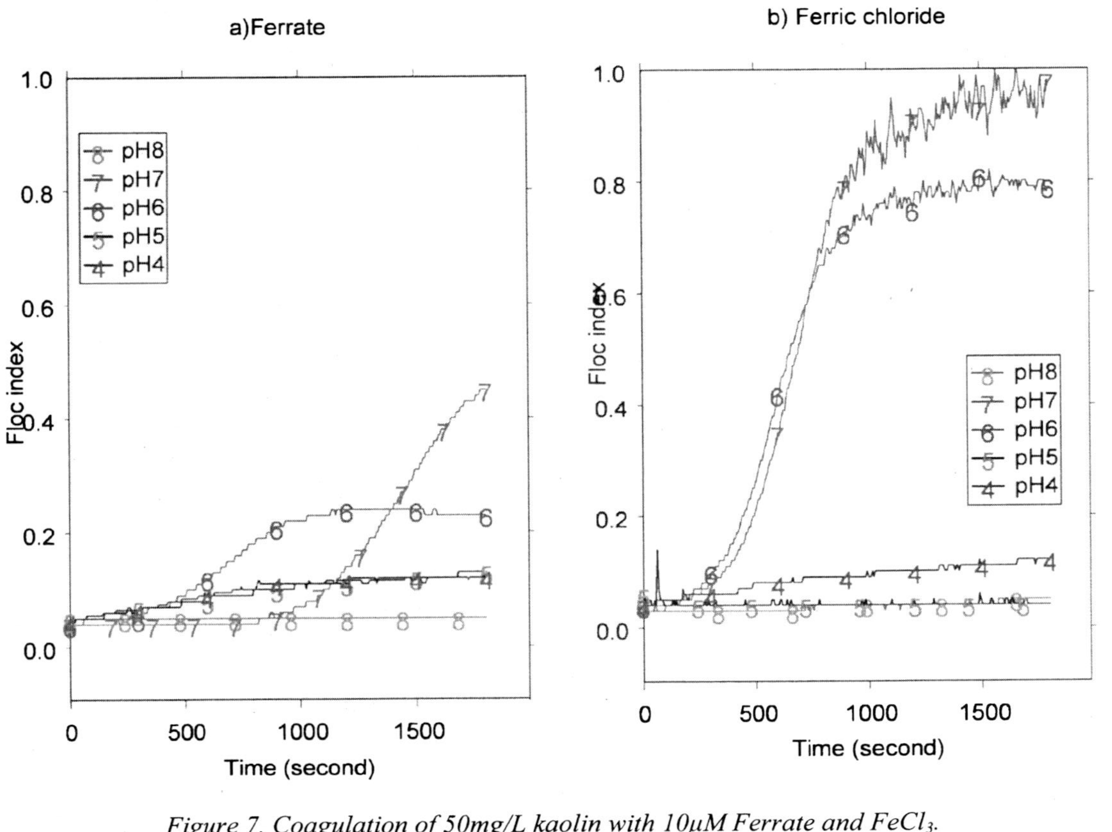

Figure 7. Coagulation of 50mg/L kaolin with 10μM Ferrate and $FeCl_3$.

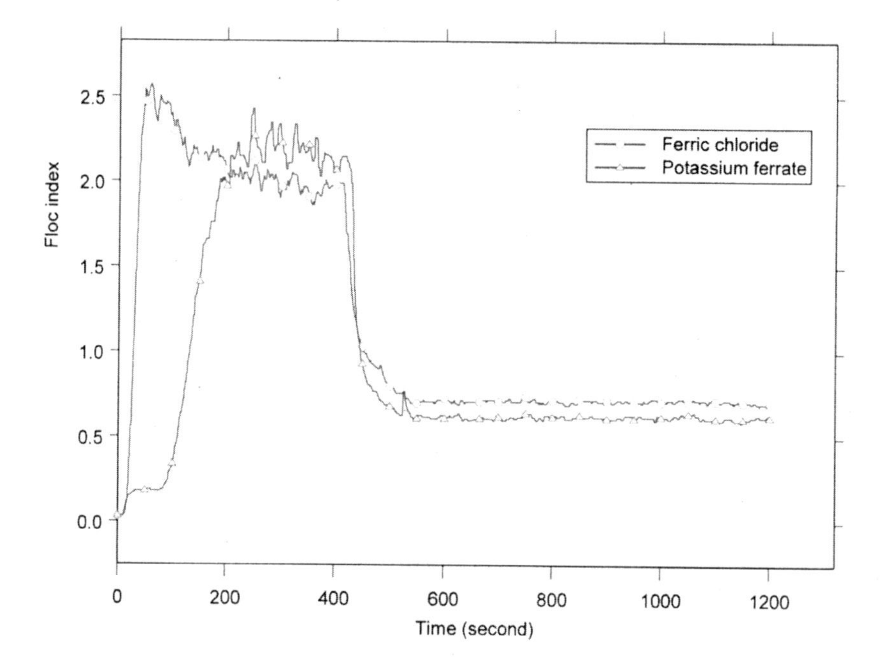

Figure 8. Variation of FI response during floc strength test (pH 7; Fe dose 200 μM).

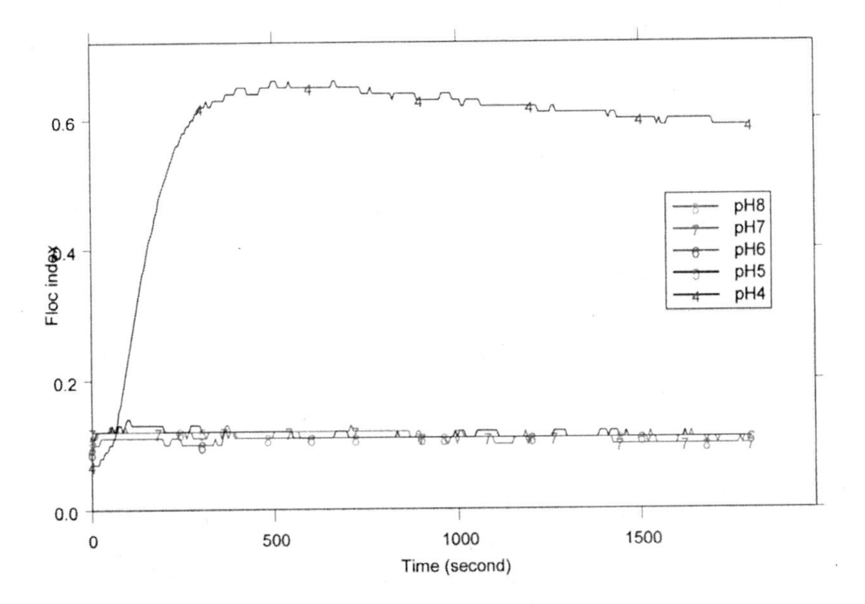

Figure 9. Coagulation of 50mg/L kaolin with 100μM Ferrate with 5mM phosphate/ 1mM borate buffer.

some solid phase precipitation is evident. A very similar behaviour was observed with the use of ferric chloride. The precise nature of the precipitate formed at pH 4 is unclear at present, as is the nature of the Fe-phosphate complex that is assumed to be formed.

Conclusions

A preliminary study has been undertaken to compare the coagulation performance of potassium ferrate in direct comparison with ferric chloride, a conventional coagulant chemical. A special test methodology was used to exclude simultaneous oxidation effects of ferrate, and to monitor the rapid formation of floc particles. The former was achieved by use of kaolin suspensions as the reacting substrate, which was assumed to be inert to the influence of ferrate oxidation. Floc formation was monitored by use of an on-line PDA light obscuration instrument.

In general, the ferrate demonstrates very similar coagulation characteristics to ferric chloride with regard to the influence of pH and Fe dose. Thus, floc formation with ferrate was relatively rapid and substantial at neutral pH and moderate Fe concentrations, corresponding to 'sweep coagulation', while at low pH there was some evidence of charge destabilisation and charge stoichiometry. In addition, there was no significant difference in the strength of flocs formed by each coagulant under optimal conditions.

The principal differences between ferrate and ferric chloride as coagulants was that the magnitude of floc formation with ferrate was always inferior to that with ferric chloride, and that in most cases the rate of floc growth with ferrate was slower than with ferric chloride. However, there was some evidence of a superior coagulation effects with ferrate at pH 4 and 5 at low Fe doses (5-15 μM). The reasons for the difference in the magnitude and rate of floc growth are not clear at present and will be considered in further work. Subsequent studies will also consider the comparative performance of ferrate and ferric chloride with humic substances, and will focus in particular on the separate roles of oxidation and coagulation in the case of ferrate.

References

1. Jiang, J.Q.; Lloyd, B.; Grigore, L. Preparation and evaluation of potassium ferrate as an oxidant and coagulant for potable water treatment. *Environ. Eng. Sci.* **2001**, *18*, 323-328.
2. Jiang, J.Q.; Wang, S. Enhanced coagulation with potassium ferrate(VI) for removing humic substances. *Environ. Eng. Sci.* **2003**, *20*, 627-633.

3. Ma, J.; Liu, W. Effectiveness of ferrate (VI) preoxidation in enhancing the coagulation of surface waters. *Water Res.* **2002**, *36*, 4959-4962.
4. Ma, J.; Liu, W. Effectiveness and mechanism of potassium ferrate(VI) preoxidation for algae removal by coagulation. *Water Res.* **2002**, *36*, 871-878.
5. Li, C.; Li, X-Z.; Graham, N. A study of the preparation and reactivity of potassium ferrate. *Chemosphere* **2005**, *61*, 537-543.
6. Lee, Y.; Yoon, J.; von Gunten, U. Spectrophotometric determination of ferrate (Fe(VI)) in water by ABTS. *Water Res.* **2005**, *39*, 1946-1953.
7. Duan, J.; Gregory, J. Coagulation by hydrolysing metal salts. *Advances in Colloid Interface Sci.* **2003**, *100-102*, 475-502.
8. Gregory, J. Monitoring floc formation and breakage. In; *Proceedings of the Nano and Micro Particles in Water and Wastewater Treatment Conference.* International Water Association, Zurich, September 2003.
9. Amirtharajah, A.; O'Melia, C.R. Coagulant Processes: Destabilization, Mixing, and Flocculation. In: Pontius, F.W., (Ed.). *Water Quality and Treatment.* McGraw-Hill, New York, pp. 269-365.
10. Gregory, J.; Yukselen, M. A. The reversibility of floc breakage. *Int J. Miner. Process* **2004**, *73*, 251-259
11. Sharma, V.K. Use of iron(VI) and iron (V) in water and wastewater treatment, *Water Sci. Technol.* **2004**, *49*, 69-74.

Chapter 18

The Use of Ferrate(VI) Technology in Sludge Treatment

Jia-Qian Jiang[1] and Virender K. Sharma[2]

[1]Centre for Environmental Health Engineering, School of Engineering, C5, University of Surrey, Guildford, Surrey GU2 7XH, United Kingdom (j.jiang@surrey.ac.uk; fax: +44 1483 450984)
[2]Department of Chemistry, Florida Institute of Technology, 150 West University Boulevard, Melbourne, FL 32901
*Corresponding author : j.jiang@surrey.ac.uk; Fax: +44 1483 450984

Sludge in large quantity is generated as byproducts of wastewater treatment processes. Various approaches have been taken to treat sludge, such as land-filling, ocean dumping, or recycling for beneficial purposes. In the USA, about 60% of sludge generated is land applied as a soil conditioner or fertilizer. Due to increasing public concern on the safety of land-applied sludge, various sludge treatment technologies are being developed or under evaluation in order to improve the quality of sludge in terms of pathogen content, odor characteristics, accumulated organic micro-pollutants. This paper summarizes the results of various reported or on-going researches on the potential use of ferrate [$Fe(VI)O_4^{2-}$] as a conditioning agent for sludge. Ferrate(VI) has high oxidizing potential and selectivity, and upon decomposition produces a non-toxic by-product, Fe(III), which is a conventional coagulant; the ferrate(VI) is thus considered to be an environmentally-friendly oxidant. Rates of oxidation reactions increase with decrease in pH. Oxidation of sulfur- and amine-containing contaminants in sludge by Fe(VI) can be accomplished in seconds to minutes with formation of non-hazardous products. Ferrate(VI) can also coagulate toxic metals and disinfect wide ranges of microorganisms including human pathogens. With its multifunctional properties, ferrate(VI) has the potential for sludge treatment.

Introduction

In the development of municipal wastewater treatment strategies, except for the efforts made to improve the quality of the effluent by using more effective treatment technologies, the problems associated with the sewage sludge produced in wastewater treatment processes have been taken into account. Toxic pollutants (e.g., heavy metals and endocrine disruptors) together with a large number of the pathogens are concentrated in the sludge, and this increases in the risks to the health and environment. Moreover, a number of organic sulfides and amines are produced in wastewater treatment which results in unpleasant odors (*1*). In the United States, 5.6 million tons of sludge is generated annually, of which 60% is applied as a fertilizer (*2*). Complaints of illness related to the land application of biosolids have been increasing (*3*), and the original application of the sludge as a fertilizer in agricultural systems has thus become increasingly under pressure. The legislation and regulations regarding the application of sludge in agriculture have changed considerably. According to Environmental Protection Agency (EPA) under 40 CFR Part 503, biosolids designated for land application are classified as Class A or Class B, depending on the presumed pathogen content. Class A biosolids are intended to have undetectable bacteria, enteric viruses, and helminthes. The revised European Union Directive on sewage sludge revises the previous one (EU Sludge Directive (86/278/EEC)) significantly. The most important new aspects are the requirement of sludge hygienization and odor reduction using advanced treatments, and the treated sludge shall not contain Salmonella spp. in 50 gram wet weighted sludge, and achieve at least a 6 Log10 reduction in Escherichia coli to less than 500 colony forming units (CFU) per gram dry solids. Also, the use of conventional treated sludge on parks, green areas, and city gardens as well as any use on forests is to be forbidden.

The revised EU Directive on sewage sludge has also proposed considerably stricter regulations for heavy metals. It has also added new limit values on organic micropollutants, such as sum of halogenated organic compounds (AOX); linear alkylbenzene sulfonates (LAS), di (2-ethylhexyl) phthalate (DEHP), nonylphenol and nonylphenolethoxylates (NPE), sum of polycyclic aromatic hydrocarbons (PAH), polychlorinated biphenyls (PCB) and dioxins. This should require water industries to apply more advanced technologies to meet the proposed treated sludge standards.

Due to all these developments the current practice of sludge management has considerably changed during the past twenty years. Improved biogas production, advanced sludge dewatering processes, controlled land filling and thermal processes are increasingly applied in practice. There is a need of innovative sludge management practices, which could not only effectively treat a wide range of contaminants and health hazardous pathogenic organisms, but could also remove unconventional contaminants (e.g., personal care products and endocrine disruptors) from sewage sludge. Due to the urgency to develop more sustainable and efficient sludge treatment technologies, an increasing growth in research is observed (*4*).

Ferrate(VI) ion has the molecular formula, FeO_4^{2-}, and is a very strong oxidant. Under acidic conditions, the redox potential of ferrate (VI) ions is greater than ozone and is the strongest of all the oxidants/disinfectants practically used for water and wastewater treatment (*5,6*). Moreover, during the oxidation/disinfection process, ferrate(VI) ions will be reduced to Fe(III) ions or ferric hydroxide, and this simultaneously generates a coagulant in a single dosing and mixing unit process. The use of ferrate(VI) in treating sludge can accomplish the important objectives of sludge management. This review aims to address currently emerging issues in sludge management and discuss the potential role of ferrate(VI) in sewage sludge treatment and management.

Sludge Management

In sludge all contaminants are present as one mixture, which includes pathogens and other harmful microbiological pollutants, odor-causing compounds, toxic heavy metals, and toxic organic micropollutants such as pesticides, endocrine disrupters. These contaminants are mixed with non-toxic organic carbon compounds (approximately 60% on dry basis), and total nitrogen and phosphorus containing components, which makes the management of sludge to be complicated. Organic carbon, phosphorus, and nitrogen containing compounds can be considered as valuable compounds and are often recovered and reused after treatment. Sludge treatment also involves the minimization of the possible adverse impact of sewage sludge on the environment and on human beings.

Sludge hygienization

Pathogens, such as bacteria, viruses, and human parasites, which may cause human diseases, exist in raw wastewaters. Common bacterial pathogens include Escherichia coli, Helicobacter pylori and Listeria montocytogenes. Of the emerging pathogens enteric viruses present the greatest risk because of their resistance to inactivation and longer survival. The literature (*7*) suggested that the concentration of enteric viruses (enteroviruses) ranged from 102 to 104 per gram dry weight of solids in raw biosolids and an average of 300 per gram in secondary biosolids. E. *coli* and L. *montocytogenes* are known capable of surviving anaerobic digestion and may regrow after land application in some cases. This is the reason that sewage sludge must be subjected to additional pathogen reduction treatment prior to land application. Aerobic and anaerobic digestion and lime treatment are the common processes for the pathogen control. The inactivation of enteric viruses by aerobic digestion is usually pH and temperature dependent. In general, a plant using mesophilic digestion (37°C) with a mean retention time of 10–20 days can reduce enterovirus concentration by 90%. Anaerobic digestion can result in a similar reduction. More recently, inactivation of vaccine-strain poliovirus and eggs from the helminth Ascaris suum in biosolids under thermophilic anaerobic digestion conditions have been reported (*8*). Interestingly, Ascaris ovum could be inactivated with the use of electron beam irradiation process (*9*). Moreover, this process may result in morphological changes in ovum as has been found in Ascaris lumbricoides ova sewage sludge water treated by gamma irradiation (*10*).

Lime treatment has been one of the most frequently applied pathogen control processes in a domestic wastewater treatment plant. The lime addition to sludge causes pH to increase to over 12 for over 2 hrs and leads to the inactivation of pathogen. However, the liming process promotes to generate ammonia and amines (e.g., trimethylamine) and odor, resulting in complaints from the public (*1*). Alternative treatment chemicals, e.g., chlorine or ozone, have been used to disinfect sludge (*11*). Ozonation has also been suggested a suitable process for minimizing the sludge production (*12*). However, due to high cost and the generation of harmful by-products, they are not commonly employed. Therefore, other chemical products which can economically favorably stabilize sludge and do not produce any harmful by-products and odors should be explored.

Odors

Odor is a problem at many urban and industrial wastewater treatment plants (*13*). Therefore, most unit processes, e.g., preliminary treatment, primary clarifiers, activated sludge basins, secondary clarifiers, sludge thickening, and conditioning, and dewatering, in a wastewater treatment plant are potential resource of odor (*14*). Especially, sludge conditioning systems, such as thickening, drying, and lime stabilization, are normally the most significant source of odors, since these unit processes are usually open to the air, and they emit odors with intensity raging from mild to nauseating. Moreover, cationic organic polymers used in enhancing thickening and dewatering processes become potential sources of strong odors (*15*). Chemical structures of these polymers are such that they are susceptible to biotic or abiotic degradation, which produce volatile odorous organic amines. Fresher sludge emits less offensive odor and septic sludge is more offensive. The odors from the sludge during solid handling processes are often carried over to the final biosolids, which is to be land applied (*14*). Since odors are considered as the most important factor affecting the operation of a wastewater treatment plant and the public acceptance of the final biosolids for land application, appropriate odor control should be designed and applied to the processes. Most frequently used odor control techniques include pH adjustment, addition of metal salts and nitrate, and addition of chemical oxidants such as hydrogen peroxide, chlorine, potassium permanganate, and ozone (*16,17*). Especially, through the addition of oxidants, the oxidation environment of the system is promoted and septic or anaerobic condition in which odorous sulfur compounds are generated can be avoided. Furthermore, some chemicals with odor potential can be destroyed through oxidation (*11*). Oxidation technologies have been applied to control odors from wastewater and sludge.

Hydrogen peroxide has been applied to oxidize H_2S, mercaptanes, thiosulfate, and sulfur dioxide in wastewater and waste activate sludge (*18*). However, hydrogen peroxide can be dangerous to handle and takes time to react, so several other oxidants also have been tested and applied. For example, potassium permanganate was added to raw dewatered sludge to reduce hydrogen sulfide production (*19*). Farooq and Akhlague (*20*) tested ozone to condition and oxidize heavy metals and organics in raw and thickened waste activate sludge. They reported a considerable reduction in odor released from the sludge, although they did not quantify their results. Gao et al. (*21*) used sodium hypochlorite to remove 95 % of hydrogen sulfide from gas emissions released from gravity thickened sludge. In the use of various chemicals for hydrogen sulfide treatment, it has been a trend since 1970s that an increase in the use of metal salts (e.g., ferric and ferrous iron) and nitrate, with a decrease in the use of lime, chlorine and oxygen.

Heavy metals

Heavy metals present in the wastewater will be removed in the treatment processes. The concentration ratio of heavy metals in the sludge to wastewater is of the magnitude of 10000:1. This means that even very small concentrations of heavy metals in wastewater will result in greater concentrations in sludge. Such high concentrations of metals in sludge may risk crop yield, long-terms soil quality, wildlife and cattle heath, and eventually human health. As stated previously that the revised EU Directive on sewage sludge has proposed considerably stricter regulations for heavy metals (see Table 1) and this should require water industries to apply more advanced technologies to meet the proposed treated sludge standards.

Table 1. Limit values of heavy metals in sludge for use on land

	Limit values		
Metals	Directive 86/278 (mg/kg-dry sludge)	Revised Directive (mg/kg-dry sludge)	Revised Directive (mg/kg P)
Cd	20–40	10	250
Cr	–	1,000	25,000
Cu	1,000–1,750	1,000	25,000
Hg	16–25	10	250
Ni	300–400	300	7,500
Pb	750–1,200	750	18,750
Zn	2,500–4,000	2,500	62,500

NOTE: The sludge producer may choose to observe either the dry matter related or the phosphorus related limit values.

Micro pollutants and endocrine disruptors

Sludge contains the components both in wastewater (domestic and industrial effluents) and that from the formation of by-products of biological and chemical treatment. The wastewater includes the expected urine and faecal matter but also synthetic organic chemicals, contributed not only by the commercial and industrial sectors, but also by residents via sinks and floor drains, and even as the pharmaceutical chemicals contained in their own wastes. With the advance of chemical/biochemical synthetic technologies, millions chemicals have been developed and manufactured but might not be considered for their complete recycle and reuse, and a significant fraction of these chemicals is sent into waste streams such as wastewater and its residual sludge. The revised EU Directive on sewage sludge has added new limit values on organic micropollutants (see Table 2), such as sum of halogenated organic compounds (AOX), linear alkylbenzene sulfonates (LAS), di (2-ethylhexyl) phthalate (DEHP), nonylphenol (NP) and nonylphenolethoxylates (NPE), sum of polycyclic aromatic hydrocarbons (PAH), polychlorinated biphenyls (PCB) and dioxins. This should require water industries to apply more advanced technologies to meet the proposed treated sludge standards.

Table 2. Limit values of organic compounds and dioxins

Compound	Limit values (mg/kg-dry matter)
AOX (*sum of halogenated organic compounds*)	500
LAS (*linear alkylbenzene sulfonates*)	2,600
DEHP (di (2-ethylhexyl) phthalate)	100
NPE (*nonylphenol and nonylphenolethoxylates with 1 or 2 ethoxy groups*)	50
PAH (*sum of various polycyclic aromatic hydrocarbons*)	6
PCB (*sum of polychlorinated biphenyls*)	0.8
PCDD (*polychlorinated dibenzodiox./dibenzofur*) /(ng-TE/kg-dry matter)	100

Endocrine disruptor chemicals (EDCs), e.g., nonylphenols (NPs) and polybrominated diphenyl ethers (PBDEs) in wastewater and sewage sludge have been detected at trace levels (*22*). The adverse effects of EDCs have been well addressed. EDCs can reach water resources indirectly through the farm runoff where biosolids has been applied. Entering of EDCs into the environment can possibly threaten the long-term viability of eco-system (*23*). NPs result from the incomplete biotransformation of domestic or industrial detergents (alkylphenol-polyethoxylate surfactants). Therefore, in the effluent of a swage treatment plant relatively high concentration of NPs can be observed (*24*). Due to their high lipophilicity, NPs can accumulate to higher concentrations in thickened sludge and the final biosolids, which can be land applied (*22*). Ahel et al. (*25*) found about 60% of total nonylphenols to a wastewater treatment plant accumulated in sludge.

PBDEs have been used as a flame retardant for furniture, televisions cast and other plastics (*26*). They can be introduced into wastewater and also accumulate in biosolids as the case with NPs (*27*). Recently, the public concerns on the land-applied biosolids have been elevated since it is thought as a potential source which delivers the endocrine disruptors to the environment, such as soil and especially aquatic ecosystem (*28*).

If these endocrine disruptors are introduced into surface water and enter the body of a fish, it is known the action of the endocrine system through various mechanisms is inhibited (*29*). Researches have shown estrogen activity of NPs in rainbow trout (*30*) and retardation in the development of secondary sex characteristics in Fathead Minnow induced by NPs. Zennegg et al. (*31*) also found PBDEs in the tissues of whitefish and rainbow trout. Although it is still under investigation, possible carcinogenic effect of the endocrine disruptors on humans is being discussed. Wren (*32*) reported the presence of endocrine disruptors at the concentrations enough to warrant concerns on the health of operators at the plant.

Therefore, it is desirable to destroy the endocrine disruptors within the treatment processes before they are released into the environment. Since endocrine disruptors in wastewater and biosolids have been recently addressed, only a few researches have been performed on their destruction (*33*).

Biological treatment alone cannot eliminate endocrine disruptors from wastewater. Lee and Paert (*34*) found that only 50% of NP could be eliminated through the secondary process in wastewater treatment. On the other hand, sonication, or peroxidase-mediated or Fenton's reagent-mediated oxidation showed more than 95 % NP removal efficiency (*33*). However, these processes require considerable capital demand. Therefore, more economically favorable alternative oxidation process should be pursued.

Ferrate as a Potential Sludge Conditioner

Properties of ferrate(VI)

Iron commonly exists in the +2 and +3 oxidation states; however, in a strong oxidizing environment, higher oxidation states of iron, +6 can be obtained. In the laboratory, ferrate(VI) can be produced from the reaction of ferric chloride with sodium hypochlorite in the presence of sodium hydroxide (*35*). This method produces sodium ferrate(VI) (Na_2FeO_4) and potassium hydroxide is added into to precipitate potassium ferrate(VI) (K_2FeO_4). The basic reactions are as follows:

$$2\ FeCl_3 + 3\ NaOCl + 10\ NaOH \rightarrow 2\ Na_2\ FeO_4 + 9\ NaCl + 5\ H_2O \quad (1)$$

$$Na_2\ FeO_4 + 2\ KOH \rightarrow K_2FeO_4 \quad (2)$$

Ferrate(VI) exhibits a multitude of advantageous properties; including higher reactivity and selectivity than traditional oxidant alternatives, disinfectant, and coagulant properties (*5*). Ferrate(VI) is one of the most powerful multi-purposes oxidizers known in treatment processes. Under acidic conditions, the redox potential of ferrate ion is the highest of any oxidant such as chlorine, ozone, hydrogen peroxide, and potassium permanganate used in treatment processes (eqs 3, 4).

$$FeO_4^{2-} + 8H^+ + 3e^- \Leftrightarrow Fe^{3+} + 4\ H_2O \qquad E^o = 2.20\ V \quad (3)$$

$$FeO_4^{2-} + 4\ H_2O + 3e^- \Leftrightarrow Fe(OH)_3 + 5\ OH^- \qquad E^o = 0.70\ V \quad (4)$$

The spontaneous oxidation of Fe(VI) in water forms molecular oxygen (eq 5), which can minimize anaerobic conditions in sludge.

$$FeO_4^{2-} + 5H_2O \rightarrow Fe^{3+} + 3/2O_2 + 10OH^- \quad (5)$$

A by-product of Fe(VI) is non-toxic, Fe(III), making Fe(VI) an environmentally friendly oxidant for sludge treatment (*36,37*). Moreover, ferric hydroxide, produced from Fe(VI), acts as a coagulant which is an excellent dewatering agent and is also suitable for removal of toxic substances such as metals in sludge. Tests show that Fe(VI) has a greater efficiency than commonly used inorganic coagulants (*38*).

Ferrate(VI)'s potential as a disinfectant

The disinfecting properties of ferrate(VI) were first observed by Murmann and Robinson (*39*) when they investigated the effectiveness of ferrate as a disinfectant to kill two pure laboratory cultures of bacteria (Non-recombinant *Pseudomonas* and Recombinant *Pseudomonas*). At a dose range of 0 - 50 ppm as FeO_4^{2-}, the bacteria could be completely destroyed. Later, another study (*40*) showed that ferrate (VI) has sufficient disinfection capability to kill *Escherichia coli* (*E. coli*). At pH 8.2 and a dose of 6 mg/l as Fe, the *E. coli* percentage kill was 99.9% when the contact time was 7 min. The results also demonstrated that the disinfecting ability of FeO_4^{2-} increased markedly if water pH was below 8.0.

In another secondary effluent disinfection study with ferrate(VI) (36), 99.9% of total coliforms and 97% of the total viable bacteria were removed at a dose of 8 mg/l as FeO_4^{2-}(Figure 1). Using ferrate(VI) to treat real drinking water and wastewater were conducted by Jiang et al. (*41-43*). Study results demonstrated that ferrate(VI) is effective in killing *Eschericha coli* (*E. coli*) and total coliforms. For example, in treating sewages, ferrate can achieve >99.99% inactivation of total coliforms at relative low dose comparing with other reagents (Table 3).

Kazama (*44*) demonstrated that Fe(VI) rapidly inactivated f2 *Coliphage* at low concentrations and a survival ratio of the virus decreased rapidly within 10 minutes after the addition of ferrate(VI). The treatment of DNA solution with micromolar concentrations of ferrate(VI) inhibits irreversibly further DNA polymerization and polymerase-chain-reaction (PCR) synthesis. Ferrate(VI) also inhibited the respiration of the bacterium *Sphaerotilus*; suggesting potential role of ferrate in treating sludge for disinfection. In a recent study (*45*) on sludge, more than 99% of indicator organisms could be inactivated at a dose of 0.2 g FeO_4^{2-} /L sludge. However, a spore formers such as *Clostridium perfringens*, which is resistant to disinfectants, required a dose of 0.8 g/L sludge.

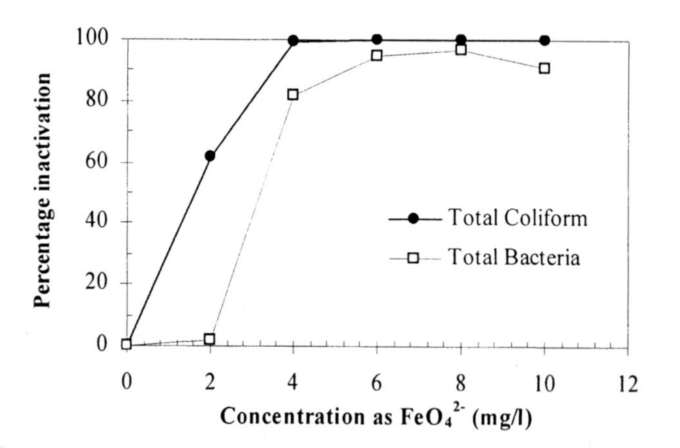

Figure 1. Percentage inactivation of bacteria with ferrate (VI)

Table 3. Comparative performance of bacteria inactivation (41)

	AS[1]	FS[1]	Ferrate(VI)[2]	
Coagulation pH	6.75-7.48	6.75-7.48	5	7
Total Coliform	87-91%	87-90%	>99.99%	>99.99%
Faecal Coliform	89-90%	90-91%	>99.99%	>99.99%

[1] Aluminium sulfate (AS) and ferric sulfate (FS) dose required was >0.50 mmol/L as either Al or Fe(III).

[2] Ferrate(VI) achieved >99.99% bacteria inactivation at doses <0.27 mmol/L as Fe(III).

Odor control with ferrate(VI)

Oxidation of sludge by ferrate(VI) to remove odor causing compounds such as hydrogen sulfide, mercaptans and amines has been studied (*46-48*). The reaction rate law and observed rate constants at pH 9 were used to determine half-lives of the oxidation processes (Fig. 2). The oxidation of odor-causing compounds with ferrate(VI) can be accomplished in seconds. The half-lives of the reactions vary from mille seconds to seconds and ferrate tends to react faster with sulfur-containing pollutants than with amines (Fig. 3). It should be pointed out that the reaction rates are pH dependent; so are the half-lives of the reactions. Half-lives of the reactions increase with increase in pH. Additional advantage of using ferrate is relatively non-toxic by-products in destruction of contaminants (49). For example, hydrogen sulfide is oxidized by ferrate to sulfate in less than a second (eq 6).

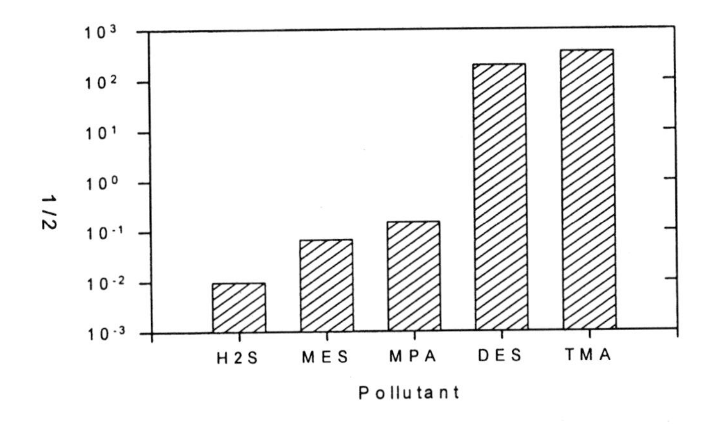

Figure 2. Half-lives of the reactions between Fe(VI) (500 mM) and pollutants (100 mM) at pH 9 and 25 °C. MES – 2-mercattoethanesulfonic acid; MPA – 2-mercaptopropionic acid; DES – diethylsulfide; TMA – trimethylamine

$$8\ HFeO_4^- + 3\ H_2S + 6\ H_2O \rightarrow 8\ Fe(OH)_3 + 3\ SO_4^{2-} + 2\ OH^- \qquad (6)$$

When an alkaline treatment is used for sludge treatment, the US-EPA requires pH of treated sludge to remain over 12.5 over two hours. As a result of oxidation with ferrate, pH tends to increase since it produces OH- as shown in eq 6. pH of various thickened sludges (solids content = 3-5%) treated with lime and ferrate were compared. With the addition of ferrate of small dosage, pH of the thickened sludge could be easily raised over 12. In fact, if the dosage was made double, the pH could be raised over 13.

De Luca et al. (*11*) carried out a study evaluating effectiveness of ferrate in improving the quality of sludge from the odor perspective. Three different kinds of sludges, i.e., waste activated sludge, anaerobically digested sludge, and primary sludge were used in this study. Ferrate(VI) successfully reduced levels of compounds producing odor for all sludge. Addition of ferrate(VI) into sludge reduced the level of hydrogen sulfide (H_2S) and H_2S removal efficiency increased with increasing ferrate dose (Fig. 3). At highest level of ferrate, reduction in sulfide was more than 90%. Sludge treatment by lime, however, provided no significant benefit in terms of sulfide reduction in the treated sludge. The oxidation of sulfide by ferrate produces sulfate in sludge (eq. 6). This was supported by experimental observations (Fig. 3). As expected, the more sulfide was oxidized, the more sulfate was produced during the sludge treatment with ferrate. The oxidation of sulfide by ferrate released OH^- ions and might also raise the pH of sludge.

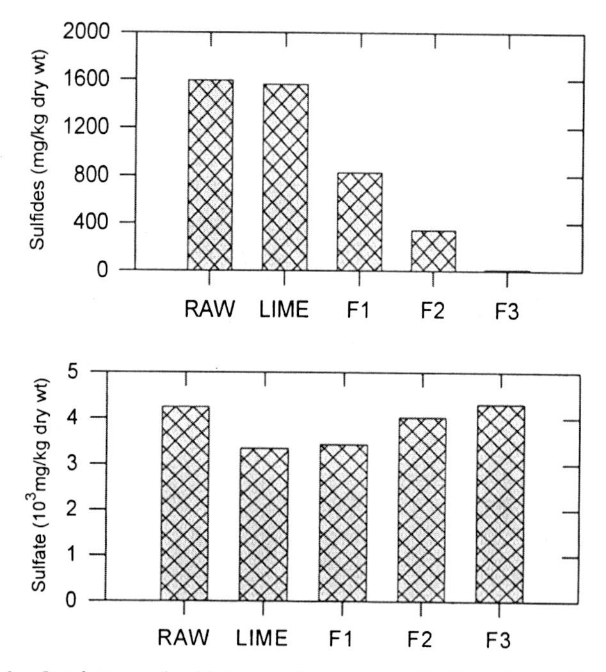

Figure 3. Oxidation of sulfide and formation of sulfate by Fe(VI) in sludge treatment (Data taken from (10))

In a study (*50*), the dose requirement for the removal of 90% sulfide from wastewater sludge with ferrous, ferric and ferrate(VI) was compared and this can be seen in Table 4. It is evident that using ferrate(VI) to replace ferrous and ferric iron, the required dose was reduced by 80% and 76%, respectively, which significantly reduce the sludge production and therefore, the sludge treatment cost.

The treatment of sludge by ferrate could also reduce the amount of ammonia-N through the oxidation process producing nitrate (eq. 7) (Fig. 4). The reductions of ammonia-N in sludge were 45.19%, 53.35%, and 59.67% for F1, F2, and F3, respectively. The changes in ammonia levels in sludge after the treatment may also be due to the formation of volatile form of ammonia ($NH_4^+ \Leftrightarrow H^+ + NH_3$; pK = 9.3). Increases in level of nitrate-N were found in ferrate treated sludge (Fig. 4). Comparatively, nitrate-N was decreased in lime treatment of sludge. Organic-N levels in sludge could also be controlled by ferrate(VI) (*11*).

$$8\ FeO_4^{2-} + 3\ NH_3 + 14\ H_2O \rightarrow 8\ Fe(OH)_3 + 3\ NO_3^- + 13\ OH^- \quad (7)$$

Table 4. The dosage required to remove 90% H_2S (50)

Chemical	Dose as Chemical to S^{2-} (g : g)
Ferrous iron	2.2
Ferric iron	1.7
Ferrate(VI)	0.4

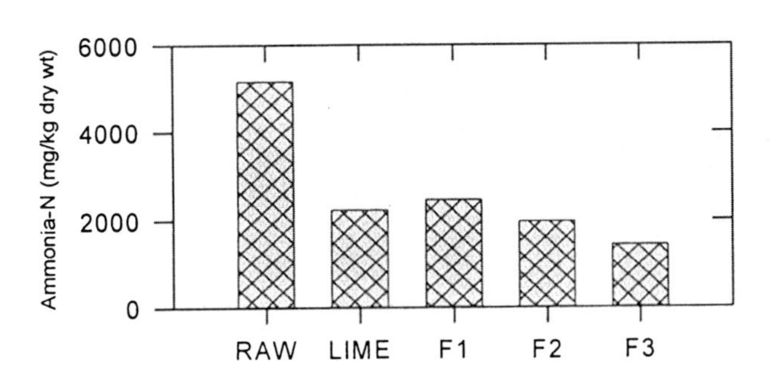

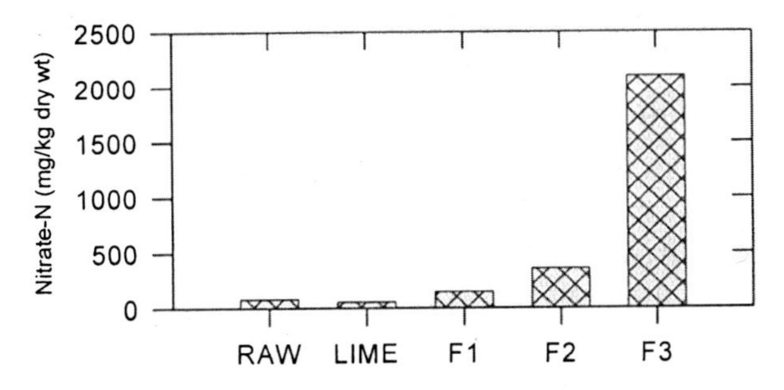

Figure 4. Removal of ammonia-N and formation of nitrate by Fe(VI) in sludge treatment (Data taken from (6))

Toxic heavy metal and radionuclides removal with ferrate(VI)

Potassium ferrate can also remove a range of metals (e.g., Fe^{2+}, Fe^{3+}, Mn^{2+}, and Cu^{2+}) and toxic heavy metals (e.g., Pb^{2+}, Cd^{2+}, Cr^{3+}, and Hg^{2+}) to a low level at a dose range of 10 - 100 mg/l as K_2FeO_4 (*51*). However, it appeared that ferrate had no significant advantages on the removal of Cr^{6+} and Zn^{2+} in comparison with ferric sulfate and aluminum sulfate (Table 5). For Moselle river water treatment (*52*), Fe(VI) effectively removes more than 99 % of metals (Figure 5).

Table 5. Removal efficiency (%) of heavy metals (51)

Metal	Initial Concentration (mg/L)	Removal efficiency (%) $FeCl_3$ 125mg/L	$Al_2(SO_4)_3$ 275mg/L	K_2FeO_4 50mg/L
Cd^{2+}	0.04-0.2	2	1	79
Pb^{2+}	0.5-0.55	15	10	63
Cu^{2+}	0.21-0.81	38	24	79
Cr^{6+}	0.22-0.43	89	56	88
Zn^{2+}	0.50-0.93	5	73	49
Total Fe	10.5-47.9	64	65	67

The common procedure for radionuclides removal is the co-precipitation of radionuclides by ferric and manganese hydroxides. However, these processes are hindered by organic ligands, which are present in wastewater and associated with radionuclides, effectively increase their solubility in water and make precipitation much more difficult. Fe(VI) as a replacement for conventional precipitating agents such as ferrous/ferric sulfate and ferric chloride can overcome these difficulties. Fe(VI) ion can possibly destroy these organic ligands, free the radionuclides, and concurrently form Fe(III) hydroxide (*53,54*).

Experiments seeking removal of ^{152}Eu in the presence of an organic (oxalate ion) using Fe(VI) have been performed. The results show that sodium Fe(VI) oxidizes oxalate ion and efficiently increases the co-precipitation of ^{152}Eu from the solution (*53*). The applicability of Fe(VI) ion for treating wastewater containing americium and plutonium has also been demonstrated (*55*). As shown in Fig. 6, the optimum pH range found was 11.5 – 12.0 where decontamination factor is the ratio of the contaminant in the untreated waste stream to that in the treated waste stream. Treating tap water spiked with americium and plutonium by Fe(VI) results in a 99.8 % gross reduction in alpha activity, from 3.0×10^6 pCi/L to 6.0×10^3 pCi/L. A dose of 5 mg as Fe in FeO_4^{2-} /L at pH 11.5-12.0 was used. In another demonstration at field site, a similar treatment with potassium ferrate(VI) resulted in a 99.9% reduction in gross alpha activity from 3.7×10^3 pCi/L to < 40 pCi/L.

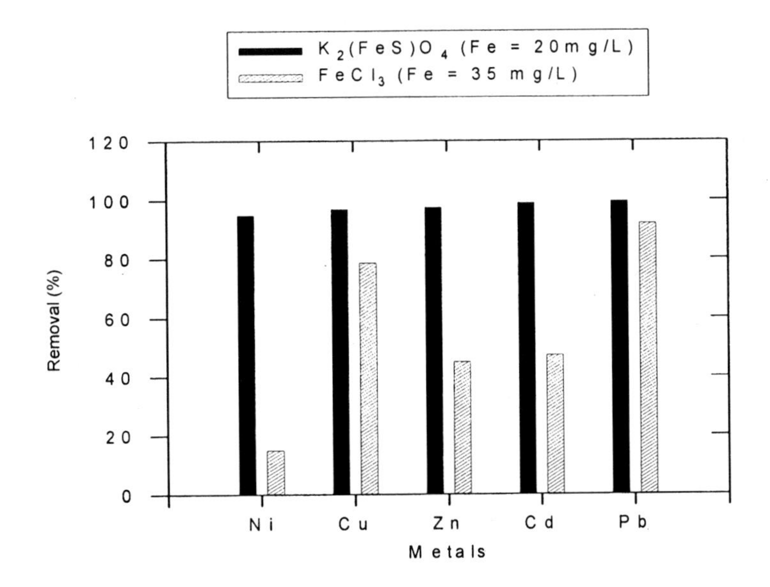

Figure 5. - Removal of metal in river water by Ferrate (Fe -20 mg/L) and Ferric Chloride (Fe - 35 mg/L) (52).

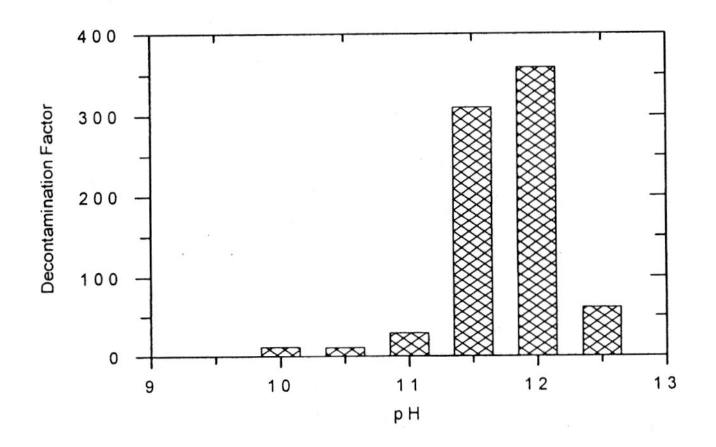

Figure 6. Decontamination of radionuclides by Fe(VI) at different pH (55).

The potential of degradation of organic micro-pollutants with ferrate(VI)

Recently, the studies on the ferrate(VI) oxidation of pharmaceuticals and EDCs in water have been carried out (*56-58*). The kinetics assessments of oxidations showed that ferrate(VI) can be effective in removing of these organic micropollutants. The reaction rate constants, k ($M^{-1}s^{-1}$) for the reaction of Fe(VI) with sulfamethoxazole (SMX) and ibuprofen (IBP) were determined as a function of pH and the rate of the reaction increases with a decrease in pH (Fig. 7). The half-lives of the reactions using observed rate constants of Fe(VI) oxidation of sulfonamides and using excess De(VI) concentration (10^{-5} M) calculated as less than five minutes at pH 7.0.

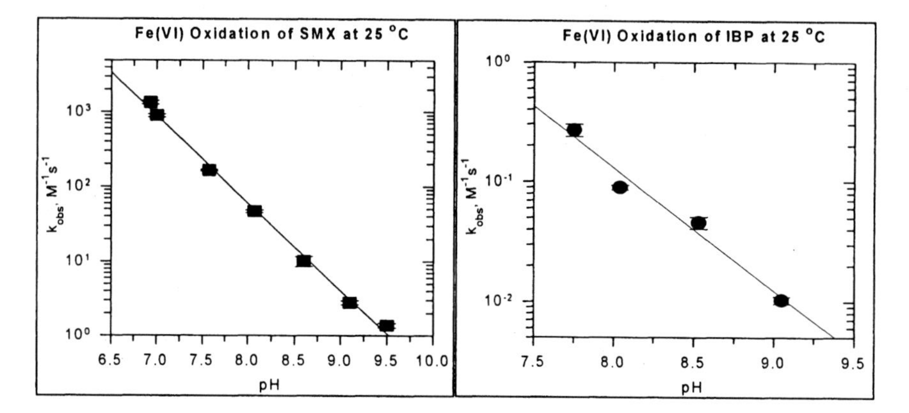

Figure 7. Ferrate(VI) oxidation of pharmaceuticals at different pH. (Reproduced with permission from reference 56. Copyright 2005 Elsevier.)

Other issues when using ferrate(VI) for waste treatment

The treatment of sludge is a major problem that water industry is facing since the resulting sludge from precipitation/primary sedimentation processes has large volumes and contains a high percentage of water. Water industries wish to use a chemical (i.e., precipitatant) with good treatment performance and less sludge production. Comparison of the sludge production is thus to be necessary.

Sludge production in treating wastewater was studied under the optimum operating conditions were studied (*43*). The optimum operating conditions used for the experiments were pre-determined. The criteria for their selection are in terms of the maximum removal of suspended solids and total chemical oxygen demand (COD). The comparative sludge production can be seen in Table 6. It is evident that ferrate removed more pollutants (e.g., organic materials as COD) and produced less wet sludge than ferric sulfate (FS) and aluminium sulfate (AS) for the equivalent doses compared; this should then make the handling of the resulting sludge easier.

Table 6. Mass balance of sewage sludge (43)

Coagulants	SS* in raw sewage	SS in supernatant		Sludge Deposit		Sludge produced from the coagulant	Contaminant removed
	g	g	mL	g	mL	g	g
AS[1]	0.382	0.0364	960	0.47	40	0.035	0.017
FS[1]	0.382	0.0204	960	0.49	40	0.042	0.046
Ferrate(VI)[2]	0.382	0.0285	965	0.54	35	0.042	0.088

* SS: suspended solids.

[1] Aluminum sulfate (AS) and ferric sulfate (FS) dose used was 0.37 mmol/L as Al and 0.36 mmol/L as Fe(III), respectively.

[2] Ferrate(VI) dose used was 0.36 mmol/L as Fe(III).

Concluding Remarks

Ferrate(VI) possesses powers for the oxidation, disinfection and coagulation, and upon its decomposition, produces a non-toxic by-product, Fe(III). Ferrate(VI), therefore, possesses great potential and an environmentally friendly chemical for sludge treatment. The release of offensive odors from sludge can be reduced, if the sludge is treated with ferrate since it can oxidize sulfur and amine compounds in sludge. The quality of ferrate(VI) treated sludge is even further improved with inactivation of pathogens and reduction of sludge volume with enhanced dewaterbility. With its oxidizing power it can destroy newly identified hazardous organic compounds like endocrine disruptors. The management practices can apply multifunctional properties of ferrate in a single dose to reduce the risks of sludge in land application.

References

1. Kim, H.; Murthy, S.; Peot, C.; Ramirez, M.; Strawn, M.; Park, C.H.; McConnell, L.L. Examination of Mechanisms for odor compounds generation during lime stabilization. *Water Environ. Res.* **2003**, *75*, 121.
2. Renner, R. *Environ. Sci. Technol.* **2002**, 338A.
3. Lewis, D.L.; Gattie, D.K. Pathogens risks from applying sewage sludge to land. *Environ. Sci. Technol.* **2002**, *36*, 287A.
4. Reimers, R.S.; Sharma, V.K.; Pillai, S.D.; Reinhart, D.R.; Boyd, G.R.; Fitzmoris, K.B. *Biosolids Tech. Bull.* **2005**, *10*, 1-4.
5. Jiang, J.Q.; Lloyd, B. Progress in the development and use of ferrate(VI) salt as an oxidant and coagulant for water and wastewater treatment. *Water Res.* **2002**, *36*, 1397-1408.

6. Sharma, V.K. Potassium ferrate(VI): an environmentally friendly oxidant. *Adv. Environ. Res.* **2002**, *6*, 143.
7. Straub, T.M.; Pepper, I.L.; Gerba, C.P. Virus survival in sewage-sludge amended desert soil. *Water Sci. and Technol.* **1993,** *27(3–4*), 421–424.
8. Aitken, M.D.; Sobsey, M.D.; Blauth, K.E.; Crunk, P.L.; Walters, G.W. *Environ. Sci. Technol.* **2005**, *39*, 5804-5809.
9. Capizzi-Banas, S.; Schwartzbrod, J. Water Res. **2001**, *35*, 2256-2260.
10. Shamma, M.; Al-Adawi, M.A.; *Rad. Phys. Chem.* **2002**, *65*, 277-279.
11. De Luca, S.J.; Idle, C.N.; Chao, A.C. Quality improvement of biosolids by ferrate(VI) oxidation of offensive odor compounds. *Water Sci. Technol.* **1996**, *33(3)*, 119.
12. Sievers, M.; Rield, A.; Koll, R. Water Sci. Technol. **2004**, *49*, 247-253.
13. Rosenfeld, P.E.; Henry, C.L. Activated carbon and wood ash sorption of wastewater, compost, and biosolids odorants. *Water Environ. Res.* **2001**, *73*, 388.
14. Kim, H.; Murthy, S.; McConnell, L.L.; Peot, C.; Ramirez, M.; Strawn, M. Characterization of wastewater and solids odor using solid phase microextraction at a large wastewater treatment plant. *Water Sci. Technol.* **2002**, *46(10)*, 9.
15. Cheng, X.; Peterlin, E.; Burlingame, G.A. *Water Res.* **2005**, *39*, 3781-3790.
16. Calvano, J.; Prakasam, T.B.S.; Sawyer, B.; Wilson, T.E. Sources and control of odor emissions from sludge processing treatment. Municipal Sewage Sludge Management. *Vol. 4,* Lancaster, PA, USA. 1992.
17. Abu-Orf, M.; Peot, C.; Ramirez, M.; Laquidara, M.; McConnell, L.L.; Kim, H.; Hunnifold, D. Inhibiting the production of odors from dewatered residuals using nitrates and anthraquinones. WEFTEC 2002 Annual Meeting, Chicago, IL, USA, October 2002.
18. Fraser, J.A.L.; Sims, A.F.E. Hydrogen peroxide in municipal, landfill and industrial effluent treatment. *Effluent and Water Treatment J.* **1984**, *24(5)*, 3.
19. Wheeler, R.A. Controlling Sewage Treatment Plant Odors. *Water and Pollution Control,* **1984**, *122(5)*, 22.
20. Farooq, S.; Akhlague, S. Oxidation of Biological Sludges with Ozone. *J. Environ. Sci. Health, Part A*, **1982**, *17(5*), 609.
21. Gao, L.; Keener, T.C.; Zhuang, L.; Siddiqui, K.F.A. Technical and economic comparison of biofiltration and wet chemical oxidation (scrubbing) for odor control at wastewater treatment plants. *Environ. Eng. Policy*, **2001**, *2(4)*, 203.
22. Planas, C.; Guadayyol, J.M.; Droguet, M.; Escalas, A.; Rivera, J.; Caixach, J. Degradation of polyethoxylated nonylphenols in a swage treatment plant. Quantitative analysis by isotopic dilution – HRGC/MS. *Water Res.* **2002**, *36*, 982.
23. Roberts, P.; Roberts, J.P.; Jones, D.L. *Soil. Biol. Biochem.* **2006**, *38*, 1812-1822.

24. Fenet, H.; Gomez, E.; Pillon, A.; Rosain, D.; Nicolas, J.C.; Casellas, C.; Balaguer, P. Estrogenic activity in water and sediments of a French river: contribution of alkylphenols. *Arch. Environ. Contam. Toxicol.* **2003**, *44*, 1.
25. Ahel, M.; Giger, W.; Koch, M. Behavior of alkylphenol polyethoxylate surfactants in the aquatic environment – I. Occurrence and transformation in sewage treatment. *Water Res.* **1994**, *28*, 1132.
26. McDonald, T.A. A perspective on the potential health risks of PBDEs. *Chemosphere* **2002**, *46(5)*, 745.
27. de Boer, J.; Wester, P.G.; van der Horst, A.; Leonards, P.E.G. Polybrominated diphenyl ethers in influents, suspended particulate matter, sediments, sewage treatment plant and effluents and biota from the Netherlands. *Environ. Pollut.* **2003**, *12*, 63.
28. Matscheko, N.; Tysklind, M.; de Wit, C.; Bergek, S.; Andersson, R.; Sellstrom, U. Application of sewage sludge to arable land-soil concentrations of polybrominated diphenyl ethers and polychorinated dibenzo-p-dioxins, dibenzofurans, and biphenyls, and their accumulation in earthworms. *Environ. Toxicol. Chem.* **2002**, *21(12)*, 2515.
29. Ahel, M.; McEvoy, J.; Giger, W. Bioaccumulation of the lipophilic metabolites of nonionic surfactants in freshwater organisms. *Environ. Pollut.* **1993**, *79*, 243.
30. Perdersson, S.N.; Christiansen, L.B.; Perdersen, K.L.; Korsgaard, B.; Bjerregaard, P. In vivo estrogen activity of branched and linear alkylphenols in rainbow trout (Oncorhynchus mykiss). *Sci. Total Environ.* **1999**, *233*, 89.
31. Zennegg, M.; Kohler, M.; Gerecke, A.C.; Schmid, P. Polybrominated diphenyl ethers in whitefish from Swiss lakes and farmed rainbow trout. *Chemosphere* **2003**, *51(7)*, 545.
32. Wren, A. Occurrence and Endocrine Disruptors in Onsite Wastewater Systems. Div. Of Environmental Science and Engineering, Colorado School of Marine, Golden, CO, USA. 2001.
33. Wagner, A.J.; Singh, H.; MacRichie, F.; Bhandari, A. An evaluation of advanced oxidation processes for the removal of 4-nonylphenol from water and wastewater. Poster at the EPA Conference on Application of Waste Remediation Technologies to Agricultural Contamination of Water Resources, Kansas City, MO, July 30-Aug. 1, 2002.
34. Lee, H.B.; Peart, T.E. Occurrence and elimination of nonylphenol ethoxylates and metabolites in municipal wastewater and effluents. *Water Qual. Res. J. of Canada* **1998**, *33(3)*, 389.
35. Thompson, G.W.; Ockerman, L.T.; Schreyer, J.M. Preparation and purification of potassium ferrate VI. *J. Amer. Chem. Soc.* **1951**, *73*, 1379.
36. Waite, T.D. Feasibility of wastewater treatment with ferrate. *J. Environ. Eng., ASCE* **1979**, *105*, 1023.
37. Waite, T.D.; Gray, K.A. Oxidation and coagulation of wastewater effluent utilizing ferrate(VI) ion. *Stud. Environ. Sci.* **1984**, *23*, 407.

38. Potts, M.E.; Churchwell, D.R. Removal of radionuclides in wastewaters utilizing potassium ferrate(VI). *Water Environ. Res.* **1994**, *66*, 107.
39. Murmann, R.K.; Robinson, P.R. *Water Res.* **1974**, *8*, 543-547.
40. Gilbert, M.B.; Waite, T.D.; Hare, C. Analytical notes - an investigation of the applicability of ferrate ion for disinfection. *J. Am. Water Works Assoc.* **1976**, *68*, 495-497.
41. Jiang, J.Q.; Lloyd, B.; Grigore, L. Preparation and evaluation of potassium ferrate as an oxidant and coagulant for potable water treatment. *Environ. Eng. Sci.* **2001**, *18*, 323-331.
42. Jiang, J.Q.; Wang, S. Inactivation of Escherichia *Coli* with ferrate and sodium hypochlorite: a study on the disinfection performance and rate constant. In: Oxidation Technologies for Water and Wastewater Treatment, Eds. C. Schroder and B. Kragert, Papiepflieger Verlag, Clausthal-Zellerfeld., 2003, pp. 447-452.
43. Jiang, J.Q.; Panagoulopoulos, A. The use of potassium ferrate for wastewater treatment. In: Innovative ferrate(VI) technology in water and wastewater treatment, Eds. V. Sharma, JQ Jiang, K. Bouzek, ICT Press, Prague. 2004, pp. 67-73.
44. Kazama, F. Viral inactivation by potassium ferrate. *Water Sci. Technol.* **1995**, *31*, 165.
45. Kim, H.; Millner, P.; Sharma, V.K.; McConnell, L.L.; Torrents, A.; Ramirex, M.; Peot, C. *Water Environmental Laboratory-Solutions* **2006**, *12*, 1-4.
46. Sharma, V.K.; Burnett, C.R.; O'Connor, D.B.; Cabelli, D. Iron(VI) and iron(V) oxidation of thiocyanate. *Environ. Sci. Technol.* **2002**, *36*, 4182.
47. Read, J.T.; Graves, C.R.; Jackson, E. The kinetics and mechanism of the oxidation of thiols 3-mercato-a-propane sulfonic acid and 2-mercaptonicotinic acid by potassium ferrate. *Inorg.. Chimica Acta* **2003**, *348*, 41.
48. Bartzatt, R.; Nagel, D. Removal of nitrosamines from waste water by potassium ferrate oxidation. *Arch. Environ. Health* **1991**, *46*, 313.
49. Sharma, V.K.; Smith, J.O.; Millero, F.J. Ferrate(VI) oxidation of hydrogen sulfide. *Environ. Sci. Technol.* **1997**, *31*, 2486.
50. Jiang, J.Q. Comparative performance of physico-chemical technologies in controlling hydrogen sulfide odors. In: Control and Prevention of Odors in the Water Industry, CIWEM and IAWQ, London, 1999.
51. Bartzatt, R.; Cano, M.; Johnson, L.; Nagel, D. Removal of toxic metals and nonmetals from contaminated water. *J. Toxicol. Environ. Health* **1992**, *35(4)*, 205-210.
52. Neveux, N.; Aubertin, R.; Gerardin, O.; Evrard, O. Chemical Water and Wastewater Treatment III, IWA Publisher, London, 1994, pp. 95-103.
53. Stoupin, D.Y.; Ozernoi, M.I. Coprecipitation of 152Eu with iron(III) hydroxide formed upon reduction of sodium ferrate(VI) in aqueous medium. *Radiochemistry* **1995**, *37*, 329-332.

54. Sylvester, P.; Jr Rutherford, L.A. A Gonzalez-Martin, J Kim, Ferrate treatment for removing chromium from high-level radioactive waste. *Environ. Sci. Technol.* **2001**, *35*, 216-221.
55. Potts, M.E.; Churchwell, D.R. Removal of radionuclides in wastewaters utilizing potassium ferrate(VI), *Water Environ. Res.* **1994**, *66*, 107-109.
56. Jiang, J.Q.; Yin, Q.; Zhou, J.L.; Pearce, P. Occurrence and treatment trials of endocrine disrupting chemicals (EDCs) in wastewaters, *Chemosphere* **2005**, *61(4)*, 544-550.
57. Lee, Y.; Yoon, J.; von Gunten, U. Kinetics of the oxidation of phenols and phenolic endocrine disruptors during water treatment with ferrate (Fe(VI)), *Environ. Sci. Technol.* **2005**, *39(22)*, 8978-8984.
58. Sharma, V.K.; Mishra, S.K.; Ray, A.K. Kinetic assessment of the potassium ferrate(VI) oxidation of antibacterial drug sulfamethoxazole, Chemosphere **2006**, *62(1)*, 128-134.

Chapter 19

Evaluation of Ferrate(VI) as a Conditioner for Dewatering Wastewater Biosolids

Hyunook Kim[1,*], Yuhun Kim[1], Virender K. Sharma[2], Laura L. McConnell[3], Alba Torrents[4], Clifford P. Rice[3], Patricia Millner[3], and Mark Ramirez[5]

[1]Department of Environmental Engineering, The University of Seoul, 90 Jeonnong-dong, Dongdaemun-gu, Seoul 130–743, Korea
[2]Department of Chemistry, Florida Institute of Technology, 150 West University Boulevard, Melbourne, FL 32901
[3]Agricultural Research Service, U.S. Department of Agriculture, 10300 Baltimore Avenue, Beltsville, MD 20705
[4]Department of Civil and Environmental Engineering, The University of Maryland, College Park, MD 20742
[5]District of Columbia Water and Sewer Authority, 5000 Overlook Avenue SW, Washington, DC 20032
[*]Corresponding author: email: h_kim@uos.ac.kr, fax: +82-2-2244-2245

Land application of sludge/biosolids is a commonly used practice for final utilization. Therefore, adequate conditioning and stabilization of wastewater solids is very critical for safe land application. The addition of ferrate (FeO_4^{2-}) has the potential to improve the dewaterablity of solids, destroy pathogenic organisms, and reduce certain endocrine disrupters. In this study, the dewaterbility and stabilization of thickened sludge treated with ferrate was evaluated under controlled laboratory conditions. To evaluate dewaterbility, three different techniques; belt-press, centrifugation, and vacuum filtration, were applied to dewater a mixture of solids. In addition, once biosoilds are generated, their safety and quality especially in terms of reduced endocrine disrupting compounds (EDCs) as well as pathogens become an important issue to the public. Therefore, the effectiveness of ferrate (FeO_4^{2-}) in disinfecting microorganisms and in oxidizing EDCs, *i.e.*, Nonylphenols and polybrominated diphenyl ethers, is briefly discussed.

Introduction

In 2005, a total of 7.5 million dry tons of biosolids were generated by municipal wastewater treatment plants (WWTPs) in the United States (*1*), and about 60 % of the biosolids were land applied. Since the management of wastewater sludge is difficult and expensive, the utilization or disposal including beneficial reuse of biosolids has become a concern of high priority (*2*). In reality, final biosolids quality is influenced by the selection, design, and operation of upstream processes (*3*), including sludge conditioning, thickening, and dewatering.

The overall cost for sludge treatment/handling is directly related to the bulk volume of sludge treated. Land application continues to be the preferred option for sludge/biosolids' final utilization, however, the production of less WWTPs solids is the most desirable and cost effective option. Minimization of water content will result in significant cost saving by reducing the overall volume of solids (*4*). Therefore, the solid-liquid separation processes, e.g., thickening and dewatering processes should be operated at the highest performance and reliability levels.

The most commonly utilized dewatering technologies are belt filter pressing and solid bowl centrifugation. Well managed operation of these technologies can increase solids content up to 25 ~ 30 % (*5*). However, most installations are not operated at optimum levels and fail to provide proper dewatering (*4*). To aid in the dewatering processes, thickened sludge is often conditioned by applying chemical additives (coagulants such as polymer, or ferric chloride, aluminum sulfate (alum), ferrous sulfate, and ferric chloro-sulfate (*6*)). Once added, the coagulant alters wastewater solids structure for better separation of solid and liquid.

Although commercially available coagulants may show outstanding solids-liquid separation and chemical oxygen demand (COD) reductions in supernatant (*7*), they do not greatly affect the bio-chemical properties of the solids. Rather pathogenic microorganisms and organic pollutants are concentrated in dewatered solids. Therefore, further stabilization and disinfection procedures are required before biosolids are land-applied (*8-,10*).

Currently, alternative chemical additives to stabilize and disinfect as well as to condition and dewater thickened solids are being evaluated. Neyens *et. al.* (*5*) oxidized wastewater solids with H_2O_2 (0.037 g H_2O_2/100 mL sludge) in the presence of Fe^{2+} (1 mg Fe^{2+}/100 mL sludge) and significantly improved the dewaterbility of the solids. In addition to improved dewaterbility, they observed higher degradable organic content and less residual heavy metals in the dissolved solids due to break-down of more recalcitrant compounds by H_2O_2.

Iron with the +6 oxidation state, ferrate (Fe(VI)) is also a potential alternative to polymer for dewatering and improving the safety of biosolids, because it is a powerful oxidizing agent. Under acidic conditions, the redox potential of Fe(VI) ion (2.2 V) is higher than that of ozone (0.72 V).

Upon decomposition, Fe(VI) in water forms molecular oxygen improving redox condition of the media, and Fe(III), a common coagulant used in water and wastewater treatment plants (Eq. 1). Ma and Liu (*11*) observed good coagulation of suspended solids in surface water preoxidized with Fe(VI). The floc size of the coagulated particles was larger in a Fe(VI) preoxidation process than those of an alum coagulated floc. They also found Fe(VI) was more effective in coagulation of organic-rich waters in which alum was less effective in reducing turbidity.

$$FeO_4^{2-} + 5\ H_2O \rightarrow Fe^{3+} + 3/2\ O_2 + 10\ OH^- \qquad (1)$$

In this study, the dewaterbility of thickened sludge treated with ferrate was evaluated in the laboratory. Simulated belt-press, centrifugation, and vacuum filtration were applied to dewater a mixture of solids collected from a gravity thickener (GT) and from a dissolved air flotation (DAF) thickener where primary-settled sludge and waste activated sludges were fed, respectively. We also briefly discuss other benefits of using ferrate (FeO_4^{2-}) as a biosolids conditioner, such as oxidizing EDCs (nonylphenol (NP) and octylphenol (OP)) and inactivtivating microbes.

Materials and Methods

Ferrate Production

In the laboratory, Fe(VI) has been produced using one of three procedures provided in the literature; wet synthesis (*12*), dry synthesis, and electrochemical synthesis. In wet synthesis, sodium ferrate(VI) (Na_2FeO_4) is produced from the reaction of ferric chloride with sodium hypochlorite in the presence of sodium hydroxide. In dry synthesis, the fusion of Na_2O_2 and Fe_2O_3 at a molar ratio of 4:1 in the presence of dry oxygen (Temp.: 370 °C) is induced to produce Na_2FeO_4 (*13*). Finally, Fe(VI) can be synthesized by applying electricity to anodic iron in NaOH solution whereby anodic iron is oxidized to Fe(VI) (*14*). Using any of the methods, Fe(VI) with relatively high purity can be produced. However, these technologies are labor and capital intensive, which has been a major obstacle to the wide application of ferrate in wastewater treatment processes.

Ferrate used in this study was produced on site, using a pilot-scale ferrate producer, FERRATOR™, supplied by Ferrate Treatment Technology, Inc. (Orlando, FL). It consists of a 100 L reactor, where industry grade NaOCl with high concentration (in this study, 16 % weight basis), NaOH of 50 % and $FeCl_3$ of 35 % are mixed to produce ferrate. A heat exchanger prevents quick decomposition of the generated ferrate. Using this reactor, which utilizes the chemical reactions described by Thompson *et al.* (*12*), sodium ferrate could be produced at about 60 % conversion yield, and the pH of the final solution was

>13. Once the ferrate solution was produced, it was utilized in the dewaterbility experiments within approximately one hour. The ferrate solution was stored at 4 ^{o}C prior to use to minimize decomposition.

Sludge Samples and Sample Treatments

Thickened sludge samples were collected from a GT and a DAF thickener of the District of Columbia Advance WWTP at Blue Plains, District of Columbia, USA. The solids contents of sludges from a GT and a DAF were 3-5 %. GTs were fed with sludge from primary sedimentation tanks. DAFs were fed with waste activated sludge from secondary process and nitrification/denitrification processes. The collected GT and DAF sludges were blended at the ratio of 1:1 in a large plastic container to prepare 20 L. In the WWTP, the GT and DAF sludges are routinely mixed at a 1:1 ratio.

Aliquots of the blended sludge (2 L) were placed in heavy duty mixers. Five treatments of ferrate solution were used in the experiment; 0, 20, 40, 100, 200 mL ferrate solution/2 L blended solids. As an additional treatment, one solids sample was treated with 25 % CaO (dry weight basis). The ferrate solution and CaO were added to the solids samples and mixed for a total of 5 min. One quarter ferrate solution or CaO was simultaneously added to each sample at 1, 2, 3, and 4 minutes, respectively. All the treatments were carried out in duplicate.

In a separate set of experiments, effects of cationic polymer addition on the dewaterbility of ferrate treated solids were also evaluated by adding cationic polymer (Stock-Housen, Willmington, DE, USA) to the blended solids.

Procedures for Dewaterbility Tests

Three conventional wastewater solids dewatering processes were tested; *i.e.*, centrifuging, belt-pressing, and vacuum filtering.

Dewatering of Sludge by Centrifugation

Blended aliquots of treated solids (45 mL) were transferred to test tubes. Then the tubes were placed in holders of a lab-scale centrifuge (Cole-Palmer, USA) (Figure 1). The sludge-dewatering performance of the chemical treatments and controls were evaluated at three different centrifuge speeds: 1000, 2000, and 3000 rpm with a centrifuge time of 3 minutes. The treatment effectiveness was also evaluated by varying the centrifuging duration at 1, 3, and 5 minutes at holding the speed constant 2000 rpm. The resulting volume and total suspended solid content of centrate were analyzed to calculating dewatering efficiency of the treatment.

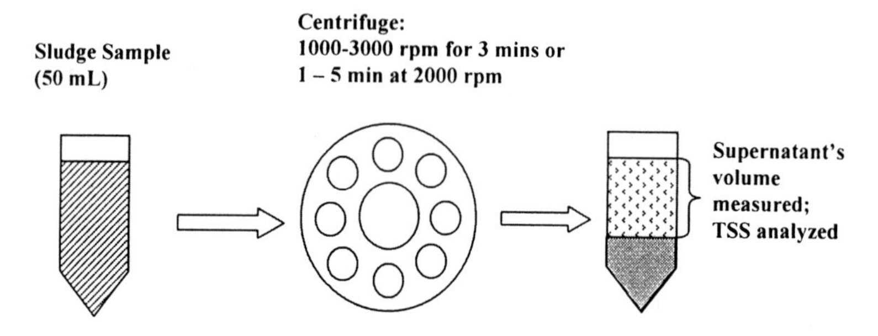

Figure 1. Procedure of sludge-dewatering using a centrifuge

Dewatering of Sludge by Belt-pressing

Sludge dewatering effectiveness of the ferrate treatments and controls were further evaluated using a Crown Press, a bench-scale belt press simulator (Cat No.: 7800-100, Phipps & Birds, Richmond, VA) (Figure 2). Before applying belt pressing, 50 mL treated solids samples were drained for 5 minutes by gravity on the pre-filter in order to simulate real practices. Then, the solids remaining on the filter were transported on the press filter, where pressure of 100 psia was manually applied for 1 min. After the pressure was applied, the dewatered solids were scrapped out on to an aluminum dish for measuring the sample mass and solids content. Total suspended solids concentration of the filtrate were also measured. Lateral migration of solids under belt-press was visually assessed.

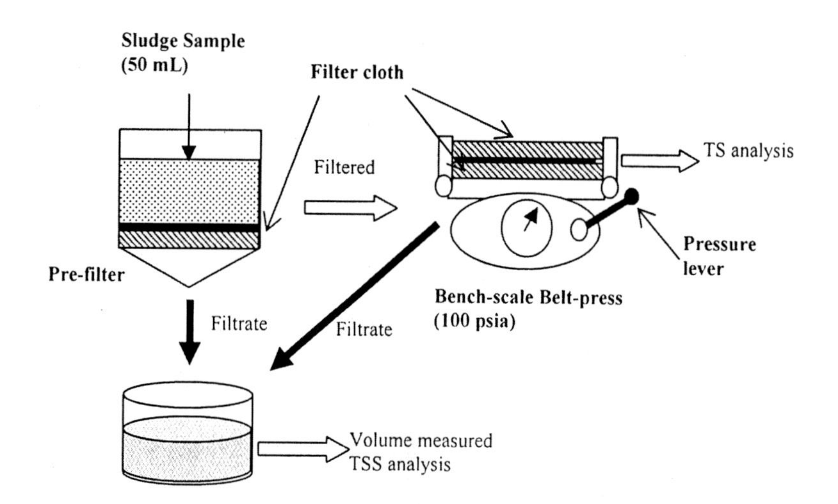

Figure 2. Procedure of sludge-dewatering using a lab-scale belt-press

Dewatering Sludge by Vacuum Filtering

Vacuum filtration was also applied to assess the effectiveness of various ferrate treatments and controls. The study was performed by placing 50 mL treated sludge on a Buckner funnel with a paper filter and by applying vacuum pressure of 49×1000 N/m^2 for 10 min (Figure 3). As with the belt-press experiments, the volume and TSS concentration of the filtrate were measured.

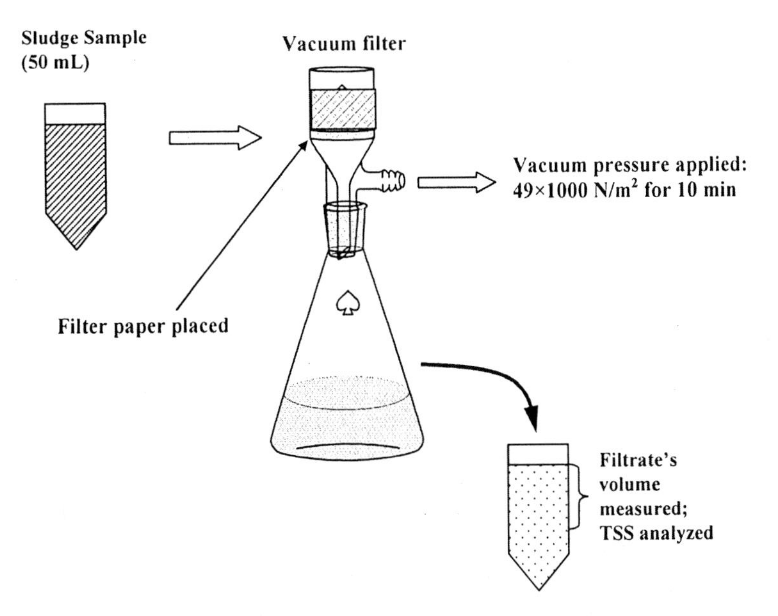

Figure 3. Procedure of sludge-dewatering using a vacuum filter

Results and Discussion

Characteristics of Solids after Ferrate Treatment

Upon the addition of ferrate solution to blended solids, the solids exhibited the consistency of a gel, probably due to the coagulation effect of Fe(III) ions generated from ferrate decomposition. This hypothesis was supported by a color change of treated solids from dark grey (control) to light reddish (typical color of iron oxide). In Figure 4, the color of ferrate treated sludge turns to lighter reddish brown as the treatment levels increases. Since the pH of the ferrate solution was high (it contains concentrated OH^-), the solids pH rose to > 12 at the lower level treatments (10 mL/L). Since the decomposition of ferrate produces OH^- as depicted in Eq. 1, it should contribute to the pH increase of sludge, too.

Figure 4. Pictures of dewatered sludge with ferrate treatments.

Effects of Ferrate Treatment on Solids Dewaterbility by Centrifuge

Effects of ferrate treatment on solids dewatering using a centrifuge were evaluated by measuring the volume and TSS of extracted water (i.e., centrate) from each solids sample. First, centrifugation was operated at three different rotational speeds, *i.e.*, 1000, 2000, and 3000 rpm and for constant operational duration, *i.e.*, 3 minutes. In general, at lower doses of ferrate solution, the dewatering efficiency of centrifuging was less than the solids without ferrate solution treatment; especially at lower rotating velocity (Figure 5(a)). At 1000 rpm and with 40 mL ferrate solution, separation between solid and liquid was not observed. It is hypothesized that organics with large molecular weights in biosolids were degraded into smaller molecules and suspended. The hypothesis was supported with the observed high TSS levels of centrates (refer to TSS of each treatment in Figure 5). Figure 6 shows the color of centrates from different ferrate solution treatments. The centrate from solid sample with 200 mL ferrate treatment is dark brown, implying higher TSS level. Ayol *et al.* (*15*) reported a similar observation in their study. In their anaerobically digested solids, they found COD increased in the centrate of solids treated with potassium ferrate along with hydrogen peroxide.

A similar result was obtained when the duration of centrifuging was varied from 1 to 5 minutes and the rotating speed of the centrifuge was fixed at 2000 rpm. It is noted that the TSS level of centrate from samples treated with ferrate was significantly higher than the control when the centrifuge duration was short. One minute of centrifuge time was not enough to force suspended solids to separate from the liquid.

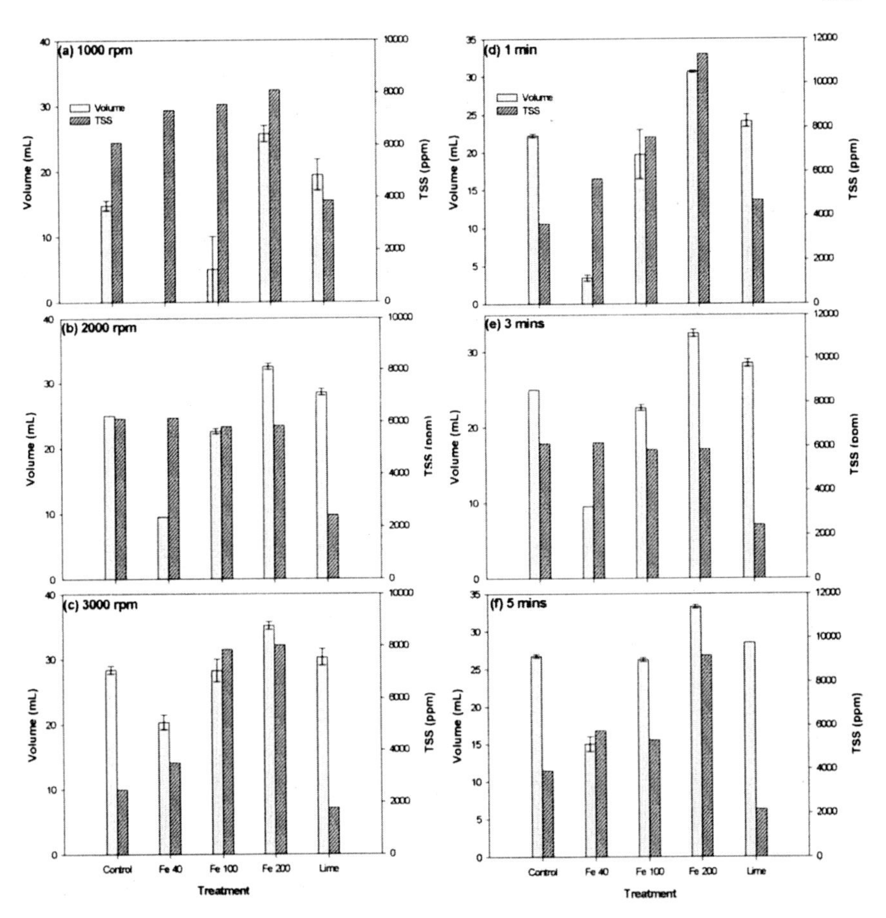

Figure 5. Dewaterbility of centrifuge at different rotating velocities ((a)~(c); duration was set at 3 minutes) and at different durations ((d) ~ (f); rotating velocity was set at 2000 rpm).

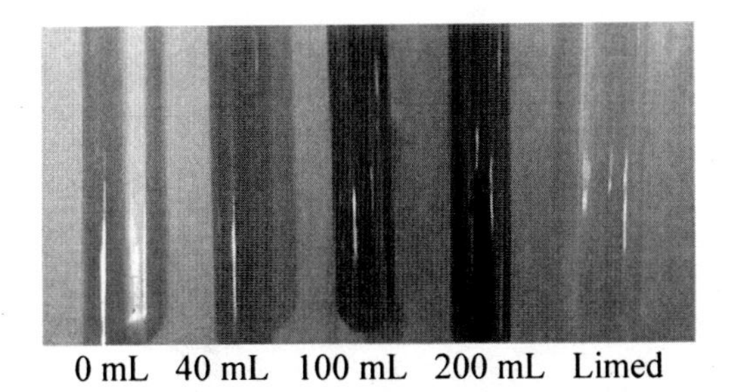

0 mL 40 mL 100 mL 200 mL Limed

Figure 6. Centrate color of dewatered sludge with ferrate treatment

Effects of Polymer Addition on Dewaterbility of Sludge Treated with Ferrate

Polymer obtained from the Blue Plains WWTP was added to blended and thickened sludge to observe its effects on dewatering of the sludge. In practice, the polymer was added at the rate of 2.25 mL/2L thickened sludge at the plant. Therefore, in this study 1.125, 2.5, 3.5 mL polymer were applied to 2 L blended sludge treated with ferrate solution. Centrifugation was performed at 2000 rpm for 3 minute in this specific experiment. In general, ferrate solution applied with polymer showed to some degree improved dewaterbility; especially when higher doses of ferrate were applied (Figure 7). 12 mL more water was extracted from the solids treated with 200 mL ferrate solution and 1.125 mL polymer, comparing to the solids with only polymer addition. Under the same operational condition, only 7 mL more water could be extracted from the solids with 200 mL ferrate treatment than the control (Figure 5(b)).

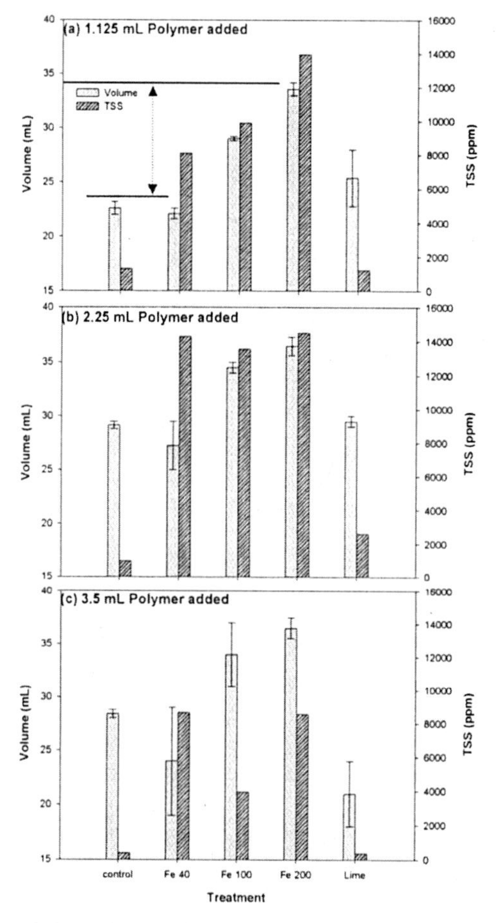

Figure 7. Dewaterbility of centrifuge at different polymer doses (Rotating velocity was set at 2000 rpm; duration: 3 minutes).

Regarding TSS concentration of centrates, polymer alone performed dramatically better than other treatments. Although it was still high, TSS levels of the centrate from ferrate treated solids were lowered with polymer addition. It is improved from 15,000 ppm to 9,000 ppm for 200 mL ferrate treatment as the polymer dose increases (Figure 7). It is hypothesized that small organic particles produced from reactions between ferrate and polymeric organic materials in sludge were coagulated by the polymer. The effects of polymer addition to ferrate treated sludge should be investigated more in detail in future study.

On the other hand, limed solids with polymer addition did not show any improvement in sludge dewatering comparing to the control.

Effects of Ferrate Treatment on Solids Dewaterbility by Belt-pressing

Dewatering of sludge treated with ferrate via belt-press was tested with a bench-scale belt-press equipped with a pre-filter and a filter-press (Figure 2).

In general, belt-press following ferrate treatment did not enhance the dewaterbility of the sludge. When the ferrate solution was applied, the sludge texture changed to liquidized gel, so more sludge was drained through the pre-filter. In addition, the lateral migration of the sludge treated with ferrate was significant, so a considerable amount of sludge was not pressurized for dewatering; the volume shown in Figure 8 includes liquid from pre-filter, from lateral migration and through the belt-filter. For the sludge treated with highest dose of ferrate (200 mL/2L sludge), around 40 mL sludge out of 50 mL was drained through the pre-filter and lateral migration during belt-pressing. Once dewatered, the sludge with ferrate treatment showed better solids contents levels, compared to the control and sludge with lime treatment; about 15 % for the sludge treated with 200 mL ferrate and about 9 % for the control and lime treated sludge.

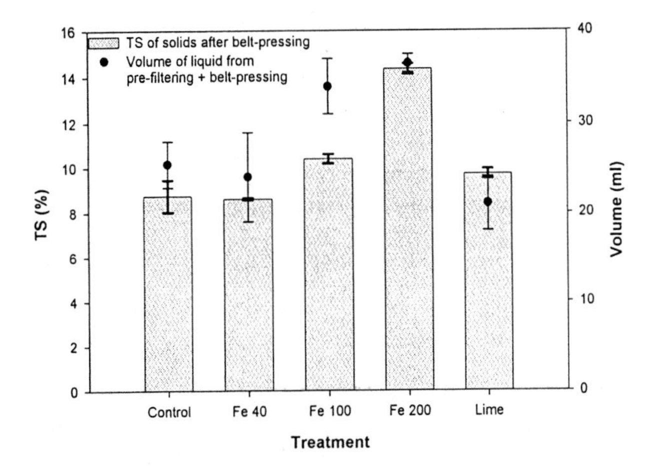

Figure 8. Solids contents and drainage volume of belt-pressed sludge with ferrate treatments

Polymer addition to ferrate treated solids before belt-pressing did not improve dewaterbility of the sludge (data not shown). Still a considerable amount of sludge was lost through prefilter and lateral migration. Since too much solids was lost through lateral migration, the use of a bench scale laboratory system to dewater ferrate conditioned solids did not appear to be the best technique to simulate full scale belt press dewatering. Additional testing on a larger scale should be evaluated.

Effect of Ferrate on the Performance of Vacuum Filtering in Sludge Dewatering

Vacuum filtering was also applied to the ferrate treated sludge as illustrated in Figure 3. Sludge with ferrate treatment could not be filtered at all, probably due to the blockage of filter pores by small particles formed through coagulation by reaction products of ferrate degradation, *i.e.*, Fe(III). Sludges with both ferrate treatment and polymer addition were not filterable with the vacuum system, either. In future studies, a robust filter assembly should be used and the surface of the filter also should be examined for blockage by ferrate-induced coagulated small particles.

Other Benefits of Using Ferrate(VI) as a Sludge Conditioner

Since ferrate is a strong oxidizer, its application as a sludge conditioner can provide several potential benefits such as destruction of endocrine disrupters and pathogens. In fact, a significant amount of EDCs like NP or OP and its ethoxylates (NPEOs and OPEOs) and PBDEs are frequently detected in sludges. We also detected 2000-10000 ng/g NP and NP(1-5)EOs and 200-1700 ng/g OP and OP(1-3)EOs from sludge samples from the Blue Plains WWTP. However, their levels were significantly reduced with the addition of ferrate. With the ferrate treatment (dose of 10 ~ 40 mL/L), up to 70 % NP and NPEOs of the biosolids were oxidized and complete removal of OP and OPEOs were observed.

Although 2-10 ng/g PBDE (i.e., BDE47, BDE99, BDE100) was also detected in the solids, their level did not change with the ferrate addition, probably due to bromide functional groups on the PBDEs which affect organic compound by making them less susceptible to oxidative reductive removal mechanisms.

Regarding microbial inactivation, ferrate was very effective in inactivating the microorganisms (Table I); even *Clostridium* sp. could be completely inactivated at the dose of 40 mL ferrate/L thickened solids. This superior disinfection efficiency can be attributed both to the oxidation power of ferrate(VI) itself and to high solids pH.

Although data was not provided, one thing should be mentioned. The ferrate treatment could not inactivate *Ascaris* ova, which should be achieved to comply with the Class A biosolids guideline (*17*). In an experiment performed with pure chemicals, the ferrate treatment could not inactivate the ova even at the dose of 0.01 g FeO_4^{2-}/mL except for causing some color change of the ova.

Table I. Microorganism removal rates (%) of treated thickened sludge(*16*)

Indicator microorganisms	N_0	*Ferrate solution dose[a], mL/L thickened solids[b]*			
		10	*20*	*40*	*80*
Total Heterotrophs	1×10^8	90 %	>99 %	>99.9 %	>99.9 %
Enterococci	5×10^5	70 %	>99.9 %	>99.9 %	>99.9 %
Total coliform	8×10^6	>99.9 %	>99.9 %	>99.9 %	>99.9 %
Fecal coliforms	5×10^6	>99.9 %	>99.9 %	>99.9 %	>99.9 %
E. coli	2×10^6	>99.9 %	>99.9 %	>99.9 %	>99.9 %
Clostridium perfringens	2×10^5	55 %	>90 %	>99.9 %	>99.9 %
Coli-phage	1×10^2	>99.9 %	>99.9 %	>99.9 %	>99.9 %

[a]Amount of Ferrate in the solution: 0.2 g of ferrate/10 mL solution.
[b]Solids content of the thickened solids: approximately 4 %.

Conclusion

In this study, the effect of ferrate on the performance of conventional bench scale laboratory simulated sludge-dewatering processes, *i.e.*, centrifugation, belt-press, and vacuum filtration were evaluated.

In short, sludge treated with ferrate could not be dewatered with a vacuum-filter at all, possibly due to the clogging caused by small particles formed by ferrate. In the case of belt-press, the texture of sludge after ferrate treatment, *i.e.*, liquidized gel, hindered the sludge from being dewatered. A significant amount of sludge were drained through pre-filter before belt-pressing. In addition, significant amount of sludge was lost through lateral migration during the pressing.

The effects of ferrate treatment on dewaterbility using centrifugation of thickened sludge, showed a slightly more promising result. In this case, dewaterbility of sludge treated with ferrate increased in proportional to the G-force or rotating speed (from 1,000 rpm to 3,000) and the time of centrifugation, duration (from 1 min to 5). However, no separation between liquid and solids could not be obtained at the lower ferrate treatment. At relatively higher ferrate level, better dewatering could be achieved. Up to 100 % more water could be extracted compared to the control. The dewaterbility of ferrate treated sludge could be further improved by combining it with cationic polymer. Although centrate from the sludge centrifugation was reddish colored and contained high TSS, it should not be a significant issue if the centrate is sent back to the headwork of the WWTP.

From these results, it appears that ferrate alone is not adequate to provide the dewatering capability required to achieve sufficient water removal from blended sludge. Considering the exceptional ability of ferrate to destroy organic compounds with oxidizable functional groups and disinfecting microorganisms, however, it appears that ferrate(VI) has the potential as a multi-purpose chemical conditioner for wastewater sludge, if it is applied as a part of a conditioning program prior to centrifuge dewatering.

Therefore, further research should be conducted to combine ferrate with other dewatering tools to gain the positive effects of oxidation and disinfection, which ferrate can provide. Research should focus on the means to tie up the Fe(III) generated from ferrate degradation to reduce the gel-like transition of the sludge material. This may drastically increase the dewatering effectiveness of ferrate for thickened sludge.

Acknowledgement

This study was funded by National Science Foundation (Award number: 0406255), which is greatly appreciated.

References

1. Krogmann, U.; Boyles, L.S.; Bamka, W.J.; Chaiprapat, S.; Martel, C.J. *Water Environ. Res.* **1999**, *71(5)*, 692-714.
2. Dentel, S. K. *Water Sci. Technol.* **2001**, *44(10)*, 9-18.
3. Oleszkiewicz, J. A.; Mavinic, D. S. *Can. J. Civ. Eng.* **2001**, *28(Suppl. 1)*, 102-114.
4. Hertle, A. *Water (Australia)* **2003**, *30(4)*, 70-73.
5. Neyens, E.; Baeyens, J.; Weemaes, M.; De Heyder, R. *Environ. Eng. Sci.* **2002**, *19(1)*, 27-35.
6. Amokrane, A.; Comel, C.; Veron, J. *Water Res.* **1997**, *31(11)*, 2775-2782.
7. Diamadopoulos, E. *Water Res.* **1994**, *28(12)*, 2439-2445.
8. Neyens, E.; Baeyens, J.; De Heyder, B.; Weemaes, M. *Management Environ. Quality* **2004**, *15(1)*, 9-16.
9. WRC "Permissible Utilization and Disposal of Sewage Sludge Edition 1." TT 85/97, Water Research Commission: Pretoria, South Africa. **1997**.
10. U.S. EPA "A plain English guide to the EPA Part 503 Biosolids Rule." EPA832-R-93-003, USA EPA DC, USA. **1994**.
11. Ma, J.; Liu, W. *Water Res.* **2002**, *36*, 4959-4962.
12. Thompson, G. W.; Ockerman, L. T.; Schreyer, J.M. *J. Am. Chem. Soc.* **1951**, *73*, 1379-1381.
13. Dedushenkol, S. K.; Perfiliev, Yu. D.; Goldfield, M. G.; Tspin, A. I. *Hyperfine Interact.* **2001**, *136/137*, 373–377.
14. Bouzek, K.; Schmidt, M.J.; Wragg, A. A. *Collect. Czech Chem. Commun.* **2000**, *65*, 133-140.
15. Ayol, A.; Dentel, S. K.; Filibeli, A. *Water Sci. Technol.* **2004**, *50(9)*, 9-16.
16. Kim, H.; Millner, P.; Sharma, V.K.; McConnell, L.L.; Torrents, A.; Ramirez, M.; Peot, C. *Water Environ. Lab. Solutions* **2006**, *12(6)*, 1-6.
17. U.S. EPA 40 CFR Part 503: The Standard for the Use or Disposal of Sewage Sludge, **1993**, *58*, 9248-9404.

Chapter 20

Ferrate(VI) Oxidation of Recalcitrant Compounds: Removal of Biological Resistant Organic Molecules by Ferrate(VI)

Virender K. Sharma, Nadine N. Noorhasan, Santosh K. Mishra, and Nasri Nesnas

Department of Chemistry, Florida Institute of Technology, 150 West University Boulevard, Melbourne, FL 32901

The oxidation of recalcitrant compounds, ethylenediamine-tetraacetate (EDTA) and sulfamethoxazole (SMX) by ferrate(VI) (Fe(VI), $Fe^{VI}O_4^{2-}$) is presented. Kinetics of the reactions were determined as a function of pH at 25 °C by a stopped-flow technique. The rate law for the oxidation of these compounds by Fe(VI) is first-order with respect to each reactant. The observed second-order rate constants for the reaction of Fe(VI) with SMX decreased non-linearly with increase in pH and are possibly related to protonations of Fe(VI) and compounds. The individual rate constants, k ($M^{-1}s^{-1}$) of Fe(VI) species, $HFeO_4^-$ and FeO_4^{2-} with protonated and deprotonated forms of compounds were estimated. The $HFeO_4^-$ species reacts much faster with these compounds than does the FeO_4^{2-} species. The results showed that Fe(VI) has the potential to serve as an environmentally-friendly oxidant for removing biologically resistant organic molecules within minutes and converting them to relatively less toxic by-products in water.

Introduction

Chemical structures that are resistant to microbiological degradation are referred as recalcitrant compounds, which cause potential problems in the environment (*1*). Such compounds have been found in groundwater, treated wastewater effluent, landfill leachate, and soils irrigated with reclaimed water (*2*). Ethylenediaminetetraacetic acid (EDTA) and sulfamethoxazole (SMX) are nitrogen-containing molecules (Figure 1) and are not readily biodegradable (*3,4*).

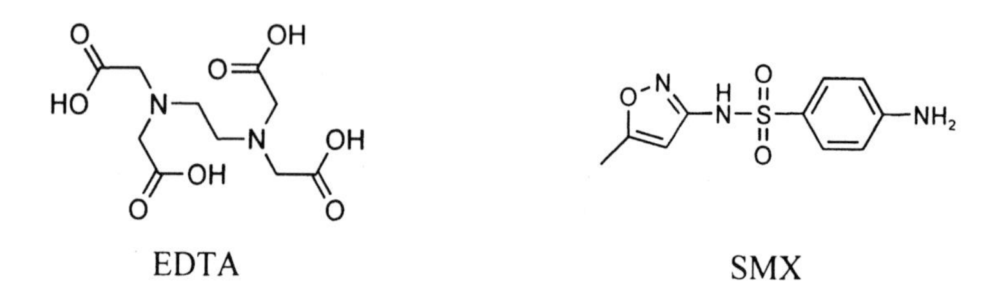

Figure 1. Molecular structures of nitrogen-containing recalcitrant compounds.

EDTA is a hexadentate chelating ligand and is widely employed in many commercial and industrial applications such as leather manufacturing, textile finishing, electroplating, and paper processes (*2,5*). The use of EDTA can be found in household items such as detergents and food items. EDTA itself does not pose significant problem, but its strong ability to complex metal ions causes an increase in the mobility of toxic metals in the aquatic environment and an adverse effect on metal removal in wastewater treatment plants (*6*). EDTA is not removed during conventional treatment methods and is thus found in the treated effluents and receiving waters. For example, a wide use of EDTA and its slow removal under many environmental conditions has led its status as an anthropogenic compound in many European surface waters (*7*). In the United States, EDTA concentrations in groundwater receiving effluent discharge have been reported as 1-72 μg/L (*7*).

Sulfamethoxazole (SMX) consists of two moieties, an aniline ring and a five membered heterocyclic group, connected to both sides of the sulfonamide linkage (-NH-SO_2-) (Figure 1). Sulfonamides are among the most frequently detected antibiotics in surface waters and are used in both human therapy and animal husbandry. Disposal of domestic and hospital waste, fields treated with animal manure, and runoff and infiltration from confined animal feeding operation result in the entry of sulfonamides into the aquatic environment (*8-10*). Among the sulfonamides, SMX has been repeatedly detected at concentrations of 200-2000 ng/L in secondary wastewater effluents and 70-150 ng/L in surface waters (*10-12*). Detection of sulfonamides is of concern due to the possibility of increased bacterial resistance (*13*). Aerobic biodegradation of sulfonamide is limited, which may be a reason explaining the finding of SMX in wastewater effluent and surface water bodies (*11,14-17*).

Various chemical oxidation processes and advanced oxidation methods have been applied to remove EDTA and SMX (*18-34*). Conventional method using chlorine may create and leave disinfection by-products (DBPs) in treated water. Ozonation has shown great potential to remove EDTA and SMX (*30,33*). However, ozone can form the potent carcinogenic bromate ion by reacting with bromide present in water. Another advanced oxidation process involves photocatalysis, which can transform both EDTA and SMX on titanium dioxide

(*29,34*). Yet another promising method is the possible application of potassium ferrate(VI) (K_2FeO_4), which can address some of the concerns of currently used methods, in treating sulfonamide antimicrobials.

Ferrate(VI) ($Fe^{VI}O_4^{2-}$, Fe(VI)) is a powerful oxidizing agent in aqueous media with reduction potential of 2.20 V and 0.70 V in acidic and alkaline solutions, respectively (*35*). The spontaneous decomposition of Fe(VI) in water gives molecular oxygen and Fe(III) thus making Fe(VI) a green chemistry chemical for coagulation, disinfection, and oxidation for multipurpose treatment of water and wastewater (*36-38*). Furthermore, the application of Fe(VI) can improve the removal of natural organic matter or DPBs precursors. Unlike ozone, Fe(VI) does not react with bromide ion; thus the carcinogenic bromate ion would not be formed in the treatment of bromide containing water by Fe(VI).

In the last few years, it has been demonstrated that the destruction of sulfur- and nitrogen-containing contaminants by Fe(VI) can be accomplished in seconds to minutes with the formation of relatively non-toxic by-products (*39-41*). In this paper, the kinetics of the oxidation of EDTA and SMX is presented. Products of the oxidation reactions are briefly discussed. The results suggest that ferrate(VI) has a potential to treat recalcitrant organic compounds in water and wastewater.

Experimental Section

Potassium ferrate (K_2FeO_4) of high purity (98% plus) was prepared by oxidizing Fe(III) with hypochlorite in an alkaline medium (*42*). A stopped-flow spectrophotometer (SX.18 MV, Applied Photophysics, U.K.) equipped with a photomultiplier (PM) detector was used to make Fe(VI) kinetic studies. The kinetic measurements were performed under pseudo-first-order conditions in which substrates were in excess. The details are given elsewhere (*43,44*). Various analytical techniques were applied to evaluate the effect of the oxidation process on the fate of EDTA and SMX in water. Experimental conditions and analytical procedures used in different techniques are reported in earlier papers from this group (*45,46*).

Results and Discussion

Ethylenediaminetetraacetate (EDTA)

Initially, the reaction rates of Fe(VI) with EDTA were measured at 25 °C at different pH in alkaline media. The rate expression for the oxidation reaction between Fe(VI) and EDTA is given by

$$-d[Fe(VI)]/dt = k[Fe(VI)]^m[EDTA]^n \quad (1)$$

where [Fe(VI)] and [EDTA] are concentrations of Fe(VI) and EDTA, respectively. The k represents the rate constant of the reaction, while the exponential variables m and n are the order of the reaction with respect to the assigned reactant. Excess EDTA was used in the kinetic studies to ensure that reactions were measured under pseudo-first-order conditions. Under these conditions equation (1) can be re-written as:

$$-d[Fe(VI)]/dt = k_1[Fe(VI)]^m \quad (2)$$

$$\text{where} \quad k_1 = k[EDTA]^n \quad (3)$$

The absorbance versus time profile for Fe(VI) gave a single-exponential decay curve, indicating the reaction was first-order with respect to Fe(VI). The k_1 values for the reaction were determined at various concentrations of EDTA. The plot of k_1 versus [EDTA] was linear with a correlation coefficient, $r^2 = 0.99$ (Figure 1). The resulting slopes of the plot of log k_1 vs. log[EDTA] were nearly one, indicating that the reaction with respect to EDTA is essentially first order. The rate law for this reaction can be expressed as

$$-d[Fe(VI)]/dt = k[Fe(VI)][EDTA] \quad (4)$$

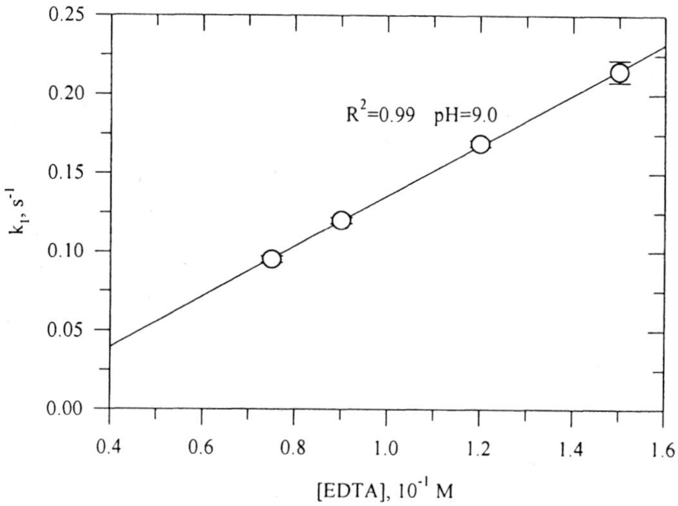

Figure 2. A plot of pseudo-first-order rate constant (k_1, s^{-1}) versus [EDTA] at 25 °C.

Next, kinetic measurements of oxidation of amines by ferrate(VI) were carried out at pH = 9.0 and at 25 °C. Rates were found to be first order with respect to each reactant. The rate constants for oxidation of different amines with ferrate(VI) are given in Table 1.

Table 1. Rate constants of Ferrate(VI) oxidation of APCs at pH = 9.0.

Compound	Structure	k, $M^{-1}s^{-1}$	$t_{1/2}$, min
Glycine	$NH_2(CH_2COO^-)$	93.7 ± 4.7	0.36
Iminodiaminetetraacetate (IDA)	$NH(CH_2COO^-)_2$	18.9 ± 1.1	1.76
Nitrilotriaminetetraacetate (NTA)	$N(CH_2COO^-)_3$	0.71± 0.05	46.9
EDTA	$CH_2N(CH_2COO^-)_2$ | $CH_2N(CH_2COO^-)_2$	1.72 ± 0.08	19.4

Glycine is a primary amine (1°), IDA is a secondary amine (2°), and NTA is a tertiary amine (3°). The order of reactivity was 1° > 2° > 3°. EDTA reacted faster with ferrate(VI) than NTA. The order of reactivity of amines with ferrate(VI) suggests that that ferrate(VI) attacks at the nitrogen atom sites of the amines. For example, successive cleavages of NTA at the nitrogen – carbon bond lead to the formation of ammonia (eq 1) (*47*). Excess ammonia can also be removed by adding additional amount of ferrate(VI) (*48*).

$$\underset{\text{NTA}}{N(CH_2COO^-)_3} \xrightarrow{FeO_4^{2-}} \underset{\text{IDA}}{NH(CH_2COO^-)_2} \xrightarrow{FeO_4^{2-}} \underset{\text{Glycine}}{NH_2CH_2COO^-} \xrightarrow{FeO_4^{2-}} \underset{\text{Ammonia}}{NH_3} \quad (5)$$

The preliminary work in the present study gave a reduction of total organic carbon (TOC) in the reaction between Fe(VI) and EDTA. The TOC reduction was approximately 20 % at a molar ratio of 6:1 ([Fe(VI):[EDTA]). It appears that similar to NTA, successive cleavages of EDTA at the nitrogen – carbon bond lead to the formation of acetate, which further reacts with ferrate(VI) to give reduction in total organic carbon (TOC).

The half-life of oxidation reactions were calculated considering concentration of ferrate(VI) (500 μM) was in excess than amine concentrations (Table 1). Amines can thus be removed by ferrate(VI) in second to minute time scales. It should be pointed out that the reaction rates are pH dependent; so are the half-lives of the reactions. Overall, the destruction of APC in water to relatively non-hazardous products by ferrate(VI) can be accomplished in minutes.

Sulfamethoxazole (SMX)

The rate law for the oxidation of SMX by Fe(VI) was found to be first-order for each reactant (*46*). The rate law may be written as

$$-d[Fe(VI)]/dt = k[Fe(VI)][SMX] \quad (6)$$

The observed second-order rate constant, k, decreased non-linearly with pH (Figure 3).

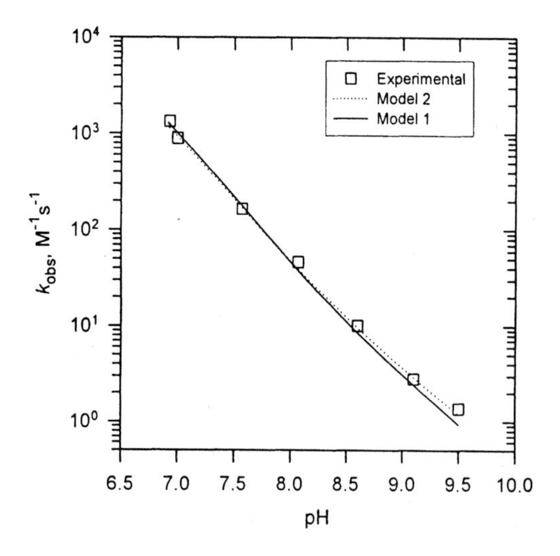

Figure 3. The rate constant, k ($M^{-1}s^{-1}$) versus pH at 25 °C.

A change in k with pH can be described by considering the equilibrium of the mono protonated Fe(VI) ($HFeO_4^-$) and SMX (SH) (SH ⇔ H^+ + S^-; $pK_{a,SH}$ = 5.7) (*49*). Two forms of Fe(VI) species ($HFeO_4^-$ and FeO_4^{2-}) can possibly react with two forms of SMX (SH and S^-) in the studied pH range (Model 1). In the non-linear regression of the data, it was determined that reaction between FeO_4^{2-}

and S^- did not contribute to the rate constant, *k* (*46*). The rate constants for other reactions were obtained as (*46*)

$$HFeO_4^- + SH \rightarrow Fe(OH)_3 + Product(s) \qquad k_{15}=3.0x10^4\ M^{-1}s^{-1} \qquad (7)$$

$$HFeO_4^- + S^- \rightarrow Fe(OH)_3 + Product(s) \qquad k_{16}=1.7x10^2\ M^{-1}s^{-1} \qquad (8)$$

$$FeO_4^{2-} + SH \rightarrow Fe(OH)_3 + Product(s) \qquad k_{17}=1.2x10^0\ M^{-1}s^{-1} \qquad (9)$$

A faster reaction rate constant of $HFeO_4^-$ with the neutral sulfonamide species (SH) (reaction 7) than with the negatively charged ionized species (S^-) (reaction 8) was obtained. This is in contrast with the reactivity of ozone and hydroxyl radical with amines and amino acids in which higher reactivity was observed as a result of deprotonation (22,30,*50*). The Hoigne and Bader (*50*) observed experimentally that ozone is almost unreactive with amines and amino acids in the highly acidic region. Contrary to this phenomena, ferrate(VI) reactivity with substrates increases in the acidic region. Reactivity of ozone and hydroxyl radicals increases with increase in pH, which is opposite to the reactivity of Fe(VI) with SMX.

In analogy with the reactivity of ozone, the pH dependence data in Figure 4 can be empirically fitted by considering interactions of ferrate species (H_2FeO_4, $HFeO_4^-$, and FeO_4^{2-}) with only deprotonated sulfonamide (S^-) (eq 10-12).

$$H_2FeO_4 + S^- \rightarrow Fe(OH)_3 + Product(s) \qquad k_{18}=4x10^6\ M^{-1}s^{-1} \qquad (10)$$

$$HFeO_4^- + S^- \rightarrow Fe(OH)_3 + Product(s) \qquad k_{19}=2x10^2\ M^{-1}s^{-1} \qquad (11)$$

$$FeO_4^{2-} + S^- \rightarrow Fe(OH)_3 + Product(s) \qquad k_{20}=1x10^{-1}\ M^{-1}s^{-1} \qquad (12)$$

According to this fitting approach (Model 2), it was assumed that ferrate species are only reactive enough with anionic SMX species (S^-) and other six reactions of three ferrate species with positively charged (SH_2^+) and neutral species (SH) had no role to play. However, the mechanism by which Fe(VI) oxidizes SMX may have proton involvement from the SMX (i.e. sulfonamide). This was the case in the oxidation of aniline by ferrate(VI) (*51*) in which involvement of substrate proton was demonstrated by performing isotopic studies. Additionally, DFT calculations modeled the abstraction of hydrogen from alcohol in reaction with ferrate species. Thus fitting of pH dependence of *k* values using Model 2 may not be appropriate. However, large difference in reactivity of $HFeO_4^-$ with SH and S^- cannot solely be from the electrostatic attraction phenomenon. A lesser degree of repulsion in the reaction of $HFeO_4^-$ with SH facilitates the involvement of the substrate proton in the reaction that ultimately increases the rate.

Next, the stoichiometry of the oxidation of SMX by Fe(VI) was determined at neutral and basic pH (Figure 4). Slopes of linear relationships between [SMX] and [Fe(VI)] were found to be 0.27 ± 0.02 and 0.24 ± 0.01, respectively, for pH 7.0 and 9.0. The formation of oxygen and Fe(III) were determined as products of the reaction. A slope between Fe(VI) consumed and oxygen formation was found to be 0.23 ± 0.01.

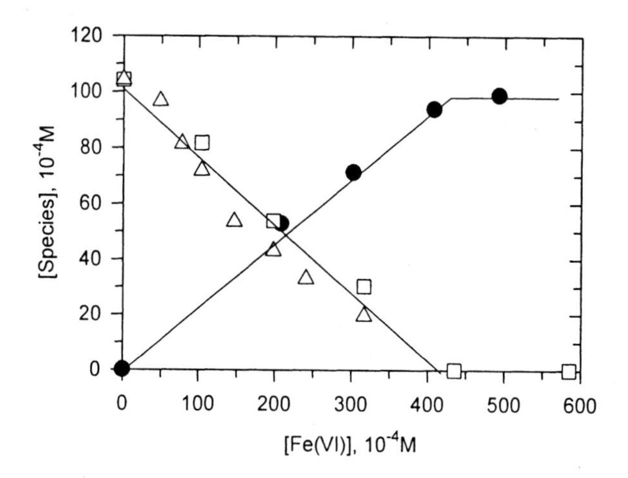

Figure 4. A plot of SMX decrease and formation of oxygen in the reaction of sulfamethoxazole with Fe(VI) under anoxic conditions. (sulfamethoxazole (△-pH 7.0, □- pH 9.0), ●-oxygen (pH 9.0)) (Data was taken from ref. 46).

The oxidation of SMX by Fe(VI) thus follows a stoichiometry of 4:1 ([Fe(VI):[SMX]) which leads to the evolution of one mole of oxygen per mole of SMX (eq 13).

$$4\ HFeO_4^- + SMX \rightarrow 4\ Fe(III) + O_2 + Product(s) \quad (13)$$

Finally, the products analysis of the oxidation of SMX by Fe(VI) at pH 9.0 gave three products A, B, and C (Scheme 1) (*46*).

Scheme 1 suggests the oxidation of both the isoxazole moiety and the aniline unit of SMX by Fe(VI). At higher molar ratios of Fe(VI) to SMX, oxidation of both the isoxazole and the aniline units may occur. We may have indeed formed some products that were oxidized at both units, however, due to the increased polarity of such products, they may have been retained on the column during chromatography. Therefore the three products described in Scheme 1 may not necessarily represent all products of ferrate oxidation, but rather the ones that were isolated and characterized. Product A indicates

Scheme I

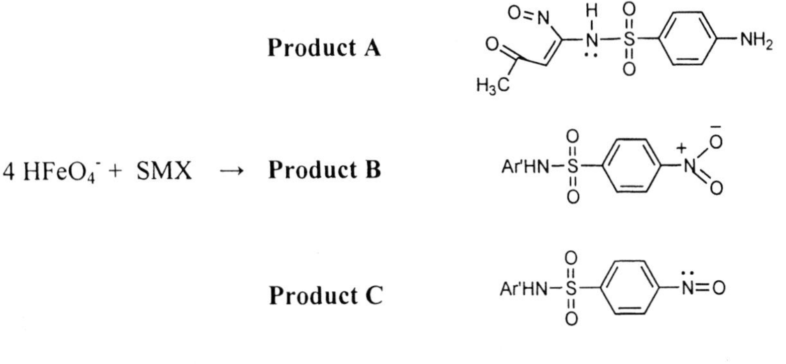

opening of the isoxazole while the presence of predominantly a nitroso and a nitro group in product C and B, respectively. Importantly, the oxidation process by Fe(VI) is the destruction of the aromatic ring, isoxazole ring, which will undoubtedly render the oxidized product a differing biological binding property. It is expected that an oxidation of the amino group and/or an oxidation of the isoxazole ring (which leads to its potential opening/destruction) will change its binding properties sufficiently rendering it less of a mimic for the important *p*-aminobenzoic acid. The latter is necessary in the synthesis of the essential vitamin: folic acid. Thus Fe(VI) not only removes SMX in water, but also produces by-products that are expected to be less toxic.

Acknowledgment

We wish to thank Dr. Y. Lee for useful comments.

References

1. Knapp, J.D.; Bromley-Challenor, K.C.A. *Recalcitrant Organic Compounds. In: Handbook of Water and Wastewater Microbiology* (Mara, D.; Horan, N. Edts). Chapter 32. Publ. Elsevier, NY. **2003**.
2. Pitter, P. and Chudoba, J. *Biodegradability of organic substances in the aquatic environment.* CRC Press, Boca Raton, Florida, USA, **1990**.
3. Bucheli-Witchel, M.; Egli, T. *FEMS Microbiol. Rev.* 2001, 25, 69-106.
4. Ingerslev, F.; Halling-Sorensen, B. *Environ. Toxicol. Chem.* **2000**, *19*, 2467-2473.
5. Korhonen, M.S.; Metarinne, S.E.; Tuhkanen, T.A. *Ozone Sci. Engg.* **2000**, *22*, 279-286.
6. Stumm, W. *Aquatic Chemistry*, 3rd ed.; John Wiley and Sons: 1996.

7. Nowack, B. *Environ. Sci. Technol.* **2002**, *36*, 4009-4016.
8. Holm, J.V.; Ruegge, K.; Bjerg, P.L.; Christensen, T.H. *Environ. Sci. Technol.* **1995**, *29*, 1415-1420.
9. Pedersen, J.A.; Soliman, M.A.; Suffet, I.H. *J. Agric. Food. Chem.* **2005**, *53*, 1625-1632.
10. Miao, X.; Bishay, F.; Chen, M.; Metcalfe, C.D. *Environ. Sci. Technol.* **2004**, *38*, 3533-3541.
11. Kolpin, D.W.; Furlong, E.T.; Meyer, M.T.; Thurman, E.M.; Zaugg, S.D.; Barber, L.B.; Buxton, H.T. *Environ. Sci. Technol.* **2002**, *36*, 1202-1211.
12. Ternes, T.A. *Trans. Anal. Chem.* **2001**, *20*, 419-434.
13. Gould, I.M. *J. Antimicrob. Chemother.* **1999**, *43*, 459-465.
14. Cunningham, V.L.; Buzby, M.; Hutchinson, T.; Mastrocco, F.; Parke, N.; Roden, N. *Environ. Sci. Technol.* **2006**, *40*, 3457-3461.
15. Diaz-Cruz, M.S.; Barcelo, D. *Anal. Bioanal. Chem.* **2006**, *386*, 973-985.
16. Holm, J.V.; Ruegge, K.; Bjerg, P.L.; Christensen, T.H. *Environ. Sci. Technol.* **1995**, *29*, 1415-1420.
17. Pedersen, J.A.; Soliman, M.A.; Suffet, I.H. *J. Agric. Food. Chem.* **2005**, *53*, 1625-1632.
18. Sillnpaa, M.; Pirkanniemi, K. *Environ. Technol.* **2001**, *22*, 791-801.
19. Leitner, N.K.V.; Guilbult, I.; Legube, B. *Rad. Phy. Chem.* **2003**, *67*, 41-49.
20. Ghielli, G.; Jardim, W.F.; Litter, M.L.; Mansilla, H.D. *J. Photochem. Photobiol. A: Chem.* **2004**, *167*, 59-67.
21. Gilbert, E.; Beyerle, M. *J. Water RT-Aqua* **1992**, *41*, 269-276.
22. Sonntag, V.V.; Hobel, B. *J. Chem. Soc. Perkin Trans. 2* **1998**, 509-513.
23. Frim, J.A.; Rathman, J.F.; Weaver, L.K. *Water Res.* **2003**, *37*, 3155-3163.
24. Chitra, S.; Paramasivan, K.; Sinha, P.K.; Lal, K.B. *J. Cleaner Prod.* **2004**, *12*, 429-435.
25. Quattara, L.; Duo, I.; Diaco, T.; Ivandint, A.; Honda, K.; Ro, T.; Fujishima, A.; Comninelli, C. *Carbon Tech.* **2003**, *13*, 97-108.
26. Emilio, C.A.; Jardin, W.F.; Litter, M.I.; Mansilla, H.D. *J. Photochem. Photobiol A: Chem.* **2002**, *151*, 121-127.
27. Chang, H-S.; Korshin, G.V.; Ferguson, J.F. *Environ. Sci. Technol.* **2006**, *40*, 1244-1249.
28. Metsarinne, S.; Tuhkanen, T.; Aksela, R. *Chemophere* **2001**, *45*, 949-955.
29. Mansilla, H.D. Bravo, C.; Ferreyra, R.; Litter, M.I.; Jardim, W.F.; Lizama, C.; Freer, J.; Fernandez, J. *J. Photochem. Photobiol. A: Chem.* **2006**, *181*, 188-194.
30. Munoz, F.; Sonntag, C.V. *J. Chem. Soc. Perkins Trans. 2* **2000**, 2029-2033.
31. Dodd, M.C.; Huang, C-H. *Environ. Sci. Technol.* **2004**, *38*, 5607-5615.
32. Huber, M.M.; Korhonen, S.; Ternes, T.A.; Gunten, U.V. *Water Res.* **2005**, *39*, 3607-3617.
33. Huber, M.M.; Goebel, A.; Joss, A.; Hermann, N.; Loffler, D.; McArdell, C.S.; Ried, A.; Siegrist, H; Ternes, T.A.; Gunten, U.V. *Environ. Sci. Technol.* **2005**, *39*, 4290-4299.

34. Calza, P.; Medana, C.; Pazzi, M.; Baiocchi, C.; Pelizzetti, E. *Applied Catal. B: Environmental* **2004**, *53*, 63-69.
35. Wood, R.H. *J. Am. Chem. Soc.* **1958**, *80*, 2038-2041.
36. Sharma, V.K. *Adv. Environ. Res.* **2002**, *6*, 143-156.
37. Jiang, J-Q.; Lloyd, B. *Water Res.* **2002**, *36*, 1397-1408.
38. Sharma, V.K. *Water Sci. Technol.* **2004**, *49*, 69-74.
39. Sharma, V.K.; Kazama, F.; Jiangyong, H.; Ray, A.K. *J. Water Health* **2005**, *3*, 42-58.
40. Sharma, V.K.; Burnett, C.R.; Yngard, R.; Cabelli, D.E. *Environ. Sci. Technol.* **2005**, *39*, 3849-3854.
41. Eng, Y.Y.; Sharma, V.K.; Ray, A.K. *Chemosphere*, **2006**, 63, 1785-1790.
42. Thompson, G.W.; Ockerman, L.T.; Schreyer, J.M. *J. Am. Chem. Soc.* **1951**, *73*, 1379-1381.
43. Sharma, V.K.; Mishra, S.K.; Ray, A.K. *Chemosphere*, **2006**, *62*, 128-134.
44. Sharma, V.K.; Mishra, S.K. *Environ. Chem. Lett.* **2006**, *3*, 182-185.
45. Sharma, V.K. and Noorhasan, N. *Prep. Pap. Natl. Meet.-Am. Chem. Soc., Div. Environ. Chem.* **2004**, 427-430.
46. Sharma, V.K.; Mishra, S.K.; Nesnas, N. *Environ. Sci. Technol.* **2006**, *40*, 7222-7227.
47. Carr, J.D.; Kelter, P.B.; Ericson III, A.T. *Environ. Sci. Technol.* **1981**, *15*, 184-187.
48. Sharma, V.K.; Bloom, J.T.; Joshi, V.N. *J. Environmental Sci. Health.* **1998**, *A33*, 635-640.
49. Boreen, A.L.; Arnold, W.A.; McNeill, K. *Environ. Sci. Technol.* **2004**, *38*, 3933-3940.
50. Hoigne, J.; Bader, H. *Water Res.* **1983**, *17*, 185-194.
51. Huang, H.; Sommerfeld, D.; Dunn, B.C.; Lloyd, C.R.; Eyring, E.M. *J. Chem. Soc. Dalton Trans.* **2001**, 1301-1305.

Chapter 21

Heterogeneous Photocatalytic Reduction of Iron(VI): Effect of Ammonia and Formic Acid

Enhancement of Photocatalytic Oxidation of Ammonia and Formic Acid in Presence of Iron(VI)

Virender K. Sharma* **and Benoit V. N. Chenay**

Department of Chemistry, Florida Institute of Technology,
150 West University Boulevard, Melbourne, FL 32901
***Corresponding author: vsharma@fit.edu**

Ammonia is a potential pollutant that can contribute to eutrophication of rivers and lakes and its removal is thus becoming an important issue. Formic acid is a byproduct of many industrial processes and its removal from wastewater is of great interest. The removal of ammonia and formic acid was sought by studying the heterogeneous photocatalytic oxidation of ammonia and formic acid in UV-irradiated TiO_2 suspensions with and without Fe(VI) ($Fe^{VI}O_4^{2-}$) at pH 9.0. The kinetics of the reactions was determined by monitoring both reduction of Fe(VI) and oxidation of ammonia/formic acid. The initial rate of Fe(VI) reduction (R) can be expressed as $R = k_{Fe(VI)}[Fe(VI)]^m$ where $m = 1.25 \pm 0.03$ and 0.70 ± 0.06 for ammonia and formic acid, respectively. The rate constant, $k_{Fe(VI)}$, depends on the concentration of ammonia and formic acid. The values of $k_{Fe(VI)}$ for the oxidation of ammonia was determined as $k_{Fe(VI)} = [Ammonia]/(a[Ammonia]+b)$, $a = 6.0 \times 10^3\ \mu M^{0.25}$ s and $b = 4.1 \times 10^6\ \mu M^{-1.25}\ s^{-1}$. The $k_{Fe(VI)}$ for the photocatalytic oxidation of formic acid showed a positive linear relationship, which can be written as $k_{Fe(VI)} = 2.41 \times 10^{-3} + 1.58 \times 10^{-7}$ ([Formic Acid]). The rates of oxidation of ammonia and formic acid in TiO_2/UV suspensions were enhanced in the presence of Fe(VI). Results suggest the photocatalytic production of a highly reactive species, Fe(V), a powerful oxidant, to oxidize ammonia and formic acid. A combination of Fe(VI) and the TiO_2 photocatalyst has the potential to enhance the oxidation of pollutants in the aquatic environment.

Introduction

Photoinduced oxidation of water and compounds at titanium dioxide (TiO_2) semiconductor surfaces has received much attention from the point of view of solar-to-chemical conversion water splitting (*1*) and environmental remediation (*2,3*). This field of using TiO_2 suspensions in heterogeneous photocatalysis process is now considered a mature field (*4*). The primary step in the process involves the generation of charge-carriers (eq 1), when a photon with enough energy ($hv \geq$ band gap energy of TiO_2) promotes an electron (e_{cb}^-) from the valence band into the conduction band of the titanium dioxide (*3*). This process leaves behind a positively charged hole (h_{vb}^+) in the valence band (eq 1).

$$TiO_2 + hv \rightarrow h_{vb}^+ + e_{cb}^- \quad (1)$$

Either electrons and holes migrate to the particle surface, becoming involved in oxidation/reduction reactions or they recombine and simply liberate heat (*5*). Electron/hole pairs escaping recombination diffuse to the oxide/water interface where they either react with electron acceptors (e.g. H^+, O_2) or electron donors (e.g. OH^-, H_2O). The transfer of an electron to the O_2 molecule produces a superoxide radical ($O_2^{-\bullet}$) (eq 2), whereas the hole can generate a hydroxyl radical ($OH^\bullet$) in water (eq 3).

$$e_{cb}^- + O_2 \rightarrow O_2^{-\bullet} \quad (2)$$

$$h_{vb}^+ + H_2O \rightarrow OH^\bullet + H^+ \quad (3)$$

The valence band holes are strong oxidizing agents (+2.53 V vs. NHE) capable of degrading pollutants (*4-6*), whereas the conduction band electrons are good reductants (-0.5 V vs. NHE) able to reduce oxidants. This work sought to understand the photocatalytic oxidation of ammonia and formic acid in TiO_2 suspensions in the aquatic environment.

Ammonia is an inorganic chemical compound widely used for industrial purposes, for the production of fertilizers and is also used as a fertilizer by itself for direct application to soils. Ammonia and ammonium compounds used as fertilizer represent 89–90% of the commercially produced ammonia, with plastics, synthetic fibers and resins, explosives, and other uses accounting for most of the remainder (*7*). Therefore, the two primary anthropogenic sources of ammonia in the environment are industrial wastewater discharge (point source) and runoff from agricultural releases (non-point source). The release of high concentrations of ammonia in the environment can lead to eutrophication of surface waters. Although nitrogen and phosphorous are two main nutrients necessary for the health of ecosystems, their overabundance can lead to excessive algal growth, which in turn depletes water bodies of oxygen, and results in fish kills. Moreover, as the concentration of ammonia in water increases, so does the rate of nitrification; a process that also consumes oxygen.

It is therefore important to find efficient methods for the removal of ammonia from wastewater.

Formic acid (HCOOH) is an organic compound formed as a byproduct in the production of other chemicals, especially acetic acid, a chemical extensively used as a condiment, preservative, in paints, textiles and the manufacture of photographic films. Formic acid is released into the environment primarily from industrial sources including textile dying and finishing, pharmaceuticals, rubber, leather and tanning treatments and catalysis (*8,9*). Levels of formic acid can be as high as 80 mgL^{-1} in wastewater and its removal from the environment is therefore of great interest (*8*).

In photocatalytic applications, higher amounts of free electrons and holes on the particle surface can make oxidation of ammonia and formic acid more efficient. However, the efficiency of decomposing pollutants using TiO_2 is low due to the fast recombination of charge carriers. Recombination can be prevented if a suitable scavenger or surface defect state is available to trap the electron or hole (*3*). Doping titanium dioxide with metal particles provides suitable scavengers that decrease the efficiency of the recombination process and enhances the photocatalytic degradation of pollutants. The use of metal ions such as silver(I), mercury(II), copper(II) and chromium(VI) has been shown to increase the photocatalytic efficiency of TiO_2 to degrade pollutants in water (*5*). An environmentally friendly oxidant, iron(VI) [$Fe^{VI}O_4^{2-}$, Fe(VI)] (*10*) may also increase the efficiency of the photocatalytic degradation by scavenging electrons.

Fe(VI) species are strong oxidizing agents which can be seen from the reduction potential of reactions 1 and 2 in acidic and alkaline solutions, respectively (*11*).

$$Fe^{VI}O_4^{2-} + 8H^+ + 3e^- \rightarrow Fe^{3+} + 4H_2O \qquad E^0 = 2.2\ V \qquad (4)$$

$$Fe^{VI}O_4^{2-} + 4H_2O + 3e^- \rightarrow Fe(OH)_3 + 5OH^- \qquad E^0 = 0.7\ V \qquad (5)$$

The reduction potential of Fe(VI) is more positive than the TiO_2 conduction band electron's potential (E_{cb} = -0.6 to 0.8 V) in basic solution and the thermodynamic driving force for the formation of OH^- ions is significantly higher than the TIO_2 photocorrosion pathway in basic media (*12*). This suggests that the photocatalytic reduction of Fe(VI) in TiO_2 suspensions is feasible. The photoreduction of Fe(VI) likely takes place through one-electron steps that would result in the sequential formation of Fe(V), Fe(IV), and Fe(III) (*13-15*). A highly reactive Fe(V) species, produced in this process, is 10^3 – 10^5 times more reactive with compounds than is Fe(VI) (16,17). Fe(V) thus has the ability to oxidize compounds which cannot be easily oxidized by Fe(VI). This paper summarizes the photocatalytic reduction of Fe(VI) in the presence of ammonia and formic acid, which react sluggishly with either Fe(VI) or illuminated TiO_2 individually. Results suggest the production of Fe(V) and thus the enhancement of the photocatalytic oxidation of ammonia and formic acid in the presence of Fe(VI).

Experimental

Chemicals

All chemicals were reagent or analytical grade and were used as received. Degussa P-25 TiO_2 (ca 80 % anatase and 20 % rutile) was used for this study. The TiO_2 concentration in the reaction vessels were fixed at 0.033 gL^{-1} for ammonia experiments and at 0.066 gL^{-1} for formic acid experiments. Doubly distilled ammonia (29.9 % ammonia) from Fisher Scientific was used as a source of ammonia. Sodium formate (99.998 %) from Sigma-Aldrich was used as a source of formic acid. A method described by Thompson et al. (*18*) was used to synthesize Fe(VI). The purity of the potassium ferrate(VI) prepared for this study was 98 % and higher, as measured spectrophotometrically (*19*).

Ammonia and formic acid oxidation

The reaction vessel was a borosilicate glass 500 ml, three-neck round-bottom flask (Figure 1). Helium gas was sparged before and during the reaction in order to create anaerobic conditions. A gas tube was inserted into one of the necks of the flask to allow excess helium to escape from the reaction vessel. A 10-ml syringe was attached to the flask for the removal of samples. The reactor vessel was illuminated with a 15-watt ultraviolet lamp (GE, T-15) (Figure 1). For each experiment the light intensity was determined using a radiometer (Cole-Parmer 9811) which measured a band of wavelengths ranging from 330-375 nm with a calibration center of 365 nm. The light intensity in the reaction vessel was calibrated with potassium ferrioxalate actinometry (*15*).

The ferrate(VI) solutions were prepared by the addition of solid samples of potassium ferrate (K_2FeO_4) to a 0.001 M borate ($Na_2B_4O_7 \cdot 10H_2O$) and 0.005 M disodium phosphate (Na_2HPO_4) solution at pH 9.0. The phosphate served as a complexing agent for Fe(III), which otherwise would have precipitated rapidly as a hydroxide that would have interfered with the optical monitoring of the reaction and accelerated the spontaneous decomposition of Fe(VI). All the experiments were conducted at pH 9.0, a condition under which Fe(VI) is most stable (*20*). The desired pH was achieved by adding phosphoric acid or sodium hydroxide. The samples were periodically withdrawn from the reaction vessel before being filtered through a Millipore (0.22μm) membrane. A molar extinction coefficient ε_{510} = 1150 M^{-1} cm^{-1} was used for the determination of Fe(VI) concentrations at pH 9.0 (*19*). The concentrations of ammonia were potentiometrically measured using an NH_3-selective electrode (ISE) (Orion model 93-18). Calibration of the ISE was performed prior to each measurement. The products of the reaction mixture were analyzed for nitrite and nitrate using high performance liquid chromatography (HPLC) . An IC-Pak HR (75x4.6 mm)

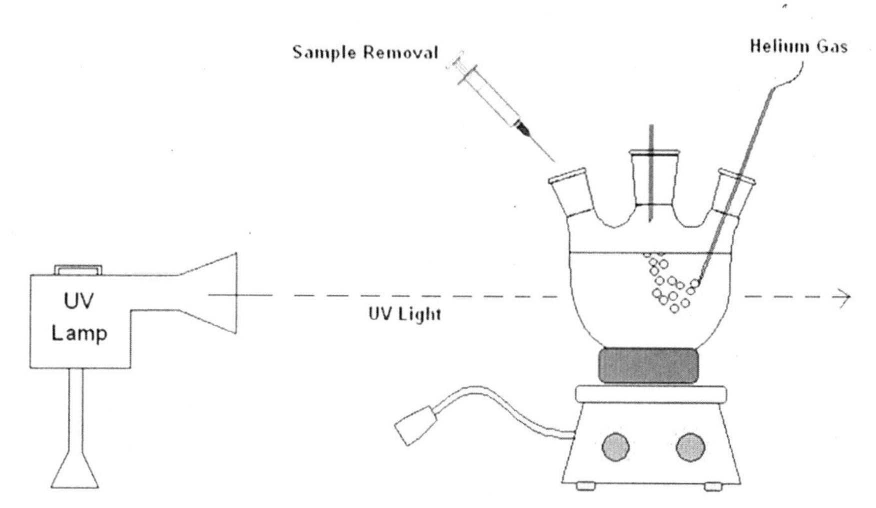

Figure 1. Setup for ammonia and formic acid photocatalytic oxidation and Fe(VI) reduction in UV-irradiated titania suspensions.

ion exchange column coupled to a conductivity detector was used for the detection of these two ions. The anion analytical eluent was a borate/gluconate buffer solution at a pH of 8.5, delivered at a flow rate of 1.0mL/min. The concentrations of formic acid present in the samples were also determined using HPLC technique.

Results and Discussion

Photocatalytic Reduction of Iron(VI)

Initially, experiments on the photocatalytic reduction of Fe(VI) were conducted at two different TiO_2 suspension doses as a function of Fe(VI) concentration at pH 9.0.

As shown in Figure 2, the photoreduction of Fe(VI) in the TiO_2 suspensions occurred at a faster rate than in the absence of TiO_2. Additionally, photoreduction was greater at higher TiO_2 suspension dose. The photoreduction of Fe(VI) to $Fe(OH)_3$ at TiO_2 surfaces can be expressed by the following equation.

$$FeO_4^{2-} + 4\ H_2O + 3\ e_{cb}^- \rightarrow Fe(OH)_3 + 5\ OH^- \qquad (6)$$

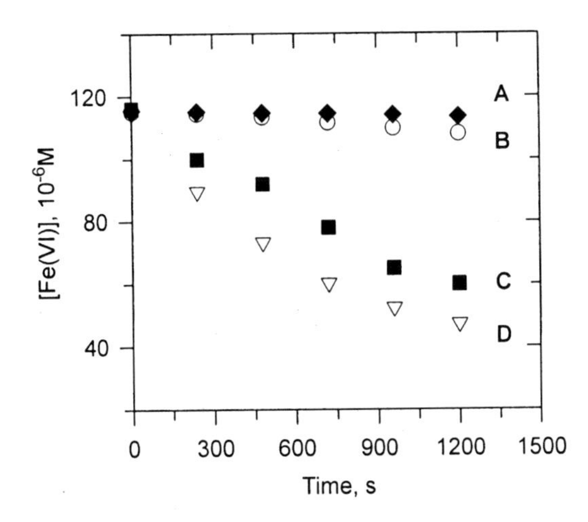

Figure 2. Photocatalytic reduction of Fe(VI) at pH 9.0. A = dark; B = UV only; C = UV + TiO_2 (0.033gL^{-1}); D = UV + TiO_2 (0.066gL^{-1}). Intensity = 1.62 × 10^{-6} einstein s^{-1} (Reproduced from (13)).

It is unlikely that the photoreduction of Fe(VI) proceeds through a multi-electron process. Therefore, it is assumed that the photoreduction of Fe(VI) occurs in successive one-electron steps with the formation of Fe(V) in the first stage (eq 7). Fe(V) has been detected in alkaline media so the assumption is made that Fe(V) was rapidly formed under existing experimental conditions (the observed rate constant for the homogeneous reaction of the hydrated electron with Fe(VI) is $k = 2.2 \pm 0.2 \times 10^{10}$ $M^{-1}s^{-1}$ (*20*)). Successive one-electron reduction steps will form Fe(IV) and Fe(III) (eqs 8, 9). In this reaction scheme, the self-decompositions of Fe(V) and Fe(IV) species have been ruled out because these reaction rate constants are of the order of $10^6 - 10^8$ $M^{-1}s^{-1}$. In comparison, reactions 8 and 9 must be occurring at the diffusion-controlled rates because Fe(V) and Fe(IV) species are much more reactive than Fe(VI) (*16,17*). The reduction of Fe(VI) to Fe(V) (eq 4) is postulated to be the rate determining step.

$$FeO_4^{2-} + e_{cb}^- \rightarrow FeO_4^{3-} \qquad (7)$$
$$FeO_4^{3-} + e_{cb}^- \rightarrow FeO_4^{4-} \qquad (8)$$
$$FeO_4^{4-} + 4H_2O + e_{cb}^- \rightarrow Fe(OH)_3 + 5\,OH^- \qquad (9)$$

The Fe(III) formed from the reaction 9 may be reduced further to Fe(II) at illuminated TiO_2 surfaces. The possibility of Fe(II) reacting with hydrogen peroxide, produced at the hole ($2\,H_2O + 2\,h_{vb}^+ \rightarrow H_2O_2 + 2\,H^+$), to initiate a Fenton-type reaction is not considered in the proposed scheme. Fe(II) would preferentially react with Fe(VI), which is a more powerful oxidant than H_2O_2 (*21,22*). Moreover, higher concentrations of Fe(VI) than H_2O_2 in the solution mixture ensure that Fe(II) is consumed by Fe(VI).

Oxidation of Ammonia

Initially, the photocatalytic reduction of Fe(VI) was studied in a TiO_2 suspension at various concentrations of Fe(VI) and ammonia at pH 9.0 (*15*). The reduction rate of Fe(VI) increased with increasing Fe(VI) concentration at all ammonia concentrations. Interestingly, the initial rates of Fe(VI) reduction linearly increased with initial concentrations of Fe(VI) (*15*). The amount of increase was also related to ammonia concentration in the solution. The initial rate (R) may be expressed by the following equation:

$$R = k_{Fe(VI)}[Fe(VI)]^{1.25} \tag{10}$$

where $k_{Fe(VI)}$ is the rate constant of the photocatalytic reduction of Fe(VI) and is related to ammonia concentration by the following expression (eq 8).

$$k_{Fe(VI)} = [\text{Amminia}]/(a[\text{Ammonia}]+b) \tag{11}$$
$$a = 6.0 \times 10^3\ \mu M^{0.25}\ s \text{ and } b = 4.1 \times 10^6\ \mu M^{-1.25}\ s^{-1}$$

Non-integer-order with respect to Fe(VI) (eq 10) is not uncommon in heterogeneous reactions where both mass transport and chemical reaction process can affect the reaction rate (*23*). The rate in a heterogeneous process thus depends on the ions in the solution and specific expression of an adsorption isotherm (*24*). A double layer (surface, Stern layer or diffuse layer) containing the Fe(VI) species surrounding the TiO_2 particle would determine the shape of adsorption isotherm (*25*). Adsorption isotherm processes in the double layer usually give non-integer order dependence on the dissolved ion concentration. If the reaction takes place in the double layer, rates would be proportional to the non-integer-order of the Fe(VI) concentration.

Finally, the photocatalytic oxidation of ammonia at two concentrations, 126 μM and 215 μM, was studied with and without Fe(VI) in the solution. The concentrations of Fe(VI) in solution mixtures were varied from 110 – 1129 μM. The significant enhancement of oxidation of ammonia was observed at [Fe(VI)]/[Ammonia] > 2. Under the studied experimental conditions, the results show an increase in the rate of ammonia oxidation due to higher molar ratios of Fe(VI):Ammonia (Figure 3). The rate of ammonia oxidation was also independent of initial ammonia concentration (Figure 3).

The faster rate of photocatalytic oxidation of ammonia in the presence of Fe(VI) compared to no Fe(VI) in the solution mixture (Figure 3) may be explained by the involvement of Fe(V) ($Fe^{V}O_4^{3-}$) (Scheme I).

Fe(V) is a highly reactive species (*16,17*), which is produced by the reduction of Fe(VI) by electrons at the conduction band (reaction 12) and the amino radical reduction of Fe(VI) (reaction 13). Recently, the formation of an intermediate Cr(V) in the heterogeneous photocatalytic reduction of Cr(VI)

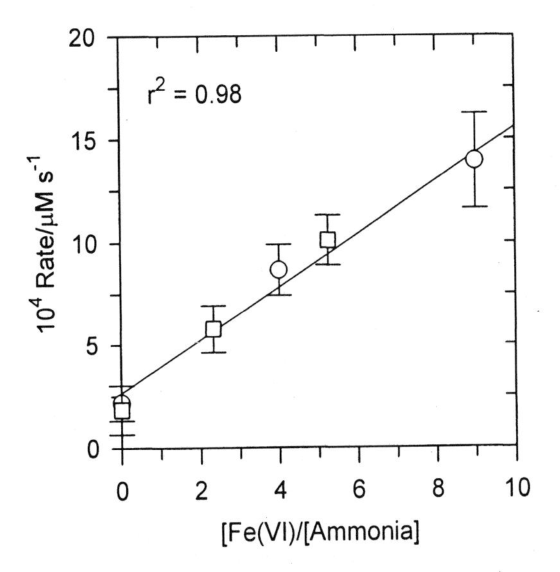

Figure 3. Rate of ammonia oxidation versus molar ratios of Fe(VI) to ammonia at pH 9.0. Initial ammonia concentration: ○ – 126 μM; □ – 215 μM (Reproduced from(14)).

Scheme I

Reduction of Fe Species

$$FeO_4^{2-} + e_{cb}^- \rightarrow FeO_4^{3-} \quad (12)$$

$$FeO_4^{2-} + NH_2^{\bullet} + OH^- \rightarrow FeO_4^{3-} + NH_2OH \quad (13)$$

$$FeO_4^{3-} + NH_3 + 3H_2O \rightarrow Fe(OH)_3 + NH_2OH + 3OH^- \quad (14)$$

$$2\ FeO_4^{3-} + NH_2OH + 4H_2O \rightarrow 2\ Fe(OH)_3 + NO_2^- + 5OH^- \quad (15)$$

$$FeO_4^{3-} + NO_2^- + 3\ H_2O \rightarrow Fe(OH)_3 + NO_3^- + 3\ OH^- \quad (16)$$

Oxidation of Nitrogen Species

$$OH^-_{ad} + h_{vb}^+ \rightarrow OH^{\bullet} \quad (17)$$

$$NH_3 + OH^{\bullet} \rightarrow NH_2^{\bullet} + H_2O \quad (18)$$

using TiO_2 in the presence of oxalate and ethylenediaminetetracetic acid (EDTA) was directly obtained by EPR spectroscopy (*26,27*). These results support the occurrence of reaction 11 to give Fe(V) in Scheme I. In homogeneous solutions, the rate constants for the Fe(VI) reduction by amino acid radicals (*28*) are of the order of 10^9 $M^{-1}s^{-1}$ and it is assumed that the rate constant for reaction 13 will be similarly fast. The amino radical in the proposed Scheme I results from the reaction of ammonia with a hydroxyl radical. The reported rate constant value of such a reaction is 1×10^8 $M^{-1}s^{-1}$ measured at pH = 11.4 (reaction 18) (*29*).

The oxidation of ammonia is enhanced in the presence of Fe(VI) because ammonia reacts with Fe(V) produced by reactions 12 and 13. The significant increase in the rate of photocatalytic oxidation of ammonia at high levels of Fe(VI) in the solution mixture suggests Fe(VI) is reacting with the amino radical to produce Fe(V), which is then oxidizing ammonia (reaction 14). Hence, the rate of NH_3 oxidation increases at higher Fe(VI) concentrations (Figure 3). The oxidation of ammonia by Fe(V) gives hydroxylamine (reaction 14). The possibility of forming hydroxylamine through a reaction, $NH_2^{\bullet} + OH^{\bullet} \rightarrow NH_2OH$, $k = 9.6 \times 10^9$ $M^{-1}s^{-1}$ (*30*), as an alternative to equation 13 is ruled out because of the much higher concentrations of Fe(VI) ($5.6 - 10.4 \times 10^{-4}$ M) than the steady state concentration of $OH^{\bullet}$ in the reaction mixture. Further reactions of Fe(V) with intermediates, hydroxylamine and nitrite, will result in nitrate (reactions 15 and 16). The formation of nitrate in the photocatalytic oxidation of ammonia has been observed by many workers (*31-34*).

Oxidation of Formic Acid

Initially, the photocatalytic reduction of iron(VI) was studied in titanium dioxide suspensions (0.066 gL^{-1}) at 970 μM formic acid concentrations. The results are given in Figure 4. The rates of Fe(VI) reduction increased with increasing Fe(VI) concentration. The initial rates of Fe(VI) reduction increased linearly with initial concentrations of Fe(VI). Interestingly, the initial reduction rates of Fe(VI) increase with increasing concentrations of formic acid (Figure 5). The initial rate versus Fe(VI) concentration plots was found linear ($r^2 = 0.99$) (Figure 5). A log-log plot of the initial rate of Fe(VI) reduction as a function of the ferrate(VI) concentration gave the order for the reaction as 0.70 ± 0.06. The rate constant, $k_{Fe(VI)}$ for reduction of Fe(VI) showed a positive linear relationship with formic acid, which can be written as $k_{Fe(VI)} = 2.41 \times 10^{-3} + 1.58 \times 10^{-7}$ ([Formic Acid]).

Next, the influence of Fe(VI) on the photocatalytic oxidation of formic acid was investigated (Figure 6). The rates of photocatalytic oxidation of formic acid followed the trend: $HCOOH+TiO_2+UV < Fe(VI)+HCOOH < HCOOH+Fe(VI)+TiO_2+UV$. The enhancement of the photocatalytic oxidation of formic acid can be explained by reactions 19-25.

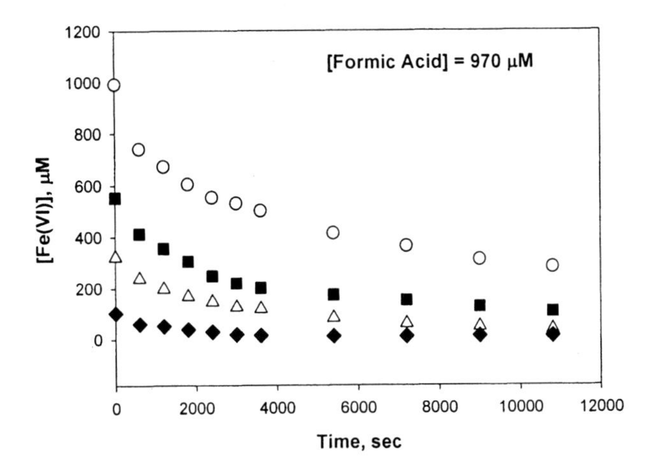

Figure 4. Photocatalytic reduction of Fe(VI) in suspensions containing 970 μM formic acid at pH 9.0. TiO_2 = 0.066 gL^{-1} and Intensity = 1.0 × 10^{-7} einstein s^{-1} (○-106 μM; ■-303 μM; △-529 μM; ◆-970 μM).

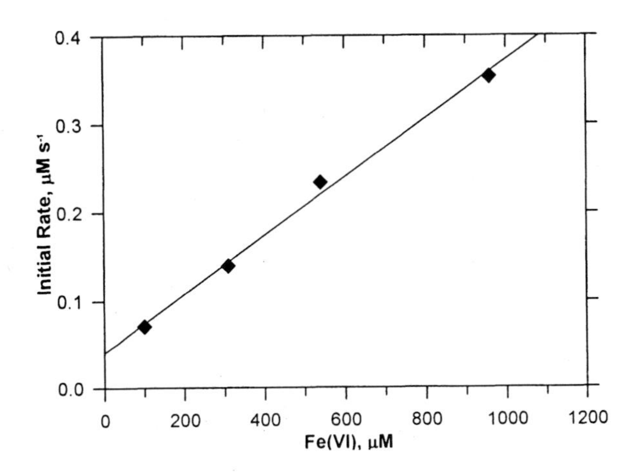

Figure 5. Initial rate of Fe(VI) reduction vs. Fe(VI) concentrations. [Formic Acid]=970 μM. A Plot was constructed from the data in Figure 4 and a line is the least squares fit.

Oxidation of Formic Acid

$$OH^-_{ad} + h_{vb}^+ \rightarrow OH^\bullet \quad (19)$$
$$HCOOH + OH^\bullet \rightarrow CO_2^{\bullet -} + H_2O + H^+ \quad (20)$$
$$HCOO^- + h_{vb}^+ \rightarrow HCOO^\bullet \quad (21)$$
$$HCOO^\bullet \rightarrow CO_2 + H^+ + e_{cb}^- \quad (22)$$

Reduction of Fe Species

$$FeO_4^{2-} + e_{cb}^- \rightarrow FeO_4^{3-} \quad (23)$$
$$FeO_4^{2-} + HCOO^\bullet + OH^- \rightarrow FeO_4^{3-} + CO_2 + H_2O \quad (24)$$
$$FeO_4^{2-} + CO_2^{\bullet -} \rightarrow FeO_4^{3-} + CO_2 \quad (25)$$

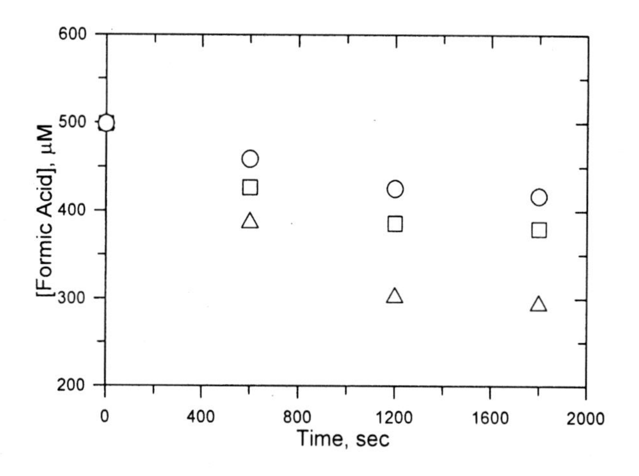

Figure 6. Photocatalytic oxidation of formic acid under different conditions (initial [HCOOH] = 1024 µM). Intensity = 1.0×10^{-7} einstein s^{-1}. ○ – no Fe(VI), UV, TiO_2 = 0.066 gL^{-1}; □ – Fe(VI) = 1000 µM, no UV, no TiO_2; △ – Fe(VI) = 500 µM, UV, TiO_2 = 0.066 gL^{-1}

The enhancement of the oxidation of formic acid in the presence of Fe(VI) can be explained by reactions 23 and 24 in which Fe(V), a highly reactive species that acts as a radical, is formed. The reduction of Fe(VI) to Fe(V) by $CO_2^{\bullet -}$ has also been studied independently (k ≈ 10^{10} M^{-1} s^{-1}) (Reaction 25) (*19*). The photocatalytic oxidation of formic acid on TiO_2 produces CO_2 without forming any long-lived intermediate (*35*), which suggests the possible occurrence of reactions 24 and 25.

Overall, results suggest the photocatalytic production of a highly reactive species, Fe(V), a powerful oxidant, to oxidize ammonia and formic acid. A combination of Fe(VI) and the TiO_2 photocatalyst thus has the potential to enhance the oxidation of pollutants in the aquatic environment.

References

1. Kato, H.; Asakura, K.; Kudo, A. Highly Efficient Water Splitting into H_2 and O_2 over Lanthanum-Doped $NaTaO_3$ Photocatalysts with High Crystallinity and Surface Nanostructure *J. Amer. Chem. Soc.* **2003**, *125*, 3082-3089.
2. Ollis, D.S., Al-Ekabi, H., Eds. *Photocatalytic Purification and Treatment of Water and Air.* Elsevier: Amsterdam, 1993.
3. Hoffmann, M.R.; Martin, S.T.; Choi, W.; Bahnemann, D.W. Environmental Applications of Semiconductor Photocatalysis *Environ. Sci. Technol.* **1995**, *95*, 69-96.
4. Prairie, M.R., Evans, L.R., Stange, B.M. and Martinez, S.L. An investigation of titanium dioxide photocatalysis for the treatment of water contaminated with metals and organic chemicals. *Environ. Sci. Technol.* **1993**, *27*, 1776-1782.
5. Fujishima A.; Rao T.N.; Tryk D.A. Titanium dioxide photocatalysis. *Journal of Photochemistry and Photobiology C: Photochemistry Reviews.* **2000**, *1*, 1-21.
6. Serpone, N.; Pelizzetti, E.; Hidka, H. Photocatalytic Purification and Treatment of Water and Air (edited by D.F. Ollis and H. Al-Ekabi, Elsevier, London, **1993**, p225.
7. Kramer DA. Nitrogen (fixed)–ammonia. U.S. Geological survey, mineral commodity summaries. 2004.
8. NTP. 1992. National Toxicology Program. NTP Technical Report Studies on Toxicity Studies of Formic Acid (CAS No. 64-18-6) Administered by Inhalation for F344/N Rats and $B6C3F_1$ Mice. NIH Publ 92-3342. NTP, Research Triangle Park, NC.
9. Selcuk, H.; Zaltner, W.; Sene, J. J.; Bekbolet, M.; Anderson, M. A. Photocatalytic and Photoelectrocatalytic Performance of 1% Pt Doped TiO2 for the Detoxification of Water. *J. Appl. Electrochem.* **2004**, *34(6)*, 653-658
10. Sharma, V.K. Use of iron(VI) and iron(V) in water and wastewater treatment. *Water Sci. Technol.* **2004**, *49*, 69-74.
11. Wood, R.H. The heat, free energy, and entropy of the ferrate(VI) ion. *J. Amer. Chem. Soc.* **1958**, *80*, 2038-2041.
12. Chenthamarakshan, C.R.; Rajeshwar, K.; Wolfrum, E.J. Heterogeneous Photocatalytic Reduction of Cr(VI) in UV-Irradiated Titania Suspensions: Effect of Protons, Ammonium Ions, and Other Interfacial Aspects. *Langmuir* **2000**, 16, 2715-2721.

13. Sharma, V.K., Burnett C.R., Rivera, W., Joshi, V.N. Heterogeneous photocatalytic reduction of ferrate(VI) in UV-irradiated titania suspensions. *Langmuir* **2001**, *17*, 4598-4601.
14. Sharma, V.K.; Winkelmann, K., Krasnova, Y., Lee, C., Sohn, M. Heterogeneous photocatalytic reduction of ferrate(VI) in UV-irradiated titania suspensions: role of enhancing destruction of nitrogen-containing pollutants. *Int. J. Photoenergy* **2002**, *5*, 183-190.
15. Sharma, V.K., Chenay, B.V.N. Heterogeneous photocatalytic reduction of Fe(VI) in UV-irradiated titania suspensions: Effect of ammonia. *J. Appl. Electrochem.* **2005**, *35*, 775-781.
16. Sharma, V.K.; Bielski, B.H.J. Reactivity of ferrate(VI) and ferrate(V) with amino acids. *Inorg. Chem.* **1991**, *30*, 4306-4311.
17. Sharma, V.K. Ferrate(V) oxidation of pollutants: a premix pulse radiolysis study. *Inorg. Chim. Acta*, **2002**, *65*, 349-355.
18. Thompson, G.W.; Ockerman, G.W.; Schreyer, J.M. Preparation and Purification of Potassium Ferrate. VI *J. Amer. Chem. Soc.* **1951**, *73*, 1379-1381.
19. Bielski, B.H.J., Thomas, M.J. Studies of hypervalent iron in aqueous solutions. 1. Radiation-induced reduction of iron(VI) to iron(V) by CO_2^- *J. Amer. Chem. Soc.* **1987**, *109*, 7761-7764.
20. Rush, J.D.; Bielski, B.H.J. Pulse radiolysis studies of alkaline iron(III) and iron(VI) solutions. Observation of transient iron complexes with intermediate oxidation states. *J. Amer. Chem. Soc.* **1986**, *108*, 5499-5501.
21. Sharma, V.K. Potassium ferrate(VI): an environmentally friendly oxidant. *Adv. Environ. Res.* **2002**, *6*, 143-156.
22. Jiang, J-Q.; Lloyd, B. Progress in the development and use of ferrate(VI) salt as an oxidant and coagulant for water and wastewater treatment. *Water Res.* **2002**, *36*, 1397-1408.
23. Sparks, D.L. Soil Physical Chemistry, second ed., CRC Press, Boca Raton, FL. 1997.
24. Gimenez, J.; Aguado, M.A.; Cervera-March, S. Comparison of TiO_2 powder suspensions and TiO_2 ceramic membranes supported on glass as photocatalytic systems in the reduction of chromium(VI). *J. Molecular Catal.* **1996**, *105*, 57-68.
25. Lyklema, J. Adsorption from Solutions at the Solid/Liquid Interface, (G.D. Parffit and C.H. Rochester, Eds.), Academic Press, New York, 1983.
26. Testa, J.J.; Grela, M.A.; Litter, M.I. Heterogeneous Photocatalytic Reduction of Chromium(VI) over TiO_2 Particles in the Presence of Oxalate: Involvement of Cr(V) Species. *Environ. Sci. Technol.* **2004**, *38*, 1589-1594.
27. Testa, J.J.; Grela, M.A.; Litter, M.I. Experimental Evidence in Favor of an Initial One-Electron-Transfer Process in the Heterogeneous Photocatalytic Reduction of Chromium(VI) over TiO_2. *Langmuir* **2001**, *17*, 3515-3517.

28. Bielski, B.H.J.; Sharma, V.K.; Czapski, G. Reactivity of ferrate(V) with carboxylic acids: a pre-mix pulse radiolysis study. *Rad. Phys. Chem.* **1994**, *44*, 479-484.
29. Alfassi, Z.B.; Hui, R.E.; Neta, P. In N-Centered Radicals; Wiley and Sons, New York, 1998.
30. Pagsberg, P.B. *RISO* **1972**, *256*, 209.
31. Zheng, G.Y.; Davies, J.A.; Edwards, J.G. *Recent Res. Devel. in Phys. Chem.* 1998, *2*, 1205.
32. Pollema, C.H.; Milosavljevic, E.B.; Hendrix, J.L.; Solujic, L.; Nelson, J.H. *Monatsh Chem.* Photocatalytic oxidation of aqueous ammonia (ammonium ion) to nitrite at TiO_2 particles. **1992**, *123*, 333-339.
33. Wang, A.H.; Edwards, J.G.; Davies, J.A. Photooxidation of aqueous ammonia with titania-based heterogeneous catalysts. *Solar Energy* **1994**, *52*, 459-466.
34. Hori, Y.; Bandoh, A.; Nakatsu, A. Electrochemical Investigation of Photocatalytic Oxidation of NO_2^- at TiO_2 (Anatase) in the Presence of O_2 *J. Electrochem. Soc.* **1990**, *137*, 1155-1161.
35. Muggli, D.S.; Backes, M.J. Two active sites for photocatalytic oxidation of formic acid on TiO_2: Effects of H_2O and temperature. *Journal of Catalysis.* **2002**, *209*, 105-113.

Chapter 22

Degradation of Dibutyl Phthalate in Aqueous Solution by a Combined Ferrate and Photocatalytic Oxidation Process

X. Z. Li[1,*], B. L. Yuan[1,2], and Nigel Graham[3]

[1]Department of Civil and Structural Engineering, The Hong Kong Polytechnic University, Hong Kong, People's Republic of China
[2]Department of Environmental Science and Engineering, Fuzhou University, Fuzhou, Fujian, 350002 People's Republic of China
[3]Department of Civil and Environmental Engineering, Imperial College London, South Kensington Campus, London SW7 2AZ, United Kingdom
[*]Corresponding author: email: cexzli@polyu.edu.hk; telephone: 852 2766 6016; fax: 852 2334 6389

The aim of the present work was to study the interaction of ferrate oxidation and photocatalytic oxidation in terms of dibutyl phthalate (DBP) degradation in aqueous solution, in which three sets of experiments were carried out, including (1) ferrate oxidation alone, (2) photocatalytic oxidation alone, and (3) the combination of ferrate oxidation and photocatalytic oxidation. Laboratory experiments demonstrated that ferrate oxidation and phtotocatalytic oxidation of DBP in aqueous solution are relatively slow processes. However, the presence of TiO_2 and ferrate together under UV illumination accelerated the DBP degradation significantly. Since ferrate was reduced quickly due to the presence of TiO_2 and UV irradiation, the DBP degradation reaction can be divided into two phases. During the first 30 min (Phase 1) the DBP was degraded by both photocatalytic oxidation and ferrate oxidation, and by interactive reactions. After 30 min (Phase 2), the ferrate residual had declined to a very low level and the photocatalytic reaction was the dominant mechanism of further DBP degradation. The influence of three main factors, ferrate dosage, TiO_2 dosage and pH, on the DBP degradation were investigated in order to understand the reaction mechanism and kinetics.

Introduction

The increasing use of "clean" reagents and photocatalytic methods in recent years has been shown to be effective for the degradation of specific organic pollutants and the general detoxification of natural and waste waters. Of current interest and concern is the presence of pharmaceutical and estrogenic compounds in effluent flows discharging into the environment.

Phthalate esters are known to be chemicals which have been linked to birth defects, organ damage, infertility and cancer. They are believed to be among the group of endocrine-disrupting compounds (*1-3*).

Dibutyl phthalate (DBP) as a representative phthalate ester is present mainly in packaging materials (papers, paperboards, etc.) for various foods (*4*). DBP is a hydrophobic and rather stable compound in the natural environment. The UV–VIS spectrum of DBP comprises two maxima, a weak one at 277 nm (ε_{277} = 1180l mol^{-1} cm^{-1}) and a more intense one at 227 nm (ε_{227} = 7700l mol^{-1} cm^{-1}), so it exhibits weak absorption of light at λ higher than 300 nm. Quite long photolysis half-lives of about 20 years (*5*) in natural waters and a very low degree of complete mineralization were observed after irradiation with artificial UV light, even at shorter wavelengths (*6*). The destruction of DBP may be achieved by photocatalytic techniques (*7*). Thus, the development of an efficient method for its removal from wastewater would be very important.

This paper considers the degradation of dibutyl phthalate (DBP), a highly toxic endocrine disrupter, by ferrate and a novel oxidation process involving a combination of ferrate and photocatalysis under UV_{365} irradiation. Of particular interest is the potential synergism in the ferrate/TiO_2/UV process via the generation of highly reactive Fe(V) species.

Experimental

Chemicals

Dibutyl phthalate (DBP) chemical was purchased from Aldrich (98.7% purity, CAS: 84-74-2). Potassium ferrate (K_2FeO_4) of high purity (>90%) was synthesized according to the method by Li *et al* (*8*). pH buffer solutions were prepared by different volumes mixed with 0.02M NaH_2PO_4/0.02M Na_2HPO_4 for pH 2-8, and 0.02M Na_2HPO_4 /0.002M $Na_2B_4O_7$ for pH 9-10. Titanium dioxide-TiO_2 (Degussa P-25) was used for the study, together with a UV lamp (8 W, Spectral output 365nm) with an irradiance of 0.40 mW/cm^2, as determined with a radiometer (Model UV365).

Photocatalysis

The initial concentration of DBP (5-7 mg L^{-1}) in aqueous solution was illuminated in the presence of ferrate (variable concentrations) and TiO_2 catalyst (40 mg/L) using a UV lamp at pH 9.0. All reactions were carried out in a quartz bottle with constant stirring. Samples were taken at different time intervals, quenched with sodium sulphite and were filtered through a Millipore (0.22 μm) membrane by syringe. DBP concentrations were determined by HPLC analysis.

Analytical Method

DBP concentrations were determined by using high performance liquid chromatography, with a high pressure pump (Spectrasystem HPLC P4000), a UV detector (UV 6000LP), and an auto sampler (AS3000). In the HPLC analysis, a pinnacle II C18 column (5 μm particle size, 250 mm × 4.6 mm i.d.) was employed and a mobile phase of acetonitrile/water (80:20, v/v) was used at a flow rate of 1.0 ml/min. An injection volume of 20 μl was used and the concentration of DBP was determined by the UV detector at 227 nm.

Results and Discussion

Preliminary Evaluation of the Ferrate/TiO_2/UV Systems

Three sets of experiments were carried out, including (1) ferrate oxidation alone, (2) photocatalytic oxidation alone, and (3) a combination of ferrate oxidation and photocatalytic oxidation as shown in Figure 1.

From Figure 1, it can be seen that the degradation of DBP in aqueous solution with ferrate alone occurred to a limited degree. A solution pH of 9 was used since ferrate is stable under these conditions over the reaction periods considered (≤120 min). Although ferrate is a very strong oxidant according to its standard potential (acid pH, E^0 = 2.20 V and basic pH, E^0 = 0.72 V), the degradation of DBP is still low only 21% degradation after 2 h reaction time. This might be because of the structural characteristics of DBP and the oxidative selectivity of ferrate.

Photocatalysis on semiconductors (TiO_2 in our case) is initiated by light absorption which promotes the excitation of valence band electrons into their conduction band. The electron deficiencies, or holes, left behind in the valence band can either be filled by adsorbed organic donors, which thus undergo direct oxidation, or react with water molecules or hydroxide anions, to form ·OH radicals. All these species can play relevant roles in the photocatalytic oxidation

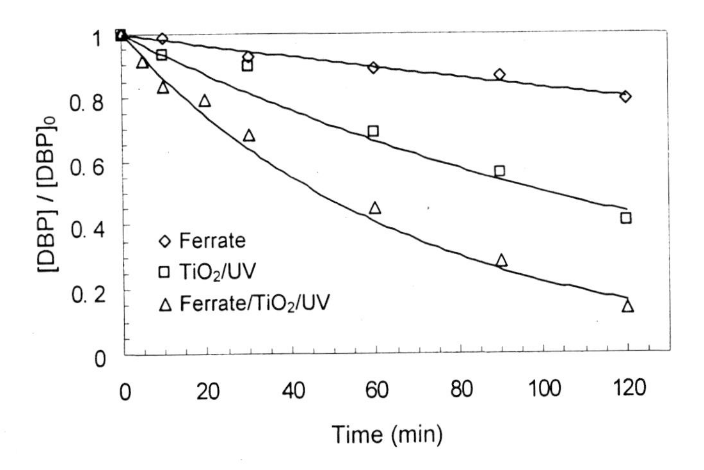

Figure 1. Degradation of DBP at pH 9.0 by ferrate/TiO_2/UV in different combinations.

of organic compounds such as DBP. From Figure 1 it can be seen that the degradation of DBP occurred in the solution with the TiO_2/UV photocatalytic process and the maximum degradation of DBP was approximately 60% after 120 min.

Finally, when photocatalysis was combined with ferrate, the overall degradation was found to be substantial (60% after 60 min) and slightly greater than the sum of the individual degradations by photocatalysis and ferrate as shown in Figure 1. This suggests the existence of a synergistic effect between the ferrate oxidation and photocatalysis. By combining the ferrate process and photocatalysis, the oxidation power of the TiO_2/UV system was significantly increased due mainly to the one-electron or two-electron reduction of Fe(VI) at the conduction band to Fe(V) (*9*) and Fe(IV), respectively; these iron species are known to be several orders of magnitude more reactive than Fe(VI) (*10*). These mechanisms are described as follows:

$$TiO_2 + h\nu \rightarrow h^+ + e_{cb}^-$$

$$Fe(VI) + e_{cb}^- \rightarrow Fe(V) \quad \text{or} \quad Fe(VI) + 2e_{cb}^- \rightarrow Fe(IV)$$

From these mechanisms, it can be seen that there are two beneficial effects of ferrate on photocatalyis. Firstly, as a sacrificial oxidant, ferrate in a UV/TiO_2

system can play the role of scavenging continuously the conduction band electrons (e_{cb}^{-}) generated on the surface of TiO_2 catalyst, which reduces the recombination of conduction band electrons and holes, and hence enhances the efficiency of the photocatalytic oxidation. Secondly, once Fe(VI) consumes the conduction band electrons (e_{cb}^{-}) on the TiO_2 surface, it is itself reduced to Fe(V) (and possibly Fe(IV)). Since Fe(V) can oxidize organics much faster than Fe(VI), the reduction of Fe(VI) to Fe(V) enhances the overall ferrate oxidation. Hence, this one-step reduction of Fe(VI) to Fe(V) on the TiO_2 surface plays an interactive, synergistic function, which is beneficial to both the photocatalytic oxidation and the ferrate oxidation simultaneously.

Effect of Ferrate concentration on DBP Degradation by the Different Oxidation Processes

The degradation of DBP by ferrate alone and the ferrate-photocatalytic system at different ferrate doses was investigated and the results are shown in Figures 2 and 3.

Figure 2 shows that ferrate can only achieve a minor degradion of DBP under these conditions. In these experiments, the original concentration of ferrate was changed from 0.04 to 0.24 mmol/L. With the increasing concentration of ferrate, the degradation of DBP was also found to increase systematically. However, the increasing effect of ferrate oxidation is not substantial. Overall the DBP degradation after a 2h reaction period was about 36% with 0.16mmol/L ferrate and only 46% with 0.24mmol/L ferrate. Therefore, it is concluded that the degradation of DBP by increasing ferrate dosage alone is not effective.

In order to enhance the degradation of DBP, we introduced the photocatalytic process to the ferrate oxidation, and the impact of the change in ferrate concentration on the photocatalytic oxidation of DBP was investigated. In these tests, as was the case with ferrate oxidation alone, the original concentration of ferrate was changed from 0.04 to 0.24 mmol/L, but the dosage of TiO_2 and the concentration of DBP were kept constant. The ferrate-photocatalysis-degradation of dibutyl phthalate with different original concentrations of ferrate are shown in Figure 3.

The results summarized in Figure 3 show that the degradation of DBP by ferrate oxidation can be enhanced substantially by combining it with TiO_2 under UV irradiation. By increasing the concentration of ferrate from 0.04 mmol/L to 0.08 mmol/L, the efficiency of DBP degradation also increased in the ferrate/TiO_2/UV system. This direct dose dependency is similar to that observed with ferrate oxidation alone. However, the maximum degradation after 2 h of 86% was observed at a ferrate concentration of approx 0.08 mmol/L. At the higher ferrate doses of 0.16 mmol/L and 0.24 mmol/L, the DBP degradation was

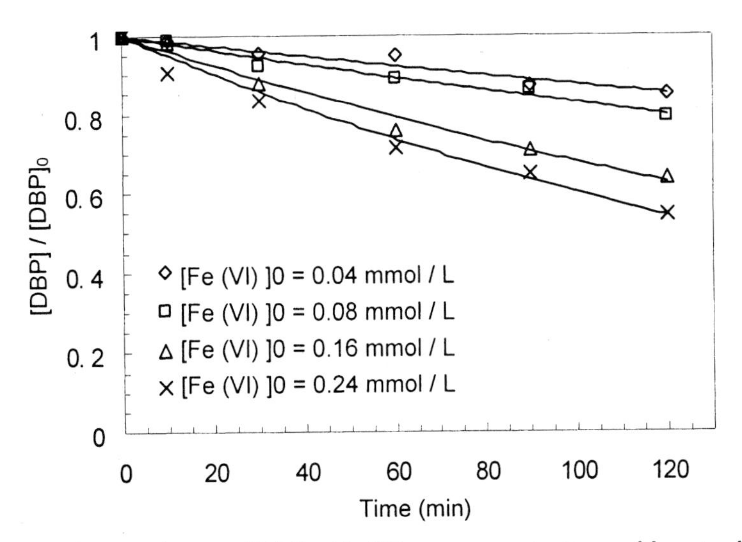

Figure 2. Degradation of DBP with different concentrations of ferrate alone (pH 9.0)

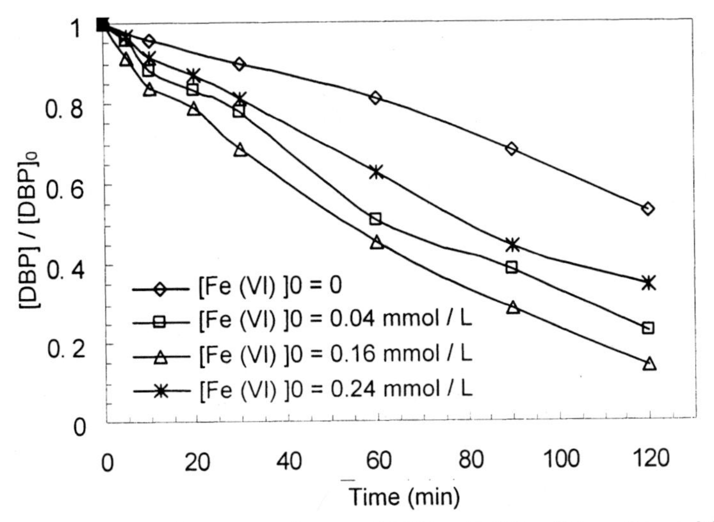

Figure 3. Photocatalytic degradation of DBP with TiO_2 (20 mg/L) at different concentrations of ferrate.

observed to decrease. It is speculated that this may have been the result of a reduction of the local UV intensity caused by its absorption by the higher ferrate concentration. Therefore, there is an optimal dosage of ferrate with TiO_2/UV system in which lower ferrate doses are not as effective in enhancing the TiO_2/UV system, and higher ferrate doses are deleterious to the adsorption of UV light.

Effect of TiO_2 Dosage on DBP Degradation in the Different Oxidation Processes

In order to define the optimum dosage of TiO_2 and further elucidate the effect of TiO_2 dosage on the relationship between the rate of ferrate decomposition and the degradation of DBP by ferrate/TiO_2/UV, different mass ratios of ferrate and TiO_2 in the ferrate-photocatalysis degradation of DBP were investigated. Figure 4 shows the efficiency of DBP degradation at different mass ratios of ferrate and TiO_2 (2:1, 1:1 and 1:2) by using three specific ferrate dosages (20 mg/L, 40 mg/L and 60 mg/L), and for each ferrate dosage varying the different TiO_2 dose under UV light irradiation.

Since ferrate was reduced quickly due to the presence of TiO_2 and UV irradiation, the DBP degradation reaction can be divided into two phases as shown in Figure 4 (a-c). During the first 30 min period (Phase 1), as the ferrate concentration declines, the DBP was degraded by both photocatalytic oxidation and ferrate oxidation, and some interactive reactions might be involved as previously described. After the initial 30 min (Phase 2), the ferrate residual has declined to a very low level and under these consitions the photocatalytic reaction is the principal process in the DBP degradation.

For the same initial concentration of ferrate, the relative dosage of TiO_2 directly affects the rate of ferrate reduction as shown in Figure 4 (d-f). Thus, it can be seen that the rate of ferrate reduction increased with the TiO_2 dosage in each case. The principal reason for this is speculated to be the loss in absorption of visible light at 505 nm (by interference) which reflects the concentration of ferrate in solution (*11*).

The maximum degradation rate of DBP under these experimental conditions occurred at the FeO_4^{2-} : TiO_2 mass ratio of 1:1 and showed almost the same removal extent of DBP after 2h reaction, whether the dosage of ferrate was 20 mg/L, 40 mg/L or 60 mg/L. Consequently it can be concluded that neither the dosage of ferrate nor TiO_2 is the dominant factor in the ferrate/TiO_2/UV system and employing a mass ratio of ferrate and TiO_2 in the range investigated is the most important factor. At the optimum FeO_4^{2-} : TiO_2 mass ratio (1:1), ferrate is reduced moderately by TiO_2/UV, thereby assisting the overall oxidation and possibly providing a synergistic effect.

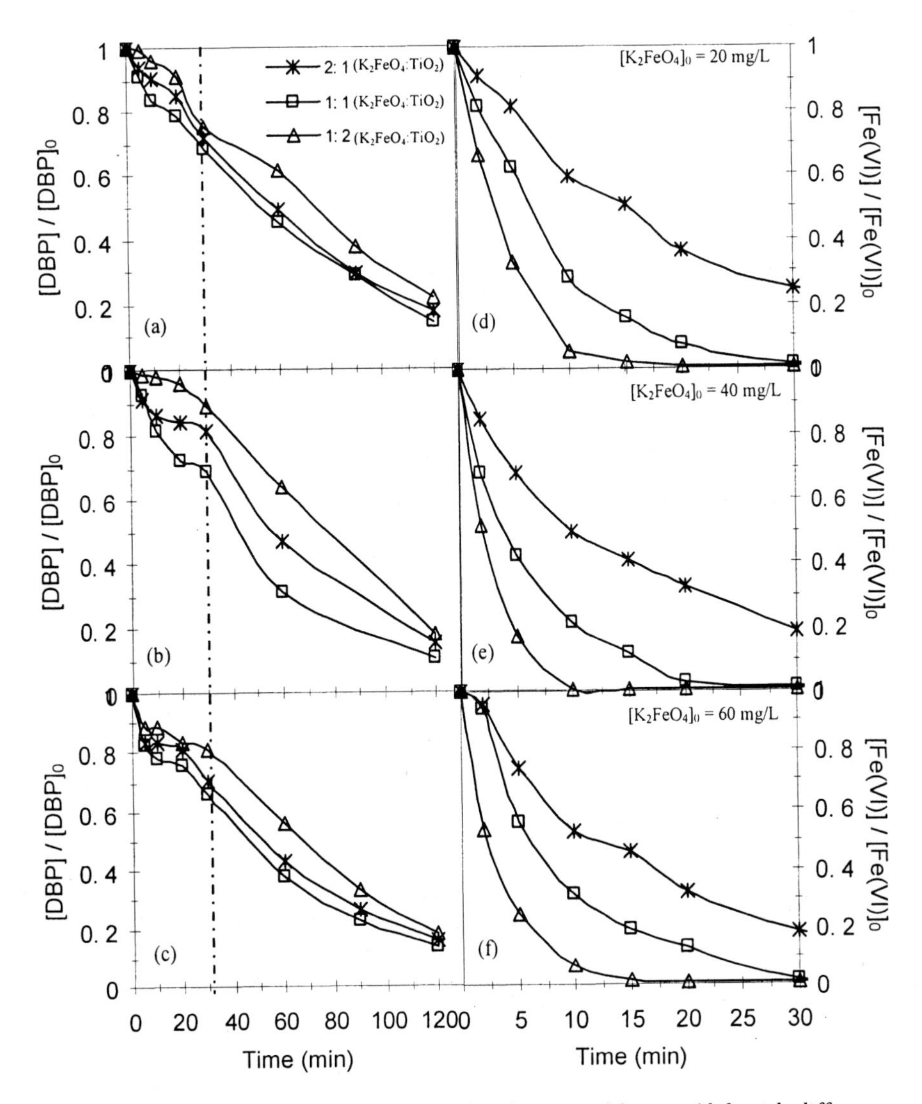

Figure 4. Degradation of DBP (a-c) and reduction of ferrate (d-f) with different weight ratios of K_2FeO_4:TiO_2 (pH 9.0)

In such a complicated reaction system it is difficult to identify the dominant phenomena in order to explain the results obtained. It can be speculated that the rapid reduction of ferrate means the the one-electron or two-electron reduction products, Fe(V) or Fe(IV), are formed and consumed indiscriminately in a very short time period, owing to their high reactivity, thereby having a limited opportunity for reaction with DBP. Another aspect is that the presence of ferrate's color during the initial period of the reaction is also deleterious for the absorption of UV light on the TiO_2 surface, which is necessary to excite the valence band electrons into their conduction band. The electron deficiencies, or holes, left behind in the valence band can either be filled by adsorbed organic donors, which thus undergo direct oxidation, or react with water molecules or hydroxide anions, to form ·OH radicals. This may possibly explain any synergism between the ferrate and photocatalysis during the compound reaction.

During Phase 2 of the reaction (t >30 min), it can be assumed that very little ferrate remains and therefore the photocatalytic oxidation of DBP is the principal process. The results indicate that the extent of the degradation of DBP after a 2h reaction time was the same, irrespective of the TiO_2 dosage, and that the DBP concentration was different at the beginning of Phase 2. Thus, for prolonged UV irradiation, DBP and its product compounds formed in Phase 1, are mostly mineralized due to the continuous formation of ·OH radicals which are known to be very reactive species and can react indiscriminately with different organic compounds.

Effect of pH on the Ferrate-Photocatalytic Degradation of Dibutyl Phthalate

The influence of pH on the ferrate-photocatalytic degradation of DBP was investigated by carrying out tests at pH 6.0, 7.0, 8.0, 9.0 and 10.0, where the pH was controlled by buffer solutions; the results are shown in Figure 5.

The solution pH can affect the oxidation efficiency of organic substances through directly influencing the form and decomposition rate of ferrate and the efficiency of photocatalysis. Previous studies have described the existence of four Fe(VI) species over a wide pH range, comprising three protonated forms, $H_3FeO_4^+$, H_2FeO_4, $HFeO_4^-$, and the de-protonated anion FeO_4^{2-} (*12*). The dissociation constants for protonated Fe(VI) species are $pK_1 = 1.6 \pm 0.2$, $pK_2 = 3.5$, and $pK_3 = 7.3 \pm 0.1$ (*12*).

At pH 6 and 7, the TiO_2/UV photocatalytic oxidation is believed to be the dominant oxidation process due to the rapid decomposition of ferrate. Thus, the extent of DBP degaradation during the first 30 minutes (Phase 1) is relatively small.

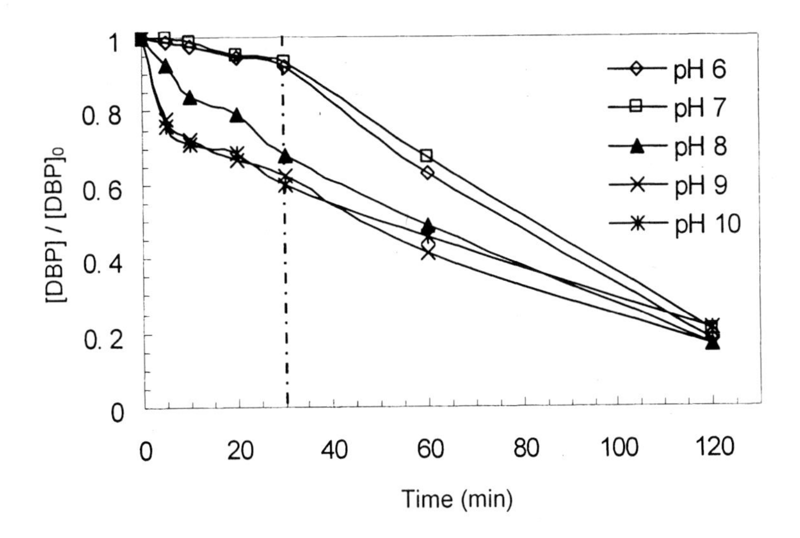

Figure 5. Influence of pH on DBP degradation by ferrate/TiO_2/UV (0.16 mmol/L ferrate and 40 mg/L TiO_2)

With increasing pH (>7), the ferrate is more stable and this leads to more rapid degradation of DBP, particularly at pH 9 and 10, where the degradation was mostly completed by about 5-10min. Associated with the greater stability of the ferrate is the speculated photoreduction of ferrate via the one-electron or two-electron transfer of Fe(VI) to Fe(V), or Fe(IV), repectively which are known to be several orders of magnitude more reactive than Fe(VI) (*10*). These processes will enable Fe(V) and Fe(IV) to contribute to the reaction with DBP and to form intermediate byproducts.

However, whilst the concentration of FeO_4^{2-} species may be higher at pH 10 than pH 9, the benefit of this in terms of the photoreduction of ferrate, may be offset by the increase in electrostatic repulsion between the negatively charged TiO_2 surface and the more negative FeO_4^{2-} species at these high pH values. Hence, there will be a balance between the benefits of greater ferrate stability and reduced Fe(VI) photo-reduction at TiO_2 surfaces and less Fe(V). This will have the effect of moderating the increased DBP degradation rate.

After the ferrate reduction in first 30 minutes, the photocatalytic oxidation of DBP is believed to be the principal process in the following 90 minutes (Phase 2). Neutral and acidic solutions will enhance the photocatalytic process because of the positively charged TiO_2 surface under these conditions. Good TiO_2 surface interaction with DBP can be inferred from the relative magnitude of Coulombic forces on the basis of negative point charge calculations. The atom bearing the highest negative point charge in DBP is the carbonyl oxygen, which of consequence is the atom that interacts the most with the positive TiO_2 surface. Thus, free surface interaction between DBP molecules and surface electron holes will lead to more rapid photo-oxidation than that in alkaline conditions (pH 8-10).

The overall effect on DBP degradation over the two phases of reaction (1 and 2), after a 2h reaction time, is that the degradation of DBP is not much different for the different pH conditions. In order to investigate the possibility of a synergistic effect between ferrate and photocatalysis, the optimum pH of 9 was selected.

Photocatalytic Degradation of DBP by Different Electron Acceptors

Some researchers have shown that the addition of metal ions to the photocatalytic system can act as electron acceptors to prevent the immediate recombination of electrons (e^-) and valence band holes (h^+), thereby enhancing the quantum efficiency and the rates of pollutant degradation (*13,14*). Other researchers have shown that the addition of chemical oxidants such as O_2, H_2O_2 or O_3 to the photocatalytic system may also act as electron acceptors resulting in enhanced rates of pollutant destruction (*15*). However, little study has been carried out on the effectiveness of ferrate as a combined oxidant and electron scavenger to increase the overall photocatalytic degradation of organic compounds in general, or of DBP in particular. In this study, a comparative evaluation of the photocatalytic degradation of DBP was carried out in the presence of enriched oxygen gas (90 vol.% of O_2, DO $\approx$ ~0.63 mmol/L (20 mg/l), hydroxyl peroxide (0.16mmol/L) and ferrate (0.16 mmol/L). An example of the results obtained are shown in Figure 6.

It is evident from Figure 6 that, compared to the case where there is no additional electron acceptor present (by bubbling N_2 in solution), the efficiency of DBP degradation increased with the addition of different chemical oxidants during the 2h reaction time. However, the extent of degradation of DBP in the presence of H_2O_2 was not as that expected from information in the literature (*12*). One reason for this is that in alkaline conditions, H_2O_2 becomes unstable and easily and rapidly decays into water and oxygen; another reason is that the concentration of H_2O_2 was relatively lower than reported studies in the literature in order to be comparable with the ferrate concentrations used. Figure 6 shows that high purity oxygen gas is an active electron acceptor, where equilibration with an oxygen-enriched gas is expected to decrease the recombination of holes and electrons on the photocatalyst surface.

The combined role of ferrate as both strong oxidant and electron acceptor corresponds to ferrate having the greatest oxidation effect compared to the other

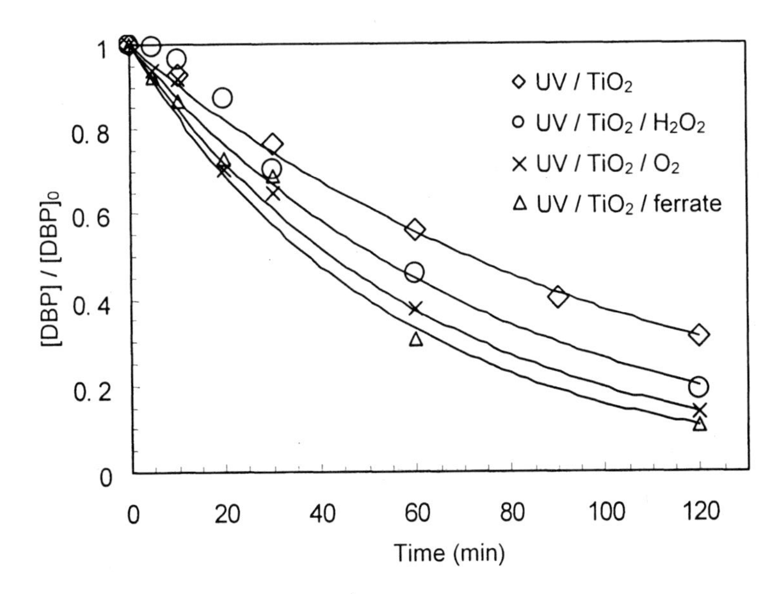

Figure 6. Photocatalytic degradation of DBP in the presence of different electron acceptors (0.16 mmol/L H_2O_2 and ferrate, 20 mg/L TiO_2 and pH 9)

chemicals (and also with permanganate and nitrate). Ferrate can act as an alternative electron acceptor to oxygen which is a thermodynamically more favorable reaction compared to oxygen reduction (*16*) (E^0 = −0.13 V for O_2 reduction, E^0 = 0.72 V for Fe(VI) reduction). The reduction of Fe(VI) by Ve^-_{cb} (-0.7 V) will take place through one-electron or two-electron steps that would result in the sequential formation of Fe(V) and Fe(IV) as iron species (electron acceptors) to enhance the photocatalytic efficiency. In addition Fe(V) and Fe(IV) have been reported to be 10^3-10^5 times more reactive with certain compounds than Fe(VI) (*17*).

Conclusions

- A substantial degradation (~80%) of DBP in aqueous solution can be accomplished by combined ferrate-photocatalysis in 120 min under our experimental conditions.
- The degradation process can be considered to occur in two sequential phases: Phase 1 (up to 30 min) of combined ferrate-photocatalysis and Phase 2 (t >30 min) of principally photocatalysis.

- The DBP degradation by ferrate-photocatalysis (Phase 1) is highly dependent on the solution pH and the greatest degradation occurred at pH 9.
- The relative concentrations of ferrate and TiO_2 influence the DBP degradation in the early stages of the reaction and an optimal mass ratio of ferrate:TiO_2 = 1:1 was evident.
- For a given TiO_2 concentration, there was an optimal ferrate concentration; above this concentration UV absorbance by ferrate is counterproductive.
- Ferrate appears to be an effective electron acceptor for the UV/TiO_2 photocatalysis; this enhances DBP degradation inherently by reducing electron-hole recombination, but also through the generation of highly reactive Fe(V) and Fe(IV) species.

Acknowledgements

The authors thank the Research Grant Committee of The Hong Kong Government for financial support of this work (RGC No: PolyU 5170/04E).

References

1. Yoshida T., Tanabe T., Miyashita Y., *et al. Chem. Lett.* **2001**, *9*, 876-877.
2. Ohtani H., Miura I., Ichikawa Y. *Environ. Health Perspect.* **2000**, *108*, 1189-1193.
3. Peters J.M., Taubeneck M.V., Keen C.L., *et al. Teratology* **1997**, *56*, 311-316.
4. Barlow N., Phillips S., Wallace D. *Toxicol. Sci.* **2003**, *73*, 431-441.
5. Staples C.A., Peterson, R.D., Parkerton T.F., *et al. Chemosphere* **1997**, *35*, 667-749.
6. Lau T.K., Chu W., Graham N. *Chemosphere* **2005**, *60*, 1045-1053.
7. Oliver B., Gilles M., Mich`ele B. *Appl. Catal. B: Environ.* **2001**, *33*, 239-248.
8. Li C., Li X.Z., Graham N. *Chemosphere* **2005**, *61*, 537-543.
9. Lee Y., Cho M., Kim J., Yoon J. *J. Ind. Eng. Chem.* **2004**, *10*, 161-171.
10. Sharma V. K., O'Connor D. B., Cabelli D. E. *J. Phys. Chem. B* **2001**, *105*, 11529-11532.
11. Wong C. C., Chu W. *Environ. Sci. Technol.* **2003**, *37*, 2310-2319.
12. Wood R. H. *J. Am. Chem. Soc.* **1958**, *80*, 2038-2044.

13. Sýkora J., Pado M., Tatarko M., Izakovič M., *J. Photochem. Photobiol. A: Chemistry* **1997**, *110*, 167-174.
14. Litter I. M. *Appl. Catal. B: Environ.* **1999**, *23*, 89-97.
15. Wong C. C., Chu W. *Environ. Sci. Technol.* **2003**, *37*, 2310-2319.
16. Wang Y., Hong C. S. *Water Res.* **1999**, *33*, 2031-2039.
17. Sharma V. K., Bielski B. H. J. *Inorg. Chem.* **1991**, *30*, 4306-4313.

Chapter 23

Preparation and Properties of Encapsulated Potassium Ferrate for Oxidative Remediation of Trichloroethylene Contaminated Groundwater

B. L. Yuan*, M. R. Ye, and H. C. Lan

Department of Environmental Science and Engineering, Fuzhou University, Fuzhou, Fujian, 350002 People's Republic of China
***Corresponding author: yuanbaoling@yahoo.com.cn; telephone: 86 591 8375 4771; fax: 86 591 2286 3070**

The oxidative degradation of chlorinated ethylenes with a metal-oxo reagent, potassium ferrate (K_2FeO_4), is investigated in this paper. The encapsulated K_2FeO_4 was prepared by a molten suspension and cooling method, with paraffin wax and K_2FeO_4 solid as the encapsulating wall and encapsulated core, respectively. The effects of three main factors, stirring time, ultrasonic time and temperature of the water bath, on the preparation of encapsulated ferrate were studied. The IR and XRD spectra showed that ferrate was stable after encapsulation. Experiments also showed that the release process of K_2FeO_4 from the encapsulated particles includes two phases: (1) the uncoated ferrate was immediately dissolved (<1min); (2) the encapsulated ferrate was sustained to release in 108 h. The release kinetics were in good agreement with the $Q = kt^n$ model. A well-defined change in the surface morphology before and after release reaction was observed by SEM. A higher amount of stable ferrate was recovered from the encapsulated paraffin when compared to pure ferrate itself. A primary research showed that trichloroethylene (TCE) degradation could be more effective in acidic medium since above 90% of the TCE was degraded after a reaction time of 60 mins in solutions with pH = 4.0 - 6.0, compared to only 60% TCE degradation after 150 mins in solutions with pH = 10.

Introduction

The common occurrence of trichloroethylene (TCE) in groundwater is largely due to its extensive use by industry and its resistance to both biotic and abiotic degradation under natural subsurface conditions. In addition to its wide spread occurrence, TCE has a drinking water limit of 0.005 mg/L (*1*) contributing further to the environmental concern for this compound. As a result, considerable research has been focused on the development of technologies for removal of TCE, as well as other chlorinated hydrocarbon compounds, from groundwater. Technologies that are suitable for removal or destruction of chlorinated hydrocarbons, occurring as dense non-aqueous phase liquids (DNAPLs) in source zones, include in-situ chemical oxidation (ISCO) and in-situ reduction. Most work had concentrated on reductive dechlorination, in particular, remedial schemes developed around zero-valent iron have shown considerable promise due to the relatively short half lives of the reactions (*2-3*).

There has been much recent work on oxidation processes even though the half life of oxidative degradation for chlorinated organic compounds is of the order of several minutes in Fenton's reagent, ozone, and O_3/H_2O_2 systems (*4-6*). An apparent limitation with this reaction is that the key reactive intermediate hydroxyl radical generated in this advanced oxidation process (AOP) strongly reacts with common inorganic species in ground water such as carbonate and bicarbonate (*7*).

Not all oxidants suffer from this limitation. A metal-oxo reagent, such as permanganate (*8-11*), can attack a double carbon–carbon bond (*12*) powerfully through direct oxygen transfer (*13*). This feature of metal-oxo reagents facilitates the degradation of chlorinated ethylenes with little scavenging by carbonate or bicarbonate.

Ferrate, the salt of the iron (VI) oxyanion FeO_4^{2-}, is a more powerful oxidant than permanganate throughout the whole pH scale, and decomposes in aqueous solution generating Fe^{3+}, hydroxide ions and molecular oxygen. As a metal-oxo reagent, ferrate may be useful to eliminate TCE through selectivity toward the alkene functional group (C=C) in chlorinated alkenes. However, in order to avoid the quick self-decomposition of ferrate in aqueous solution and increase the oxidative efficiency with target compounds, the controlled release techniques were realized by encapsulating solid potassium ferrate (*14*).

The purpose of the present work was to prepare and characterize the encapsulated K_2FeO_4 particles with commercially available paraffin wax as a coating material and matrix, which will not only ensure the stability of ferrate but also the sustained release of ferrate to oxidize hydrophobic organic contaminants in groundwater and soil. A further purpose was to evaluate the release characteristics from the coated K_2FeO_4 particles into aqueous solutions and degradation efficiency of the organic contaminant TCE.

Experimental

Chemicals

Standard trichloroethylene (TCE) was purchased from Aldrich. Potassium ferrate (K_2FeO_4) of high purity (>90%) was synthesized according to the Li's method (*15*). Section paraffin wax with a melting (congealing) range of 52-54°C (Sigma) was used as received.

Encapsulation

Solid paraffin wax was introduced into a 3-neck-flask which was heated in the water-bath. The temperature of the water-bath was maintained at 80°C. The K_2FeO_4 solid was added to the molten wax with continuous stirring and ultrasonic shaking to ensure uniform dispersion of the K_2FeO_4 into the wax phase for 30 mins. The mass ratios of added solid paraffin wax : K_2FeO_4 solid were 10:1, 5:1 and 3:1 respectively. The molten suspension was rapidly transferred into ice water (the mixture of water and ice to keep the temperature at 0°C) by using a burette. The molten suspension was rapidly dropped into the cool pure water. The encapsulated potassium ferrate was collected after desiccation. The particles were formed by milling and stored in the vacuum desiccator.

Loading efficiency

Ferrate-encapsulated paraffin (0.1000 g) were dissolved in 100 ml volumetric flask containing 20 ml of cyclohexane. The flask was placed on a rotary wheel shaker for 10 min at 60°C in the water-bath to ensure complete dissolution of the paraffin wax. 6 M KOH solution was added to extract the K_2FeO_4. The aqueous and organic phases were vigorously mixed for 1 h to make sure that the organic and aqueous phases were separated. The ferrate content was assayed by measuring the absorbance at 510 nm using a UV spectrophotometer (VIS-7220 visible spectrometer, Ruili Analytical Instrument Corporation, Beijing). The molar absorptivity at 510 nm has been determined as 1150 M^{-1} cm^{-1} by Bielski and Thomas in 1987 (*16*). Experiments were performed in triplicate (n = 3) and loading efficiencies were calculated using equation (1).

$$\text{Loading efficiency(\%)} = \frac{\text{Calculated ferrate concentration}}{\text{Theoretical ferrate concentration}} \times 100 \qquad (1)$$

Release Experiments

Because of the instability of the K_2FeO_4 solution in acidic even netural environment, the 6 M KOH solution was chosen as the reagent to extract the K_2FeO_4. The ferrate-encapsulated paraffin samples (0.55, 0.30, and 0.20 g for the mass ratios of 10:1, 5:1, and 3:1 paraffin : K_2FeO_4, respectively) were introduced into 200 mL of 6 M KOH at room temperature under constant stirring (200 rpm) with a mechanical stirrer to ensure a well-mixed solution. The aqueous solution were withdrawn at different intervals every a few minutes and then the concentration of ferrate in KOH solution was measured with a UV-Visible spectrophotometer at 510 nm. Experiments were performed in triplicate (n = 3) and releasing efficiencies were calculated using equation (2).

$$\text{Releasing efficiency}(\%) = \frac{\text{Released ferrate concentration}}{\text{Loaded ferrate concentration}} \times 100 \quad (2)$$

Stability study

The ferrate levels in encapsulated paraffin were monitored periodically. The stable ferrate present in encapsulated paraffin was extracted by the method described above. The stable ferrate present in encapsulated paraffin and pure ferrate in 6M KOH were determined by UV spectroscopic method at 510 nm. The analysis were carried out after exposed in air for 1 d, 3 d, 5 d, 7 d, 10 d, 15 d, 20 d, 25 d, 30 d and 40 d.

TCE oxidation by encapsulated ferrate

The initial concentration of TCE (6-7 mg/L) in an aqueous solution was oxidized by variable amounts of ferrate at different pH. All reactions were carried out in a series of sealed glass volumetric flasks with constant stirring. Samples were taken at different time intervals and quenched with sodium sulphite to stop ferrate oxidation of TCE. Analyses for dissolved TCE were conducted using a GC-ECD (HP 6890 gas chromatograph, Hewlett-Packard Co.). Samples of the headspace in the sample vials were injected automatically with an injection volume of 10 μL. A capillary column (HP-5MS, 30 m × 0.25 mm × 0.25 μm) was used with nitrogen carrier gas at a flow rate of 1.0 mL/min. The oven temperature was programmed from 60°C (3 mins) to 240°C(15 mins) at a ramp rate of 10°C / min, and then to 280°C (3 mins) at ramp rate of 15°C/min. The injection and detector temperature were 250 and 280°C, respectively.

Instrumental Analysis

The surface morphology of ferrate-encapsulated paraffin was examined by means of an XL30 scanning electron microscope (SEM, Philips, The Netherlands); FTIR spectra of pure ferrate, paraffin and ferrate-encapsulated paraffin were obtained by using a AVAT-AR360 Fourier Transform Infrared Spectrometer (Nicolet corporation, USA); X-ray powder diffraction patterns of pure ferrate, paraffin and ferrate-encapsulated paraffin were obtained at room temperature using a Philips X'Pert MPD diffractometer (Philips, The Netherlands).

Results and Discussion

Preparation of ferrate encapsulated paraffin

A molten suspension and cooling (MSC) method according to Kang (*14*) was employed for the encapsulation of K_2FeO_4 (the oxidant) with a paraffin wax as the coating material (matrix). Stirring and ultrasonic methods were used to ensure uniform dispersion of the K_2FeO_4 in the paraffin phase. The speed of the stirring and the duration of the ultrasonic method had a significant impact on the effectiveness of the encapsulation.

The effect of stirring speed on the loading efficiency is presented in Figure 1(A). The loading efficiency increases with stirring speed. For example, the loading efficiency doubles as the stirring rate is increased from 400 rpm to 500 rpm. When the speed increases to 800 rpm, the efficiency reaches 44.2%.

The effect of ultrasonication time on the loading efficiency is presented in Figure 1(B). The loading efficiency increases with prolonging the ultrasonication time from 5 to 35 mins and afterwards the loading efficiency remains constant at approximately 40%. Neither mechanical stirring or ultrasonication can yield adequate loading efficiency, through combining mechanical mix (800 rpm) with ultrasonication (30 mins) to disperse ferrate into melten paraffin results in a high loading efficiency (85.5%). This combined method of dispersion was used for subsequent encapsulated ferrate samples described below.

Fourier transform infrared spectroscopy

FTIR spectra of pure ferrate, paraffin and ferrate-encapsulated paraffin are presented in Figure 2. The characteristic peaks of pure K_2FeO_4 crystals were obtained at 807, 1106, 1383, 1569, 2346 and 3433 cm^{-1}. The peaks at 807 and 1106 cm^{-1} are assigned to stretching and bending vibrations of Fe-O bond $v_3(F_2)$ and Fe-O bond $v_1(A_1)$ in the K_2FeO_4 crystals, respectively (*17*). The spectra of ferrate-encapsulated paraffin also shows these characteristic peaks, confirming the stability of K_2FeO_4 during its encapsulation into paraffin.

X-ray diffraction

The XRD studies help to understand the nature of core material (crystalline or amorphous) in the polymeric matrix (*18*). The X-ray diffraction spectra of pure K_2FeO_4 crystals, paraffin and ferrate-encapsulated paraffin are presented in Figure 3. From Figure 3(A), it is evident that pure K_2FeO_4 crystals exhibited characteristic crystalline peaks at 2θ of 17.20°, 20.79°, 23.17°, 25.70° and 30.17°,

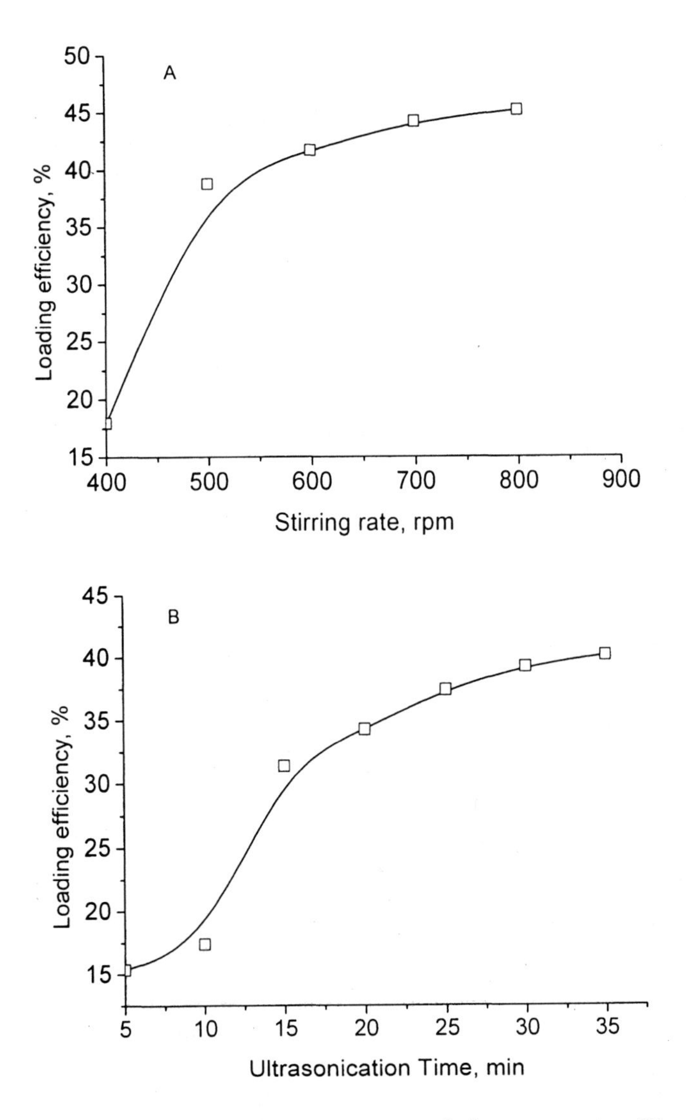

Figure 1. Effect of mechanical mixing (A) and ultrasoniccation (B) on loading efficiency.

indicating the presence of crystalline K_2FeO_4 (*15*). The characteristic crystalline peaks of K_2FeO_4 were little changed after encapsulation in paraffin (Figure 3(B)). This indicates that the encapsulation of K_2FeO_4 in paraffin is only a physical process, and, hence, the reactivity and stability of K_2FeO_4 crystals are not affected even after the ferrate is encapsulated within paraffin.

Release study

The release profile of ferrate from encapsulated paraffin is shown in Figure 4, along with pure ferrate itself. The release studies of ferrate from encapsulated paraffin were carried out in a 6 M KOH solution. The dissolution of pure ferrate in 6 M KOH was rapid and 100% of ferrate dissoluted within several seconds. This is due to the high solubility of ferrate in the aqueous medium. However, the release process of ferrate from encapsulated paraffin includes two phases: (1) dissolution of the uncoated ferrate was immediate (<1 min); (2) dissolution of the encapsulated ferrate was significantly retarded: 15-50% of the K_2FeO_4 was released in 108 h. The ferrate that dissolved in phase 1 is not suitable for controlled release. Figure 6 also shows the influence of mass ratios of paraffin : K_2FeO_4 on the release rate of ferrate. It can be seen that the release rate of ferrate from ferrate-encapsulated paraffin increased with the decreasing mass ratio of paraffin : K_2FeO_4. The increased releasing rate of ferrate from encapsulated paraffine could be due to increased ferrate dosage of encapsulated paraffin.

To analyse the release mechanism of ferrate from encapsulated paraffin, the release data obtained were fit to the Sinclair and Peppas equation (*3*):

$$Q = kt^n \qquad (3)$$

where Q is the amount of ferrate released at time *t*, *k* is a rate constant incorporating the characteristics of the ferrate and paraffin matrix system, and *n* is the diffusional exponent.

This equation indicates that the release of ferrate from encapsulated paraffin did not follow Fick's law of diffusion with constant pseudoconvection due to the stress within the paraffin matrix (*19*). The rapid dissolution of K_2FeO_4 particles on the paraffin surface leaves crevices and pores. It is hypothesized that these pores may possibly generate inter-particulate openings or channels within the paraffin matrix. Consequently, the aqueous phase may diffuse in and contact the remaining K_2FeO_4 in the matrix.

The release mechanism was also studied by scanning electron microscopy. Figure 5(A) shows a micrograph of uncoated or incompletely coated K_2FeO_4 particles which are embedded into or near the surface of encapsulated paraffin. These particles are readily accessible to water when the products were introduced into solution.

After a period of continuous contact with 6 M KOH, empty crevices appeared on the surface of the paraffin medium (Figure 5(B)). KOH solution can penetrate into the interior of the matrix through these crevices so that additional K_2FeO_4 further released.

Stability study

The amount of stable ferrate (%) remaining in unencapsulated ferrate and encapsulated paraffin *vs.* time at room temperature is shown in Figure 6. The stable ferrate remaining in pure ferrate sample decreases rapidly and less than

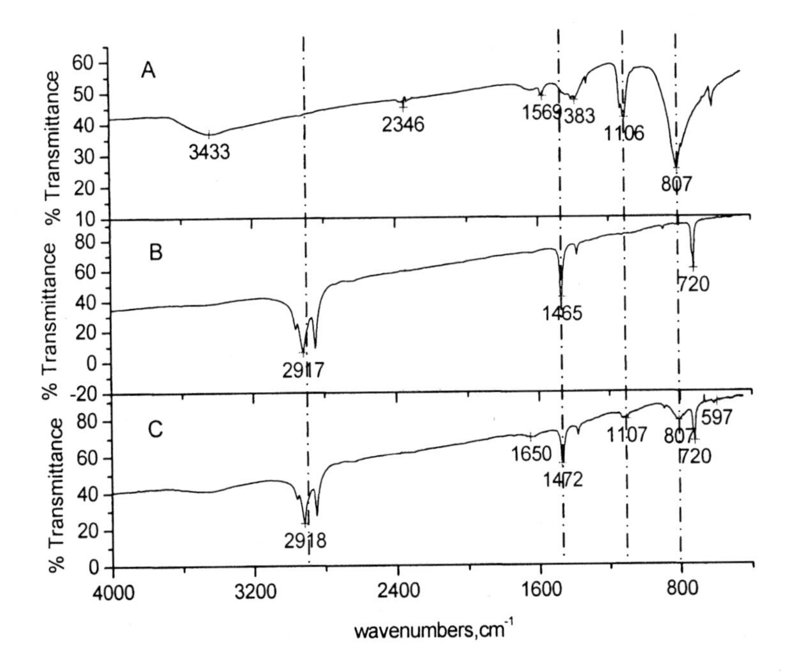

Figure 2. FTIR spectra of (A) pure K_2FeO_4 crystals, (B) pure paraffin and (C) ferrate-encapsulated paraffin.

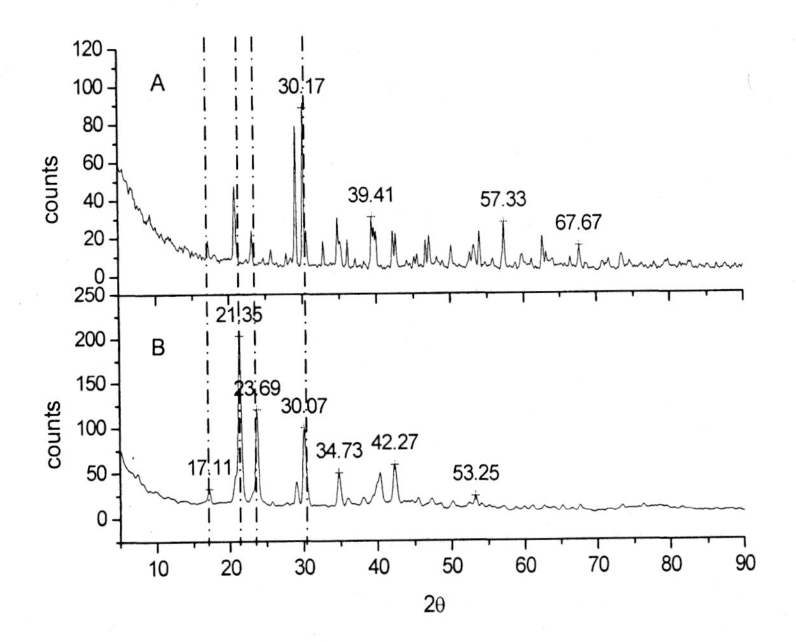

Figure 3. XRD spectra of pure ferrate (A) and ferrate-encapsulated paraffin (B).

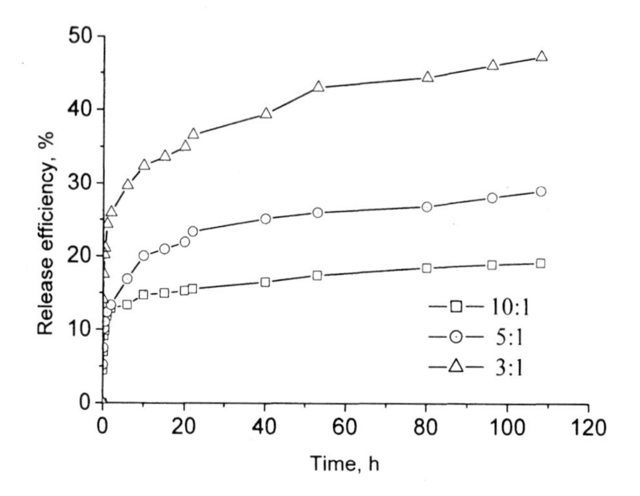

Figure 4. Cummulative percentage of ferrate released over time from the encapsulated paraffin matrix for 10:1, 5:1 and 3:1 paraffin to K_2FeO_4 into 6 M KOH.

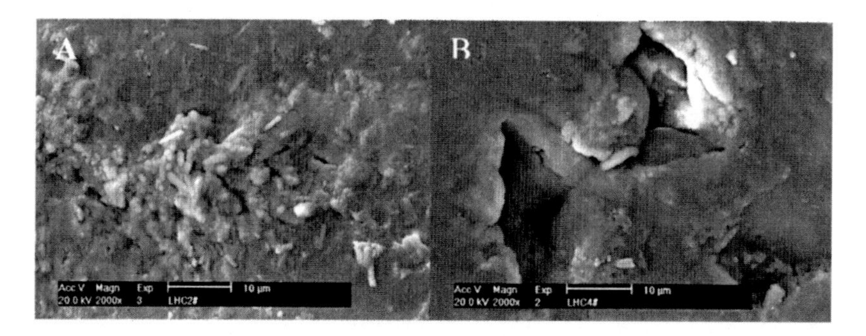

Figure 5. Scanning electron micrographs of ferrate-encapsulated paraffin surface before (A) and after (B) ferrate release.

20% of stable ferrate remained in the sample within 30 days. More than 90% of the ferrate encapsulated in parrafin remained in paraffin after 30 days. The 10% loss of ferrate is due to the uncoated or incompletely coated ferrate on the paraffin. It is consistent with the SEM analysis of ferrate encapsulated paraffin. Therefore, better storage stability of the ferrate could be obtained by the encapsulation process.

TCE oxidation by ferrate

In the experiments for identifying products, the reaction was initiated with 6.12 mg/L TCE reacting with 0.12 g ferrate-encapsulated paraffin for the mass

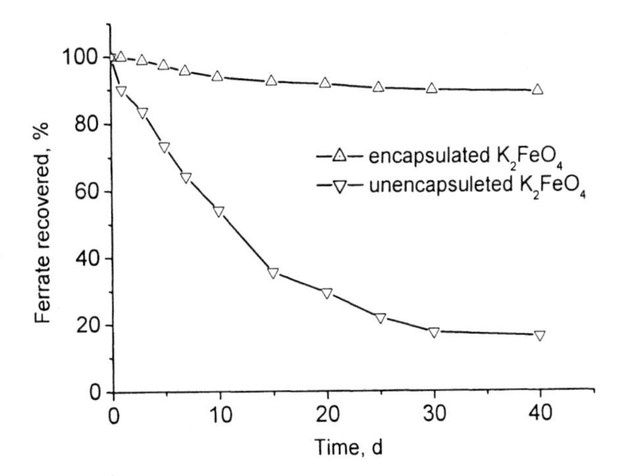

Figure 6. Cummulative amount of stable ferrate (%) recovered from unencapsulated ferrate and ferrate-encapsulated paraffin over time.

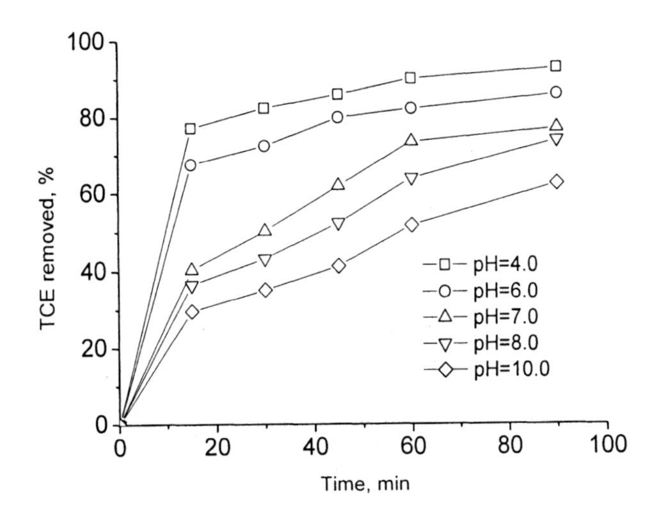

Figure 7. Percentage of TCE removed by ferrate-encapsulated paraffin in solutions at various pH values.

ratio of 3:1 paraffin to K_2FeO_4) in a phosphate-buffered solution having an ionic strength of 0.02 M at pH 4, 6, 7, 8 and 10, respectively, as showed in Figure 7. From Figure 7, the TCE oxidation is more effective in acidic medium than alkaline medium. At pH = 4, the reaction is very fast and the efficiency of TCE oxidation can reach to 90% after 30 mins. However, increasing the pH of the solution slows the TCE oxidation rate. At pH = 10, the removal efficiency of TCE is only 51% TCE after 30 mins and 60% after 150 mins.

Conclusions

- The optimum encapsulated conditions were achieved with a stirring speed of 800 rpm and 30 mins of ultrasonication at 80°C.
- FTIR and XRD spectra of ferrate-encapsulated paraffin showed that the stability and reactivity of K_2FeO_4 is not changed after its encapsulation into paraffin.
- The release process of K_2FeO_4 from the encapsulated particles includes two phases: (1) the uncoated ferrate was immediately dissolved (<1 min) and (2) the encapsulated ferrate was released much more slowly during the proceeding 108 h. The release kinetics were in good agreement with the $Q = kt^n$ model. The SEM results confirm the biphasic release patterns of ferrate-encapsulated paraffin.
- TCE degradation is more effective in an acidic medium. Over 90% of TCE was removed after 60 mins under pH = 4.0-6.0 while only 60% of TCE was removed after 150 mins at pH = 10 through the controlled release of encapsulated K_2FeO_4.

Acknowledgements

The authors are grateful to Fok Ying Tung Education Foundation (Grant No. 104003) and the Natural Science Foundation of Fujian Province (Grant No. 2006J0134) for their support of this project.

References

1. United States Environmental Protection Agency, EPA570/9-91-012FS; U.S. EPA: Washington, DC, 1991.
2. Gillham, R.W., O'Hannesin, S. F. *Ground Water* **1994**, *32*, 958-967.
3. Matheson, L. J., Tratnyek, P. G. *Environ. Sci. Technol.* **1994**, *28*, 2045-2053.
4. Gates, D. D, Siegrist, R L. *J. Environ. Eng.* **1995**, *121*, 639-644.
5. Glaze, W. H., Kang, J. K. *J. AWWA* **1988**, *80*, 57-63.
6. Khan, M. D. A. J., Watts, R. J. *Water Air Soil Pollut.* **1996**, *88*, 247-260.
7. Hoigne, J., Bader, H. *Water Res.* **1983**, *17*, 173-183.
8. Gardner, K. A., Mayer, J. M. *Science* **1995**, *269*, 1849-1851.
9. Schnarr, M. *J. Contam. Hydrol.* **1998**, *29*, 205-224.
10. Seol, Y., Schwartz, F. W. *J. Contam. Hydrol.* **2000**, *44*, 185-201.
11. Yan, Y. E., Schwartz, F. W. *Environ. Sci. Tech.* **2000**, *34*, 2535-2541.
12. Stewart, R. *Oxidation Mechanisms.* Benjamin, New York, **1964**, pp. 58-76.
13. Wiberg, K. B., Saegebarth, K. A. *J. Am. Chem. Soc.* **1957**, *79*, 2822-2824.
14. Kang, N., Hua, I., Rao, P. S. C. *J. Ind. Eng. Res.* **2004**, *43*, 5187-519
15. Li, C., Li, X. Z., Graham, N. *Chemosphere* **2005**, *61*, 537-543.
16. Bielski, B. H. J., Thomas, M. J. *J. Am. Chem. Soc.* **1987**, *109*, 7761-7764.
17. Audutte, R. J., Quai, J. W. *J. of. Solid. State. Chem.* **1973**, *8*, 43-49.
18. Palmieri, G. F., Bonacucina, G., Martino, P. D., Martelli, S. *Drug Development and Industrial Pharmacy* **2001**, *27*, 195-204.
19. Sinclair, G. W., Peppas, N. A. *J. Membrane Sci.* **1984**, *17*, 329-334.

Chapter 24

Oxidation of Nonylphenol Using Ferrate

Myongjin Yu, Guisu Park, and Hyunook Kim*

The University of Seoul, Department of Environmental Engineering, 90 Jeonnong-dong, Dongdaemun-gu, Seoul 130-743, Korea
***Corresponding author: h_kim@uos.ac.kr**

Public concerns on nonylphenol (NP) and nonylphenol ethoxylates (NPEOs) are growing because they are frequently detected in the aquatic environment and proven endocrine disrupter compounds (EDCs). Since these compounds cannot be biologically completely degraded, chemical oxidation has been frequently applied to degrade NP and NPEOs. In this study, ferrate(VI) (Fe(VI)) was used to oxidize NP and its oxidation kinetics was evaluated. It should, however, be noted that the first order rate was evaluated using data collected only after the initial degradation phase, in which 50-70 % Np was degraded. In fact, the NP and Fe(VI) concentrations during the ID phase could not be quantified since the oxidation was too fast. The effect of hydrogen peroxide (H_2O_2) presence on the NP oxidation by Fe(VI) was also evaluated. In general, the initial destruction of NP by Fe(VI) at lower pH was more significant than higher pH (i.e., 26% at pH 9.0 and 71% at pH 6.0). H_2O_2 addition did not have much impact on the NP oxidation. When applied to oxidation of NP in natural water, Fe(VI) showed less removal efficiency possibly due to the presence of dissolved organics in the water.

Introduction

Currently, a large number of chemicals have been emitted into the environment through human activities. Some of these chemicals can cause serious effects on wildlife species and human health. Among the chemicals, nonylphenol (NP) and its ethoxylates (NPEOs) are of more interest due to their anti-androgenic potential in the environment and their possible adverse effects on human health (*1-3*). For these reasons, utilization of NPEOs is under strict regulation in some countries in Europe and North America (*4*). In addition, NP is on the second priority list of substances drawn up under the European Union's Existing Substances Regulations (793/93/EEC).

Since the first synthesis of these chemicals in the 1940s, their production has been increased yearly due to their wide range of application as an ingredient of detergents, emulsifiers, wetting and dispersing agents. The world production of NPEOs in 1995 was about 520,000 tones (*5*).

Several researchers have pointed out that the effluent from sewage treatment plants (STPs) is one of the major sources of the NPEOs' metabolites detected in the aquatic environment (*6-8*). Incomplete biodegradation of commonly used NPnEO (usually n = 8-12) in STPs results in the production of NPEOs with shorter ethoxylate chains (1-3 ethoxylate units) and NP, which exhibit higher toxicity than their parent compounds (*9,10*). Therefore, it is necessity to decompose or remove the metabolites in the STPs before treated water is discharged into a stream or a river, which can be a drinking water source in downstream. Table I shows level of NP and its ethoxylates in the effluent from STPs in Canada.

Table I. Concentration of Nonylphenol and Its Ethoxylates in Sewage Treatment Plant Effluents

STP level of treatment		*(μg/L)*	
	NP	*NP1EO*	*NP2EO*
Primary	< 0.02 - 62.1	0.07 - 56.1	0.34 - 36.3
Secondary	0.12 - 4.8	< 0.02 - 43.4	< 0.02 - 32.6
Tertiary	< 0.02 - 3.2	0.30 - 26.4	0.25 - 12.5

SOURCE: Reproduced with permission from Reference 11, Copyright 2000 CEPA

Advanced oxidation processes seem promising for breakdown of the NPEOs' metabolites as the post-treatment of biological processes in STPs since a variety of aromatic compounds can be efficiently decomposed by these treatments. In fact, the ozone (O_3) has been successfully applied to oxidize various endocrine disrupter compounds (EDCs) (e.g., 17 β-estradiol (*12*), NP and bisphenol-A (*13*), and NPEOs (*13,14*)).

Ferrate(VI) is the oxidant which recently attracts more attention from researchers with its versatility as a multi-purpose water and wastewater treatment agent for disinfection, oxidation, and coagulation (*15,16*). Fe(VI) has very strong oxidizing power in the entire pH range (*17*). Under acidic conditions, the redox potential of ferrate ion is higher than that of any other oxidants, such as Cl_2, O_3, H_2O_2, and $KMnO_4$.

Due to its strong oxidizing nature, Fe(VI) has been applied to decompose various organic and inorganic compounds; for examples, hydrogen sulfide (*18*), 1,4-thioxane (*19*), thioacetamide (*20*), phenol and chlorophenols (*21*). Nearly all reactions of Fe(VI), with a variety of compounds, were reported to be first-order for each reactant. Table II shows second-order reaction rate constants of Fe(VI) with selected organic and inorganic compounds. The reported second-order reaction rate constants of Fe(VI) range from about 10^{-2} to 10^{5} $M^{-1}s^{-1}$ in the pH range of 8.0-9.0. The reactivity of Fe(VI) is highly dependent on the type of compound involved, indicating that Fe(VI) is a very selective oxidant. High reactivity of Fe(VI) was found for reduced sulfur- and nitrogen compounds (i.e., hydrogen sulfide and hydroxylamine) and aromatic compounds (i.e., phenol). On the other hand, Fe(VI) showed low reactivity toward tertiary alkylamine, ammonia, carboxylic acid, aldehydes and alcohols.

Table II. Second-order Rate Constants of Ferrate with Various Compounds at 25°C

Compounds	*pH*	*k* ($M^{1}s^{-1}$)	*Ref.*
Sulfur compounds			
Hydrogen sulfide	9.0	7.4×10^{5}	*18*
Cysteine	9.0	1.0×10^{5}	*22*
Thiourea	9.0	3.4×10^{3}	*23*
Thiosulfate	9.0	7.2×10^{2}	*24*
1,4-Thioxane	9.0	5.8×10^{1}	*19*
Nitrogen compounds			
Hydrazine	9.0	5.6×10^{3}	*25*
Hydroxylamine	9.0	4.8×10^{3}	*26*
Aniline	9.0	3.9×10^{2}	*27*
Ammonia	9.0	1.7×10^{-1}	*28*
Aldehydes, carboxylic acid, and alcohols			
Formaldehyde	8.0	5.0×10^{-1}	*29*
Formic acid	8.0	4.0×10^{-1}	*29*
Methyl alcohol	8.0	3.0×10^{-2}	*29*
Other compounds			
Superoxide ion	9.0	3.0×10^{5}	*30*
Phenol	9.0	1.0×10^{2}	*31*
Hydrogen peroxide	9.0	3.0×10^{1}	*32*

In this study, NP oxidation by Fe(VI) was performed in organic-free water and in natural water. The oxidation efficiency was evaluated at different FeO_4^{2-} doses, solution pHs or molar ratios of H_2O_2/FeO_4^{2-}.

Material and Methods

Reagents

NP (>85% purity) was obtained from Aldrich (USA) and diluted with methanol before use as necessary. The chemical structures of NP and NPEOs are shown in Figure 1. Methanol was chosen as solvent because it was miscible with water and has a low reactivity with Fe(VI) (k = 3.0×10^{-2} $M^{-1}s^{-1}$, pH 8.0) (*29*). H_2O_2 solution (35%) was purchased from Junsei (Japan) and diluted to the desirable concentration with the Milli-Q water as necessary. High-performance liquid chromatography (HPLC) grade solvents were used for all the HPLC works. Potassium ferrate (K_2FeO_4) (>94% purity) was prepared by modifying the method proposed by Thompson *et al.* (*33*). Fe(VI) solution with desirable concentration prepared by adding appropriate amount of K_2FeO_4 to 0.005 M Na_2HPO_4/0.001 M borate water buffered at pH 9.0. Fe(VI) is known the most stable at pH 9.0 (*34*).

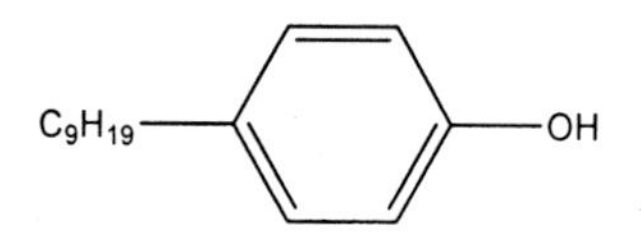

Figure 1. Chemical structure of nonylphenol.

Experimental Procedures

NP oxidation with Fe(VI) was performed in a 5 L Pyrex glass reactor in which buffered sample was agitated with a magnetic stirrer. Oxidation experiments were conducted with three different FeO_4^{2-} doses (i.e., 2, 5 and 10 mg/L) and at four different pHs (i.e., pH 6.0, 7.0, 8.0 and 9.0). In addition, H_2O_2 was injected at different H_2O_2/FeO_4^{2-} molar ratios (i.e., 0, 0.15 and 0.30) in order to investigate the effect of H_2O_2 presence on the NP oxidation by Fe(VI).

Rate constant for Fe(VI) decomposition was estimated using a UV/VIS spectro-photometer (at 510 nm) with a flow cell attached. Once an aqueous sample was taken for NP analysis during an oxidation study, it was immediately quenched with sodium sulfite (1.25×10^{-4} M) to remove the residual Fe(VI). Subsequently, the sample was centrifuged to remove particle at 2,500 rpm for 30 min. In our study, the first order rate was evaluated with data collected only after the initial decomposition (ID) phase in which 50-70% NP were removed, since decomposition rates of Fe[VI] and NP were too fast to measure.

Analytical Methods for Nonylphenol

For the analysis, NP was separated and concentrated in water by modifying the solid phase extraction (SPE) method proposed by Marinez *et al.* (*35*). SPE method for NP analysis is provided in Figure 3. SPE cartridge was obtained from Waters (Oasis, HLB 6 cc, 200 mg), USA). All glassware was washed by 0.1 M HCl and distilled water and baked at 550°C for at least 4 hours before used for an experiment.

The analysis of the NP was performed with a HPLC with fluorescence detection (HPLC/F) and mass spectrometry detection (HPLC/MS). The ring chromophore in NPEO molecules enables direct UV or fluorescence detection possible. The instrument used in the study was a Dionex Summit HPLC System (Dionex, USA) with an Inertsil PH column, 150 mm × 4.6 mm ID (GL Sciences Inc., Japan) at 40°C. A 50 μL sample was chromatographed at a flow rate 1 mL/min. A mobile-phase gradient was made to separate the compounds: solvent A was pure methanol and solvent B was distilled water (*36*). Initial conditions were 60% A; a gradient was started immediately after injection until 85% A was reached; these conditions were maintained for 25 min and then the percentage of B was gradually increased over 10 min to 40%. Finally, the column was stabilized for 15 min with 60% A; total run time was 60 min. Fluorescence detection (RF 2000, Dionex, USA) was achieved by 275 nm excitation and 300 nm emission wavelengths. Table III shows analytical condition of HPLC/MS for NP analysis.

Results and Discussion

Decomposition of Ferrate

Figure 3 shows the decomposition of Fe(VI) at different pHs and at different H_2O_2/FeO_4^{2-} molar ratios. In general, the decomposition rate of Fe(VI) strongly depended on the initial ferrate concentration, pH, and solution temperature (*37*). Under our experimental condition, the Fe(VI) decomposition at < pH 6.0 was too fast to determine, possibly due to instability of Fe(VI) at lower pH. Fe(VI) decomposition rate was significantly increased from 0.5×10^{-3} to 10.7×10^{-3} s^{-1} as

Table III. Analytical Conditions of HPLC/MS for Nonylphenol

Items	*Conditions*
Separation column	C_{18} reverse phase column; 2.1 mm × 150 mm, 5 μL
Mobile phase	Ammonium acetate gradient 40-100%; 0-15 min
Flow rate (mL/min)	0.4 mL/min
UV wavelength (nm)	228 nm
MS detection ionization	API-ES, NI mode
Mass (Mr)	219, 133
Drying gas temp. (°C)	350
Fragment voltage (V)	100
Capillary voltage (V)	3,500
Operation mode	SCAN and/or SIM mode

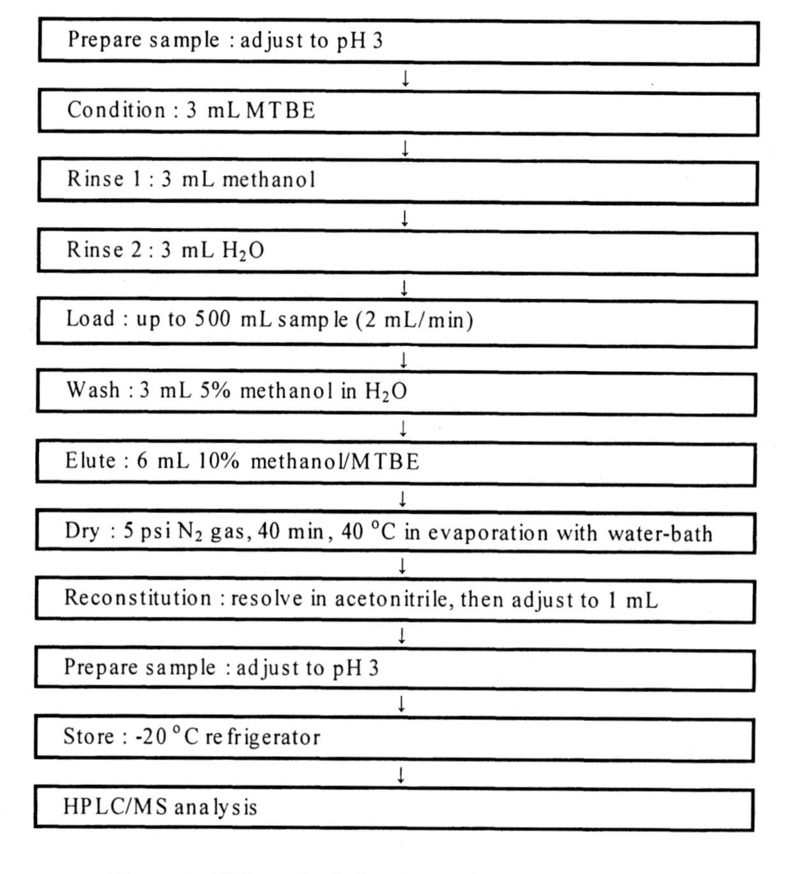

Figure 2. SPE methods for the analysis of nonylphenol.

pH decreased from 8.0 to 6.0. Enhanced Fe(VI) decomposition could be observed with the addition of H_2O_2. However, the magnitude of the enhancement decreased as the solution pH decreased.

Figure 4 shows effect of organic compounds present in water on Fe(VI) decomposition. This study was carried out with water sample collected from the Han River in Korea. The pH, alkalinity, and total organic carbon concentration of the water were 7.15, 38 mg/L as $CaCO_3$, and 2.35 mg/L, respectively. Before an experiment, pH of the DI water (organic free water) and Han River water were adjusted with phosphate and borate buffer (i.e., pH 6.0, 7.0, and 8.0). In short, Fe(VI) decomposition at lower pH was significantly affected by the organic material while it was not much affected at higher pH (pH 8.0).

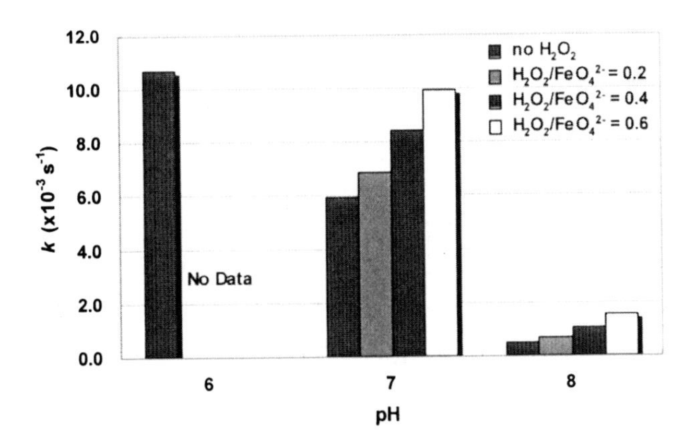

Figure 3. Decomposition of ferrate at different pHs and H_2O_2/FeO_4^{2-} molar ratios in DI-water. (FeO_4^{2-} = 10 mg/L, T = 20±1 °C)

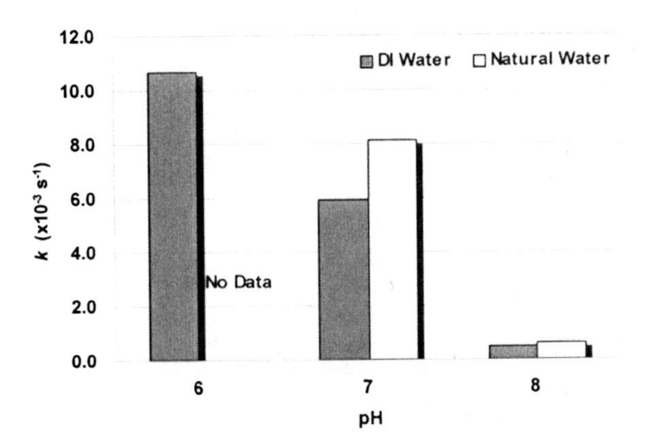

Figure 4. Decomposition of ferrate at DI-water and natural water (Han River water). (FeO_4^{2-} = 10 mg/L, T = 21±1 °C)

Effect of Ferrate Concentration on the Degradation of Nonylphenol

The decomposition of Fe(VI) in water is characterized with two phases of kinetics as the case of O_3. The first one is called instantaneous ferrate demand (ID) that occurs within 1-2 min. During this period, an instantaneous or very fast consumption of Fe(VI) takes place. In this study, the ID phase was defined as the initial decomposition of Fe(VI) within first 30 seconds.

NP of 1.03 mg/L was oxidized with Fe(VI) of different concentrations; molar ratios of 3.7-18.5 (FeO_4^{2-}/NP). Figure 5 shows the effect of initial FeO_4^{2-} concentration on the NP degradation. The oxidation of NP strongly depended on the amount of FeO_4^{2-} added. When the FeO_4^{2-} concentration was increased from 2 to 10 mg/L, the rate constants for NP oxidation after the ID phase increased from 2.2×10^{-3} to 3.1×10^{-3} s^{-1} at pH 7.0. More than 49% of NP could be oxidized in the ID phase when molar ratio of FeO_4^{2-} to NP was greater than 3.7 (Table IV); at the maximum molar ratio of FeO_4^{2-} to NP (18.5), more than 85% of NP could be removed within 5 min. As shown in the insert of Fig 5, Fe(VI) decomposition rate after ID phase could be estimated with the first order. However, for evaluating Fe(VI) decomposition more accurately, Fe(VI) concentration profile also should be monitored even during the ID phase; as mentioned earlier in the current experimental condition, we could not evaluate Fe(VI) or NP during the ID phase in which 50-70 % NP was oxidized.

Effect of pH on Nonylphenol Degradation

The reaction rate constants of NP oxidation by Fe(VI) at different pH value (6.0-9.0) were determined. As stated before, the Fe(VI) decomposition rate at pH less than 6.0 could not be carried out since the reaction was too fast to monitor. The fraction of protonated ferrate increases at lower pH (eq 1). Therefore, the reaction rates are expected to increase at lower pH because protonated ferrate species are unstable and more reactive, while unprotonated ferrate is not very reactive but relatively stable (eq 2).

$$H_2FeO_4 \leftrightarrow H^+ + HFeO_4^- \qquad pK_1 = 3.5\ (30) \qquad (1)$$

$$HFeO_4^- \leftrightarrow H^+ + FeO_4^{2-} \qquad pK_2 = 7.3\ (30) \qquad (2)$$

In alkaline pH (8-9), Fe(VI) does not decompose rapidly. Therefore, if sufficient amount of Fe(VI) is provided (high Fe(VI)/NP molar ratio), satisfactory degree of NP degradation can be achieved.

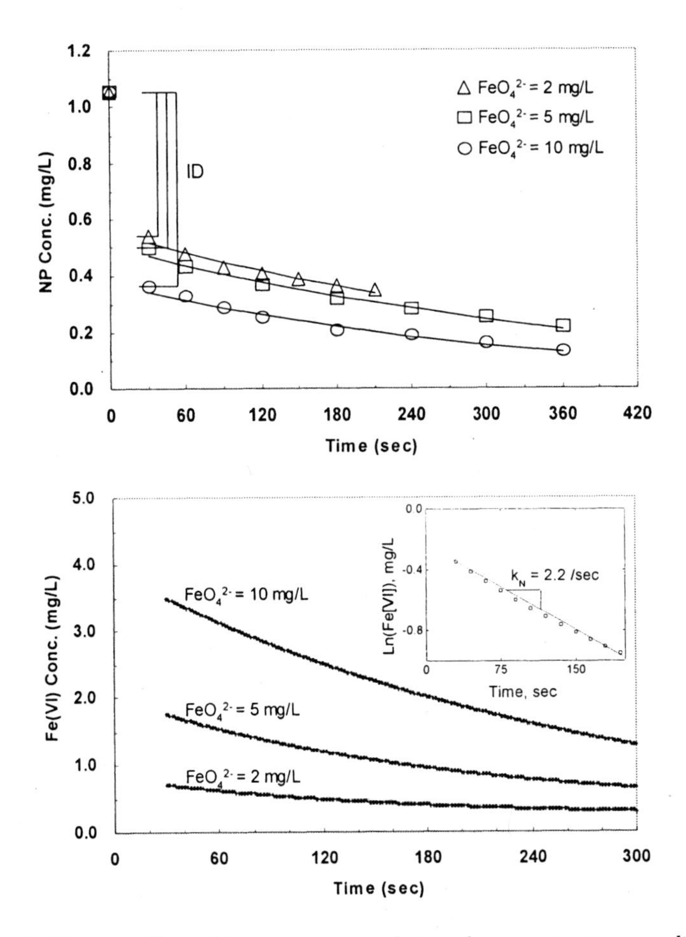

Figure 5. Time profiles of ferrate and nonylphenol concentrations at different FeO_4^{2-} doses. (NP = 1.03 mg/L, pH = 7.0, T = 20±1°C); insert = ln(Fe(VI)) at the initial Fe(VI) of 2 mg/L.

Table IV. Rate Constant and Removal Efficiency of Nonylphenol Oxidation by Ferrate at Different Ferrate Doses

	FeO_4^{2-} = 2 mg/L	FeO_4^{2-} = 5 mg/L	FeO_4^{2-} = 10 mg/L
ID (mg/L)	1.29	3.25	6.50
k_N ($\times 10^{-3}$ s^{-1})	2.2	2.4	3.1
$Removal_{5min}$ (%)	70* (49)	76 (53)	85 (66)

ID : Instantaneous ferrate demand within first 30 sec.

k_N : First order rate constant of NP oxidation was determined after ID phase.

( ) : Removal efficiency within first 30 sec.

* : Removal efficiency at 3.5 min.

Given that pK_a (= 10.7) for NP is well outside the pH range under study (pH 6.0-9.0), it is assumed that the results obtained reflect principally the redox behavior of Fe(VI) over the pH range studied. Comparing to NP degradation during ID phase at pH 6.0 (about 71%), minor degree of NP degradation during the ID was observed at pH 9.0 (26%) (Table V). The high NP degradation at pH 6.0 in the ID phase relates with the high oxidation power of protonated ferrate and low residual Fe(VI) after ID.

In our study, therefore, reaction rate constant for NP oxidation after ID was the lowest at pH 6.0, since oxidation was limited by the residual Fe(VI) concentration in water (Figure 6); for example, as pH decreased from 9.0 to 6.0, rate constant decreased from 5.0×10^{-3} to 3.3×10^{-3} s^{-1}. Nonetheless, similar level of overall removal efficiencies (at 5 min) was observed within all the pH range under study (Table V).

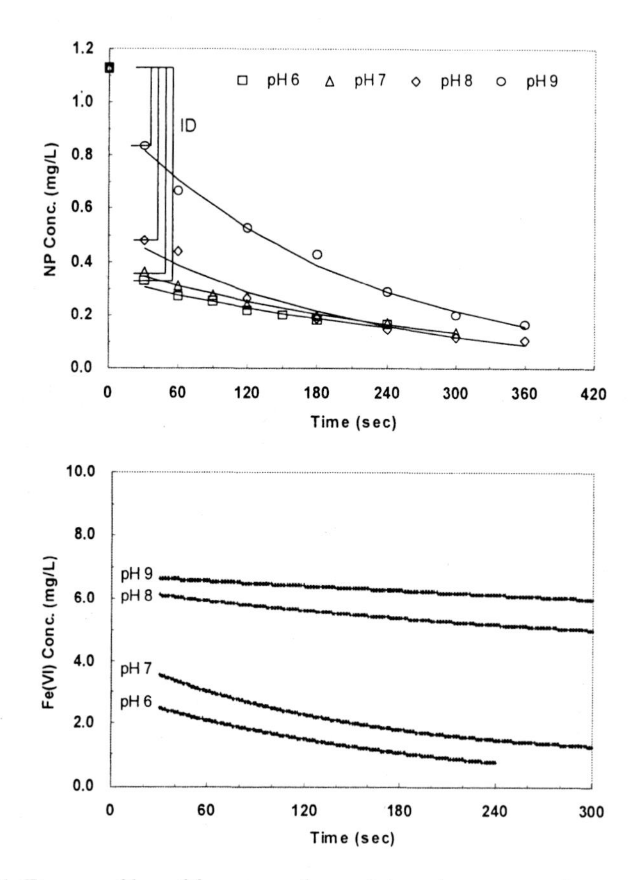

Figure 6. Time profiles of ferrate and nonylphenol concentrations at different pH values. (FeO_4^{2-} = 10 mg/L, NP = 1.129 mg/L, T = 21±1°C)

Table V. Rate Constant and Removal Efficiency of Nonylphenol Oxidation by Ferrate at Different pHs

	pH 6.0	*pH 7.0*	*pH 8.0*	*pH 9.0*
ID (mg/L)	7.53	6.46	3.86	3.34
k_N ($\times 10^{-3}$ s^{-1})	3.3	3.5	4.9	5.0
$Removal_{5min}$ (%)	86* (71)	88 (68)	90 (57)	83 (26)

* : Removal efficiency at 4 min.

Effect of Hydrogen Peroxide on Nonylphenol Oxidation by Ferrate

In our other study, ozone, when used together with H_2O_2, showed more effective NP oxidation (data not shown). However, Fe(VI) is known to rapidly react with H_2O_2 (*30*). This may indicate that NP may not be effectively oxidized at all if Fe(VI) is applied with H_2O_2. In this section, therefore, effect of H_2O_2 addition on the NP oxidation by Fe(VI) is evaluated.

In this specific experiment, H_2O_2 at different doses (molar H_2O_2/FeO_4^{2-} of 0, 0.15, and 0.3) were added into the water along with Fe(VI) of 10 mg/L (= 0.0833 mM) was applied at pH 8.0. No pH change was observed after the addition of H_2O_2 in phosphate buffer solution. In fact, we expected H_2O_2 addition would significantly enhance Fe(VI) decomposition. However, its decomposition rate was not significantly enhanced by the H_2O_2 addition (Figure 7); for all H_2O_2/FeO_4^{2-} ratios of 0-0.3, similar rate constant for NP oxidation was obtained (i.e., 4.7×10^{-3} to 4.0×10^{-3} s^{-1}). Similarly, H_2O_2/FeO_4^{2-} ratios did not much affect the NP destruction during ID phase, either; 44-56% NP removal (Table VI).

Oxidation of Nonylphenol in Natural Water Using Ferrate

An experiment to oxidize NP in water samples from the Han River was performed. The pH, alkalinity, TOC, and UV_{254} of the water were 7.3, 37 mg/L as $CaCO_3$, 2.45 mg/L, and 0.032 cm^{-1} respectively. Once collected, the raw water was pre-chlorinated and filtered with a 0.45 μm membrane filter prior to oxidation experiments. 1 mg/L NP was, on experiment, spiked in the filtered water. During this specific experiment, raw water was not buffered, so the water pH was increased from 7.3 to 8.1 during the reaction. In spite of organic material present in the raw water, Fe(VI) decomposition rate was almost similar to that obtained from the experiment with DI water at pH 8.0 (i.e., rate constants of Fe(VI) decomposition in raw water and in DI water were 0.69×10^{-3} and 0.61×10^{-3} s^{-1}, respectively). NP decomposition rate constant from the experiment with raw water was similar to that with DI water. However, NP removal efficiency

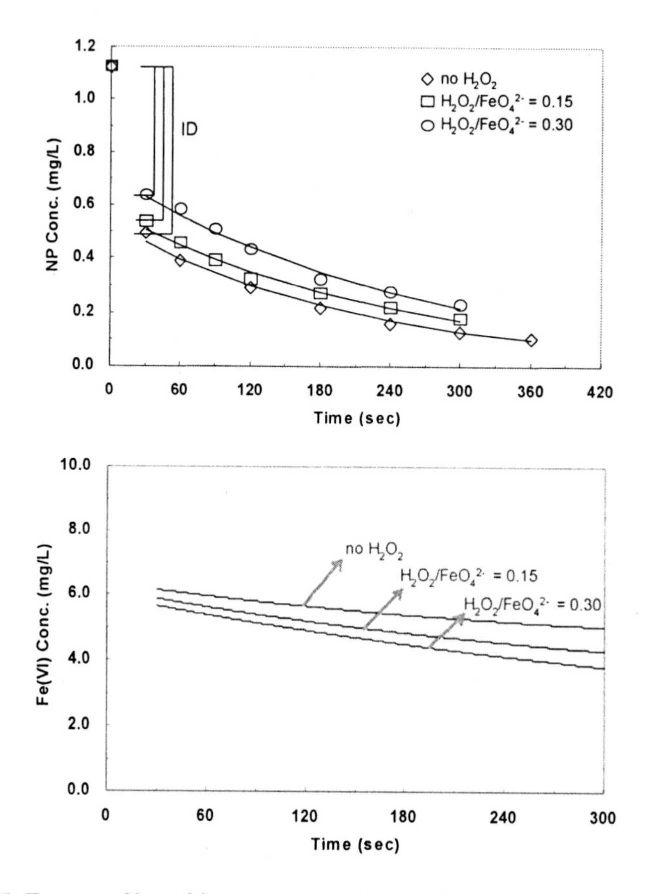

Figure 7. Time profiles of ferrate and nonylphenol concentrations at different H_2O_2/FeO_4^{2-} molar ratios. (FeO_4^{2-} = 10 mg/L, NP = 1.126 mg/L, pH = 8.0, T = 19±1°C)

Table VI. Rate Constant and Removal Efficiency of Nonylphenol Oxidation by Ferrate at Different Molar H_2O_2/FeO_4^{2-} Ratios

	no H_2O_2	*H_2O_2/FeO_4^{2-} = 0.15*	*H_2O_2/FeO_4^{2-} = 0.30*
ID (mg/L)	3.85	4.14	4.36
k_N ($\times 10^{-3}$ s^{-1})	4.7	4.1	4.0
$Removal_{5min}$ (%)	89 (56)	85 (52)	80 (44)

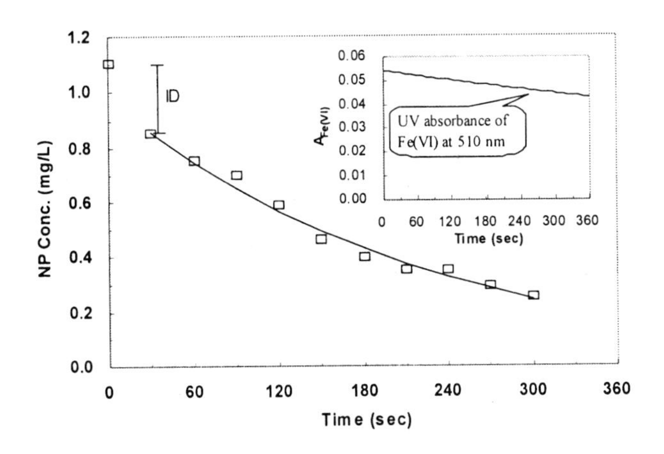

Figure 8. Time profile of nonylphenol concentration in natural water (insert; shows the first-order decomposition for Fe(VI)). (NP = 1.102 mg/L, T = 19±1°C)

during the ID phase decreased from approximately 57% (DI water and pH 8.0) to 23% (natural water). On the other hand, 78% overall removal of NP was obtained over the reaction time of 5 min (Figure 8).

Conclusions

In this preliminary study, oxidation of NP using by Fe(VI) has been performed. The percentage removal of NP strongly depended on the initial FeO_4^{2-} concentration. At the maximum molar ratio of FeO_4^{2-} to NP in this study, more than 85% removal of NP removal could be achieved within 5 min.

NP oxidation by Fe(VI) at different pHs is characterized as followings: at neutral or acidic pHs, most NP is removed during ID phase. Therefore, NP oxidation kinetics was not properly estimated with the second order rate; the fore, first order rate constant was estimated with the data collected after the ID phase. Currently, a new experimental set-up is being designed to quantify NP and Fe(VI) during ID phase for better estimation of kinetic order. At higher pHs, the NP oxidation apparently follows first order rate. Unlike O_3, the addition of H_2O_2 did not enhance the oxidation of NP by Fe(VI).

When Fe(VI) was applied to oxidize NP in Han River water sample, Fe(VI) showed low ID stage unlike the case with the experiment with NP in DI water. Probably because dissolved organic compounds in the natural water, only 23% NP could be oxidized within ID phase. Nonetheless, significant amount of NP (about 80%) could be removed if sufficient reaction time is provided. Based on the result of the current study, it can be said Fe(VI) be applied to treat water with

NP. However, it should be mentioned that due to its characteristic color (reddish yellow) after oxidation, Fe(VI) may not be used in downstream processes; water with color cannot be discharged or distributed. It means that location of its application should be carefully determined.

Acknowledgement

Financial support from Korean Science Foundation (R01-2006-000-10284-0) is greatly appreciated.

References

1. Jobling, S.; Sumpter, J. P. *Aquatic Toxicol.* **1993**, *27*, 361-372.
2. Routledge, E. J.; Sumpter, J. P. *Environ. Toxicol. Chem.* **1996**, *15*, 241-248.
3. Yadetie, F.; Arukwe, A.; Goksøyr, A; Male, R. *The Science of the Total Environment* **1999**, *233*, 201-210.
4. Servos, M. R. *Water Qual. Res. J. Can.* **1999**, *34*(1), 123-177.
5. Darnerud, P. O. *FEMS Micro. Ecol.* **2002**, *46*, 159-170.
6. Ahel, M.; Giger, W.; Koch, M. *Water Res.* **1994**, *28*, 1131-1142.
7. Giger, W.; Brunner, P. H.; Schaffner, C. *Science* **1984**, *225*, 623-625.
8. Rudel, R. A.; Melly, S. J.; Geno, P. W.; Sun, G.; Brody, J. G. *Environ. Sci. Technol.* **1998**, *32*, 861-869.
9. Manzano, M. A.; Perales, J. A.; Sales, S.; Quiroga, J. M. *Water Res.* **1999**, *33*, 2593-2600.
10. Maki, H.; Okamura, H.; Aoyama, I.; Fujita, M. *Environ. Toxicol. Chem.* **1998**, *17*, 650-654.
11. *Assessment of NP and its ethoxylates*; CEPA, 2000.
12. Kim, S.; Yamada, H.; Tsuno, H. *Ozone Science and Engineering* **2004**, *26*, 563-571.
13. Deborde, M.; Rabouan, S.; Duguet, J.; Legube, B. *Environ. Sci. Technol.* **2005**, *39*, 6086-6092.
14. Ike, M.; Asano, M.; Belkada, F. D.; Tsunoi, S.; Tanaka, M.; Fujita, M. *Wat. Sci. Technol.* **2002**, *46*(11-12), 127-132.
15. Schink, T.; Wait, T. D. *Water Res.* **1980**, *14*, 1705-1717.
16. Waite, T. D.; Gray, K. A. *Stud. Environ. Sci.* **1984**, *23*, 407-420.
17. Wood, R. H. *J. Am. Chem. Soc.* **1958**, *80*, 2038-2041.
18. Sharma, V. K.; Smith, J. O.; Millero, F. J. *Environ. Sci. Technol.* **1997**, *31*, 2486-2491.
19. Read, J. F.; Boucher, K. D.; Mehlman, S. A.; Watson, K. J. *Inorg. Chem. Acta.* **1998**, *267*, 159-163.

20. Sharma, V. K. *Water Sci. Technol.* **2004**, *49*(4), 69-74.
21. Graham, N; Jiang, C.; Li, X.; Jiang, J.; Ma, J. *Chemosphere* **2004**, *56*, 949-956.
22. Read, J. F.; Bewick, S. A.; Graves, C. R.; MacPherson, J. M.; Salah, J. C.; Theriault, A.; Wyand, A. E. H. *Chem. Acta* **2000**, *303*, 244-255.
23. Sharma, V. K.; Rivera, W.; Joshi, V. N.; Millero, F. J.; O'Connor, D. *Environ. Sci. Technol.*. **1999**, *33*, 2645-2650.
24. Johnson, M. D.; Read, F. J. *Inorg. Chem.* **1996**, *35*, 6795-6799.
25. Johnson, M. D.; Hornstein, B. H. *Inorg. Chem. Acta* **1994**, *225*, 145-150.
26. Johnson, M. D.; Hornstein, B. H. *Inorg. Chem.* **2003**, *42*, 6923-6928.
27. Huang, H.; Sommerfeld, D.; Dunn, B. C.; Lloyd, C. R.; Eyring, E. M. *J. Chem. Soc. Dalton Trans.* **2001**, 1301-1305.
28. Sharma, V. K.; Bloom, J. T.; Joshi, V. N.; J. *Environ. Sci. Health* **1998**, *A33*, 635-650.
29. Carr, J. D.; Kelter, P. B.; Tabatabai, A.; Spichal, D.; Erickson, J.; McLaughin, C. W. *Proc. Conf. Water Chlorin. Chem. Environ. Impact Health Effects* **1985**, *5*, 1285-1298.
30. Rush, J. D.; Zhao, Z.; Bielski, B. H. J. *Free Rad. Res.* **1996**, *24*, 187-198.
31. Rush, J. D.; Cyr, J. E.; Zhao, Z.; Bielski, B. H. J. *Free Rad. Res.* **1995**, *22*, 349-360.
32. Rush, J. D.; Zhao, Z.; Bielski, B. H. J. *Free Rad. Res.* **1996**, *24*, 187-198.
33. Thompson, G. V.; Ockerman, L. T.; Schreyer, J. M. *J. Am. Chem. Soc.* **1951**, *73*, 1379-1381.
34. Bielski, B. H. J.; Thomas, M. J. *J. Am. Chem. Soc.* **1987**, *109*, 7761-7764.
35. Marinez, E.; Gans, O.; Weber, H.; Scharf, S. *Water Sci. Tech.* **2004**, *50*(5), 157-163.
36. Kawamura, S. *Integrated design and operation of water treatment facilities, 2nd ed.*; Wiley-Interscience: New York. 2000.
37. Jiang, J. Q.; Lloyd, B. *Wat. Res.* **2002**, *36*, 1397-1408.

Chapter 25

Preparation of Potassium Ferrate for the Degradation of Tetracycline

Shih-fen Yang and Ruey-an Doong*

Department of Biomedical Engineering and Environmental Sciences, National Tsing Hua University, Hsinchu, 30013, Taiwan

Tetracycline antibiotics are widely used in veterinary medicine and growth-promoting antibiotics for treatment and/or prevention of infectious disease because of their broad-spectrum activity and cost benefit. Tetracycline is rather persistent and a large fraction, which can be up to 75 %, of a single dose can be excreted in non-metabolized form in manures. In addition, potassium ferrate (Fe(VI)) is a powerful oxidant over a wide pH range and can be used as an environmentally friendly chemical in treated and natural waters. Therefore, the ability of ferrate (VI) to oxidize tetracycline in aqueous solution was examined in this study. The stability of Fe(VI) was monitored by the 2,2'-azinobis(3-ethylbenzothiazoline-6-sulfonate) (ABTS) method. Results showed that the decomposition of Fe(VI) is highly dependent on pHs. A minimum decomposition rate constant of 1×10^{-4} s^{-1} was observed at pH 9.2. A stoichiometry of 1.4:1 was observed for the reaction of tetracycline with Fe(VI) at pH 9.2. In addition, tetracycline was rapidly transformed in the presence of low concentrations of Fe(VI) in the pH rang 8.3-10.0. The degradation efficiency of tetracycline is affected by both pH and initial Fe(VI) concentration. The degradation of tetracycline by Fe(VI) was also confirmed by ESI-MS and total organic carbon (TOC) analyses.

Introduction

The widespread occurrence of pharmaceutical and personal care products (PPCPs) has recently received a great concern. PPCPs are released into aquatic environments as a result of industrial, agricultural, and sewage runoff *(1-3)*. PPCPs such as hormones, antimicrobials, and fragrances have been detected in sewage or wastewater treatment effluents at concentrations in the range of monogram per liter to microgram per liter *(1)*. Recent studies have indicated that several treatment technologies such as coagulation, sedimentation, adsorption, photo catalytic degradation and chemical oxidation can result in the decrease in concentrations and risks of PPCPs in the environments *(4-6)*. In addition, advanced oxidation processes (AOPs) using ozone, hydrogen peroxide and the combination with UV light have shown to effectively degrade PPCP contaminants in waters and wastewaters *(5, 7)*.

Tetracyclines (TCs) such as tetracycline, chlortetracycline, oxytetracycline and doxycycline are broad-spectrum antibiotics which show activity against gram-positive and gram-negative bacteria including some anaerobes *(8, 9)*. These compounds are widely used in veterinary medicine and growth-promoting antibiotics for treatment and/or prevention of infectious disease. In general, TCs are added in animal foodstuffs and present in animal waste products. These antibiotics are rather persistent and a large fraction, which can be up to 75 %, of a single dose can be excreted in non-metabolized form in manures, and then entered into the aquatic environments. Biodegradation of PPCPs such as tetracycline under aerobic conditions is limited. Several studies, however, have demonstrated that tetracycline can be removed with respect to the chemical oxidation. Reyes et al. *(10)* reported a rapid tetracycline degradation in the presence of 0.5 g/L of TiO_2 and UV light. About 50% of the initial tetracycline concentration was photodegraded after 10, 20 and 120 min when tetracycline was irradiated with a UV lamp, a solarium device and a UV-A lamp, respectively. In addition, significant mineralization was also obtained when the UV lamp and solarium were used for photocatalysis. Dodd et al. *(11)* investigated the oxidation of 14 antibacterial molecules by aqueous ozone and found that tetracycline reacted rapidly with ozone over a wide range of pHs at 20 °C. The degradation of tetracycline by ozone followed second-order kinetics and a second-order rate constant of 1.9×10^6 $M^{-1}s^{-1}$ at pH 7.0 was observed *(11)* These results clearly depict that oxidation processes are suitable technologies for controlling tetracycline in aquatic environments.

Ferrate (Fe(VI)) is a powerful oxidant over a wide pH range and can be used as an environmentally friendly chemical in treated and natural waters *(12, 13)*. The redox potentials of ferrate(VI) are 2.20 V and 0.72 V under acidic and alkaline conditions (Eq.(1) and (Eq.(2)), respectively *(12, 14)*. During the oxidation processes, ferrate(VI) is reduced to Fe(III) ions or ferrihydrite ($Fe(OH)_3$), resulting in the simultaneous coagulation in a single unit process. In addition, ferrate(VI) is an efficient coagulant for removing toxic contaminants after oxidation. Therefore, ferrate can serve as a dual-function chemical reagent in water and wastewater treatment.

$$FeO_4^{2-} + 8H^+ + 3e^- \rightarrow Fe^{3+} + 4H_2O \qquad E^0 = 2.20\ V \tag{1}$$

$$FeO_4^{2-} + 4H_2O + 3e^- \rightarrow Fe(OH)_3 + 5OH^- \qquad E^0 = 0.72\ V \tag{2}$$

Potassium ferrate(VI) has been used to oxidize many priority pollutants including reduced sulfur- and nitrogen-containing compounds, chlorophenols, and some toxic inorganics such as arsenic, cyanides and ammonia *(11-16)*. More recently, ferrate(VI) has been used to decompose endocrine disrupting chemicals *(17, 18)*, sulfonamide antimicrobials *(19, 20)* and microorganisms *(14)*. However, the use of ferrate for the degradation of tetracycline has received less attention.

When the potassium ferrate is dissolved in water, oxygen is evolved and ferric oxide is precipitated, making this oxidant unstable in water. The decomposition rate of ferrate is highly dependent on initial ferrate concentration, pH value and temperature *(12, 14)*. Usually the change in aqueous ferrate concentration is determined by a simple and convenient method which monitors the absorbance at 510 nm wavelength *(12, 14, 21)*. However, the inherent limitation of this simple UV-Vis spectroscopic method is that the molar absorbance coefficient is strongly pH dependent. The molar absorption coefficient ($\varepsilon_{ferrate,\ 510\ nm}$) of aqueous ferrate(VI) at pH 9.1, the most stable condition of the aqueous ferrate(VI), is 1150 $M^{-1}cm^{-1}$, and the molar absorbance coefficient decreases to 520 $M^{-1}cm^{-1}$ at pH 6.2. In addition, the molar absorption coefficient is too low to measure the Fe(VI) concentration when it is lower than 100 μM.

Recently, another method for monitoring the concentration of aqueous ferrate(VI) has been developed. This method is based on the reaction of ferrate(VI) with 2,2'-azinobis(3-ethylbenzothiazoline-6-sulfonate) (ABTS) *(22)*. The colorless ABTS will react with ferrate(VI) to form a green radical $ABTS^{\bullet +}$, which has a maximum absorbance at 415 nm. The change in molar absorbance for the ABTS method at 415 nm is reported to be $3.40 \times 10^4\ M^{-1}cm^{-1}$ per mol L^{-1} of added ferrate(VI) *(22)*. This method is highly sensitive and simple for low concentration of aqueous ferrate(VI), which has been used to determine the apparent second-order rate constant for the reaction of Fe(VI) with phenolic endocrine disruptors in aqueous solution *(17)*. However, the use of ABTS method for the determination of the reaction of tetracycline with Fe(VI) has received less attention.

The objective of this study was to determine the potential of Fe(VI) for the degradation of tetracycline at various pHs ranging from 7.0–10.0. The potassium ferrate was prepared using wet method by oxidizing Fe(III) salt under strong alkaline conditions. The stability of the prepared potassium ferrate at various pH values was also examined using ABTS method. The stoichiometric relationship between ferrate(VI) and tetracycline was determined. In addition, the oxidative degradation rates of tetracycline by Fe(VI) as a function of pH were determined. Electrospray ionization-mass spectrum (ESI-MS) was also used to identify the degradation of tetracycline by Fe(VI).

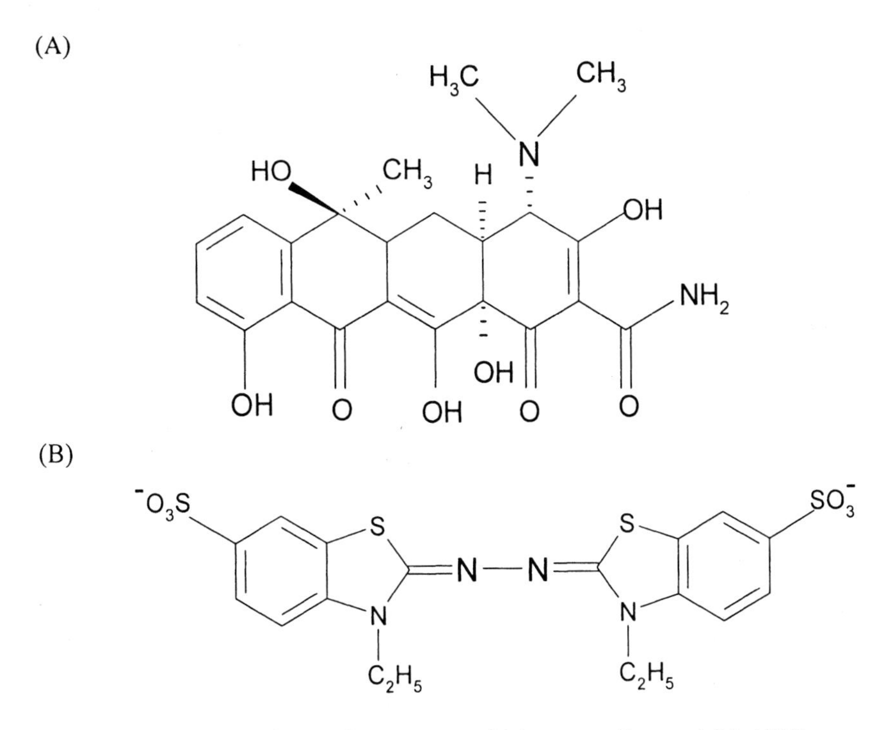

Figure 1 The chemical structures of (a) tetracycline and (b) ABTS.

Materials and Methods

Chemicals

Tetracyclines ($C_{22}H_{24}N_2O_8$, 98 %), N-(2-hydroxyethyl)-piperazine-N'-(2-ethanosulfonic acid) (HEPES) (99.5%), and 2,2'-azinobis(3-ethylbenzo-thiazoline-6-sulfonate) (ABTS) were purchased form Sigma-Aldrich Co. (Milwaukee, WI). Figure 1 shows the chemical structures of tetracycline and ABTS. Ascorbic acid (97 %) was purchased from Mallinckrodt Co. (Phillips-burg, NJ). Solutions were prepared with distilled water that was passed through a Milli-Q water purification system (> 18 MΩ cm). All chemicals were of analytical grade and were used as received without further treatment

Preparation of potassium ferrate

Solid phase of potassium ferrate (K_2FeO_4) with a purity up to 90% was synthesized by modifying the method of reaction between OCl^- and $Fe(NO)_3$ in strong basic solutions *(12, 14)*. Sodium hypochlorite ($NaOCl^-$) was first reacted with hydrochloric acid (HCl) to produce chlorine gas, and then the chlorine gas was flowed into the vacuumed vessel containing $Fe(NO)_3$ and KOH to produce potassium ferrate. The resulting purple slurry was filtered and then the precipitate was washed with 1 M aqueous KOH solution. The precipitate was then flushed with n-hexane (four times × 25 mL), n-pentane (four times × 25 mL), and methanol (four times × 25 mL) in sequence *(22)*. The resulting potassium ferrate solid was obtained and stored in the glove box prior to use.

The concentration of potassium ferrate was determined using UV-Vis spectroscopy at 510 nm. Aqueous Fe(VI) solution was prepared by dissolving solid samples of K_2FeO_4 in a 5 mM phosphate/1 mM borate buffer at pH 9.1 ± 0.1, which is well-known to be most stable for Fe(VI) solution *(12, 14)*. A molar absorption coefficient of aqueous Fe(VI) $\varepsilon_{ferrate,\ 510\ nm}$ of 1150 $M^{-1}cm^{-1}$ at pH 9.0 was used for the calculation of FeO_4^{2-} concentration at 510 nm *(21)*.

Stability of ferrate(VI) at various pH values

To understand the stability of potassium ferrate in aqueous solution at various pH values, the prepared solid potassium ferrate was dissolved in various pH buffer solutions. The aqueous ferrate solution was buffered with a 5 mM phosphate buffer in the pH range of 6-8, while 5 mM phosphate/1 mM borate buffer was used in the pH range of 9-11. The stability of ferrate was monitored using ABTS method. An appropriate volume of stock aqueous ferrate(VI) was added into solution containing ABTS at various pHs ranging from 7.0 to 11.0. The final concentration of ferrate(VI) and ABTS were 200 and 100 μM, respectively. After the complete reaction of ABTS with ferrate(VI), a green color was formed *(22)*. The solution was then transferred to the cuvette and the absorbance of green radical ($ABTS^{\bullet+}$) was monitored at 415 nm by using Hitachi U-3010 UV-Vis spectrophotometer.

Degradation of tetracycline

Reactions between tetracycline and ferrate were conducted using a series of 10 ml serum bottles. The ferrate and tetracycline concentrations were on the order of 10 – 80 μM and 200 μM, respectively. 5 mM phosphate buffer was used to maintain the pH at 7-8, while 5 mM phosphate/1 mM borate buffer was used to control the pH range of 8.5-10.0. All the experiments were performed at room temperature (23 – 25 °C). Samples were periodically withdrawn from the reaction vessels, and immediately treated with 0.5 mL of ascorbic acid (25 mM) or Na_2SO_3 (6 mM) to quench the reaction. The 0.45 μm PTFE filters were used to remove the precipitate, and then the absorption of supernatants were monitored at 359 nm to determine the residual concentration of tetracycline using Hitachi U-3010 UV-Vis spectrophotometer.

Results and Discussion

Sensitivity of ABTS method

Ferrate(VI) is a powerful oxidant with the high reduction potential and is unstable in water. The aqueous decomposition rate of ferrate is strongly dependent on the pH and temperature. In this study, the effect of pH on the ferrate stability at room temperature was evaluated by the ABTS method. Lee et al. *(22)* have depicted that ABTS method is a highly sensitive method for the determination of low concentration of aqueous ferrate over a wide pH range. In this study, the ABTS method was also used to determined decomposition of Fe(VI) and degradation of tetracycline at various pH values.

Figure 2 shows the calibration curves of Fe(VI) concentration ranging from 0 - 200 μM at pH 7.0 and 9.2. A good linearity (r^2 = 0.979, n = 10) of the calibration curve of Fe(VI) at pH 7.0 with the slope of 3.76×10^4 M^{-1} cm^{-1} was obtained. Similar result was also observed for the calibration curve of Fe(VI) at pH 9.2. The absorbance increased linearly upon increasing Fe(VI) concentration from 0 to 154 μM and then gradually leveled off to saturation when further increasing the Fe(VI) concentration to 220 μM. A correlation coefficient of 0.995 with the slope of 3.68×10^4 M^{-1} cm^{-1} was obtained at pH 9.2. Lee et al. *(22)* depicted that the reaction of Fe(VI) with ABTS has a stoichiometry of 1:1 when ABTS concentration was in excess (73 μM). The absorbance coefficient was found to be $(3.4 \pm 0.05) \times 10^4$ M^{-1} cm^{-1} per mole L^{-1} of added Fe(VI) ranging between 0.03 and 35 μM. In this study, we also found that the absorbance of Fe(VI) determined by excess ABTS (100 μM) at various pH values (7.0 – 9.2) were in the same range. It is noted that the linearity of Fe(VI) used in this study can be up to 150 μM in the presence of excess ABTS, which shows a wide dynamic range than that determined by Lee et al. *(22)*. In addition, the absorbance ($\sim 3.7 \times 10^4$ M^{-1} cm^{-1}) is much higher than that used at 510 nm (1150 M^{-1} cm^{-1}), showing that ABTS method is highly sensitive and stable for Fe(VI) determination.

Stability of Fe(VI) at various pH values

The stability of Fe(VI) at various pH values ranging between 7.0 and 11.0 was further examined using ABTS method. Figure 3 shows the concentration profiles of 200 μM Fe(VI) solutions at various pH values. Results showed that the stability of ferrate(VI) solution is highly dependent on pH values. The ferrate(VI) solution is more stable under alkaline conditions. Only 22 % of the original ferrate(VI) solution remained at pH 7.0 after 540 sec (9 min), while 49 and 93 % of those remained at pH 8.0 and 9.2, respectively. Further increase in

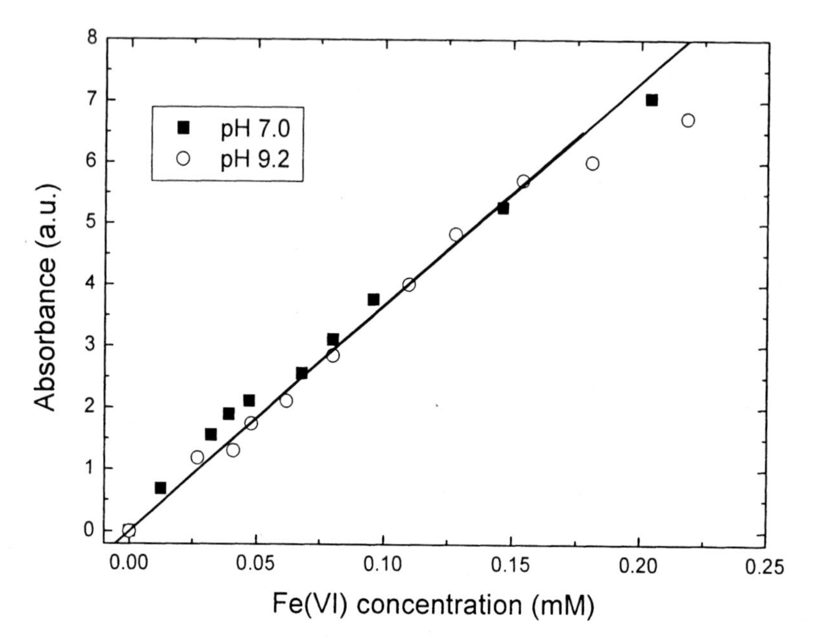

Figure 2. The calibration curves of Fe(VI) solution determined by ABTS method at pH 7.0 and 9.2.

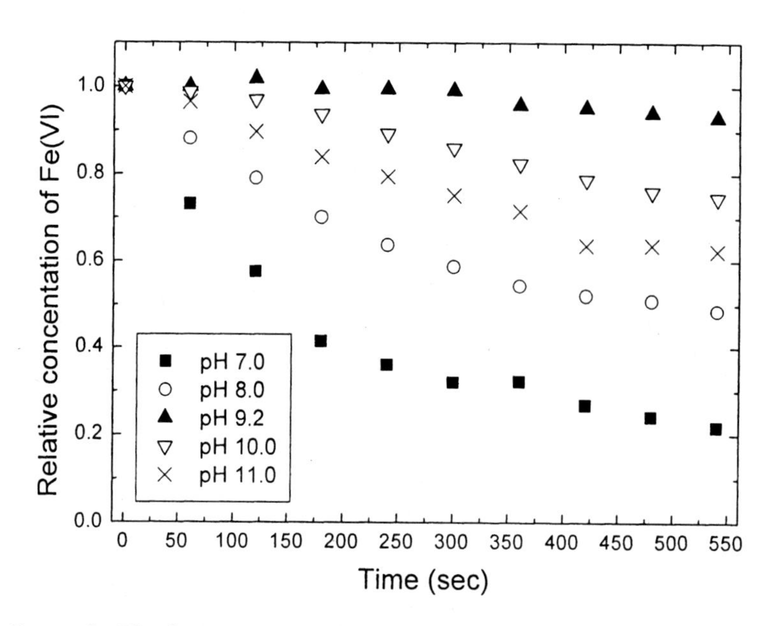

Figure 3. The decomposition of ferrate(VI) in aqueous solution at various pH values. The initial ferrate concentration was 200 μM.

pH value slightly decreased the stability of ferrate solution, and 63-75 % of the original ferrate remained in solutions when pH was higher than 10. Several sduies have reported that Fe(VI) solution was stable in the pH range 9.0-10.0 *(14, 15, 24, 25)*, which is consistent with the result obtained in this study. The reason for the slight increase in decomposition ratio of Fe(VI) at pH > 10 is not clear. Lee and Gai *(25)* depicted that the lowest initial reeduction rate of ferrate by water occurred at pH 9.4-9.7, below which decomposition occurred by a reaction that was second-order with respect to ferrate concentration and approximately first-order above pH 10. In addition, Graham et al. *(15)* showed that the stability of the ferrate is highly pH dependent and there appeared to be a maximum stability at approximately pH 10. They proposed that the descrease in ferrate stability at pH > 10 may be attributed to the association with a different reduction pathway leading to the formation of anionic ion species instead of $Fe(OH)_{3(s)}$.

The decomposition of ferrate in aqueous solution can be explained by a first-order reaction kinetics with respect to ferrate(VI) concentration *(23)*. Table 1 shows the first-order rate constants (k_1) for ferrate decomposition at various pH values. The k_1 for ferrate decomposition decreased significantly from 3.9×10^{-3} s^{-1} at pH 7.0 to 1×10^{-4} s^{-1} at pH 9.2. Further increase in pH value slightly increased the decomposition rate of ferrate, and the k_1 for ferrate decomposition was in the rang $5.0\text{-}9.7 \times 10^{-4}$ s^{-1} at pH 10-11. Li et al. *(23)* reported that the decomposition rate constant of ferrate ranged between $(1 - 37) \times 10^{-4}$ s^{-1} in the pH range of 7.1 – 11.9, and the rate constant has a minimum value between pH 9.2 and 9.4. In this study, a minimal rate constant for ferrate decomposition was found to be 9.2, which shows that the potassium ferrate prepared in this study is most stable under alkaline conditions. It is also noted that the decomposition rate of ferrate(VI) in solution can be neglected during the experiment because the reaction of tetracycline with ferrate can be complete within 3 min.

Table 1. The first-order rate constant (k_1) for ferrate decomposition at various pH values.

pH	k_1 (s^{-1})	r^2
7.0	3.9×10^{-3}	0.967
8.0	1.8×10^{-3}	0.988
9.2	1.0×10^{-4}	0.989
10.0	5.0×10^{-4}	0.969
11.0	9.7×10^{-4}	0.988

The ferrate solution was purple initially and gradually changed to a yellowish color when decomposition occurred. Under acidic conditions, ferrate has a high oxidation potential which may lead to a rapid redox reaction with water, and thus results in the instability of ferrate(VI) solution at low pH.

$$H_2FeO_4 \rightarrow H^+ + HFeO_4^- \quad pK_a = 3.5 \ (12, 26) \quad (3)$$

$$HFeO_4^- \rightarrow H^+ + FeO_4^{2-} \quad pK_a = 7.23 \ (12, 13) \quad (4)$$

$$4\ K_2FeO_4 + 10\ H_2O \rightarrow 4\ Fe(OH)_3 + 8\ KOH + 3O_2 \quad (5)$$

On the contrary, the dominant ferrate species is FeO_4^{2-} when the pH is higher than 8.3 *(5)*, which is more stable and persists longer time in solution. It is believed that ferrate could undergo a different reaction pathway at pH > 10, leading to the formation of anionic iron species ($Fe(OH)_4^-$ and $Fe(OH)_6^{3-}$) instead of the formation of ferrihydrite ($Fe(OH)_3$) *(15, 27, 28)*. Therefore, the pH value of 9.2 was selected to prepare the ferrate(VI) solution for further experiments.

Stoichiometry

The stoichiometric experiments were conducted by fixing the initial concentration of aqueous ferrate(VI) at 70 μM and varied the initial concentration of aqueous tetracycline ranging from 0 - 120 μM in the solution at pH 9.2. The residual concentration of ferrate(VI) after reaction was measured by ABTS method. Figure 4 illustrates the residual ferrate concentration as a

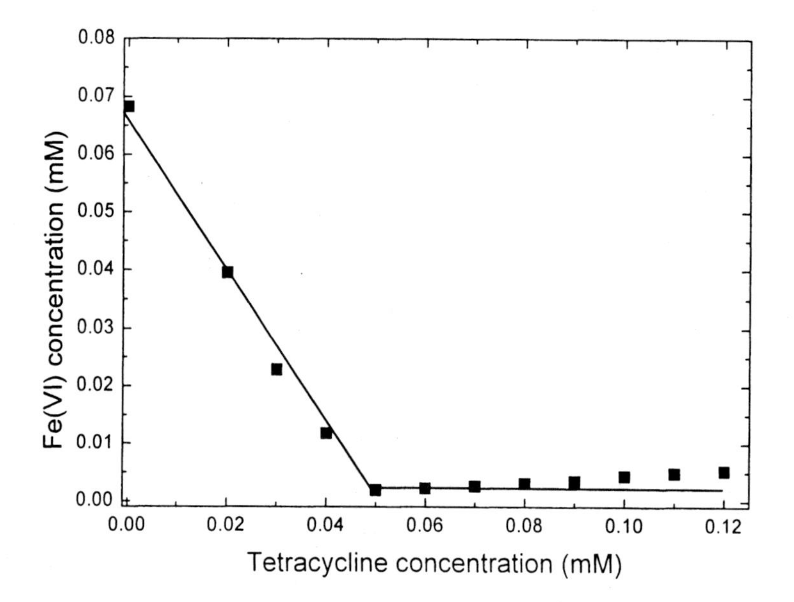

Figure 4. The stoichiometry of ferrate to tetracycline during the reaction of tetracycline with Fe(VI). The initial concentrations of Fe(VI) was 70 μM and the Fe(VI) concentration in solution was determined by ABTS method.

function of tetracycline concentration. The ferrate concentration decreased progressively with the increase in tetracycline concentration. When the tetracycline concentration was higher than 50 μM, the residual ferrate concentration in solution was below to the detection limit. The slope of linear relationship between ferrate and tetracycline in the concentration range of lower than 50 μM was found to be 1.37. In addition, a yellow color after the reaction of tetracycline with ferrate was clearly observed, suggesting that the final product of Fe(VI) was Fe(III) ions. Therefore, a stoichiometry of 1.4 (Fe(VI):tetracycline) was proposed in this study for the degradation of tetracycline by Fe(VI).

$$1.4\ FeO_4^{2-} + \text{tetracycline} \rightarrow 1.4\ Fe(III) + O_2 + \text{product(s)} \qquad (6)$$

Degradation of tetracycline

The mixing of aqueous tetracycline and ferrate(VI) leads to a series of solution color changes. The solution starts out purple, rapidly turns yellowish brown, and then slowly faded to colorless. Therefore, the UV-Vis spectra of Fe(VI) and tetracycline solutions were determined first. The UV-Vis spectra of tetracycline showed two peaks at 278 and 359 nm, while a strong absorption peak at about 510 nm was observed for ferrate. Since 278 nm has a strong interference after the reaction of tetracycline with ferrate, the wavelength of 359 nm was used to monitor the change in tetracycline concentration for further experiments.

The effect of solution pH on the degradation of tetracycline was first studied. Figure 5 shows the concentration profile of tetracycline as a function of pH value. The degradation efficiency of tetracycline increased upon increasing pH value (Figure 6). The degradation efficiency of tetracycline at pH 7.5 was only 35%, while the oxidative removal efficiencies of tetracycline increased to 53-64% when the pH were higher than 8.3. The low degradation efficiency at pH 7.5 may probably be due to that ferrate(VI) is unstable at pH < 8. This result clearly depicts that pH plays a crucial role in controlling the degradation of tetracycline.

The degradation experiments were further undertaken in terms of initial ferrate(VI) concentration where an excess tetracycline (200 μM) was used. The effect of ferrate concentration on the degradation of 200 μM tetracycline at pH 10 was examined. Figure 7 shows the effect of initial Fe(VI) concentration on the degradation of tetracycline. The degradation efficiency increased upon increasing initial ferrate(VI) concentrations ranging from 10 to 75 μM, showing that ferrate(VI) can be used as an environmentally friendly reagent to efficiently oxidative remove tetracycline in aquatic environments.

Figure 8 shows the electrospray ionization-mass spectrometry (ESI-MS) spectra of tetracycline and the products after the reaction of tetracycline with

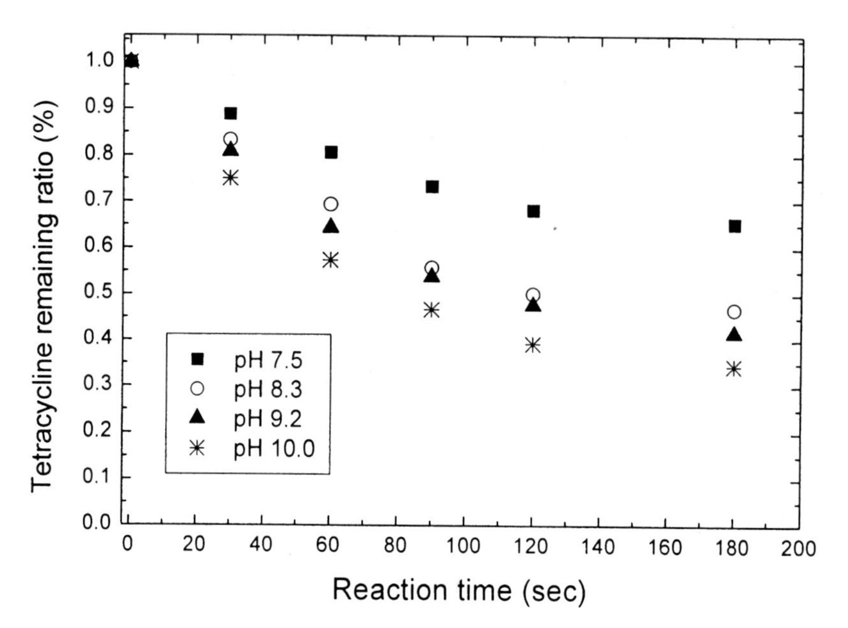

Figure 5. The degradation of tetracycline by Fe(VI) at various pH values ranging from 7.5 to 10.0. The concentration of tetracycline and Fe(VI) were 200 and 30 μM, respectively.

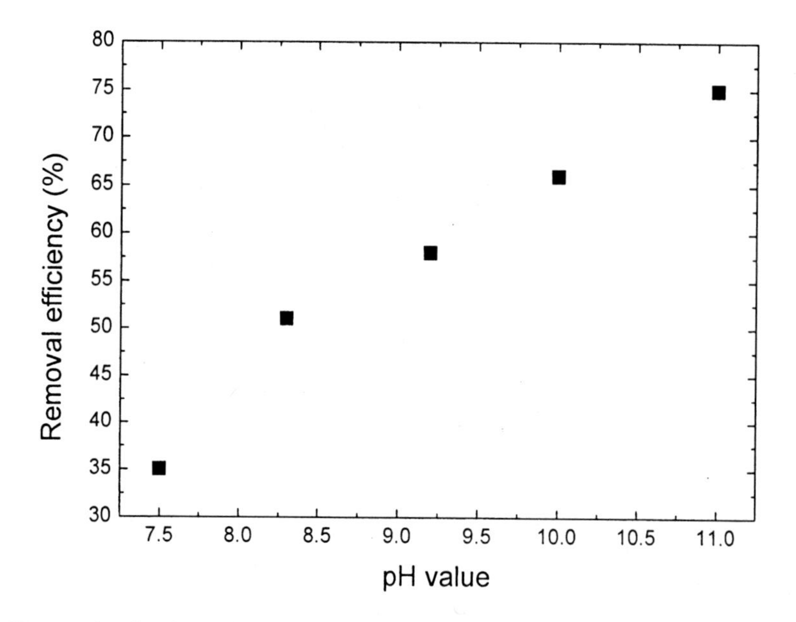

Figure 6. The degradation efficiency of tetracycline as a function of pH values.

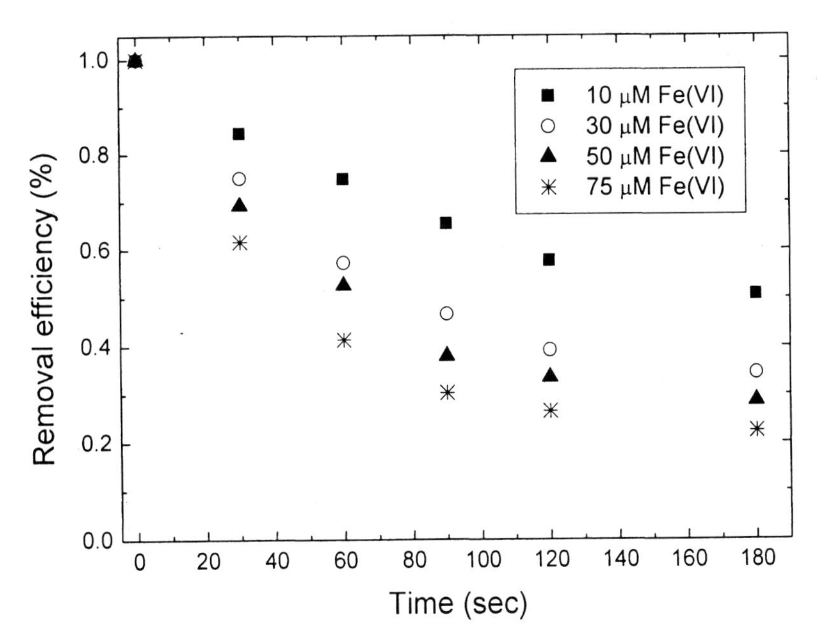

Figure 7. Effect of initial ferrate concentration on the degradation of tetracycline.

Fe(VI). The samples were directly injected into the ESI-MS to generate sufficient readable signal. An abundant peak at *m/z* 410.0 and a small peak at m/z 427 appeared in tetracycline standard. Several studied have demonstrated that tetracycline gives $[M + H]^+$, $[M+H-H_2O]^+$, and $[M+H-NH_3-H_2O]^+$, which correspond to m/z 445, 427, and 410, in the ESI-MS spectra measured at the high capillary voltage *(29, 30)*. This clearly depicts that the spectrum obtained can represent the parent compound of tetracycline. After the reaction of tetracycline with Fe(VI), a substantial decrease in peaks at m/z 445 and 410 was observed, clearly showing the degradation of tetracycline by Fe(VI). Total organic carbon (TOC) analysis also showed that the TOC decreased from 17.65 mg-C/L at initial to 15.65 mg-C/L after the reaction. However, it is difficult to identify the possible products from the ESI-MS spectra and TOC analysis from the results obtained in this study. Further experiments are necessary to clarify mechanisms and reactive pathways of tetracycline by Fe(VI).

Conclusions

In this study, the degradation of tetracycline by Fe(VI) was investigated. Results show that potassium ferrate exhibits good potential to be an effective

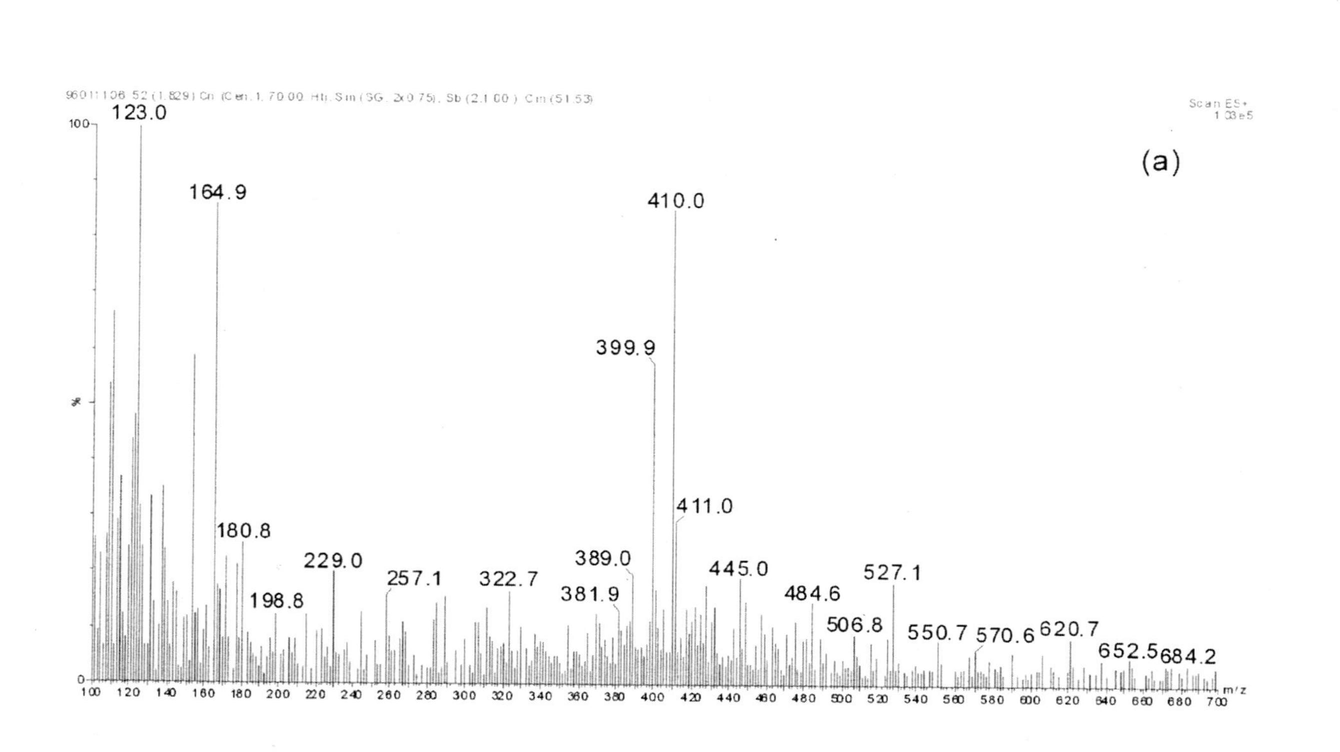
Scan ES+
1.03e5
(a)
123.0
164.9
180.8
198.8
229.0
257.1
322.7
381.9
389.0
399.9
410.0
411.0
445.0
484.6
506.8
527.1
550.7
570.6
620.7
652.5
684.2
%
100
0
m/z
100 120 140 160 180 200 220 240 260 280 300 320 340 360 380 400 420 440 460 480 500 520 540 560 580 600 620 640 660 680 700

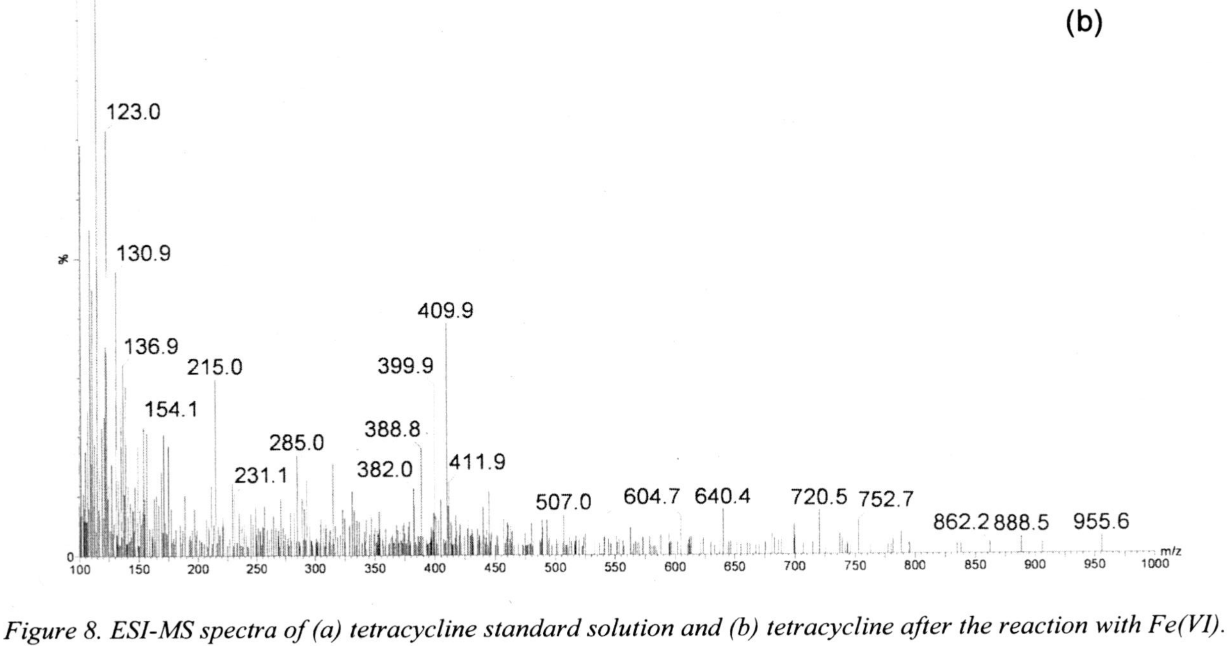

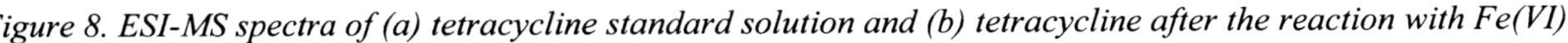
Figure 8. ESI-MS spectra of (a) tetracycline standard solution and (b) tetracycline after the reaction with Fe(VI).

oxidant for the degradation of tetracycline in aqueous solutions. The stability of ferrate in aqueous solution is highly dependent on pH value. The prepared Fe(VI) is rather stable under alkaline condition at pH 9.2. A stoichiometry of 1.4 for the reaction of Fe(VI) with tetracycline was also obtained. In addition, the oxidative degradation of tetracycline by Fe(VI) is affected by pH value and Fe(VI) concentration. Tetracycline can be successfully degraded by different dosages of Fe(VI) at various pHs. The degradation efficiency of tetracycline was linearly correlated to both pH and initial Fe(VI) concentration. ESI-MS spectra showed that tetracycline was degraded by Fe(VI) and transformed to low-molecular-weight compounds. Additional experiments are needed to clarify mechanisms and reactive pathways of tetracycline by Fe(VI).

Acknowledgements

The authors thank National Science Council, Taiwan for financial support under contract No. NSC95-2221-E-007-077-MY3.

References

1. Westerhoff, P.; Yoon, Y., Snyder, S.; Wert, E. *Environ. Sci. Technol.* **2005**, *39*, 6649-6663.
2. Daughton, C. H.; Ternes, T. A. *Environ. Health Perspect.* **2000**, *108*, 598-598.
3. Kolpin, D. W.; Furlong, E. T.; Meyer, M. T.; Thurman, E. M.; Zaugg, S. D.; Barber, L. B.; Buxton, H. T. *Environ. Sci. Technol.* **2002**, *36*, 1202-1211.
4. Adams, C.; Wang, Y.; Loftin, K.; Meyer, M. *J. Environ. Eng.-ASCE*, **2002**, *128*, 253-260.
5. Huber, M. C.; Canonica, S. Park, G. Y.; Gunten, U. V. *Environ. Sci. Technol.* **2003**, *37*, 1016-1024.
6. Acero, J. L.; Von Gunten, U. *J. Am. Water Works Assoc.* **2001**, *93*, 90-100.
7. Von Gunten, U. *Wat. Res.* **2003**, *37*, 1443-1467.
8. Goodman, G.A.; Goodman, L.S.; Rall, T.W.; Murad, F. (Eds.), *The Pharmacological Basis of Therapeutics*, seventh ed., Mac-Millan, New York, **1985**.
9. Cherlet, M.; Croubels, S.; De Becker, P. *J. Chromatogr. A* **2006**, *1102*, 116-124.
10. Reyes, C.; Fernandez, J.; Freer, J.; Mondaca, M. A.; Zaror, C.; Malato, S.; Mansilla, H. D. *J. Photochem. Photobiol. A-Chem.* **2006**, *184*, 141-146.
11. Dodd, M. C.; Buffle, M. O.; Von Guten U. *Environ. Sci. Technol.* **2006**, *40*, 1969-1977.
12. Sharma, V. K. *Adv. Environ. Res.* **2002**, *6*, 143-156.

13. Sharma, V. K. *Wat. Sci. Technol.* **2004**, *49*, 69-74.
14. Jiang, J. Q.; Lloyd, B. *Wat. Res.* **2002**, *36*, 1397-1408.
15. Graham, N.; Jiang, C. C.; Li, X. Z.; Jiang, J. Q.; Ma, J. *Chemosphere* **2004**, *56*, 949-956.
16. Sharma, V. K.; Burnett, C. R.; Yngard, R. A.; Cabelli, D. E. *Environ. Sci. Technol.* **2005**, *39*, 3849-3854.
17. Lee, Y.; Yoon, J.; Gunten, U. V. *Envieron. Sci. Technol.* **2005**, *39*, 8978-8984.
18. Jiang, J. Q.; Yin, Q.; Zhou, J. L.; Pearce, P. *Chemosphere*, **2005**, *61*, 544-550.
19. Sharma, V. K.; Mishra, S. K.; Ray, A. K. *Chemosphere* **2006**, *62*, 128-134.
20. Sharma, V. K.; Mishra, S. K.; Nesnas, N. *Environ. Sci. Technol.*, **2006**, *40*, 7222-7227
21. Bielski, B. H. J.; Thomas, M. J. *J. Am. Chem. Soc.* **1987**, *109*, 7764-7791.
22. Lee, Y.; Yoon, J.; Gunten, U. V. *Wat. Res.* **2005**, *39*, 1946-1953.
23. Li, C.; Li, X.Z. ; Graham, N. *Chemosphere* **2005**, *61*, 537-543.
24. Wagner, W. F.; Gump, J. R.; Hart, E. N. *Anal. Chem.* **1952**, *24*, 1497-1498.
25. Lee, D.G., Gai, H. *Can. J. Chem.* **1993**, *71*, 1394-1400.
26. Rush, J.D., Zhao, Z., Bielski, B.H.J. *Free Rad. Res.* **1996,** *24*, 187-198.
27. Li, C., Li, X. Z. Graham. N. *Chemosphere*, **2005**, *61*, 537-543.
28. Sharma, V. K.; Rendon, R. A.; Millero, F. J.; Vazquez, F. G. *Mar. Chem.* **2000**, *70*, 235-242.
29. Oka, H.; Ito, Y.; Ikai, Y.; Kagami, T.; Harada, K. I. *J. Chromatogr. A.* **1998**, *812*, 309-319.
30. Cherlet, M.; Schelkens, M.; Croubels, S.; De Becker, P. *Anal. Chim. Acta* **2003**, *492*, 199-213.

Chapter 26

Removal of Estrogenic Compounds in Dairy Waste Lagoons by Ferrate(VI)

Jarrett R. Remsberg[1], Clifford P. Rice[2,*], Hyunook Kim[3], Osman Arikan[4], and Chulhwan Moon[3]

[1]Brook-Lodge Farms, 7229 Holter Road, Middletown, MD 21769
[2]Environmental Management and Byproduct Utilization Laboratory, Agricultural Research Service, U.S. Department of Agriculture, Beltsville, MD 20705
[3]Department of Environmental Engineering, The University of Seoul, 90 Jeonnong-dong, Dongdaemun-gu, Seoul 130–743, Korea
[4]Department of Environmental Engineering, Istanbul Technical University, Istanbul 34469, Turkey

Ferrate(VI) was used to remove steroidal estrogens (SE) from dairy waste lagoon effluent (DWLE). Dairy lagoon sites were sampled for estrogenic content (EC) and assayed using high performance liquid chromatography coupled to triple quadrupole mass spectrometry. Effects of varying amounts of ferrate(VI) and ferric chloride treatments on the EC of these DWLE samples were determined. Of the compounds measured, 17β-estradiol, 19.7 μg/L, was the most abundant and estriol, 2.10 μg/L the least abundant. When DWLE was treated with a high concentration (0.84%) of ferrate(VI) there was a significant decrease (>50%) in 17β-estradiol content. Ferrate(VI) treatment of DWLE may be an environmentally sound approach to reduce estrogenic compounds.

The widely accepted meaning for endocrine disruptor compounds (EDCs) are any exogenous substances that cause adverse biological effects by interfering with the endocrine system and disrupting the physiologic function of hormones. At the present time a number of compounds are known and/or suspected EDCs. They can be either naturally or artificially produced. For example, steroidal estrogens (SEs) and androgens are naturally produced by female and male organisms. Some common ones are E1 - estrone; E2 - 17β-estradiol; E3 – estrol; 17β-keto testosterone - testosterone; and progesterone. Some industrial chemicals that are potential EDCs are tributyltin, akylphenol polyethoxylates (APEOs) and their intermediates such as nonylphenol (NP) and octylphenol (OP), bisphenol A (BPA), phthalate esters, brominated flame retardants, polychlorinated biphenyls, dioxins, furans, and more. Typical environmental concentrations and potencies of three extensively studied synthetic EDCs are summarized in Table 1 and these values are compared to similar data for some of steroidal estrogens (SEs) including the synthetic birth control agent, 17α-ethylnyl estradiol (EE2).

The potential environmental impacts of each EDC can be determined by converting its concentration in the environment to an E2 concentration equivalency (*1-3*). This approach is useful in determining what compounds are of most environmental concern. Table 1 shows the predicted potential impacts of some of EDCs. Based on this table, some SEs appear to be of greater concern than industry-derived EDCs; EE2, especially, has a relatively high *in-vivo* potency. These compounds can exert estrogenic activity on aquatic organisms; they can cause changes in development, growth, and reproduction of exposed organisms (*4-8*). The effects of SEs are particularly pronounced in aquatic organisms, with feminization of male fish observed in Europe, the UK, Japan, Canada and the USA, including Maryland waterways (*6-9*).

Table 1. The potencies of known and suspected EDCs

Compound	General effluent conc. range (ng/l)	*In vitro* potency	Predicted *in vitro* potency E2 equiv*	*In vivo* potency*	Predicted *in vivo* potency E2 equiv*
NP(branched)†	100-14,000	10^{-3}-10^{-6}	2×10^{-8}-6.9×10^{-2}		
4t-NP or 4t-OP‡	2,000	10^{-4}	0.2	1×10^{-2} to 6×10^{-3}	1.2-2
BPA‡	0.0001-4.5	10^{-4}-10^{-5}	1×10^{-9}-4.5×10^{-4}		
Estrone (E1)†	5	0.5	2.5	0.5	2.5
17β-estradiol; (E2)†	1.5	1	1.5	1	1.5
17α-ethylnyl estradiol (EE2)†	0.5	1-2	0.5-1	25	12.5
Estrol (E3)†	20	0.5×10^{-2}-0.04	0.1-0.8	0.001	0.02

† adapted from Johnson and Sumpter (*10*)
‡ adapted from Johnson and Jürgens (*11*)
* calculated based on the general effluent concentration and in vitro/in vivo potency

Municipal wastewater treatment plants, which handle estrone and 17α-ethinyl estradiol excreted by those using oral contraceptives or estrogen replacement therapy have been documented as major entry points for SE into waterways (*12-15*). Farm waste lagoon effluents, which are used as field fertilizers may be a significant secondary source of aquatic SE contamination. A rapid, inexpensive and safe means of breaking down and/or removing SE from waste lagoon effluents, prior to their use as fertilizer, is needed to minimize introduction of SE into the environment.

SE compounds are easily biodegradable and sorb onto digested sludge. A recent study by Anderson et al., (*16*) quantified the removal efficiency of SE compounds in an activated sludge system with a series of anoxic and aerobic tanks. Over 98% of E1 (initial concentration: 65.7 ng/L) and E2 (initial concentration: 15.8 ng/L) was removed while 90% removal efficiency was achieved for EE2. Additionally they observed E1 and E2 degradation both in the nitrification and denitrification tanks. On the other hand, EE2 could be degraded in the nitrification tank due to the presence of its conjugated form in the denitrification tank, where conjugated estrogens of EE2 were firstly transformed to easily biodegradable unconjugated estrogens.

Chemical oxidation can be applied to remove SEs from water or wastewater (*17-22*). In particular, ozone, an effective oxidant and disinfectant, has frequently been studied as an oxidizer of EDCs in water and wastewater treatment. Huber et al. (*19*) oxidized EE2 to less estrogenic compounds (i.e., 5,6,7,8-tetrahydro-2-naphthol and 1-ethinyl-1-cyclohexanol) by ozone and observed decreased overall estrogenicity mainly due to the destruction of the phenolic ring. Despite its effectiveness, ozone is rarely adopted for practical application because it requires a large intial capital investment and high operational cost.

Recently, ferrate(VI) has been considered as an alternative oxidant for wastes treatment, because it is a very strong oxidant (the redox potential of ferrate(VI) is 2.20V at acidic condition and 0.57V at basic condition, and it can effectively destroy organic pollutants and disinfect pathogenic organisms (*23-27*). Ferrate(VI) treatment of waste can also yield "environmentally friendly by-product", Fe(III) (14,17) (*24,27*), which has been used as a coagulant in some waste operations (*28*). Because of these beneficial qualities, ferrate has been tested to oxidize or destroy various organic compounds; e.g., sulfurous (*29*), phenolics (*30*), and nitrogen compounds (*31*) in water. Oxidation of OP, NP, BPA, E1, and E2 using Fe(VI) were also reported (*20, 32-34*)). Kim et al (*35*) applied on-site produced ferrate to thickened and dewatered biosolids and successfully disinfected pathogenic microorganisms.

This project investigates the use of ferrate(VI) oxidation-coagulation as a means of breaking down and/or removing SE from dairy waste lagoon effluent (DWLE), to minimize the amount of SE entering groundwater/watersheds.

The SE compounds monitored in this study include E1, E2, E3, EE2, progesterone, and testosterone (Figure 1). All of these compounds consist of three hexagonal rings, A-ring, and one pentagonal ring, D-ring. Each SE compound is characterized with its functional group at carbon positions C-3, C-16, and C-17, Fig 1 (estrone). E1, E2, E3, and EE2 have hydroxyl group at C-3, whereas, an oxyl group occupies the C-3 position in the case of progesterone. Estradiol can be further divided into α- and β-configurations depending on the direction of functional group at C-17 downward or upward, respectively. High affinity of these compounds to an estrogen receptor is known due to the phenolic and hydroxyl groups.

Methods

Reagents: Estrone (E1), 17 β-estradiol (E2), 17 alpha-ethinylestradiol (EE2), 17-keto testosterone, and progesterone were all obtained from Amersham Searl (Amherst, MN) while ^{13}C- 17 β-estradiol was obtained from Sigma-Aldrich (Miamisburg, OH). The solvent, diethyl ether, was pesticide grade from Burdick and Jackson (Honeywell International, Muskegon, MI, USA) and methanol was UV grade from Fisher Scientific (Pittsburg, PA). Syringe filters were 13-mm GHP brand Acrodisc Pall Gelman Lab 0.45 μ pore size syringe filters (Ann Arbor MI.)

The ferrate(VI) utilized in this experiment was prepared as described by Park et al. (*34*) and handled as potassium ferrate (K_2FeO_4). Ferric chloride ($FeCl_3$) and sodium borate ($Na_2B_4O_7$), which was used in the buffer solution to test the purity of the ferrate(VI) were obtained from Fisher Scientific (USA).

Sample Collection: To obtain the dairy waste lagoon effluent (DWLE) for these experiments, a dairy lagoon near Frederick, Maryland was sampled on two occasions in March 2006. The samples were collected by dipping an open bucket and the lagoon slurry material decanted into sealed 1-L plastic jars. The samples were then placed in a standard refrigerator and subsequently delivered to the laboratory walk-in refrigerator, which was maintained at 4 C°. In order to minimize losses during storage, these chilled samples were extracted within one week of collection.

Sample Processing: Each experimental test was run at least in duplicate. Three levels of ferrate were tested (0.07, 0.34 and 0.84%) which were crudely set to give 0.1, 0.5 and 2 g of dried ferrate powder per 125 ml of each experimental sample. To account for the possibility that ferric(III) precipitation might also be occurring in these trials due to incidental ferric(III) formation during reduction of the ferrate, separate $FeCl_3$ tests were carried out by adding ferric(III) amounts to equal the molar concentrations of ferrate, e.g., 0.14 g $FeCl_3$ and 1.8 g $FeCl_3$. Distilled water was also tested after spiking it with a mixture of the hormones (E1, E2, E3, EE2, testosterone and progesterone) and reacting this mixture with a medium level of ferrate (0.034% FeO_4^{-2}). Since all the samples had labeled 13C-estradiol added as internal standard just prior to extraction, this compound was used to represent spike recovery of the method. The ferrate powders were supplied to us from samples employed in another experiment (*36*). Since the purity varies with each batch it was necessary to deterimine the exact ferrate

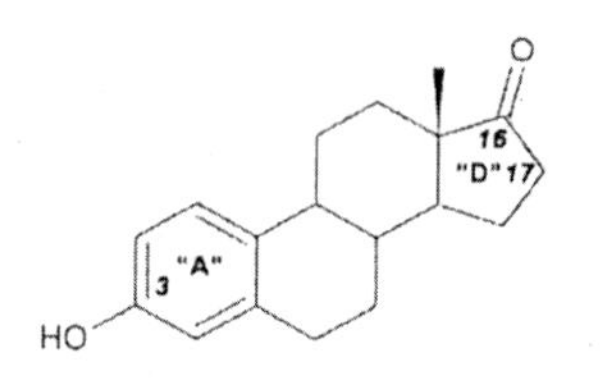

Estrone (E1)

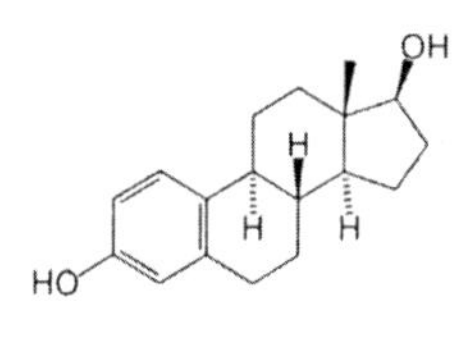

17β-estradiol (E2)

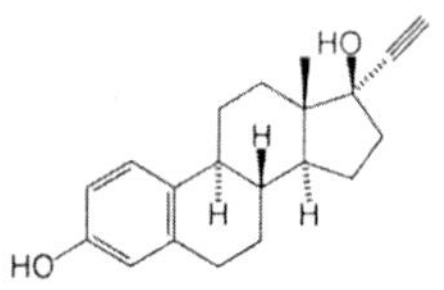

17α-ethinyl estradiol (EE2)

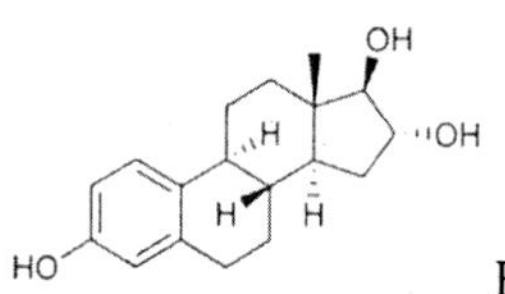

Estriol (E3)

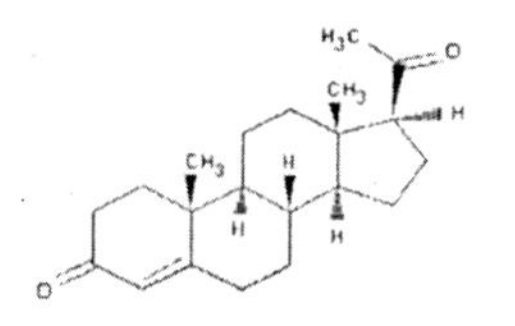

Progesterone

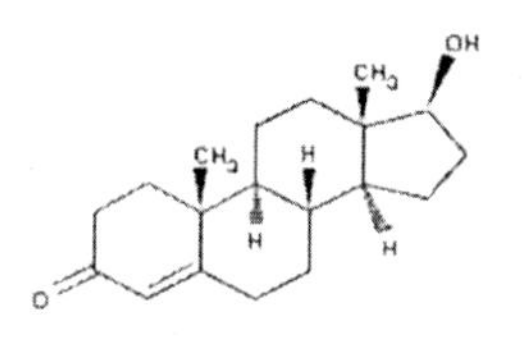

17β-keto testosterone (Testosterone)

Figure 1. The structure of EDCs selected in this study.

concentration using the method described by Jiang and Lloyd, (*26*) and Arispe et al., (*36*). An exact portion of the crude ferrate(VI) mixture is dissolved in borate buffer solution and evaluated by spectrometry at an absorption of 510 nm. This determination disclosed that one preparation was 83 % pure vs. a second batch that had a purity of 54%, These percentages were used to correct the weighed quantities of each mixture to the true concentrations of ferrate (VI) used in each experimental tests.

Each reaction was carried out using 125 ml of solution placed in 250 ml plastic centrifuge tubes with ferrate(VI) or ferric chloride solution (0 to 2 grams). These tubes were capped and rapidly shaken (200 rpm) initially for 5 minutes with the reagents being added in three equal amounts and at three equally spaced intervals of 30 seconds each. These three 30-second-spaced additions improved the efficiency of the process by providing maximum contact between the reagents and the solids contained in each sample. The reaction was then allowed to continue for 30 minutes with slow shaking (10 rpm) and finally a static incubation was continued for another 30 minutes. At this point the internal standard (13C-estradiol) was added and the reaction immediately stopped by addition of 125 ml of diethyl ether and shaken rapidly by hand to start the extraction. Then each sample was further shaken for 5 minutes at 200 rpm on a Glas-Col mechanical shaker. This diethyl ether extraction method was adapted from a procedure reported by Hanselman and associates (*37*) where diethyl ether was used to isolate 17 β-estradiol from biosolids in their immunoassay determinations. The rapid shaking caused a dense region of emulsion to form that was reduced by centrifuging the entire bottle, still capped, using a Glas-Co R brand centrifuge operated at 3000 RPM for 10 minutes. The upper diethyl ether layer was decanted and evaporated to dryness under nitrogen. The residue from this blow down was first dissolved in 1 mL of methanol, and then 1ml of distilled water was added. Just prior to introduction of this mixture into the LC column of the LC/MS-MS instrument the samples were syringe filtered using 13-mm GHP brand Acrodisc Pall Gelman Lab 0.45 μ pore size syringe filters.

Hormone Analysis: The hormones were analyzed by LC/MS-MS using an adaption of a procedure reported by Laganá, et al., (*38*). The LC instrument was a 2695 XE separations module (Waters Corp., Milford, MA) equipped with an Xterra MS C18 column (150 mm x 2.1 mm i.d., 5 μm) (Waters Corp.) and operated at a temperature of 45 °C. The injection volume was 10 μl. The mobile-phase solvents used to carry out the separation were gradient mixtures of solvent "A", 1% formic acid-methanol (70:30, v/v); "B", water; and solvent "C", methanol. The solvents were mixed as follows: 0-1 min. (50:50, A:B); 1-12 min. was a linear gradient from the previous mixture to (70:30, A:C); then 12-30 min. to (7:93) A:C); and finally the instrument was returned to starting conditions and allowed to stabilize for 10 min to the initial 50:50 mixture of solvent A and B. The total run time was 42 min. The flow of the column was set at the rate of 0.25 mL/min. The analytes were detected using atmospheric pressure ionization-tandem mass spectrometry. The instrument was a benchtop triple quadrupole mass spectrometer (Quattro LC from Micromass Ltd., Manchester, U.K.)

operated in electrospray ionization mode. The source parameters were as follows: capillary voltage at 3.0 kV and extractor voltage at 3 V, respectively; rf lens set at 0.1 V; source and desolvation temperatures were 150 and 450 °C. Liquid nitrogen was used to supply the nebulizer and desolvations gas (flow rates were approximately 80 and 600 L/h, respectively). Argon was used as collision gas to fragment the parent ions; the typical pressure of the collision cell was 2.6 x 10^{-3} mbar. High and low mass resolutions were set at 12.0 for both quadrupoles. Acquisition was done in the multiple-reaction monitoring mode (MRM) in electrospray positive (ES+) mode. The parent and daughter ions used for compound identification and quantitation are listed in Table 1 along with the optimum cone voltages and collision energies for each analyte. The detector was a photomultiplier set at 650 V. A typical LC/MS-MS run is shown in Figure 2. Quantitation of the hormones were calculated by internal standard method using ^{13}C-17β-estradiol. Quantitation for the non-estradiol compounds was confirmed using the method of standard additions as described by Lindsey et al.,(*42*) since matrix effects were possibly different for those compounds whose retention times differed from estradiol.

Table 1. Parent and daughter ions used for quantitation and the MS parameters used to produce them.

Compound	Parent ion, Da	Daughter ion, Da	Retention time, min	Cone, V	Collision, eV
E3	271	133	6.4	22	18
E2	255	159	6.8	14	21
EE2	279	133	6.8	24	16
E1	271	253	6.4	20	15
Testosterone	289	109	7.5	25	21
Progesterone	315	109	12.7	14	23
^{13}C-17β-Estradiol	258	159	6.8	22	18

Quality Assurance: For concentrations of greater than 0.2 µg/L, the relative standard deviations of duplicates were < 25%. Recovery was followed by adding ^{13}C-labelled estradiol just prior to extraction of each sample. The apparent recovery of this compound varied from 11.6 to 38.8% with an average of 28.8% overall. This apparent recovery occurred because of matrix suppression due to the high amount of coextractants, which was estimated to be approximately 65%; therefore correcting for this suppression resulted in an average recovery of 81% for the internal standard. All the samples were corrected using the method of internal standard calculations based on ^{13}C-17β-estradiol. For the natural estradiol this internal standard method of quantitation is termed isotope dilution and produces very accurate data (*41*).

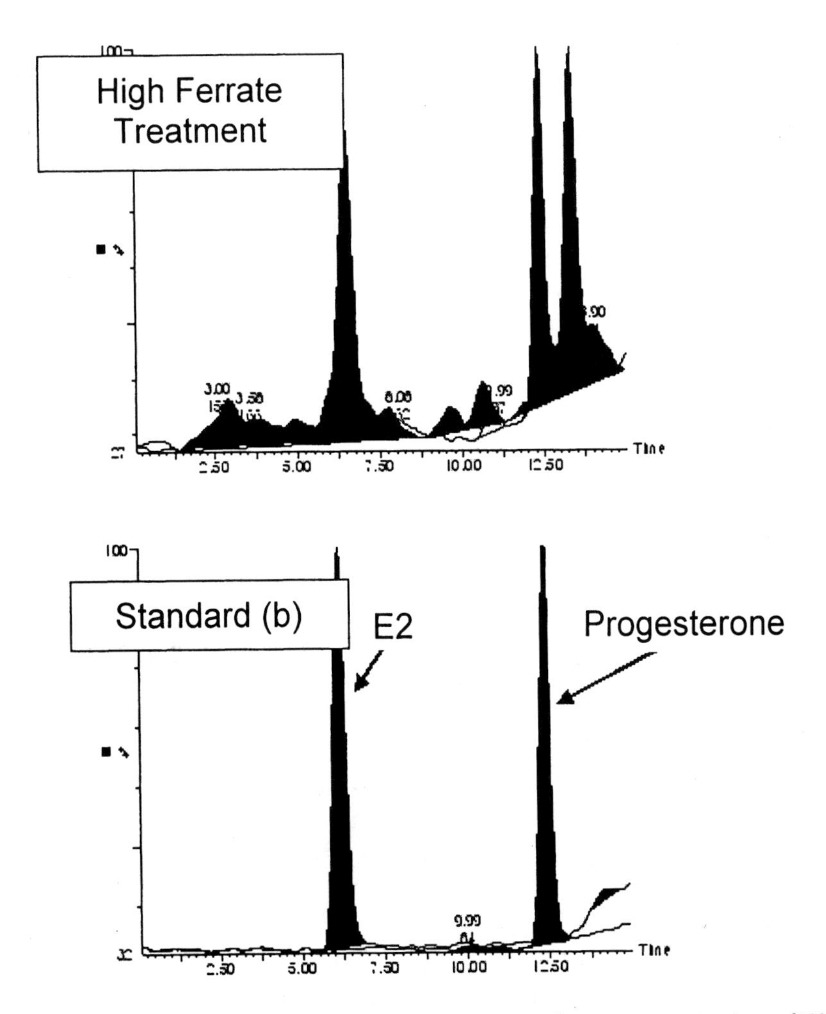

Figure 2. A typical extracted ion chromatogram showing quantitation of E2 and progesterone in one of the high ferrate lagoon sample treatments (a) versus the standard (b) for these two compounds.

Results and Discussion

Controlled treatment of a standard aqueous mixture of estrogenic compounds with ferrate(VI) (0.34%) significantly decreased the estrogenic content ($P < 0.05$) while producing a slight, but statistically non-significant increase, 12.5% in progesterone (Table 2.). The dairy waste liquid effluent (DWLE) contained measurable amounts of most of the estrogenic compounds (Table 2), except for testosterone, which is not surprising since the waste came predominantly from female cows. Initial average total concentration of extractable estrogenic compounds in these DWLE was 27.70 μg/L. 17β-estradio, 19.67 μg/L, was the most abundant estrogenic form and estriol, 2.10 μg/L and testosterone (< detection limits) the least abundant forms. Treatment of DWLE with ferrate(VI) decreased the concentration of all the extracted estrogenic compounds, except for progesterone, which actually increased with increasing ferrate(VI) treatment, Figure 2. Even though not shown in Table 2, trace levels of testosterone, <0.01 μg/L, were detected and a slight reduction was noted, Figure 3. There was a significant and dose-dependent reduction in estradiol (Table 2, Figure 4). Comparison of ferrate(VI) treatment of DWLE with low to medium and medium to high concentrations of Ferrate(VI) using a paired t-test revealed a significant increase in progesterone ($P < 0.05$); this is shown graphically in Figure 2. A paired t-test showed a significant decrease in estradiol content ($P < 0.05$) when DWLE was treated with a high concentration (0.84%) of ferrate(VI). Treatment of DWLE with a low (0.11%) level of ferric chloride, a by-product of ferrate(VI) had little effect on the estrogenic content. Treatment of DWLE with a high (0.30%) level of ferric chloride did appear to increase the total extractable estrogenic content: total 79.26 μg/L, 41.17 μg/L of 17 *alpha*-ethinylestradiol and 1.25 μg/L estriol (Table 2).

Treatment of spiked distilled water with ferrate (intermediate concentration 0.34%) removed most of the spiked compounds, e.g., 100% estriol, 97% estradiol, 100% 17 α-ethinylestradiol, 96% estrone, 82% testosterone, but did not remove any of the progesterone.

Table 2. Mean levels of estrogenic compounds in DWLE before and after treatment with varying concentrations of ferrate (Fe(VI)) or ferric chloride ($FeCl_3$).

Treatment	E3 (μg/L)	E21 (μg/L)	EE2 (μg/L)	E1 (μg/L)	Progesterone (μg/L)
DWLE	2.1	19.67	2.36	3.57	0.37
DWLE +$FeCl_3$ Low	2.25	16.13	2.17	3.17	0.5
DWLE +$FeCl_3$ High	1.25	29.33	41.17	7.51	1.3
DWLE+ Fe(VI) Low	2.17	21.48	2.25	3.21	0.48
DWLE+ Fe(VI) Medium	1.5	12.22	0.68	1.33	1.42
DWLE +Fe(VI) High	0.92	6.94	0.24	1.4	3.82

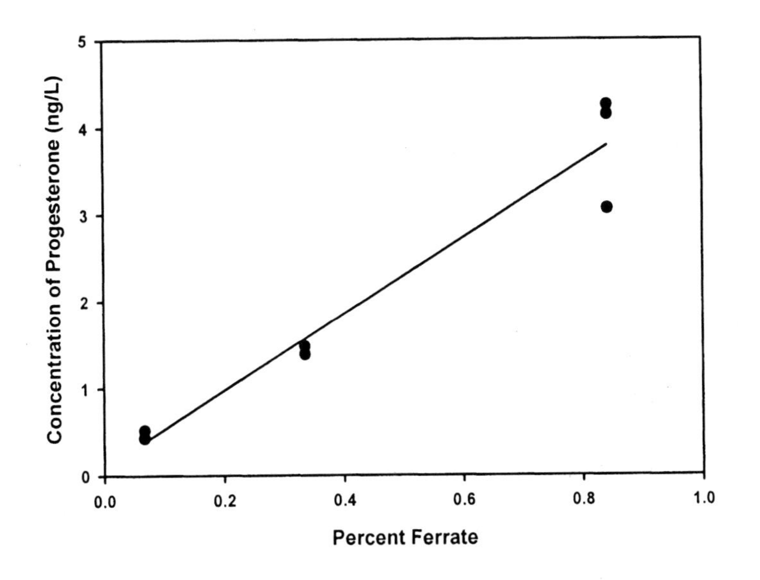

Figure 2. Change in concentration of progesterone in dairy lagoon waste upon reaction with increasing concentration of ferrate (FeVI).

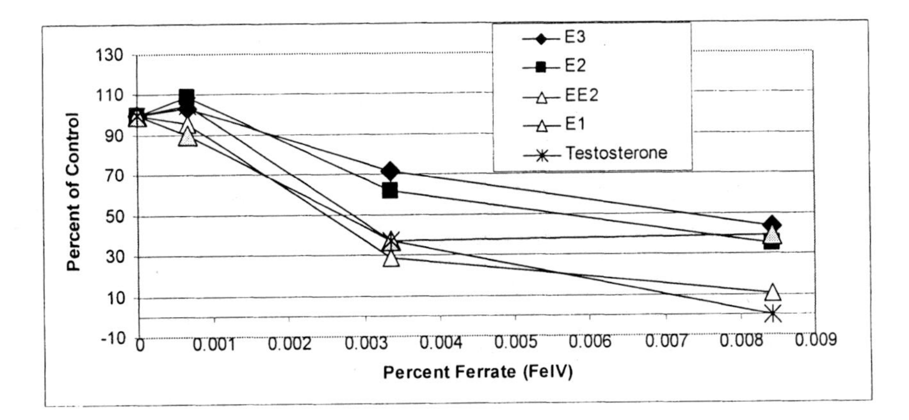

Figure 3. The concentration of each hormone expressed as percent of its starting concentration versus the concentration of Ferrate (FeVI) used.

In the dairy waste lagoon effluent sample tests with ferrate, there was significant destruction of all of the naturally occurring hormone compounds except progesterone. The most abundant hormone, 17 β-estradiol went from an average of 20.6 μg/L in the control to 6.9 μg/L for the highest ferrate treatment. This was a 66% removal of this compound; the other hormonal compounds

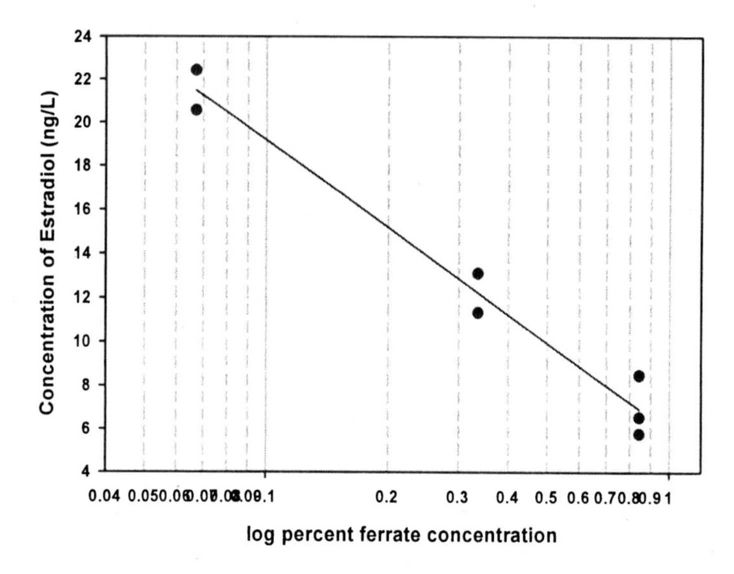

Figure 4. Destruction of estradiol versus the log of the percent concentration of ferrate (FeVI).

losses varied from 57% to 90%. Figure 3. Their levels were low and subject to high experimental error in these rather crude extracts. As shown by the distilled water control, progesterone is not destroyed by ferrate treatment and it may actually be formed in the lagoon slurry samples from some unknown precursor(s) that produce this compound by the ferrate treatment.

Ferrate(VI) treatment of DWLE decreased the concentration of 17 β-estradiol logarithmically with increasing concentration of ferrate(VI), Figure 4. The logarithmic relationship ($R^2 = 0.9724$) can be used to estimate the amount of ferrate necessary to remove this compound from a given amount of DWLE. The equation for this curvilinear relationship is ($y = -5.7471\text{Ln}(x) - 20.516$) with an R^2 curve fit value of 0.9724.

In no cases did the treatments with ferric chloride decrease the levels of any of these hormones. In fact in two cases, estradiol and ethinylestradiol, the concentration of these compounds were highest with the highest ferric chloride treatment. This was significantly higher for ethinylestradiol, which possibly occurred through release of sorbed materials by the ferric chloride treatments.

The linear increase in concentration of progesterone upon reaction with increasing concentration of ferrate(FeVI), Fig 2, was not expected. We do not have any explanation for this; however, several possibilities exist. The increases in progesterone suggests, possibly, that the ferric chloride and ferrate(VI) treatments disassociated the bio-solids in the sample, allowing bound progesterone to be released. A second possibility could be that this progesterone resulted from the oxidation of precursor compounds in the lagoon waste. And finally, a third possibility might be that the Ferric chloride and ferrate(VI) treatments caused cleavage of conjugated molecules either from progesterone conjugates directly or cleavage of these conjugates from progesterone precursors, which were then oxidized to progesterone. Finally all of these possibilities may be acting together to yield the higher progesterones observed after Ferrate(VI) and the high Ferric chloride treatments. Futher experiments must be performed to validate any of these hypotheses.

Raman et al. (*41*) examined dairy holding ponds in Tennessee for estrogenic compounds and found that estrone to be the most abundant of the three, estrone, 17β-estradiol and 17α-estradiol reported in their study. Their concentrations were all lower than ours, e.g., < 10 μg/L. Their extraction methods were similar to ours, however, they used GC/MS to quantify these analytes.

Further research is warranted to investigate the economic feasibility of using Ferrate(VI) treatment to minimize the estrogenic content of DWLE as well as the effects of the resulting progesterone upon aquatic organisms and the environment.

References

1. Jobling, S.; Sumpter, J. P. Fate of estrogens in a municipal sewage treatment plant. *Aquat. Toxicol.* **1993**, *27*, 361-372.
2. Brian, J. V.; Harris, C. A.; Scholze, M.; Backhaus, T.; Booy, P.; Lamoree, M.; Pojana, G.; Jonkers, N.; Runnalls, T.; Bonfa, A.; Marcomini, A.; Sumpter, J. P. Accurate Prediction of the Response of Freshwater Fish to a Mixture of Estrogenic Chemicals. *Environ. Health Perspect.* **2005**, *113*, 721-728.
3. Nimrod, A. C.; Benson, W. H. Environmental estrogenic effects of alkylphenol ethoxylates. *Crit. Rev. Toxicol.* **1996**, *26*, 335-364.
4. Scruggs, C.; Hunter, G.; Snyder, E.; Long, B.; Snyder, S. What's the Next Step? *Water Environ Technol.* **2005**, *17*, 25-31.
5. Jobling, S.; Casey, D.; Rodgers-Gray, T.; Oehlmann, J.; Schulte-Oehlmann, U.; Pawlowski, S.; Baunbeck, T. A.; Turner, A. P.; Tyler, C. R. Detergent components in sewage effluent are weakly oestrogenic to fish: An in vitro study using rainbow trout (Oncorhynchus mykiss) hepatocytes. *Aquatic Toxicol.* **2004**, *66* , 207-222.
6. Hanselman T.A.; Graetz, D. A.; Wilkie, A. C. Manure-borne estrogens as potential environmental contaminants: A review. *Environ. Sci. Technol.* **2003**, *37*, 5471-5478.
7. Hermanowics, S. W. Biodegradation of estrogenic compounds and its enhancement in a membrane bioreactor. in California Water Resources Center – Annual Report 2001-2002, **2002**, 63-65.
8. Roepke, T. A.; M.J. Snyder, M. J.; Cherr, G. N. Estradiol and endocrine disrupting compounds adversely affect development of sea urchin embryos at environmentally relevant concentrations. *Aquatic Toxicol.* **2005**, *71*, 155-173.
9. Purdom, C. E.; Hardiman, P. A.; Bye, V. J.; Eno, N. C.; Tyler, C. R.; Sumpter, J. P. Estrogenic effects of effluent from sewage treatment works. *Chemistry and Ecology* **1994**, *6*, 275-285.
10. Johnson, A.; Jürgens, M. Endocrine active industrial chemicals: Release and occurrence in the environment. *Pure Appl. Chem.* **2003**, *75*, 1895-1904.

11. Johnson, A. C.; Sumpter, J. P. Removal of Endocrine-Disrupting Chemicals in Activated Sludge Treatment Works. *Environ. Sci. Technol.* **2001**, *35*, 4697-4703.
12. Shore, L. S. ; Shemesh, M. Naturally produced steroid hormones and their release into the environment. *Pure Appl. Chem.* **2003**, *75*, 1859-1871.
13. Cargouet, M.; Perdiz, D.; Mouatassim-Souali, A.; Tamisier-Karolak, S.; Levi, Y. *Sci. Total Environ.* **2004**, *324*, 55-66.
14. Servos, M.; Bennie, D. T.; Burnison, B. K.; Jurkovic, A.; Mcinnis, R.; Neheli, T.; Schnell, A.; Seto, P.; Smyth, S. A.; Ternes, T. A. Assessment of river contamination by estrogenic compounds in Paris area (France). *Sci. Total Environ.* **2005**, *336*, 155-170.
15. Johnson, A. C.; Aerni, H.-R.; Gerritsen, A.; Gilbert, M.; Giger, W.; Hylland, K.; Jurgens, M.; Nakari, T.; Pickering, A.; Sutter, M. J-F.; Svenson, A.; Wettstein, F. E. Comparing steroid estrogen, and nonylphenol content across a range of European sewage plants with different treatment and management practices. *Water Res.* **2005**, *39*, 47-58.
16. Anderson, H. ; Siegrist, H.; Halling-SŘrensen, B.; Ternes, T. A. Fate of estrogens in a municipal sewage treatment plant. *Environ. Sci. Technol.* **2003**, *37*, 4021-4026.
17. Snyder, S., Proceedings International Water Authority, **2002**.
18. Schäfer, A. I.; Waite, T. D. 2002 International Water Authority.
19. Huber, M. C. ; Ternes, T. A.; von Gunten, U. Removal of estrogenic activity and formation of oxidation products during ozonation of 17-ethinylestradiol. *Environ Sci Technol.* **2004**, *38* , 5177-5186.
20. Jiang, J. Q. ; Yin, Q.; Zhou, J. L.; Pearce, P. Occurrence and treatment trials of endocrinedisrupting chemicals (EDCs) in wastewaters. *Chemosphere* **2005**, *61*, 544-440.
21. Barron, E.; Deborde, M.; Rabouan, S.; Mazellier, P.; Legube, B. Kinetic and mechanistic investigations of progesterone reaction with ozone. *Water Res.* **2006**, *40*, 2181-2189.
22. Buffle, M.-O.; Schumacher, J.; Salhi, E.; Jekel, M.; Von Gunten, U. Measurement of the initial phase of ozone decomposition in water and wastewater by means of a continuous quench-flow system: Application to disinfection and pharmaceutical oxidation. *Water Res.* **2006**, *40*, 1884-1894
23. Sharma, V. K. Potassium Ferrate (VI): An Environmentally Friendly Oxidant. Adv. *Environ. Res.* **2002**, *6*, 143-156.
24. Sharma, V. K.; Kazama, F.; Hu, J.; Ray, A. K. Ferrates (iron(VI) and iron(V)): Environmentally friendly oxidants and disinfectants. *J. Water Health* **2005**, *3*, 45-58.
25. Sharma, V.K.; Mishra, S.K.; Nesnas, N. *Environ. Sci. Technol.* **2006**, *40*, 7222-7227..
26. Jiang, J. Q.; Lloyd, B. Progress in the development and use of ferrate (VI) salt as an oxidant and coagulant for water and wastewater treatment. *Water Res.* **2002**, *36*, 1397-1408.
27. Reimers, R. S.; Sharma, V. K.; Pillai, S. D.; Reinhart, D. R.; Boyd, G. R.; Fitzmorris, K. B. *Biosolids Tech. Bull.* **2005**, *10*, 1-2.

28. Powers, W. J.; Flatow, L. A. Iowa State University, 2001.
29. Sharma, V.K.; Smith, J.O.; Millero, F.J. Ferrate(VI) oxidation of hydrogen sulfide. *Environ. Sci. Technol.* **1997**, *31*, 2486-2491.
30. Huang, H.; Sommerfeld, D.; Dunn, B. C.; Eyring, E. M.; Lloyd, C. R. J. Ferrate(VI) Oxidation of Aqueous Phenol: Kinetics and Mechanism. *Phys. Chem.* A. **2001**, *105*, 3536-3541.
31. Sharma, V.K.; Burnett, C.R.; Millero, F.J. Dissociation constants of the monoprotic ferrate(VI) ion in NaCl media. *Phys. Chem. Chem. Phys.* **2001**, *3*, 2059-2062.
32. Deborde, M.; Rabouan, S.; Gallard, H.; Legube, B. Aqueous chlorination kinetics of some endocrine disruptors. *Environ. Sci. Technol.* **2004**, *38*, 5577-5583.
33. Lee, Y.; Yoon, J.; von Gunten, U. Kinetics of the Oxidation of Phenolsand Phenolic Endocrine Disruptorsduring Water Treatment with Ferrate (Fe(VI)). *Environ. Sci. Technol* **2005**, *39*, 8978-8984.
34. Park, G.; Yu, M.; Go, J.; Kim, E.; Kim, H. Oxidation of geosmin and MIB in water using O_3/H_2O_2: kinetic evaluation. *Water Sci. Technol.* **2006**, *6*, 43-69.
35. Kim, H.; Milner, P.; McConnell, L. L.; Torrents, A.; Sharma, V. K.; Ramirez, M.; Peot, C. Disinfection of thickened sludge and biosolids using ferrate (FeO_4^{2-}). WEF/AWWA/IWA Disinfectant Conference Proceedings. WEF Publishing, Alexandria **2005**.
36. Arispe, S. Measuring and Developing a Control Strategy for Odorous Gases from Solids Handling Processes of a Large Wastewater Treatment Plant. Master's Thesis.Univ. of Maryland, College Park, Maryland, USA. 2005.
37. Hanselman, T. A.; Graetz, D. A.; Wilkie, A. C. J. Comparison of three enzyme immunoassays for measureing 17ß-estradiol in flushed dairy manure wastewater. *Environ. Quality* **2004**, *33*, 1919-1923
38. Laganá, A. ; Bacaloni, A.; De Leva, I.; Faberi, A.; Fago, G.; Marino, A. Analytical methodologies for determining the occurrence of endocrine disrupting chemicals in sewage treatment plants and natural waters. *Analytical Chimica Acta* **2004**, *501*, 79-88.
39. Lindsey, M. E.; Meyer, M.; Thurman, E. M. Analysis of trace levels of sulfonamide and tetracycline antimicrobials in groundwater and surface water using solid-phase extraction and liquid chromatography/mass spectrometry. *Anal. Chem.* **2001**, *73*, 4640-4646.
40. In Guidelines for achieving high accuracy in isotope dilution mass spectrometry (IDMS).; Sargent, M., Harte, R., and Harrington, C.; Royal Society of Chemistry: 2002; pp. 3.
41. Raman D.R.; Williams, E. L.; Layton, A. C.; Burns, R. T.; Easter, J. P; Daugherty, A. S.; Mullen, M. D.; Sayler, G. S. Estrogen content of dairy and swine wastes. *Environ. Sci. Technol.* **2004**, *38*, 3567-3573.

Chapter 27

Use of Ferrate(VI) in Enhancing the Coagulation of Algae-Bearing Water: Effect and Mechanism Study

Wei Liu and Yong-Mei Liang

Department of Environmental Engineering, Sun Yat-Sen University, Guang Zhou, People's Republic of China

This study found that ferrate preoxidation significantly enhanced the algae removal in alum coagulation. A very short preoxidation time, e.g. several minutes, was enough to achieve substantial enhancement of algae removal by ferrate. It was also found that ferrate preoxidation was much more powerful than pre-chlorination in enhancing the coagulation of algae-bearing water. Ferrate oxidation left obvious impacts on surface architecture of algal cells. Upon oxidation with ferrate, the cells were inactivated and some intracellular and extracelluar components were released into the water, which act as coagulant aid. The coagulation was also improved by increasing particle concentration in water, because of the formation of the intermediate forms of precipitant iron species during preoxidation. In addition, it was also observed that ferrate preoxidation caused algae agglomerate formation before the addition of coagulant, the subsequent application of alum resulted in further coagulation. Ferrate preoxidation also improved the reduction of residual organic matters in algae-bearing water.

Introduction

The eutrophication of lakes or reservoirs is a worldwide problem, and it is increasing significantly in China. Eutrophication is caused by excessive inputs of nutrients that stimulate nuisance growth of algae. To control the massive growth of algae in lakes and reservoirs, the impact of some chemicals such as copper sulphate, potassium permanganate, on algae were studied. Copper sulphate has been used to control algae growth in lakes and reservoirs for more than 80 years. It has been considered to be a cost-effective algicide available *(1)*. Potassium permanganate was also studied specifically to use as an algicide for reservoirs *(2, 3)*.

The omnipresence of algae in surface water is also the current and growing problem in the drinking water treatment. Coagulation-sedimentation is still the main treatment process to remove algae from drinking water. Meanwhile, other treatment processes, like dissolved air flotation *(4)*, sand filtration *(5)*, and direct filtration *(6)* have also been investigated. Massive growths of algae in drinking water sources cause many problems. Some algae cause uncomfortable tastes and odors, some algae cause filter clogging, some can penetrate the filter and lead to the deterioration of water quality. Algae are also a precursor of disinfection by-products. Algae removal from water treatment process is difficult because of their small size and the low specific gravity.

The effects of chlorine, ozone, and chlorine dioxide on *Scenedesmus sp.* cultures were studied *(7)*. Algal cells activity and chlorophyll concentration decreased and the concentration of dissolved organic substances increased with increasing applied oxidant concentration. It was found that pretreatment with chlorine dioxide (1, 3 or 5 mg/L) or ozone (2.6, 4.6 or 8.1 mg/L) on algal cultures enhanced algal coagulation with aluminium sulphate, while prechlorination with 10 or 20 mg/L chlorine increased the required dosage of alum by 15%. However, the negative effect of using chlorine and chlorine dioxide resulting from the formation of by-products are limiting the use of these chemicals as pre-oxidants. In addition, it was recently recognized that the ozonation of waters containing bromide may lead to the formation of bromate at a level suspected of being hazardous to health, which is a negative aspect for using ozone as a preoxidant.

Potassium permanganate has been investigated as an alternative preoxidant for the direct filtration of impounded surface water. The experiments of modified jar test apparatus and pilot plants showed that permanganate pre-treatment followed by coagulation with dual coagulants (ferric sulphate and cationic polymer) distinctly improved the particle and algae removal commonly achieved in direct filtration *(6)*. It was suggested that the basic mechanism of algae removal by oxidant is the destruction of the algae structure.

Ferrate (VI) has a very strong redox potential through the entire pH range, ranging from –2.2V in acid to –0.7V in base *(8)*. Several investigations have been conducted in applying ferrate as a promising alternative to chlorine for the disinfection of water and wastewater *(9, 10)*. It is found that ferrate ion appeared to be an effective antifoulant *(11)*, as only short contact times were required for ferrate concentration of 10^{-5}M to control the biofilm growth. In addition, recent study found that ferrate treatment did not produce any mutagenic by-products during the treatment process *(12)*.

The objectives of this paper were to investigate the effects of pre-treatment with ferrate ion on algae by alum coagulation. The influence of ferrate preoxidation on surface architecture of algae cell and dissolved organic materials (DOM) were also studied.

Materials and methods

Waters

Raw water from shallow lake in northeast China was collected. The water is in deep green color and containing high concentration of algae. Observation results by microscopy shows that the lake water principally contains green algae, such as *Chlorella, Spirogyra, Chlorococoam, Scenedesmus* etc. The raw water quality was as follows: Turbidity of 10~30 NTU; pH of 7.5~7.7; temperature of 15~18 °C; algal concentration of 8×10^6~2×10^7 cells/L.

Cultured water containing green algae species was also used in this study in order to overcome the influence of other materials except algae in natural surface water on experimental results. Algae species in cultured process are *Chlorococoum* and *Scenedesmus*, because they are commonly found in natural waters and are typical green algae. They are also easily available and cultured in the laboratory.

The algae seeds were cultured in the plastic culture tank containing total 280L modified inorganic nutrient solution. The inorganic salts included KNO_3, $Ca(NO_3)_2$, $MgSO_4$, KH_2PO_4, and $FeCl_3$. The nutrient solution mixed with 200 mL soil exudation liquid. The starting cell concentration in the water was 4×10^6 cells/L at a constant temperature of 15 °C, pH 7.3. Continuous light was provided by incandescent lamp and daylight lamp. A gas mixture (1% CO_2 in air) was bubbled into the medium for a period of 15 min every other day. After cultured for 25 days, algal concentration in nutrient solution achieved 3.5-4.2$\times10^8$ cells/L, with pH of 9.1 and turbidity of 20-40 NTU.

Chemicals

Potassium ferrate solid was prepared by the modification of the method of reaction between OCl^- and $Fe(OH)_3$ (gel) in strongly basic media and isolated from saturated KOH solution (13), and stored in the desiccator. Potassium ferrate solution (0.3 g/l, calculated in K_2FeO_4) was prepared by dissolving potassium ferrate solid in distilled water just before use in order to minimize the loss of ferrate as a result of rapid decomposition rate in solution.

Jar tests

The Standard jar tests were carried out with a six-unit stirrer apparatus. Before tests, the pH of each 0.5L cultured sample was adjusted to 7.0 with 1N HCl. A carefully calculated amount of ferrate solution was injected into beakers a certain time before subjection to alum coagulation. Water samples were first rapidly stirred at 200 rpm for 1 min and then at 45 rpm for 10 min. After that, samples were allowed to settle quiescently for 20 min. Then, the upper 100 mL of the water sample was taken for counting of residual algae cell concentration. In the case of lake water, the settled samples were further filtered with filter paper (1-2 μm pore size), and the residual algal concentration after filtration was also determined. Residual algal concentration after coagulation test was determined by microscopically counting of cells.

Scanning electron microscopy

A blob of treated algae by potassium ferrate preoxidation and the control algae without ferrate preoxidation were dried for 2h in the drying table, and they were gold-coated to a calculated coating thickness of 150 nm by Eiko IB-3 ion emitting apparatus. Then, they were examined in a Hitachi S-520 scanning electron microscope operated at 15KV.

UVA at 254 nm and UVA scanning

The scanning of ultraviolet absorbance using wavelength scan from 200 nm to 320 nm of untreated and treated cultured samples after filtered with a 0.45 μm cellulose acetate membrane filter, the absorbance of ultraviolet absorbance at 254 nm was also determined. The absorbance of ultraviolet at 254 nm by natural waters is a semi-quantitative indicator of the concentration of natural organic materials (NOM) in water. In water treatment practice, the use of absorbance at

254 nm has been found to be useful for monitoring the concentration of DOC *(14)*. UV absorbance was also used to characterize NOM by the degree of its aromaticity. UVA at 254 nm and UVA scanning were used in this study to indicate the variation of dissolved organic materials (DOM) in cultured waters during different treatment processes.

Results and Discussion

Enhanced algae removal

The enhanced algae removal from lake and cultured water in alum coagulation by ferrate concentration were exhibited in Figure 1. The removal efficiency is expressed as the ratio of the initial algal concentration of raw algae water to the concentration measured after treatment. Alum coagulation can partially remove the algal cells from lake water. Alum coagulation at dosage of 20-50 mg/L removed 20-30% of algae, and 50% removal was observed at higher alum dosage of 80 mg/L. In the case of cultured water, relative high algae removal was achieved at all alum dosages. It is worth to notice that there was a sharp increase of removal efficiency when the alum dosages were between 50-60 mg/L in lake water and 40-50 mg/L in cultured water by solely alum coagulation. This indicated that there exists optimum dosage range of alum for effective coagulation of algae-bearing water. This was attributable to the isoelectric point between the alum and algal cells.

The data also showed that ferrate preoxidation significantly enhanced the coagulation of either lake water or cultured water. At any alum dosages, the algae removal of settled samples pretreated with ferrate is higher than that without ferrate pretreatment. Considerable enhancement in algae removal can be obtained by ferrate preoxidation even at small dosage. The removal of algae gradually increased with the increase of alum dosage when pretreated by ferrate. There were not obvious isoelectric point in the removal curves for ferrate preoxidation, especially in cultured water, and lead to a relative broader optimum alum dosage range. Therefore, the alum dosage required for achieving certain algae removal efficiency can be reduced.

Algal cells always carry negative surface charges at natural water conditions *(15)*. Normally, the basic mechanism of coagulation consists in mutual attraction and neutralization of the negative surface charge by the positively-charged hydroxide coagulant. Since NOM has a very strong influence on coagulation effectiveness, with the presence of NOM, the coagulant reacts first with the free natural organic acids, e.g. humic acids and fulvic acids in waters. Only when the coagulant dosages are high enough to neutralize the surface charges of the organic materials, the coagulant can take part in the neutralization coagulation of algal cells *(16)*.

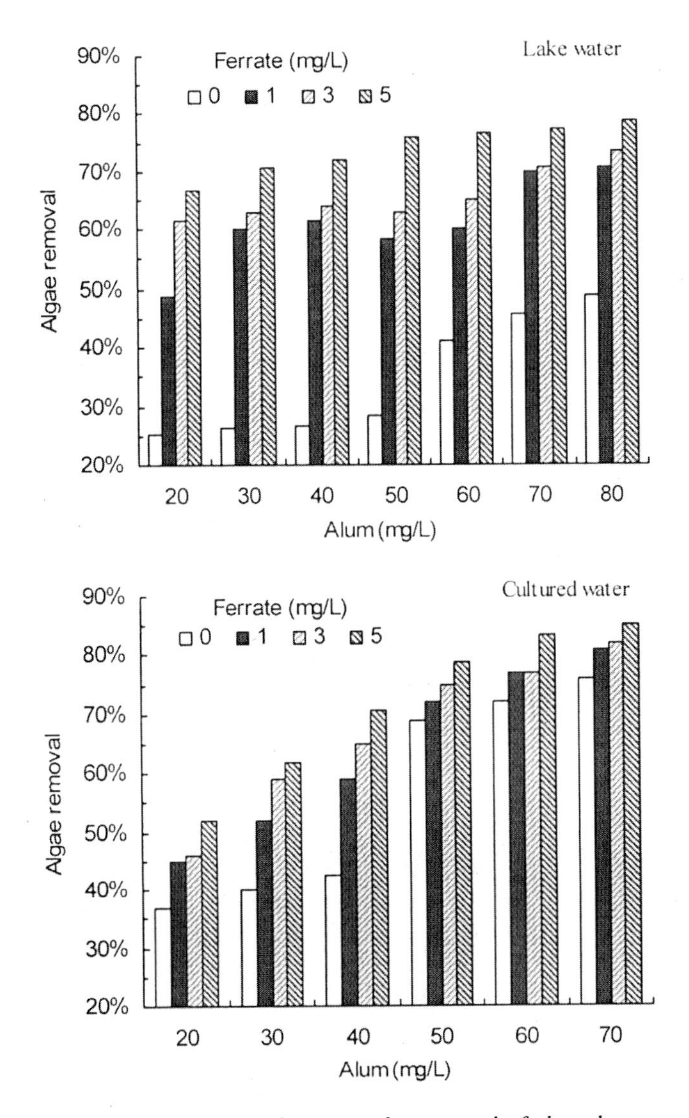

Figure1. Effect of ferrate preoxidation on the removal of algae by coagulation.

It could be concluded that the existence of NOM in surface water results in different algae removal between lake water and cultured water which is lack of natural organic materials, when treated with solely alum. The effects of ferrate preoxidation on algae removal in the case of lake water were more significant than that in cultured water. It is believed that these effects were attributable to ferrate preoxidation, since ferrate preoxidation acted as aid to coagulation processes *(17)*.

The effects of various preoxidation time on the removal of algae by ferrate preoxidation were showed in Figure 2. It was observed that algae removal efficiency increased obviously in a short preoxidation time and the removal efficiency further increases with the continuing extension of the preoxidation time. Considering the slight increase of removal efficiency with the extension of pretreatment times (longer than 1 min), it was suggested that pretreatment with ferrate can influence the surface characteristics of algal cells in a very short time of oxidation.

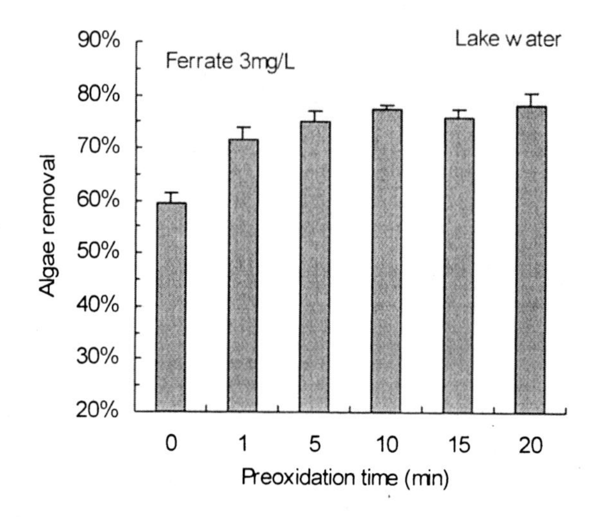

Figure 2. Enhanced removal of algae by ferrate preoxidation with respect to preoxidation time

The algae removal of the water samples preoxidized with ferrate is further increased after filtered with filter papers (1~2 μm pore size). The data was shown in Figure 3. Apparently, the filtration process strengthened the enhancement of ferrate preoxidation on algae removal. Some algal cells after ferrate preoxidation and alum coagulation are easily trapped by the filter, if did not precipitated during sedimentation stage.

The effects of ferrate and chlorine on removal of algae were compared and the results were illustrated in Figure 4. It was clear that the enhancement in algae removal by ferrate preoxidation was much more significant than that achieved by prechlorination.

Only 1 mg/L of ferrate can improved the removal of algae from less than 30% to more than 80%. 2 mg/L of chlorine only increased the algae removal to

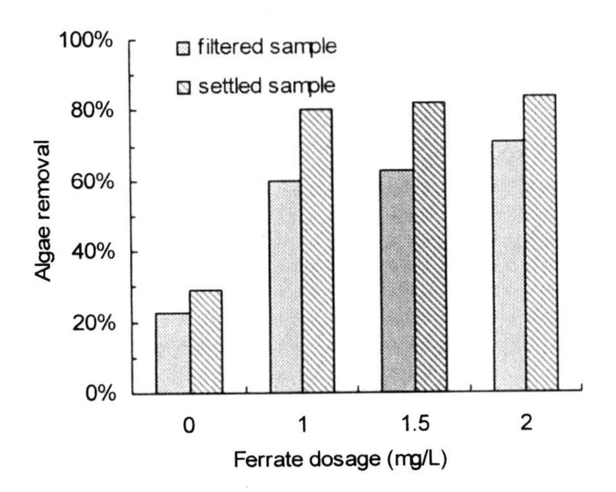

Figure 3. Comparative results of ferrate preoxidation on algae removal of settled and filtered samples of lake water

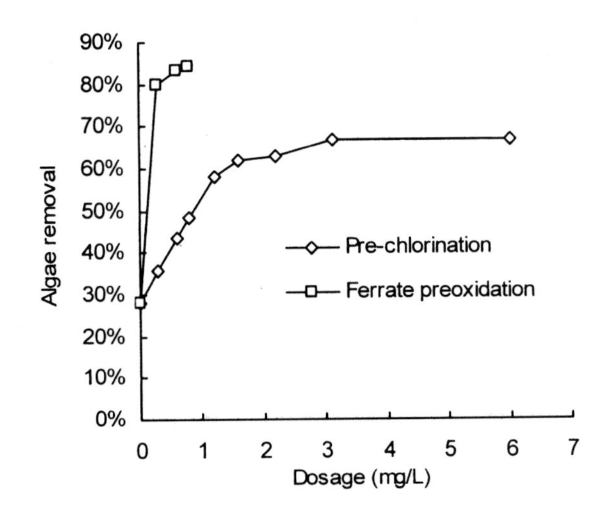

Figure 4. Comparative results of ferrate preoxidation and pre-chlorination on algae removal

approximate 60%, where the removal of algae reached the plateau value even with higher dosage up to 6 mg/L. It seemed that there is a limitation of algae removal by prechlorination. Apparently, there are advantages for ferrate preoxidation in practice comparing with chlorine, considering the disinfection by-products formation of prechlorination.

Influences on surface architecture of algae cells

Comparative results of scanning electron microscopy (SEM) micrographs between algal cells before and after pretreatment with ferrate demonstrated that the ferrate preoxidation induced a number of clearly discernible effects on algal behavior and cell architecture *(18)*.

First, ferrate caused excess release of intracellular component into the surrounding medium. This phenomenon possibly caused by ferrate stimulation on algal cell or cleaved sheath by ferrate oxidation. Generally, algae may release organic compounds into water, which is species-specific and growth phase-specific. Extracellular organic matters (EOM) from cultures of green and blue-green algae and diatoms behave like anionic and non-ionic polyelectrolytes *(19)*. Hence, it was suggested that algal biopolymers secreted in response to ferrate oxidation behave as a coagulant aid.

Second, ferrate preoxidation caused intense sheath convolutions on the surface of algal cells. It was observed that the cell surface architecture was eminent damaged, row organization of warts of *Scenedesmus* were not remained.

Third, ferric hydroxide, which is generated from the decomposition of ferrate, precipitated onto the surface of algae cells. These precipitates can obviously change algal surface properties. Once it is attached to the algal surface, the weights of the algal cells is increased and the algae settling character is improved. In addition, the $Fe(OH)_3$ colloids increased the concentration of particles in water. All these effects led to better coagulation and consequent better removal of algae.

And last, ferrate preoxidation also led to conglomeration of algal cells. It was suggested that preoxidation with ferrate enhanced the coagulation through the modification of algae envelope and their behavior as well, thus reducing the stability of algae colloids. Details are given in reference *(18)*.

Changes of dissolved organic matters (DOM)

Figure 5 shows the changes in other parameters of cultured water samples under various treatment processes. Parameters included the UVA at 254 nm, permanganate index (COD_{Mn}), and number of organic matters in water samples, which is detected by GC/MS. In order to make a clear comparison, three parameters of cultured water in the figure were plotted at equal level labeled with different values. Correspondingly, the level for parameters of treated water samples was proportional to the changes of exact value.

Alum coagulation could reduce these parameters to some extent. Ferrate oxidation increased three parameters from UVA_{254} of 12, COD_{Mn} of 2.1 and number of 40 to 13, 2.7 and 73, respectively. After following alum coagulation, the increment of three parameters dropped to 9, 1.5 and 45, even lower than that of solely alum coagulation.

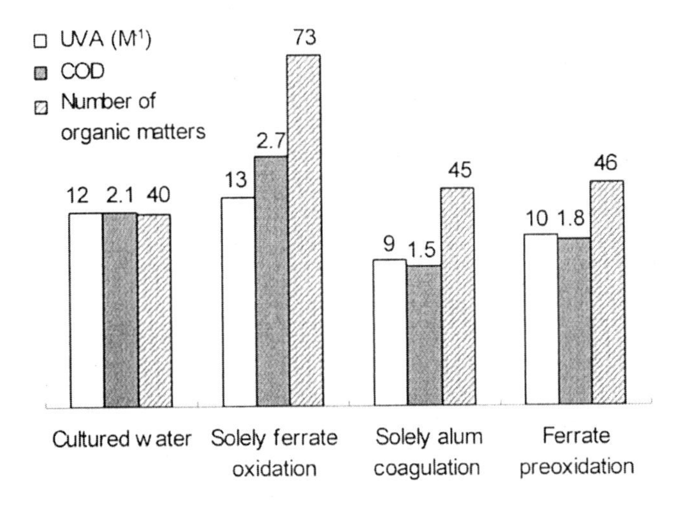

Figure 5 Effect of ferrate preoxidation on scanning of UVA of cultured water after filtration.

Comparative results of UV absorbance spectra from 200 nm to 320 nm of water samples that solely alum coagulation, solely ferrate oxidation, and ferrate preoxidation followed by alum coagulation showed the similar information *(18)*. Alum coagulation decreased UVA spectrum to some extent; indicating the removal of organic matters from water. UVA spectrum significantly increased after ferrate oxidation at a very short contacting time. It was also interesting to notice that the UVA spectrum for ferrate oxidation decreased after following alum coagulation, and the spectrum was even lower than that for solely alum coagulation.

Considering the variation of above parameters, it could be concluded that ferrate preoxidation increased the dissolved organic concentration of cultured water, and the increment can be removed by following alum coagulation. This was in consistence with the observation by SEM, in which the secretion of intracellular component of algal cells caused by ferrate oxidation was identified. This process increased the concentration of organic matters in the water and then caused the increases of UVA spectrum. These results possibly support that algal biopolymers secreted in response to ferrate preoxidation, which may behave as a coagulant aid.

These results may demonstrate the mechanisms for enhancement of algae by ferrate preoxidation as follows: ferrate preoxidation inactivated the algal cells in a very short contacting time and then the algae architecture was destroyed. Consequently, the cellar components were released to act as coagulant aid, which largely enhanced the following coagulation. Induced coagulant aid and the ferric hydroxide colloids derived from the decomposition of ferrate caused the conglomeration of algal cells in prolonged preoxidation time, leading to the primary coagulation of algae, which also enhanced the following algae coagulation by alum.

Conclusions

Laboratory studies using algae-bearing lake water and cultured algae water demonstrated that pretreatment with potassium ferrate obviously enhanced the algae removal by alum coagulation. Algae removal increased remarkably even at a short period of preoxidation time, and the efficiency was further increased at a prolonged contact time. Ferrate preoxidation exhibited much more powerful than prechlorination in enhancing algae removal.

Efficient removal of algae caused by potassium ferrate preoxidation and coagulation with alum is suggested to be a consequence of several process mechanisms. Ferrate preoxidation inactivated algae, and also induced coagulant aid secreted by algal cells. Meanwhile, ferric hydroxide derived from the decomposition of ferrate improved the coagulation condition by increasing particles concentration in water. In addition, ferrate preoxidation caused algae agglomerate formation before the addition of coagulant, and the subsequent application of alum resulted in further coagulation.

References

1. Elder, J. F.; Horne, A. J. Copper cycles and copper sulphate algicidal capacity in two Californian lakes. *Environ. Manage.* **1978**, *2*, 17-30.
2. Fitzgerald, G. P. Laboratory evaluation of potassium permanganate as an algicide for water reservoirs. *J. South West Water Works Assoc.* **1964**, *45*, 16-25.
3. Kemp, H. T.; Fuller, R. G.; Davidson, R. S. Potassium permanganate as an algicide. *J. Amer. Water Works Assoc.* **1966**, *58*, 255-263.
4. Bare, W. F. R.; Jones, N. B.; Middleebrooka, E. J. Algae removal using dissolved air flotation. *J. Water Pollut. Control Feder.* **1975**, *47*, 153-169.
5. Borchardt, J. A.; O'melia, C. R. Sand filtration of algae suspension. *J. Amer. Water Works Assoc.* **1961**, *53*, 1493-1508.
6. Fetrusevski, B.; Van Breemen, A. N.; Alaerts, G. Effect of permanganate pre-treatment and coagulation with dual coagulants on algae removal in direct filtration. *J. Water Supply: Res. Technol.-Aqua* **1996**, *45*, 316-326.
7. Sukenik, A.; Teltch, B.; Wachs, A. W.; Shelef, G.; Nir, I.; Levanon, D. Effect of oxidants on microalgal coagulation. *Water Res.* **1987**, *21*, 533-539.
8. Wood, R. H. The heat, free energy and entropy of the ferrate(VI) ion. *J. Amer. Chem. Soc.* **1958**, *80*, 2038-2041.
9. Gilbert, M. B.; Waite, T. D.; Hare, C. An investigation of the applicability of ferrate ion for disinfection. *J. Amer. Water Works Assoc.* **1976**, *68*, 495-497.
10. Schink, T.; Waite, T. D. Inactivation of f2 virus with ferrate(VI). *Water Res.* **1980**, *14*, 1705-1717.
11. Fagan, J.; Waite, T. D. Biofouling control with ferrate(VI). *Environ. Sci. Technol.* **1983**, *17*, 123-125.

12. Decula, S. J.; Chao, A. C.; Smallwood, C. Ames test of ferrate treated water. *J. Environ. Eng.* **1983**, *109*, 1159-1167.
13. Goff, H.; Murmann, R. K. Studies on the mechanism of isotopic oxygen exchange and reduction of ferrate(VI) ion (FeO_4^{2-}). *J. Amer. Chem. Soc.* **1971**, *93*, 6058-6065.
14. Owen, D. M.; Amy, G. L.; Chowdhury, Z. K.; Paode, R.; McCoy, G.; Viscosil, K. NOM characterization and treatability. *J. Amer. Water Works Assoc.* **1995**, *87*, 46-63.
15. Ives, K. J. Electrokinetic phenomena of planktonic algae. *Proc. of the Society for Water Treatment and Examination* **1956**, *5*, 41-53.
16. Narkis, N,; Rebhun, M. The mechanism of flocculation processes in the presence of humic substances. *J. Amer. Water Works Assoc.* **1975**, *67*, 101-108.
17. Ma, J.; Liu, W. Enhanced coagulation of low temperature and low turbidity water by ferrate composite chemicals. *Water Wastewater Eng.* **1997**, *23*, 9-11 (in Chinese).
18. Ma, J.; Liu, W. Effectiveness of ferrate (VI) preoxidation in enhancing the coagulation of surface water. *Water Research* **2002**, *36(20)*, 4959-4962.
19. Bernhart, H.; Clasen, J. Flocculation of micro-organisms. *J. Water Supply: Res. Technol.-Aqua* **1991**, *40*, 76-87.

Chapter 28

Combined Process of Ferrate Preoxidation and Biological Activated Carbon Filtration for Upgrading Water Quality

Jun Ma*, Chunjuan Li, Yingjie Zhang, and Ran Ju

School of Municipal and Environmental Engineering, Harbin Institute of Technology, Harbin 150090, People's Republic of China
***Corresponding author: email: majun@hit.edu.cn; fax: 8645182368074**

The preoxidation of polluted surface water with ferrate was conducted with respect to its impact on the following biofiltration. It was found that preoxidation with ferrate promoted the biodegradation of organics with substantial reductions of chemical oxygen demand (COD_{Mn}), UV_{254}-absorbance. It was also found that the removal of NH_4^+-N in biological activated carbon (BAC) process was also substantially improved as compared with the cases without ferrate preoxidation and with ozone preoxidation. In addition, the experiments were conducted related to the effect of potassium ferrate oxidation of raw water of Songhua River on its changes of molecular weight distribution in order to investigate further the enhancement of ferrate preoxidation on the removal of organics. The results indicated that the concentration of organics with molecular weight (MW) of 10k-100k and less than 0.5k were substantially increased after the raw water was coagulated with ferrate preoxidation, which suggested that these oxidation products are readily removed by subsequent biofiltration.

Introduction

One of the principal methods for enhancing the biodegradation is preoxidation, which is generally aimed at destroying the structure of organics and then the formation of some products readily removed by the following biological treatment. Preoxidation partially oxidizes the dissolved organic carbon (DOC), which results in an increase of assimilable organic carbon (AOC) and a decrease of organic micro-pollutants. The partially oxidized DOC is better biodegraded in BAC process than that in GAC process (The oxidant is immediately used before granular activated carbon column and no biocoenosis is detected on the column). Preoxidation processes have been widely studied in drinking water treatment to remove pollutants and to improve the biodegradability of organic constituents (*1-6*). Ozonation is the most commonly applied process for preoxidation. The O_3/BAC combination proved to be very efficient in reducing DOC because of the biodegradability enhanced upon ozonation (*7-12*). However, ozone oxidation has the disadvantages of high cost, inconvenient operation and maintenance. For example, ozone can reduce the levels of THMs and halo acetic acids (HAAs), but it can form the potent carcinogenic bromate ion by reacting with bromide present in water (*13,14*).

Ferrate is environmental friendly water treatment agent that can meet these new challenges confronted in the water industry. Ferrate has been proved in previous studies as a selective and effective oxidant in degrading various synthetic and natural organic pollutants, inactivating microorganisms, and coagulating colloidal particles and removing heavy metals (*15-20*). More importantly, it is an environmental friendly treatment chemical, which does not produce any harmful by-products in water treatment process.

In this paper, the potential role of Fe(VI) as preoxidants to enhance the following biological process in drinking water treatment was investigated through long-term experiment. The performance of ferrate was compared with ozone preoxidation in terms of removing COD_{Mn}, UV_{254}-absorbance, ammonia-nitrogen. And the variation of molecular size of raw water induced by ferrate preoxidation was analyzed in a batch study.

Experiments

Materials and Equipment

Potassium ferrate (K_2FeO_4) of high purity (98 % plus) was prepared by the method described by Thompson et al. (*21*). Potassium ferrate solution was prepared by dissolving potassium ferrate solid in distilled water just before use in order to minimize the loss of ferrate. Ozone was produced with an ozone generator (Tongli XFZ-58I, Tsinghua) that used dried oxygen. Distilled water was continuously bubbled with gaseous ozone. The aqueous ozone was monitored with ultraviolet spectrometer at λ=258nm (ε=3000 $M^{-1}cm^{-1}$)(*22*) until a constant concentration was reached (i.e. saturated ozone containing solution). This ozone containing solution was prepared just before addition. Aluminum Sulfate ($Al_2(SO_4)_3 \cdot 18H_2O$, Tianjin chemical Inc., Tianjin, China) is selected as the coagulant.

A polluted water from down stream part of Songhua River was selected in this study. A summary of the raw water quality is shown in Table I.

Table I. Characteristics of the raw water from Songhua River

Parameters	*Maximum*	*Minimum*	*Average*
pH	7.5	7.0	7.1
Turbidity (NTU)	144.00	1.85	23.8
UV_{254}-abs (cm^{-1})	0.260	0.093	0.147
COD_{Mn} (mg L^{-1})	13.30	3.33	7.42
Ammonia-nitrogen (NH_4^+-N)	2.30	0.19	0.72

Experiment Procedures

Dynamic Experiments

The effectiveness of micro-pollutants removal was studied by dynamic experiments in these processes, such as chemical preoxidation, coagulation and precipitation, filtration and BAC process. Potassium ferrate and ozone were selected as the preoxidant. The dosage of preoxidants was 1mg L^{-1}. Preoxidation was conducted for 10 min at 300 rpm. The dosage of Aluminum Sulfate was 50 mg L^{-1}. The parameters such as turbidity, chemical oxygen demand (COD_{Mn}), UV_{254}-absorbance and NH_4^+-N were measured according to the standard methods(Water and Wastewater Monitoring and Analysis Method, the Fourth Edition, China, 2002).

Batch Test

In order to have a deeper understanding of the mechanism of ferrate preoxidation to enhance the biological process for treating the polluted surface water, a batch study of preoxidation of raw water was further conducted. The experiment was carried out with a magnetic stirrer. 1 mg L^{-1} of potassium ferrate was added and mixed at a speed of 300 rpm for 10 min. Then, all the samples were coagulated by adding Aluminum Sulfate, followed by filtration through a 0.45 μm filter to separate any particulate matter. These samples were collected and fractionated for the molecular weight distribution using a stirred ultra-filtration cells (Amicon; model 8400). The samples were subsequently measured with a TOC (total organic carbon) analyzer (TOC-VCPH, Shimadzu).

Results and Discussion

Dynamic Experiments

The variation of parameters in different processes with ferrate is shown in Table I. It can be seen that turbidity, COD_{Mn}, UV_{254}-absorbance and NH_4^+-N substantially decreased by 99.4%, 82.9%, 74.9%, 65.3%, respectively, after BAC, with ferrate preoxidation. Furthermore, BAC outlet values of turbidity and NH_4^+-N absolutely meet drinking water standards. It must be noted that NH_4^+-N concentration is not decreased, but rather increased after preoxidation and coagulation with ferrate. This phenomenon can be explained by the fact that organics containing nitrogen in raw water were partly oxidized to inorganic nitrogen by ferrate preoxidation.

In the experiments, it was found that preoxidation with a low dose of potassium ferrate (1mg L^{-1}) could promote the degradation of organics in the water, with substantial reduction of COD_{Mn}, UV_{254}-absorbance and NH_4^+-N in BAC reactor as compared to the case without ferrate preoxidation. There were an additional average removal 5%, 12%and 47%, respectively with ferrate preoxidation compared to the case without ferrate preoxidation in BAC process. Obviously, the biodegradability of organics in raw water was improved by ferrate preoxidation. This might be due to the enhanced nitrification of BAC by oxidation products with ferrate and the transformation of higher molecular weight compounds into lower molecular weight compounds induced by ferrate, which could be readily removed by BAC process. Previous studies indicated that higher biological activity in GAC filter was achieved with preozonation of humic or fulvic solutions as compared with the case without ozone preoxidation (*23,24*).

The slightly higher removal of COD_{Mn} and UV_{254}-absorbance was observed with ferrate preoxidation than that with ozone, as shown in Figure 1. It is necessary to note the higher removal of NH_4^+-N with ferrate than with ozone, in Figure 1. Such information could be seen from Figure 2. The amount of NH_4^+-N removed by GAC with ferrate was up to 57% on average. Ferrate/GAC was always effective in reducing NH_4^+-N, when NH_4^+-N values of raw water were highly differential (0.19-4.12mg L^{-1}). Major factors to effect the nitrification process were dissolved oxygen (DO), organic load, temperature and ammonia concentration (*25*). Nitrification was a biological nitrogen removal process, nitrosifying-bacteria (e.g. *Nitrosomonas*) oxidized ammonia to nitrite and nitrate. In this process, which required a reduced nitrogen species, an increasing in NH_4^+-N after ferrate preoxidation was an un-omitted factor (see Table II).

On the other hand, due to the inhibition of heterotrophic bacteria on nitrification, the competing for DO was of more concern in the system. In our study, organic load was lower and DO was enough under aerobic condition. Therefore, heterotrophic bacteria had little competitiveness over nitrifying bacteria with both ferrate and ozone preoxidation. In contrast to ozone, ferrate/BAC resulted in an additional decreasing of 36% in NH_4^+-N. Under the same performance (DO, temperature, hydraulic load, etc), this additional decrease might be due to trace amount of iron existed, which was possibly favorable for *Nitrosomonas* bacteria and *Nitrobacter* bacteria. Unfortunately, the detailed study related to this was not performed deeply in this paper and will be done in the future work.

The results of the long-term tests were presented to estimate the effect of both ferrate and ozone pretreatment on GAC running time. The data allowed us

Table II. Variation of parameters in different processes with ferrate preoxidation*

Parameters	*Raw water*	*Preoxidation and coagulation*	*Filtration*	*BAC*
Turbidity (NTU)	23.80	0.72	0.13	0.13
UV_{254}-abs(cm^{-1})	0.147	0.063	0.060	0.025
COD_{Mn}(mg L^{-1})	7.42	3.36	2.97	1.86
Ammonia-nitrogen (mg L^{-1})	0.72	0.80	0.66	0.25

*Average values of long-term performance

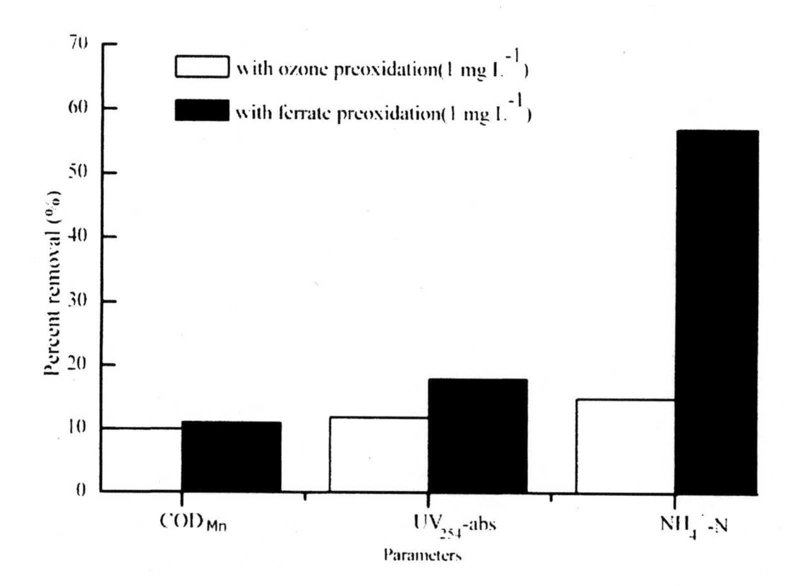

Figure 1. The effect of different preoxidants on COD_{Mn}, UV_{254}-absorbance and NH_4^+-N in BAC reactor

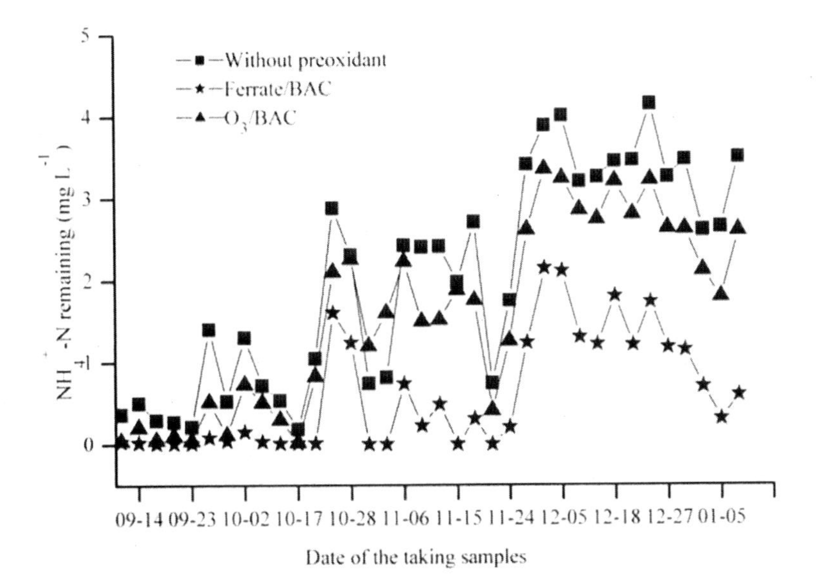

Figure 2. The effect of different preoxidants on the variation of NH_4^+-N after BAC

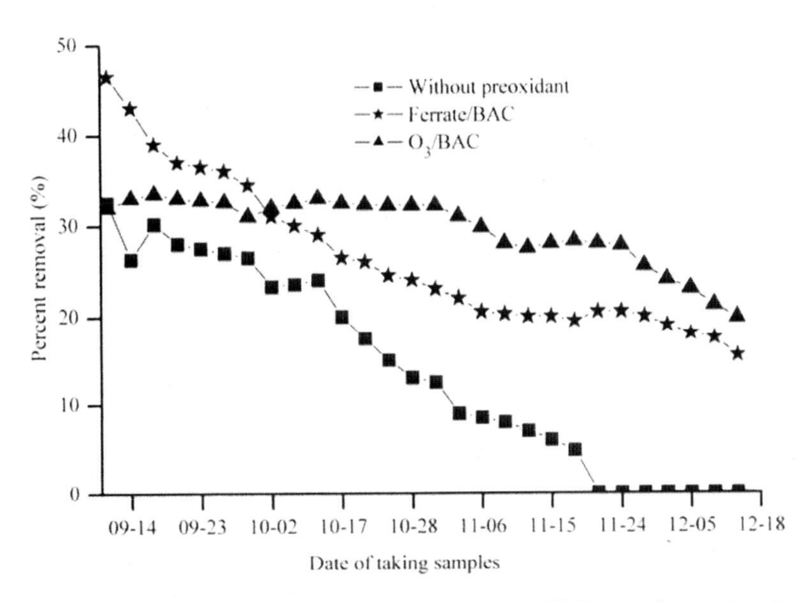

Figure 3. The effect of different preoxidants on the COD_{Mn} with running time after BAC process

to see if the BAC running time would be prolonged. As shown in Figure 3, the removal of COD_{Mn} was decreasing as the running time was increasing in three different tests. When the experiments had been conducted over 60 days, the removal of COD_{Mn} was 0%, 15.5% and 19.8% in GAC process, respectively, with none, ferrate and ozone preoxidation. These data indicated that the BAC running time was prolonged by either ferrate or ozone pretreatment.

Batch Study

As shown in Figure 4, ferrate preoxidation caused an additional removal of about 13% and an absolute removal of 27% after coagulation, which were mainly due to structural transformations of waters. The decrease in organics with molecular size of 100k-0.45μm and 0.5k-10k was observed. However, it was also noted that the organics with molecular size of 10k-100k and less than 0.5k were increased with ferrate preoxidation during coagulation. The decrease of both 100k-0.45μm and 0.5k-10k as well as the increase of less than 0.5k indicated that some high molecular weight organic substances were broken into smaller ones during oxidation. It was more interesting to note that the molecular size of 10k-100k was increased, which might be the result of polymerization induced by oxidation.

It has been realized by some researchers that NOM affected the adsorption of trace organic compounds not only by directly competing for adsorption sites but also by blocking carbon pores (*26,27*). The decrease of the organic concentration of molecular size of 100k-0.45μm reduced the pore blockage, so heterotrophic bacteria and nitrifying bacteria obtained enough space to propagate. In the GAC filter, the filtration mainly reduced the amount of intermediate and low MW (*28*). The decrease of 0.5k-10k helped BAC reactor prolonged the running time. Molecular size less than 0.5k having lower molecular weight, such as aldehydes and carboxylic acids, could be easily removed by using a BAC process (*29*).

In general, larger molecules were microbially resistant, whereas smaller molecules were microbially labile fractions of DOC, and ferrate preoxidation produced more available molecules for BAC process and reduced non-available ones.

Conclusion

Preoxidation of organics with ferrate produced lower molecular weight compounds that could serve as substrate for microorganisms. In spit of the highly differential of NH_4^+-N concentration in raw water, its removal rate in BAC with ferrate preoxidation was up to 57% on average. It was the most cost-effective way to use biological nitrification with ferrate preoxidation to reduce ammonia load. Compared with preozonation, ferrate preoxidation has the

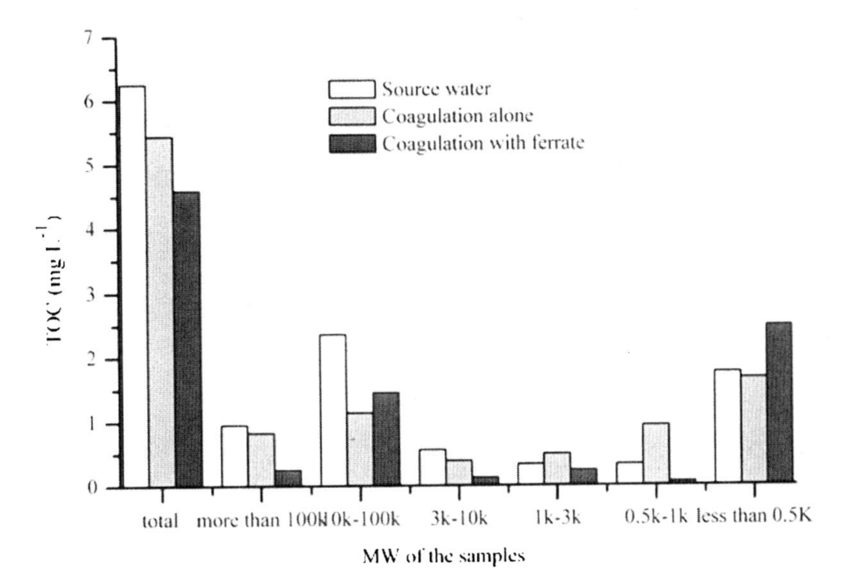

Figure 4. The change of TOC in terms of different MW DOM with $Al_2(SO_4)_3$ of 50 mg L^{-1} and ferrate of 1 mg L^{-1}

advantages of low cost, easy operation and maintenance. Ferrate preoxidation might be an economical method to enhance the conventional water treatment process on occasions with limited funds for capital investment. Combined process of ferrate preoxidation and BAC could efficiently purify drinking water.

References

1. Hoigne, J. Chemistry of aqueous ozone and transformation of pollutants by ozonation and advanced oxidation processes. *In the Handbook of Environmental Chemistry*. **1998**, *5*, p83-143.
2. Yu, Y.H.; Hu, S.T. Preoxidation of chlorophenolic wastewaters for their subsequent biological treatment. *Water Sci. Technol.* **1994**, *29(9)*, 313-320.
3. Kaludjerski, M.; Gurol, M.D. Assessment of enhancement in biodegradation of dichlorodiethyl ether (DCDE) by preoxidation. *Water Res.* **2004**, *38(6)*, 1595-1603.
4. Rivas, F.J.; Beltrán, F.J.; Gimeno, O.; Alvarez, P. *J. Environ. Eng.* **2001**, *127(7)*, 611-619.
5. Manzano, M.A.; Perales, J.A.; Sales, D.; Quiroga, J.M. Enhancement of aerobic microbial degradation of polychlorinated biphenyl in soil microcosms. *Environ. Toxicol. Chem.* **2003**, *22(4)*, 699-705.

6. Zhang, Z.S.; William, A.A.; Murray, M.Y. Photocatalytic pretreatment of contaminated groundwater for biological nitrification enhancement. *J. Chem. Technol. Biotechnol.* **2002**, *77(2)*, 190-194.
7. Lawrence, J.; Tosine, H.; Onuska, F.I.; Comba, M.E. The ozonation of naturel waters: product identification. *Ozone-Sci. Eng.* **1980**, *2*, 55-64.
8. Rice, R.G. The use of ozone to control trihalomethanes in drinking water treatment. *Ozone-Sci. Eng.* **1980**, *2*, 75-99.
9. Yamada, H.; Uesugi, K.; Myoga, H. Study on byproducts of ozonation during the control of trihalomethanes formation. *Ozone-Sci. Eng.* **1986**, *8*, 129-150.
10. Van, L.H. Preliminary investigation into the improvement of the biodegradability of organic substances in surface waters and effluents through ozonation. *Water Sci. Technol.* **1987**, *19*, 931-937.
11. Takahashi, M.; Nakai, T.; Satoh, Y.; Katoh, Y. Ozonolysis of humic acid and its effect on decoloration and biodegradability. *Ozone-Sci. Eng.* **1995**, *17*, 511-525.
12. Lefebvre, E.; Crouue, J.P. Change of dissolved organic matter during conventional drinking water treatment steps. *Revue Sci. Eau.* **1995**, *8*, 463-479.
13. Gunten, U.V. Ozonation of drinking water: Part II. Disinfection and by-products formation in presence of bromide, iodide or chlorine. *Water Res.* **2003**, *37*, 1469-1487.
14. Richardson, S.D. Water analysis: emerging contaminants and current issues. *Anal. Chem.* **2003**, *75*, 2831-2857.
15. Jiang, J.Q.; Lloyd, B. Progress in the development and use of ferrate (VI) salt as an oxidant and coagulant for water and wastewater treatment. *Water Res.* **2002**, *36*, 1397-1408.
16. Jiang, J.Q.; Lloyd, B.; Grigore, L. Preparation and evaluation of potassium ferrate as an oxidant and coagulant for potable water treatment. *Environ. Eng. Sci.* **2001**, *18*, 323-331.
17. Jiang, J.Q. Potassium ferrate (VI), a dual functional water treatment chemical. *Leading Edge Water and Wastewater Treatment Technologies.* Noordwijk/Amsterdam, The Netherlands, **2003**, p26-28.
18. Jiang, J.Q.; Wang, S. Enhanced coagulation with potassium ferrate (VI) for removing humic substances. *Environ. Eng. Sci.* **2003**, *20*, 627-635.
19. Jiang, J.Q.; Wang, S. *Oxidation Technologies for Water and Wastewater Treatment.* Eds. C. Schroder and B. Kragert, Papiepflieger Verlag, Clausthal-Zellerfeld, **2003**, p447-452.
20. Sharma, V.K. Potassium ferrate (VI): an environmental friendly oxidant. *Adv. Environ. Res.* **2002**, *6*, 143-156.
21. Thompson, G.W.; Ockerman, G.W.; Schreyer, J.M. Preparation and purification of potassium ferrate(VI). *J. Am. Chem. Soc.* **1951**, *73*, 1279-1281.
22. Elovitz, M.S.; Gunten, U.V. Hydroxyl radical/ozone ratios during ozonation process. I. the Rct concept. *Ozone-Sci. Eng.* **1999**, *21*, 239-260.

23. De, L.J.; Dore, M.; Mallevialle, J. Effects of preozonation on the adsorbability and the biodegradability of aquatic humic substances and on the performance of granular actived carbon filters. *Water Res.* **1991**, *25*, 151-164.
24. Kainulainen, T.K.; Tuhkanen, T.A.; Vartainen, T.K.; Kalliokoski, P.J. Removal of residual organics from drinking water by ozonation and activated carbon filtration: a pilot plant study. *Ozone-Sci. Eng.* **1995**, *17*, 449-462.
25. Sharma, B.; Ahlert, R.C. Nitrification and nitrogen removal. *Water Res.* **1977**, *11*, 897-925.
26. Carter, M.C.; Weber, W.J. Modeling adsorption of TCE preloaded by background organic matter. *Environ. Sci. Technol.* **1994**, *28(4)*, 614-623.
27. Kilduff, J.E.; Karanfil, T.; Weber, W.J. Competitive effects of nondisplaceable organic compounds on trichloroethylene uptake by activated carbon. I. Thermodynamic predictions and model sensitivity analyses. *J. Colloid Interf. Sci.* **1998**, *205(2)*, 271-279.
28. Matilainen, A.; Lindqvist, N.; Korhonen, S.; Tuhkanen, T. Removal of NOM in the different stages of the water treatment process. *Environ Int.* **2002**, *28*, 457-465.
29. Schechter, D.S.; Singer, P.C. Formation of aldehydes during ozonation. *Ozone-Sci. Eng.* **1995**, *17*, 53-69.

Chapter 29

Enhanced Removal of Cadmium and Lead from Water by Ferrate Preoxidation in the Process of Coagulation

Jun Ma*, Wei Liu, Yingjie Zhang, and Chunjuan Li

School of Municipal and Environmental Engineering, Harbin Institute of Technology, Harbin 150090, People's Republic of China
***Corresponding author: email: majun@hit.edu.cn; fax: 8645182368074**

This paper discussed the effect of ferrate preoxidation on enhanced removal of cadmium and lead from water in the process of coagulation. Some factors affecting the removal of heavy metals were discussed such as pH value, the dosage of ferrate and the water quality condition etc. The results showed that ferrate preoxidation could effectively increase the removal efficiency of lead, whilst a little increase of removal efficiency of cadmium; the removal efficiency increased with the increase of pH. The presence of humic acid greatly affected the removal efficiency of lead in the process of coagulation, but hardly affected the removal efficiency of cadmium. The combined effect of adsorption by intermediate iron species formed in the process of ferrate oxidation and the enhanced coagulation of iron colloids co-precipitated with heavy metals might be responsible for the effective removal of heavy metals.

Introduction

Heavy metals are very harmful to the health, they could accumulate in the living bodies with different forms and could not be degraded by some microorganisms, which made the removal of heavy metal important in drinking water treatment. Lead and Cadmium are common heavy metals in raw water, they are harmful to the human body. Lead and its compounds would lead to acute and chronic lead poisoning. The symptom of acute lead poisoning was abdomen angina, hepatitis, high blood pressure, around neuritis, toxic cephalitis and anaemia; the symptom of chronic lead poisoning is neurasthenia. Cadmium in the body may accumulate in stomach, liver, pancreas and hypothyroid as wll as in cholecyst and bones. Cadmium in the bones would replace calcium in bones and make the bones loose and soft, the sufferer would be in pain and finally dead. So it is very important to effectively remove heavy metals in drinking water.

Coagulation has been used for heavy metal removal due to the formation of hydrolyzed species of aluminum which have adsorption ability to heavy metals. However, coagulation with aluminum alone has limited effectiveness for heavy metal removal at neutral pH ranges. Although higher pH condition is suitable for the removal of heavy metals but trace amount of heavy metals removal in drinking water treatment might increase the cost of operation by adjusting pH.

In recent years, ferrate oxidation has received much attention because of its high oxidizing power, good selective reactivity, and non-toxic decomposition by-products of ferric ion. Fe(VI) has been known as a green oxidant, coagulant, disinfectant, and antifoulant, therefore a promising multi-purpose water treatment chemical (*1-4*). Since the newly formed intermediate constituents in the process of ferrate oxidation might have high adsorption ability (*5,6*), it is the objectives of this research to investigate the effect of ferrate preoxidation on the enhancement of heavy metals removal during aluminum coagulation.

Experimental Methods

Materials

Potassium ferrate (K_2FeO_4) of a high purity (98 % plus) was prepared by the method described by Thompson et al. (*7*). Potassium ferrate solution was prepared by dissolving potassium ferrate solid in distilled water just before use in order to minimize the loss of ferrate. Aluminum Sulfate ($Al_2(SO_4)_3 \cdot 18H_2O$, Tianjin Chemical Inc., Tianjin, China) was selected as the coagulant.

Humic acid was from England, humic acid solution was prepared by dissolving humic acid in distilled water, then shattered the solution by ultrasonic instrument, then dissolved it in water for 12h at 50°C, followed by filtration through a 0.45 μm filter to separate any particulate matter, finally determined the TOC of the sample.

The raw water was obtained from River Songhua, and the raw water quality is shown in Table I. To this raw water, 250μgL^{-1} of Pb^{2+} or 50μgL^{-1} of Cd^{2+} were added respectively to simulate the heavy metal polluted water.

Table I. Typical raw water quality

Parameters	*Turbidity (NTU)*	*Colour (CU)*	*pH*	*COD_{Mn} (mgL^{-1})*	*Alkalinity (mg$CaCO_3L^{-1}$)*
Concentrations	15~80	30~40	8.1~8.2	10~12	50~60

Experiment Procedures

The evaluation of the performance for heavy metal removal was conducted through a jar test apparatus (Model DBJ-621). The raw water was transferred into 6 beakers with 500ml volume. Potassium ferrate was added into the water and stirred for 1min. followed by the addition of alum, and then stirred at 300rpm for 1min. The water was then subject to slow stirring at 60rpm for 10min. The solution was settled for 20min and the supernatant was withdrawn at 1cm below the surface and then the water sample was filtrated by a membrane with a pore size of 0.45 μm before analyzing cadmium and lead concentration by atomic absord spectrometer.

Results and Discussion

Effect of ferrate preoxidation on removal of heavy metal

As shown in Figure 1, alum alone could remove some of lead, with the increase of dosage of alum, the removal percentage increases, but when the dosage of alum was more than 40mgL^{-1}, the removal rate would decrease, which might be due to the bad coagulation with excess amount of alum as a result to lead to lower removal rate of lead.

Ferrate preoxidation could improve the removal of lead, at the conditions, of alum dosage 40mgL^{-1}, ferrate 1mgL^{-1}, the removal rate was 88.6%, which achieve additional average removal of 28.6% compared with the case of alum alone, also the removal rate of lead increases with the increase of alum dose. The cause of removal of heavy metals by ferrate may due to higher adsorption of the by-products, $Fe(OH)_3$, but the higher removal rate would reach only at the conditions of good coagulation.

In addition, the results also showed that the removal rate would not increase with the increase of ferrate dosage, which implied that not all of $Fe(OH)_3$ adsorbed Pb^{2+}, some of $Fe(OH)_3$ would adsorb another substance (organic matter), there might be competitive adsorption of many substances in the reactions.

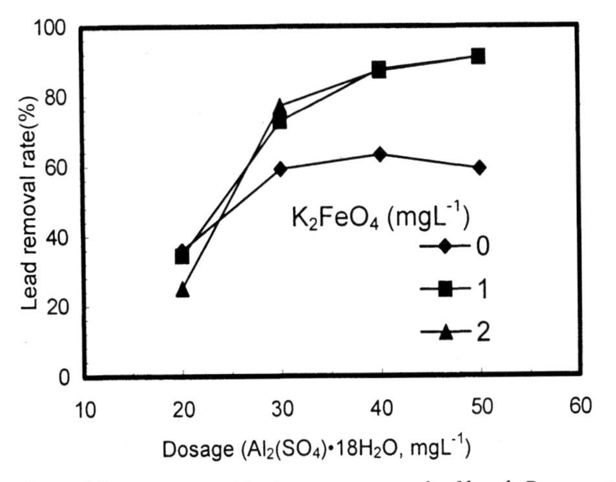

Figure 1. Effect of ferrate preoxidation on removal of lead. Raw water quality: Turbidity 15~80NTU; Colour 30~40CU; pH 8.1~8.2; COD_{Mn} 10~12mgL^{-1}; Alkalinity 50~60 (mgCaCO$_3$L^{-1}) ; Pb(II); 250 μgL^{-1}

As indicated in Figure 2, alum alone hardly remove cadmium, even though the dosage of alum was increased, the removal rate was still not increased, which suggested that alum alone could not effectively remove cadmium.

Ferrate preoxidation can effectively increase the cadmium removal percentage, for example, the removal rate of cadmium was 20%, 25% at the dosage 0.5, 1mgL^{-1} FeO_4^{2-}, respectively; the removal rate of cadmium increased with the incease of ferrate dosage, but did not increase with the increase of alum dosage, which was different from the case of lead. It could concluded that the newly formed hydroxide iron might be more prone to adsorb cadmium than hydroxide aluminium.

Compared with the data in Figure 1 and Figure 2, it is seen that the removal rate of lead and cadmium is totally different, which might be attributed to the different forms of Pb(II)and Cd(II) in water. pH value could affect the forms of heavy metals, so let us to discuss the effect of pH value on the removal rate of heavy metals.

Effect of ferrate dosage on heavy metals removal

Figure 3 showed the percentage removal of lead as a function of ferrate dosage. It indicated that at pH=3, the removal rate of lead increased with the

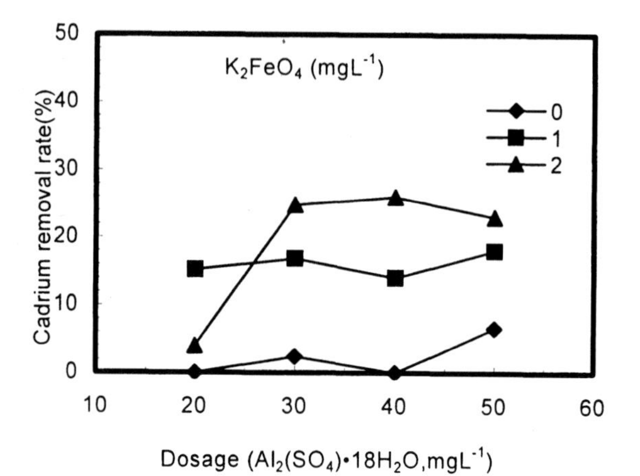

Figure 2. Effect of ferrate preoxidation on the removal of cadmium. Raw water quality: Turbidity 15~80NTU; Colour 30~40CU; pH 8.1~8.2; COD_{Mn} 10~12mgL^{-1}; Alkalinity 50~60 (mgCaCO$_3$L^{-1}) ; Cd(II) 50 μgL^{-1}

increase of ferrate dosage, but the removal rate was less than 40%; at pH=7, 9, the removal rate of lead obviously increased, the increase of pH from 3 to 7 caused 60% higher removal of lead.

Figure 4 showed the variation of the percentage cadmium removal with respect to ferrate dosages at several selected pH values. It is seen that even though adjusting pH to alkaline condition could increase the removal rate of cadmium by coagulation alone without the dose of ferrate, but the percentage removal of cadmium is limited to a certain extent. However, ferrate preoxidation achieved obvious increase of the removal of cadmium when the water was subjected to ferrate preoxidation. Higher dosage of ferrate achieved higher degree of removal of cadmium but after the dose of ferrate increased beyond 3mgL^{-1}, the extent of removal of cadmium was decreased.

Effect of pH on heavy metals removal

The condition of pH determine the speciation of heavy metals, it is important to evaluate the effect of pH on the removal of heavy metals during ferrate preoxidation.

Figure 5 shows the effect of pH on ferrate preoxidation for lead removal during coagulation with alum. In the case of coagulation with alum alone, the percentage removal of lead was increased as the increase of pH. When pH was below 6, the increase of pH caused sharp increase of the percentage removal of lead. But further increased pH beyond 6, the percentage removal of lead maintained at certain level. However, with ferrate preoxidation, the percentage removal of lead was always higher than the case with alum coagulation alone.

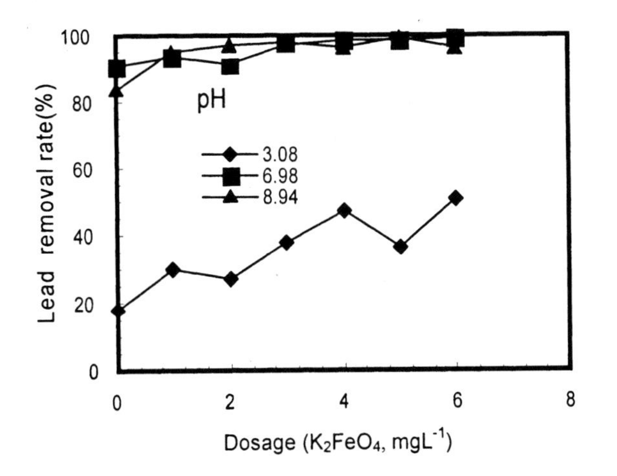

Figure 3. Effect of ferrate dosage on the removal of lead during coagulation. Pb(II) spiked; 250 μgL^{-1}. Alum dose: 10mgL^{-1}. Raw water quality: Turbidity 15~80NTU; Colour 30~40CU; pH 8.1~8.2; COD_{Mn} 10~12mgL^{-1}; Alkalinity 50~60 ($mgCaCO_3L^{-1}$)

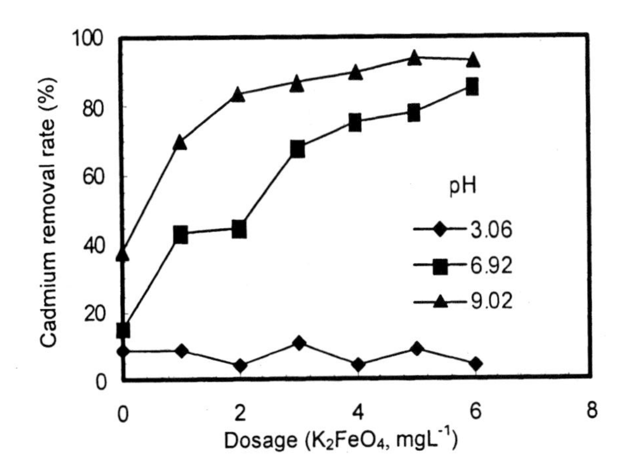

Figure 4. Effect of ferrate dosage on the removal of cadmium during coagulation by ferrate preoxidation. Cd(II) spiked; 250 μgL^{-1}. Alum dose: 10mgL^{-1}. Raw water quality: Turbidity 15~80NTU; Colour 30~40CU; pH 8.1~8.2; COD_{Mn} 10~12mg L^{-1}; Alkalinity 50~60 ($mgCaCO_3L^{-1}$)

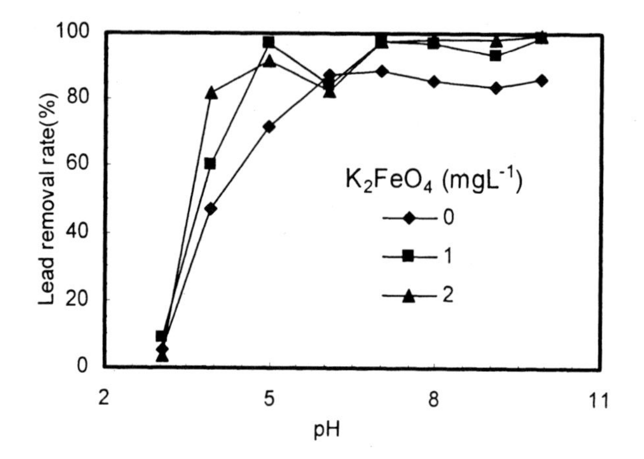

Figure 5. Effect of pH on the removal of lead by ferrate preoxidation during coagulation. Pb(II) spiked; 250 μgL^{-1}. Alum dose: $10mgL^{-1}$. Raw water quality: Turbidity 15~80NTU; Colour 30~40CU; pH 8.1~8.2; COD_{Mn} 10~12mgL^{-1}; Alkalinity 50~60 ($mgCaCO_3L^{-1}$)

pH 5 was the optimum condition for achieving the highest lead removal. In addition, preoxidation with ferrate also widened the optimum range of alum dosages. This results indicated that the intermediate species formed at weakly acidic condition might also have strong adsorption.

As seen in Figure 6 that with the increase of pH, there was obvious increase in the percentage removal of cadmium. However, coagulation under neutral condition had limited percentage removal of cadmium when coagulate with alum alone. In contrast, the dose of ferrate caused substantial increase in cadmium percentage removal. At ferrate dose of 1-2mgL^{-1}, the percentage removal of cadmium was increased about one time compared to the case without ferrate preoxidation.

From the experiments above, the variation of pH had a great effect on the removal of heavy metals, the higher pH of water, the higher the removal rate. But there was existing great difference between lead and cadmium, even at neutral pH conditions, the removal rate of lead was high, the lead concentration after filtration could meet with the drinking water standards; but for cadmium, only when the pH value was more than 11, the removal rate was high, so in order to achieve the higher removal rate of cadmium, there are two ways, one is to raise pH value, the other is to increase the dosage of ferrate.

Effect of ferrate preoxidation on heavy metals removal in the presence of humic acid

Natural organic matter (NOM) in water, such as humic acid, have large numbers of active groups on their surface such as carboxyl, phenolic hydroxyl

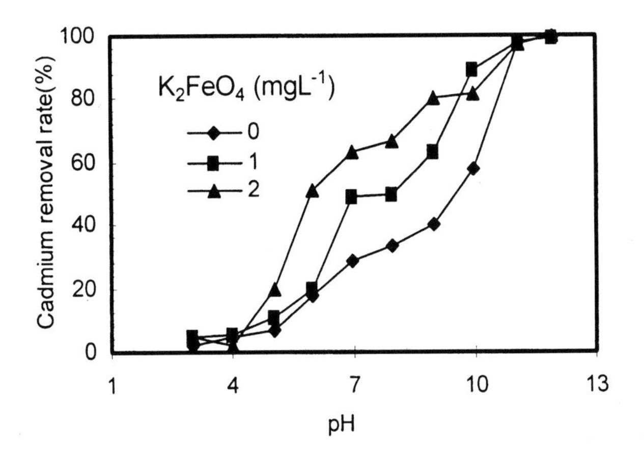

Figure 6. Effect of pH on the removal of cadmium by ferrate preoxidation during coagulation. Cd(II) spiked; 250 μgL^{-1}. Alum dose: 10mgL^{-1}. Raw water quality: Turbidity 15~80NTU; Colour 30~40CU; pH 8.1~8.2; COD_{Mn} 10~12mgL^{-1}; Alkalinity 50~60 ($mgCaCO_3L^{-1}$)

group, amido etc., which make the surface more negative charged. Those active groups could chelate with the heavy metals in water, so the humic acid could affect the forms of heavy metals in water and thus affected the removal rate of heavy metals.

As indicated in Figure 7and Figure 5, the removal rate of lead obviously decreased in the presence of humic acid with alum alone, the optimal removal rate was at pH >10.

The removal rate of lead by ferrate preoxidation was still higher than that of alum alone, but the removal rate of lead was lower in the presence of humic acid than that without the presence of humic acid in the case of ferrate preoxidation, the removal rate of lead increased with the increase of ferrate dosages in the presence of humic acid.

The presence of humic acid could greatly affect the removal percentage of lead in the process of coagulation, because it could form complex with lead, which was hardly removed by coagulation. Ferrate could oxidize some functional groups of humic acids and could reduce the chelation between humic acid and lead or the newly formed ferric hydroxide could adsorb the complex with lead to raise the removal rate of lead.

As illustrated in Figure 8 and Figure 6, the presence of humic acid hardly affected the removal percetage of cadmium, the removal rate at neutral pH and the optimal pH was still the same, which suggested that the presence of humic acid did not affect the removal of cadmium.

Humic acid chelated with metal ions according to Irving Williams Serial: Pb^{2+} > Cu^{2+} > Ni^{2+}> Co^{2+} > Zn^{2+} > Cd^{2+} > Fe^{2+} > Mn^{2+} > Mg^{2+} (*8*). It was obvious that the chelating ability of lead was stronger than that of cadmium, so the presence of humic acid did affect the removal of lead but did not affect the removal of cadmium.

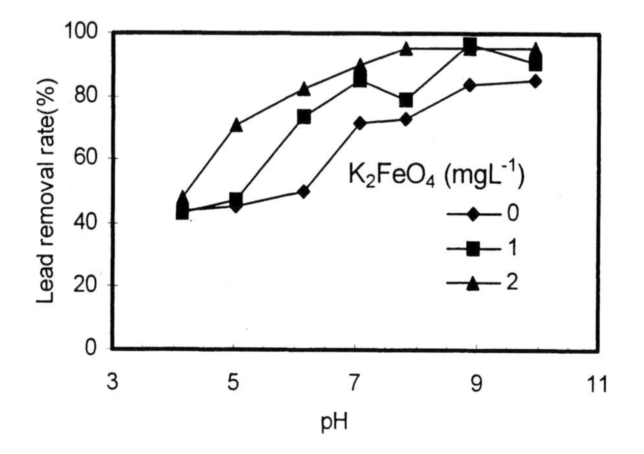

Figure 7. Effect of pH on lead removal in the presence of humic acid. Raw water quality: Turbidity 15~80NTU; Colour 30~40CU; pH 8.1~8.2; COD_{Mn} 10~12mgL^{-1}; Alkalinity 50~60 (mgCaCO$_3$L^{-1}) ; Pb(II); 250 μgL^{-1}; Humic acid :1.75 mgL^{-1}(DOC).

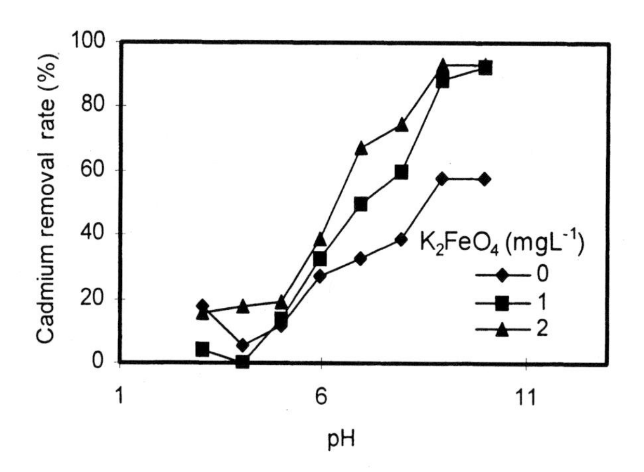

Figure 8. Effect of pH on cadmium removal in the presence of humic acid. Raw water quality: Turbidity 15~80NTU; Colour 30~40CU; pH 8.1~8.2; COD_{Mn} 10~12mgL^{-1}; Alkalinity 50~60 (mgCaCO$_3$L^{-1}) ;Cd(II); 250 μgL^{-1}; Humic acid :1.75 mgL^{-1}(DOC)

Conclusion

Alum alone hardly removed Pb^{2+} and Cd^{2+} in water during the coagulation, especially for Cd^{2+}. Ferrate preoxidation could effectively increase the removal rate of lead, which made the effluent meet with the drinking water quality standards, but it did not obviously increase the removal rate of cadmium.

There were obvious increases in the percentage removal of cadmium with the increase of pH. The water pH was an important factor to affect the removal rate. With the increase of pH, the remove rate increased. Ferrate preoxidation widened the optimum pH ranges for alum to remove heavy metals. At weakly acidic condition, the removal rate was still high.

The adsorption ability of hydrolyzed species and the final product $Fe(OH)_3$ was the main cause to remove heavy metals. The combined effect of adsorption by intermediate iron species formed in ferrate oxidation and the enhanced coagulation of iron colloids co-precipitated with heavy metals might be responsible for the effective removal of heavy metals.

The humic acid could greatly affect the removal rate of lead in the process of coagulation, because it could form complex with lead, which was hardly removed by coagulation. Ferrate preoxidation could oxidize the active groups of humic acid and reduce the extent of chelation, so ferrate preoxidation could enhance the removal of lead in the presence of humic acid. The humic acid hardly affected the removal rate of cadmium, for the chelation between humic acid and cadmium was very weak.

References

1. Waite, T.D.; Gray, K.A. Oxidation and coagulation of wastewater effluent utilizing ferrate (VI) ion. *Stud. Environ. Sci.* **1984**, *23*, 407-420.
2. Carr, J.D.; Kelter, P.B.; Tabatabai A.; Splichal D.; Erickson J.; McLaugh-lin, C.W. *Properties of ferrate (VI) in aqueous solution: an alternate oxidant in wastewater treatment.* In: Jolley RL (Ed.), Proceedings of Conference on Water Chlorination Chem Environment Impact Health Effects, Lewis Chelsew.**1985**, p1285-1298.
3. Sharma, V.K. Potassium ferrate (VI): an environmentally friendly oxidant. *Adv. Environ. Res.* **2002**, *6*, 143-156.
4. Jiang, J.Q.; Lloyd, B. Progress in the development and use of ferrate (VI) salt as an oxidant and coagulation for water and wastewater treatment. *Water Res.* **2002**, *36*, 1397-1408.
5. Murmann, R.K.; Robinson, P.R. Experiments utilizing ferrate ion for purifying water. *Water Res.* **1974**, *8(8)*, 543-547.
6. Potts, M.E.; Churchwell, D.R. Removal of redionuclides in wastewater utilizing potassium ferrate (VI). *Water Environ. Res.* **1994**, *66(2)*, 107-109.
7. Thompson, G.W.; Ockerman, G.W.; Schreyer, J.M. Preparation and purification of potassium ferrate(VI). *J. Am. Chem. Soc.* **1951**, *73*, 1279-1281.
8. Bowe, C.A.; Martin, D.F. Extraction of Heavy Metals by 2-Mercaptoethoxy Groups Attached to Silica Gel. *J. Environ. Sci. Heal. A* **2004**, *39(6)*, 1479-1485.

Chapter 30

Potential of Ferrate(VI) in Enhancing Urban Runoff Water Quality

Umid Man Joshi[1], Rajasekhar Balasubramanian[1,*], and Virender K. Sharma[2]

[1]National University of Singapore, 21 Lower Kent Ridge Road, Singapore 119077
[2]Chemistry Department, Florida Institute of Technology, 150 West University Boulevard, Melbourne, FL 32901
***Corresponding author: cherbala@nus.edu.sg**

Urban development and increasing water demand are putting a lot of stress on existing water resources around the world. A great deal of attention is now paid to alternative sources of water such as stormwater catchment systems as they serve multi-purpose functions. However, human activities introduce a variety of contaminants into the stormwater catchments, which affect the quality of the water to be used for both potable and non-potable purposes. Environmentally friendly treatment technologies are needed to treat and to use urban runoff without having negative impacts on the environment. Ferrate (VI) technology has the potential to be one of the most environmentally friendly water treatment technologies of the twenty-first century. Ferrate(VI) has advantages in treating heavy metals (e.g., Pb^{2+}, Cd^{2+}, Cr^{3+}, Hg^{2+}, Cu^{2+}, and Mn^{2+}), suspended particles, synthetic/natural organic matter (present as TOC, BOD and COD), microorganisms (e.g., bacteria and virus), without producing chlorinated by-products. Ames test on ferrate(VI) treated water demonstrated negative results, suggesting no mutagenic by-products. Uniquely, Ferrate(VI) performs distinctly different treatment functions (oxidation, coagulation, flocculation, and disinfection) from the application of a single dose, thus providing a simplified and cost-effective process.

Introduction

Water is indispensable for human health and well-being. Global consumption of water is doubling every 20 years, more than twice the rate of human population growth. Two-thirds of the world's population is predicted to be living with serious water shortages, or absolute water scarcity by 2025, if the current growth rate in global water demand persists. More than 30 countries already face water stress/scarcity. The Earth's water system can support, at most, only one more doubling of demand, estimated to occur in less than 30 years. Limited fresh water resources, growing demand of water, environmental concern, and ever stringent water quality standards by the regulatory agencies have encouraged the scientific community to look for non-conventional water sources, and more effective, environmentally friendly treatment technologies. Technologies are now available by virtue of which wastewater and even seawater can be converted to potable water, however, the excessive cost of these treatment processes, environmental concern of their by-products (concentrate from reverse osmosis process) and psychological barrier for the use of treated wastewater are the major constrains for implementing these technologies.

The demand for potable water is centralized in and around urban areas. About half of the world's population (47%) now lives in urban areas, compared to little more than one-third in 1970s. To add on to the already existing water scarcity, the forecast that urban population is expected to grow by 2% per year during 2000–15, and to reach an overall 65% by 2050 indicates the crunching demand on fresh water supplies in the days ahead. Urban areas are characterized by imperviousness, where a substantial fraction of the land is made impervious by construction of structures such as buildings, roads, pavements etc. Thus, there is a potential to use urban runoff as a substitute and/or supplementary source of much needed potable water.

The origin of urban runoff is the rainfall. Urban runoff is excess water, not absorbed by soil after heavy rains. It flows over surfaces such as roads, parking lots, building roofs, driveways, lawns, and gardens. Precipitation chemistry plays an important role in the quality of stormwater as it scavenges soluble gases and particles from the atmosphere (*1*). During its movement through soils and impervious surfaces, stormwater can carry a wide range of potentially harmful environmental contaminants, such as metals, oil and grease, pesticides, and fertilizers. For this reason, even when there is scarcity of water in urban areas, the conventional practice has been inclined towards discharging the urban stormwater as soon as possible through a network of drainage systems. Now that there is greater interest in utilizing the runoff from urban areas, the same system can be used to collect the urban runoff. However, use of water from urban runoff is not straight-forward. Even though the quantity of urban runoff makes it justifiable to consider it as an alternative/supplementary source, the quality is questionable. Accurate knowledge of the quality and quantity of runoff is required to develop appropriate treatment technologies.

There have been several attempts to collect and treat urban stormwater. Selecting a specific treatment technology is one of the most difficult steps in the stormwater management plan due to the presence of a number of pollutants derived from a variety of sources (*2, 3, 4*). The main reason for this difficulty is the lack of accurate information on the kind of pollutants being introduced into the runoff. By characterizing the runoff and quantifying the pollutant load, it would be possible to predict the overall quality of urban runoff entering into the treatment plant. It would then be easier to decide on technology to be used to meet the required water quality standards.

Contradicting the traditional belief that the stormwater is contamination-free, recent studies have proved that the stormwater could be a potential source of water pollution and the concentration of pollutants varies according to the location through which it passes (*5, 6, 7*). Stormwater even from a visibly clean residential catchment can be a significant source of pollution (*4*). Anthropogenic activity introduces chemical and biological constituents to the catchments. Trace metals, suspended solids, nutrients, pesticides, petroleum products, and E. coli and fecal coliform bacteria are generally found in higher concentrations in urbanized and urbanizing areas than in natural systems, due to increased numbers of people, vehicles, roads, and building materials introduced into the landscape.

Among the chemical pollutants in urban runoff, trace metals, introduced into catchments through anthropogenic activities, are a cause for concern due to their potential toxicity. Sources of trace metals in urban stormwater runoff are numerous and complex. Once they are present, they cannot be chemically transformed, or destroyed, as other constituents such as organic matter may be (*6*). The International Agency for Research on Cancer (IARC) classifies As and its compounds, Cd and its compounds, Cr (VI) compounds, Ni compounds as Class I agents (carcinogenic to humans), Pb compounds as Class 2A agents (probably carcinogenic to humans), and Pb and Ni as Class 2B agents (possibly carcinogenic to humans). Some metals are necessary for vital metabolic processes, for example, Fe and Cu are required for the synthesis of hemoglobin; Mn and Fe are constituents of some enzymes; Zn is required for proper functioning of enzymes, growth of tissue and healing of wounds; Cr^{3+} is necessary for production and functioning of insulin. However, even the essential metals, if present in higher concentration, are toxic.

The contribution of PAHs in any water system could vary (site-specific). For instance, Hoffman et al. (*8*) have reported that urban runoff was the main contributor to local water bodies, with atmospheric fallout and asphalt abrasion that contributed small and large particles, respectively. Several studies also suggested that urban runoff could act as a major source of PAHs to the coastal environment (*8,9,10*). In the United States, urban runoff was the second most frequent source of PAHs that contaminate the surface water. The concentrations of PAHs in urban runoff are generally much higher than those of surface water (*11*). Nationwide Urban Runoff Program (NURP) reported concentrations of PAHs in the range of 300-10,000 ng/l, with the average concentrations of individual PAHs being mostly above 1,000 ng/l (*12*). Furthermore, Pitt et al. (*13*) reported elevated concentration of Flt, which was 130 μg/l in urban runoff samples of Birmingham, Alabama.

Runoff waters are formed when rain or melting snow washes the ground surfaces. In general, ground surfaces can be grouped into impervious and pervious surfaces. Impervious surfaces enlarge the quantity of runoff as well as increase the level of PAHs in the urban runoff, which would dissipate once they attenuate on pervious surfaces (*14*). The examples of impervious surfaces are paved parking lots, streets, driveways, roofs, and sidewalks. Among the types of impervious surfaces, runoff from highway areas has been reported in various studies (*15, 16*). This concern might be due to the runoff loading of highway areas in contributing PAHs to adjacent water bodies. A case study in the US river indicates that over 50% of the total PAHs in the river originated from a highway runoff. The runoff-loading factor per vehicle was as high as 24 mg/kg (9). However, the relative contribution of urban runoff to adjacent water bodies is site-specific, and dependent heavily on the magnitudes of the wet and dry deposition as well as sources of PAHs in the urban runoff (*17*).

This chapter gives the assessment of contamination of urban runoff in Singapore, caused by metals and PAHs. The potential of ferrate (VI) to improve the urban runoff water quality is discussed. Future research directions in terms of using ferrate (VI) for decontamination of storm water in Singapore are highlighted.

Metals

The 13 metals in urban runoff water were analyzed. Analyzed metals were aluminum (Al), arsenic (As), cadmium (Cd), cobalt (Co), chromium (Cr), copper (Cu), iron (Fe), manganese (Mn), nickel (Ni), lead (Pb), titanium (Ti), vanadium (V), and zinc (Zn). These metals were selected based on their toxicity to ecosystem and human health. Figure 1 shows the average concentration of trace metals for the urban runoff. Since the variation in concentrations among various trace metals is quite large, the data are presented in log-graph. It can be seen that Al, Fe, and Zn were in higher concentration than other trace metals. Various factors are known to influence the concentration which include the number of days antecedent to rainfall, intensity of rainfall, atmospheric deposition, surface over which the runoff passes, anthropogenic activity in the area considered etc (*6, 18, 19*). One of the significant observations made is that the average values lie closer to the lowest values, which suggest that most of the time the concentrations of the metals are fairly constant.

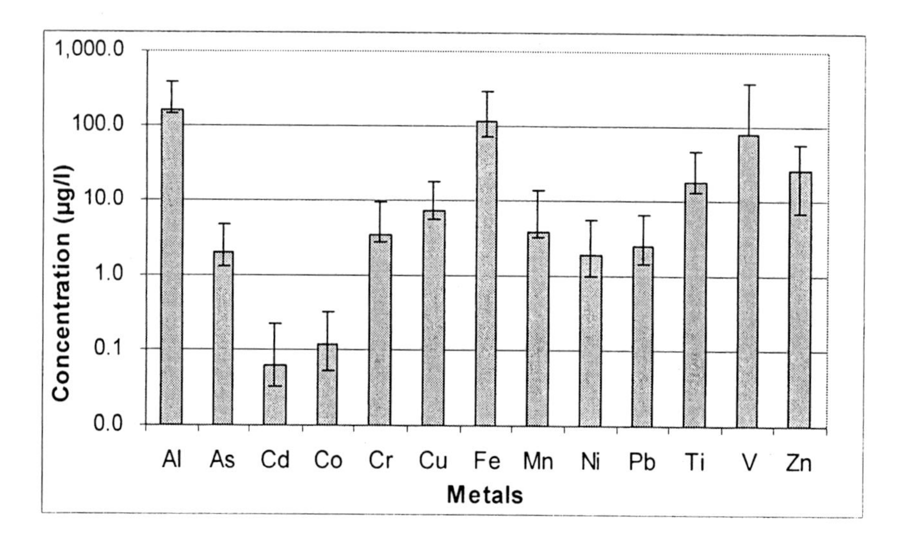

Figure 1. Average concentration of trace metals in urban runoff.

Comparing the average concentration of trace metals with international standards of water quality, it was found that for most of the times, the concentrations of majority of metals were below World Health Organization (WHO) guidelines. However, it is to be noted that the standards are getting stringent and the pollution levels are increasing over time. It is likely that their maximum concentration levels in drinking and recreational waters would be reduced by the concerned authority in the days ahead.

Polycyclic Aromatic Hydrocarbons (PAH)

Fifty-five stormwater samples were collected from a drain located at the shoulder of Ayer Rajah Expressway (AYE) during the period of October 2005 to March 2006. AYE is one of the main expressways in Singapore, with eight lanes

Table 1. Concentration of PAHs in Dissolved phase of stormwater

Compound	Dissolved Phase	
	Range (µg/L)	Average (µg/L)
Naphthalene	20-1,600	1,991
Acenaphthylene	7-1,300	155
Acenaphthene	6-62	51
Fluorene	1-41	49
Phenantherene	10-77	132
Anthracene	5-1,470	59
Fluoranthene	25-560	35
Pyrene	2-430	27
Benzo[a]anthracene	29-430	33
Chrysene	2-430	28
Bezo[b]fluoranthene	1-470	61
Benzo[k]fluoranthene	2-200	26
Benzo[a]pyrene	1-240	25
Indeno(1,2,3-cd)pyrene	3-960	118
Dibenz[a,h]anthracene	9-3,840	443
Benzo[g,h,I]perylene	4-328	205

stretching across the road. The average concentrations of the PAHs in the dissolved phases of the stormwater are given in Table 1.

Concentrations of total PAHs in the dissolved phase ranged from 506 to 1,940 ng/L, with an average value of 1,143 ng/l. The concentrations of $\sum 6PAH_{CARC}$ ranged from 49 to 887 ng/l, with an average value of 318 ng/l, while the concentrations of $\sum 6PAH_{ECstd}$ ranged from 64 to 881 ng/l with an average value of 232 ng/l. $\sum 6PAH_{CARC}$ and $\sum 6PAH_{ECstd}$ were 32 and 20% of the total PAHs.

Although these concentrations were relatively lower than values reported in the literature and the European Community standard for untreated waters (i.e. 1000 ng/l), they were higher than the sum permitted for drinking water, which is 200 ng/l. Moreover, as these carcinogenic compounds are in the dissolved phase, they are more readily taken up by the biota than those in the particulate phase (*20, 21*). Thus, there is a potential health hazard resulting from the levels of carcinogenic PAHs present in the dissolved phase in stormwater, particularly if it is used for the drinking water purposes.

Ferrate(VI) Technology for Storm Water Quality Enhancement

Ferrate(VI) is made from iron which is the third most common element on earth. Ferrate(VI) is an iron atom with a plus-six charge. Ferrate(VI) is highly reactive and unstable when exposed to air or water because of its high oxidation/reduction potential. As Ferrate(VI) is reduced from 6+ to 3+, it delivers superior oxidation, disinfection, coagulation, dewatering, and deodorization. The by-product of ferrate Treatment is Fe(III) or ferric oxide or rust, which is non-toxic in nature, thus making the use of ferrate very environmentally friendly. Sodium ferrate(VI) (Na_2FeO_4) and potassium ferrate(VI) (K_2FeO_4) are two types of ferrates, which are considered to be used for water treatment.

The metal levels in urban runoff can be reduced by treating with ferrate(VI). Previous studies have demonstrated effectiveness of ferrate(VI) to remove a range of metals (e.g., Fe^{2+}, Fe^{3+}, Mn^{2+}, and Cu^{2+}) and toxic heavy metals (e.g., Pb^{2+}, Cd^{2+}, Cr^{3+}, and Hg^{2+}) to a low level at a dose range of 10–100 mg/L as K_2FeO_4 (*22, 23*). Table 2 shows removal of toxic metals and nonmetals from contaminated water, prepared in the laboratory by the addition of salts into water. The results in Table 2 indicate removal of a broad category of toxic elements. Removal of these elements was accomplished by co-precipitation in the ferric oxide gel formed by the action of potassium ferrate. During the same process, turbidity is reduced. The removal of these toxic elements is highly desirable, as it will beneficial for human health and ecosystem.

Manganese also affects the flavor and color of food and water. In addition, manganese may react with tannins in coffee, tea, and some alcoholic beverages to produce a black sludge, which also affects both taste and appearance. Manganese(II) is usually removed by the oxidation of soluble Mn(II) to a colloidal precipitate followed by solid/liquid separation via sedimentation and filtration. Ferrate(VI) ion can readily oxidize Mn^{2+} to insoluble MnO_2 (equation 1) and Figure 2 shows that ferrate(VI) is very effective for removing manganese from water (24). Ferrate(VI) initially oxidizes Mn(II) to Mn(IV), which can then easily be removed from water by flocculation through iron(III) hydroxide (eq 1)

$$3\ Mn^{2+} + 2\ Fe^{6+} \rightarrow 3\ MnO_2 + 2\ Fe^{3+} \qquad (1)$$

Ferrate(VI) acts as an oxidation and coagulation (precipitation with $Fe(OH)_3$) agent which can also remove organic pollutants and pesticides from wastewater. For example, Fe(VI) was shown to be efficient in removing five priority pollutants; these include bromodichloromethane, nitrobenzene, naphthalene, 1,2-dichlorobenzene, and trichloroethylene (*25*). As shown in Figure 3, Ferrate(VI) can also treat a number of other organic priority pollutants. Subsequent treatments or larger doses would remove these pollutants further.

Table 2. Metals Removed by Potassium Ferrate

Constituent	Amount[a] Metal mg/cm^3	Amount[b] K_2FeO_4 mg/cm^3	Constituent	Amount[a] Metal mg/cm^3	Amount[b] K_2FeO_4 mg/cm^3
Ag(I)	14.18	0.9300	Ag(I)	0.01418	1.020
Cd(II)	10.99	0.9300	Cd(II)	0.01099	1.020
Al(III)	0.01420	0.9300	Cu(II)	0.001360	0.7600
Ce(III)	3.678	0.7600	Sn(II)	2.959	0.7600
Pb(II)	0.7798	1.067	Ba(II)	0.7670	1.067
Co(II)	0.1539	1.067	Mg(II)	0.2403	2.220
Mn(II)	0.1365	2.220	Hg(II)	0.7705	2.220
Mg(II)	0.002403	2.220	Ce(III)	0.1819	2.410
Pb(II)	0.005000	1.500	Tl(III)	0.6909	1.045

[a] Quantity of ion present initially in test solution.
[b] Quanity of potassium ferrate used in test.
SOURCE: Reproduced with permission from reference 23. Copyright 1992.

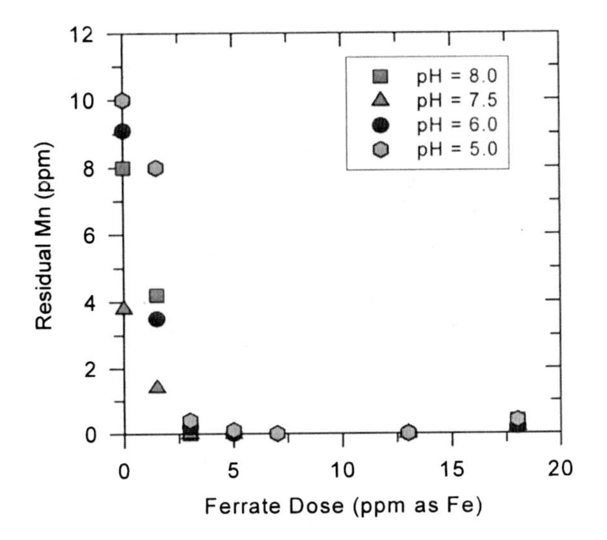

Figure 2. Effect of Ferrate dosage on Mn removal. (Reproduced with permission from reference 24. Copyright 1998.)

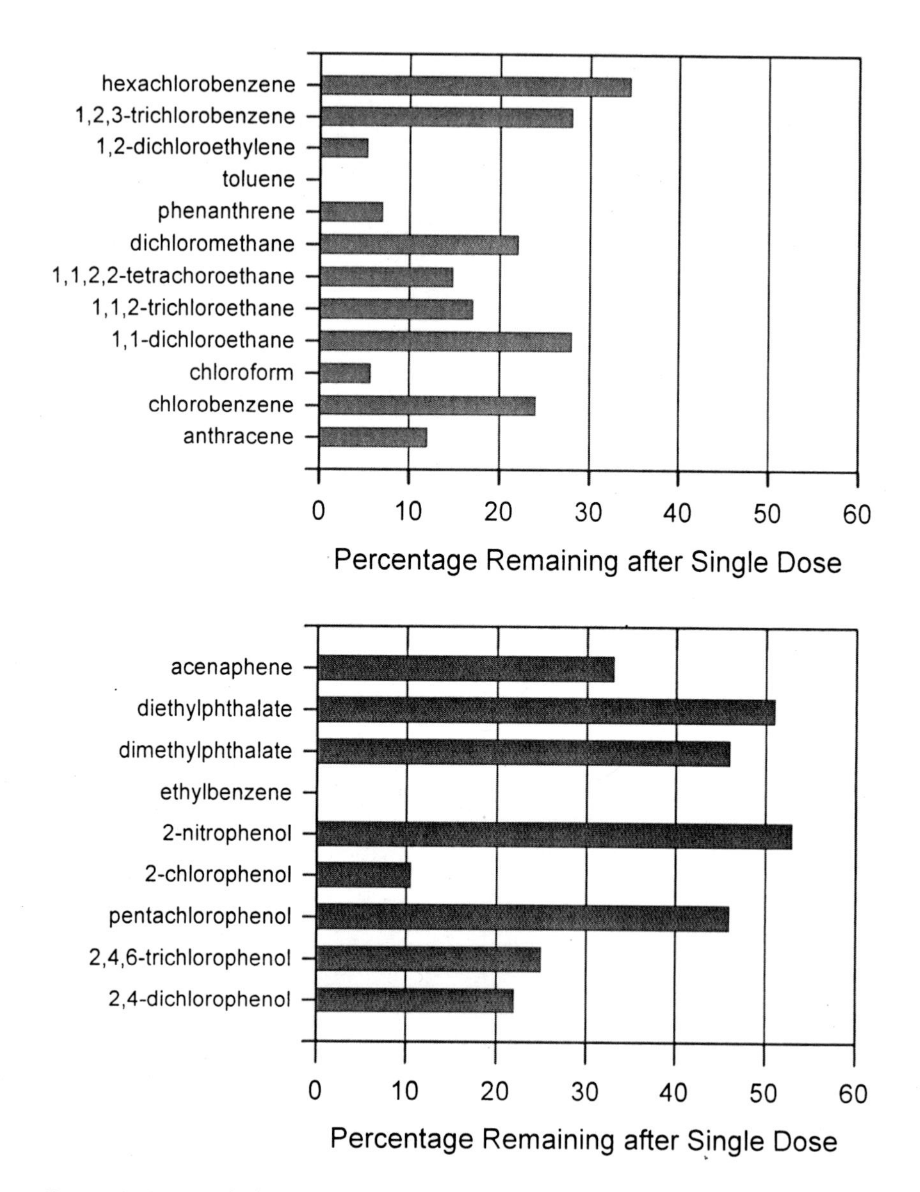

Figure 3. Removal of organics by treatment with ferrate(VI) (data taken from (25)).

Mutagenic tests

As ferrate(VI) is proposed to be used in water treatment works, public health concern regarding mutagenicity is also being addressed simultaneously. Thus, attempts were made to determine whether the ferrate(VI) treated water contains/produces any mutagenic substances. The Ames test was used to screen for the existence of mutagenic compounds. It is claimed that about 90% of known carcinogens can also be shown to be mutagens by use of Ames test. Special strains of *Salmonella typhimurium* are used for this test. The mutagenicity of the chemicals tested is usually determined based on the number of bacterial revertants observed for a certain concentration or the slope of the linear dose response. If a positive result is obtained, mutagenic compounds are considered to be present. The Ames test applied to ferrate(VI) treated water demonstrated negative results (*26*), suggesting that ferrate(VI) reagents do not produce mutagenic by-products.

Conclusion

As the demand for potable water is increasing at unprecedented rate, search for new sources of water on. Urban runoff, which used to be considered as a nuisance is now getting attention as it has the potential to be used as alternative/supplementary source of potable water. Pollutants generated by anthropogenic activities in urban areas are transferred to urban runoff. The levels of pollutants have to be quantified, and appropriate treatment technologies have to be applied to treat the urban runoff to be used for potable purposes.

Emerging ferrate(VI) technology has the potential to treat urban runoff. As Ferrate(VI) is reduced it falls back down from iron 6+ to 5+ to 4+ to 3+ (ferric iron), in the process acting as an oxidant, then coagulant, and finally as a flocculent. Further, ferrate(VI) performs these three distinct treatment functions from the application of a single dose, thus making it cost effective and environmentally friendly treatment technology. Using ferrate(VI) technology, it is possible to remove metals and PAHs. The same technology can also be used to remove endocrine disrupting chemicals, pharmaceuticals, and pesticides. Lastly, this technology is suitable for disinfecting water without the production of disinfection-by-products. Therefore, ferrate(VI) has the potential to be used in water treatment plants worldwide. Bench and pilot scale studies on the ferrate technology are currently underway, which will form the basis for designing full-scale plants for treatment of urban runoff.

References

1. Hu, G.P.; Balasubramanian R.; Wu, C.D. *Chemosphere* **2003**, *51*, 747.
2. Boving, T.B.; Zhang, W. *Chemosphere* **2004**, *54*, 831.
3. Jang, A.; Seo, Y.; Bishop, P.L. *Environ. Pollut.* **2005**, *133*, 117.
4. Gromaire-Mertz, M.C.; Garnaud, S.; Gonzalez, A.; Chebbo, G. *Water Sci Technol,* **1999**, *39(2)*, 1.
5. Barrett, M.E.; Irish, Jr. L.B.; Malinia, Jr. J.F.; Charbenuea, R.J. *J. Environ. Eng. - ASCE* **1998**, *124*, 131.
6. Davis, A.P.; Shokouhian, M.; Ni, S. *Chemosphere* **2001**, *44*, 997.
7. Metre, P.C.V.; Mahler, B.J. *Chemosphere* **2003**, *52*, 1727.
8. Hoffman, E.J.; Mills, G.L.; Latimer, J.S.; Quinn, J.G. *Environ. Sci. Technol.* **1984**, *18*, 580
9. Hoffman, E.J.; Latimer, J.S.; Hunt, C.D.; Mills, G.L.; Quinn, J.G. *Water Air Soil Pollut.* **1985**, *25*, 4349.
10. Barrick, R.C.; Prahl, F.G. *Estuar. Coast. Shelf Sci.* **1987**, 25(*2)*, 175
11. ATSDR (Agency for Toxic Substances and Disease Registry), U.S. Department of health and human services **1995**.
12. Cole, R.H.; Frederick, R.E.; Healy, R.P.; Rolan, R.G. *J. Water Pollut. Control Fed.* **1984**, *56*, 898
13. Pitt, R.; Lalor, M.; Field, R.; Brown, M. *Water Sci. Technol.* **1993**, *28*, 271.
14. Ötvös, E.; Kozák, I.O.; Fekete, J.; Sharma, V.K.; Tuba, Z. *Sci. Total Environ.* **2004**, *33*, 89.
15. Bomboi, M.T.; Hernandez A. *Water Res.* **1991**, *25*, 5, 557.
16. Hewitt, C.N.; Rashed, M.B. *Water Res.* **1992**, *26*, 3311
17. EHC (Environmental Health Criteria) 202, World Health Org. Geneva, **1998**. http://www.inchem.org/documents/ehc/ehc/ehc202.htm
18. Granier, L.; Chevreuil, M.; Carru, A.M.; Letolle, R. *Chemosphere* **1990**, *21*, 1101.
19. Hatje, V.; Rae, K.; Birch, G.F. *J. Environ. Monitor.* **2001**, *3*, 251.
20. Roesijadi, G.; Woodruff, D.L.; Anderson, J.W. *Eniron. Pollut.* **1978**, *15(*3), 223
21. Neff, J.M. Polycyclic Aromatic Hydrocarbons in the Aquatic Environment - Sources, Fates and Biological Effects. **1979**.
22. Murmann, R.K.; Robinson, P.R. *Water Res* **1974**, *8*, 543.
23. Bartzatt, R.; Cano, M.; Johnson, L.; Nagel, D. *J. Toxicol. Environ. Health* **1992**, 35, *4*, 205
24. White, D.A.; Franklin, G.S. *Environ. Technol.* **1998**, *19*, 1157.
25. DeLuca, S.J. In: Removal of Organic Compounds by Oxidation-Coagulation with Potassium Ferrate. Ph.D. Thesis **1981**.
26. DeLuca, S.J.; Chao, A.C.; Smallwood, J.C. Jr. *J. Environ. Eng,* **1983**, *109(5)*, 1159.

Indexes

Author Index

Arikan, Osman, 421
Bakac, Andreja, 167
Balasubramanian, Rajasekhar, 467
Baum, J. Clayton, 145
Benová, M., 68
Bouzek, Karel, 52, 68
Cabelli, Diane E., 158
Cai, Shengmin, 81
Carr, James D., 189
Chen, Liu, 94
Chenay, Benoit V. N., 350
Chengchun, Jiang, 94
Criado, C., 102
Deng, HongBo, 81
Doong, Ruey-an, 405
Evrard, O., 102
Filip, Jan, 124
Graham, Nigel, 292, 366
Híveš, J., 68
Homonnay, Zoltán, 124, 255
Hornstein, Brooks J., 177
Jeannot, C., 102
Jiang, Jia-Qian, 292, 306
Johnson, Michael D., 177
Joshi, Umid Man, 467
Ju, Ran, 448
Kanari, N., 102
Kim, Hyunook, 326, 390, 421
Kim, Yuhun, 326
Kuzmann, Ernő, 255
Lan, H. C., 379
Li, Chunjuan, 448, 458
Li, X. Z., 366
Liang, Yong-Mei, 436
Licht, Stuart, 2, 197, 268
Liu, Wei, 81, 436, 458
Ma, Jun, 448, 458
Machala, Libor, 124
Mácová, Zuzana, 52
Madarász, János, 124
McConnell, Laura L., 326
Millner, Patricia, 326
Mishra, Santosh K., 339
Moon, Chulhwan, 421
Nesnas, Nasri, 339
Neveux, N., 102
Ninane, L., 102
Noorhasan, Nadine N., 145, 255, 339
Park, Guisu, 390
Perfiliev, Yurii D., 112
Pestovsky, Oleg, 167
Pokol, György, 124
Qu, Deyang, 268
Ramirez, Mark, 326
Remsberg, Jarrett R., 421
Rice, Clifford P., 326, 421
Schneeweiss, Oldrich, 124
Sharma, Virender K., 52, 68, 81, 112, 124, 145, 158, 255, 306, 326, 339, 350, 467
Shichao, Wang, 94
Szilágyi, Petra Á., 255
Tien, Khoi Tran, 292
Torrents, Alba, 326
Wischnewsky, Jacob, 177
Wu, Feng, 81
Yang, Kai, 81
Yang, Shih-fen, 405

Ye, M. R., 379
Yngard, Ria, 124
Yu, Myongjin, 390
Yu, Xingwen, 2, 197, 268
Yuan, B. L., 366, 379
Zboril, Radek, 124
Zhang, Cun Zhong, 81
Zhang, Yingjie, 448, 458
Zhao, Tingting, 81

Subject Index

A

Activated carbon
removal of oxidation products of aniline, phenol and hydrazine, 284–287
water purification configuration, 272
See also Biological activated carbon (BAC)
Alcohols, second-order rate constants of ferrate with, 391*t*
Aldehydes, second-order rate constants of ferrate with, 391*t*
Algae-bearing water. *See* Coagulation of algae-bearing water
Alkali ferrates
Cs_2FeO_4 and $Rb(K)FeO_4$, 7–9
Fourier transform infrared (FTIR) spectra, 9, 11*f*, 12
IR absorption of pure $Na_{1.1}K_{0.9}FeO_4$ and K_2FeO_4, and Na_2FeO_4, 13*f*
$Na(K)FeO_4$, 9
powder x-ray diffraction (XRD) of K_2FeO_4 and Cs_2FeO_4, 10*f*
pure Li_2FeO_4, 13–14
solid state stability of Cs_2FeO_4, K_2FeO_4, $Rb_{1.7}K_{0.3}FeO_4$, and $Na_{1.1}K_{0.9}FeO_4$, 13, 14*f*
solution precipitation synthesis, 7–14
Alkali ferrate synthesis
drinking water treatment, 110
fluid bed with chlorine, 109–110
granulation fluid bed process, 108–109
industrial waste water, 110–111
new dry process, 105, 106*f*
objectives of recent EEC program, 105–108
semi-dry synthesis, 108
titration methods for testing ferrate, 107–108
urban waste water, 110
See also Waste alkali
Alkaline earth ferrates
infrared (IR) absorption, 11*f*
powder x-ray diffraction patterns, 17*f*
room temperature solubility, 17*f*
solid synthesis of $BaFeO_4$, 19, 21
solution precipitation synthesis, 15–16
Alkaline hypochlorite synthesis, K_2FeO_4, 5–7
Alkaline thin film cathodes. *See* Rechargeable alkaline Fe(VI/III) thin film cathodes
Ammonia
commercial uses, 351
oxidation by Fe(VI), 356–358
oxidation of aqueous solution, 192–193
oxidation procedure, 353
oxidation rate vs. molar ratios of Fe(VI) to ammonia, 357*f*
rate constants of oxidation by pH, 193*f*
release into environment, 351–352
removal by Fe(VI) in sludge treatment, 317*f*
second-order rate constants of ferrate with, 391*t*
setup for photocatalytic oxidation, 354*f*
See also Heterogeneous photocatalytic reduction of iron(VI)

Anilines
  formation of ferrate(VI)-aniline intermediate, 182*f*
  loss of ferrate(VI)-aniline intermediate, 183*f*
  measurement, 282–283
  oxidation by Fe(VI) and removal of oxidation products by activated carbon, 286*f*
  oxidation of, and phenol by Fe(VI), 285*f*
  oxidative treatment and removal from wastewater, 284–287
  proposed reaction for *cis*-azobenzene formation, 183*f*
  quantitative analysis with UV/vis spectroscopy, 283*f*
  reaction of ferrate(VI) with *N,N*-dimethylaniline, 184*f*
  reactions of ferrate(VI) with, 181–184
  second-order rate constants of ferrate with, 391*t*
  simultaneous oxidation of, phenol and hydrazine, 285, 287, 288*f*
  See also Ferrate(VI) oxidation of nitrogen compounds
Anode dissolution kinetics
  anode material composition, 53–54
  batch electrolyses, 62–65
  behavior of silicon rich steel (SRS) and white cast iron (WCI), 57
  charge transfer, 62
  constant phase element (CPE), 58
  CV curves of anode materials, 56*f*
  CV (cyclic voltammetry), 55–57
  dependence of ferrate current yield on anode current density for WCI and SRS, 64*f*
  dependence of iron in oxidation state lower than Fe^{6+} content in analyte on anode current density, 65*f*
  EIS of pure iron anode, 60*f*, 61*f*
  electrical circuit to evaluate EIS, 60*f*
  electrochemical impedance spectroscopy (EIS), 57–62
  electrochemical preparation of ferrate (Fe(VI)), 53
  experimental, 54–55
  influence of electrolyte composition, 56–57
  positive effect of hydroxide concentration, 53
  resistance of internal protective oxide layer, 59, 61
  value of CPE on internal surface layer, 59, 61
Aqueous ferryl(VI) ion. *See* Ferryl(VI) ion, aqueous
Aqueous phase reactants, $BaFeO_4$ synthesis, 4
Aqueous solution, ferrate(V) and ferrate(IV) generation, 158–159
Aromatic compounds, reactivity of ferrate(V) with, 162*t*, 163
Arsenite
  contaminated water treatment, 281
  cyclic voltammetry of Pt electrode in, solution, 276*f*
  inorganic contaminant analyses, 275–276
  remediation by on-line Fe(VI) treatment, 281*f*
Azide
  oxidation of aqueous solution, 192–193
  oxidation products, 194
  rate constants of oxidation by pH, 193*f*
2,2'-Azinobis(3-ethylbenzothiazoline-6-sulfonate) (ABTS)
  chemical structure, 407*f*
  sensitivity of ABTS method, 408–409, 411
  stability of ferrate(VI) by pH, 408, 410*f*, 411
  stability of Fe(VI), 406, 408

B

Bacteria inactivation
 comparative performance, 314*t*
 percentage with ferrate(VI), 314*f*
 See also Sludge treatment
Barium ferrate(VI)
 chromite analysis, 42
 discharge, 203*f*
 energy capacity, 203*f*
 Mössbauer spectroscopy, 138, 140*f*, 140
 phase diagram of products from heating, at various temperatures and oxygen pressures, 137*f*
 stability at room temperature, 139–140, 140*f*, 141*f*
 synthesis, 134, 136
 TGA/DTA curves of, under nitrogen, 138*f*, 139*f*
 thermal decomposition, 134–140
 See also Electrochemical synthesis (in situ); Super-iron batteries; Thermal stability of solid ferrates(VI)
Batch electrolyses, anode dissolution, 62–65
Battery, super-iron
 ferrate Fe(VI), 69, 82, 146, 198
 See also Super-iron batteries
Belt filter pressing
 dewatering technique, 327, 330
 effects of ferrate treatment on solids dewaterability, 332, 334–335
 See also Dewatering wastewater biosolids
Biological activated carbon (BAC)
 oxidizing dissolved organic carbon (DOC), 446–447
 See also Activated carbon; Ferrate preoxidation and biological activated carbon filtration
Biosolids. *See* Dewatering wastewater biosolids

C

Cadmium
 ferrate dosage and, removal, 460, 461*f*
 ferrate preoxidation and humic acid for removal, 463, 464*f*
 health concerns, 457
 pH effect on removal, 462, 463*f*
 removal by potassium ferrate, 472, 473*t*
 See also Heavy metals; Lead
Carboxylic acids
 reactivity of ferrate(V) with, 162*t*, 163
 second-order rate constants of ferrate with, 391*t*
Cathodes. *See* Rechargeable alkaline Fe(VI/III) thin film cathodes
Centrifugation
 dewaterability of centrifuge at different polymer doses, 334*f*
 dewatering technique for sludge, 327, 329
 effects of ferrate treatment on solids dewaterability by, 334–335, 333*f*
 effects of polymer addition on dewaterability of sludge with ferrate, 334
 See also Dewatering wastewater biosolids
Cesium ferrate. *See* Alkali ferrates
Cesium ferrate(VI)
 thermal decomposition, 142
 See also Thermal stability of solid ferrates(VI)
Charge storage
 Fe(VI) compounds, 269
 Fe(VI) salts as alkaline cathode, materials, 200
 theoretical 3-electron charge capacities of Fe(VI) salts, 201*t*
 See also Super-iron batteries

Charge transfer
activation of $BaFeO_4$ and K_2FeO_4 composite cathode super-iron batteries, 214, 215*f*
alkaline Fe(VI) K_2FeO_4 or $BaFeO_4$, 202–220
alkaline Fe(VI) $Na(K)FeO_4$, $Rb(K)FeO_4$, Cs_2FeO_4, $SrFeO_4$ or Ag_2FeO_4, 220–224
$BaFeO_4/MnO_2$ composite super-iron batteries, 210–212
$BaFeO_4$/Mn(VII) composite super-iron batteries, 210, 211*f*
cell potential and energy capacity of alkaline cells with K_2FeO_4 composite cathodes, 208*f*
chemical mediation of Fe(VI), 206–207
co-cathode electronic and redox mediated Fe(VI), by Ag(II) addition, 213*f*
co-cathode redox mediated Fe(VI), 206*f*
comparison of discharge behavior of alkaline AAA cells of cathode composite, 209*t*
conductive matrix facilitated Fe(III/VI), 230–234
constant comparison of MnO_2, $BaFeO_4$, and AgO/K_2FeO_4 cathodes, 214–216
discharge capacity of K_2FeO_4, 204*f*
discharge efficiency of super-iron batteries with $SrTiO_3$, $Ba(OH)_2$ additives, 216
inhibition of Fe(VI), 205–206
K_2FeO_4/Mn(VII or VI) composite super-iron batteries, 207–210
Mn(IV), Co(III) modifiers for super-iron batteries, 216, 219–220
passivation of thick Fe(III/VI), on smooth platinum, 228, 230
platinized titanium substrates for Fe(III/VI), 236, 237*f*
simultaneous chemical and electronic mediation of Fe(VI), 212–214
titanate, Ba(II), Mn(VI) and Co(III) modifiers, 216–220
zirconia's hydroxide mediation of Fe(VI/III), 202, 204
See also Rechargeable alkaline Fe(VI/III) thin film cathodes; Super-iron batteries
Chemical mediation
Fe(VI) charge transfer, 206–207
inhibition of Fe(VI) charge transfer, 205–206
See also Charge transfer
Chlorinated ethylenes. *See* Trichloroethylene (TCE)
Chlorination
comparison to ferrate preoxidation, 440, 441*f*
pretreatment for algal coagulation, 435
water treatment, 268
See also Coagulation of algae-bearing water
Coagulation. *See* Heavy metals
Coagulation of algae-bearing water
changes of dissolved organic matters (DOM), 442–443
comparing ferrate preoxidation and pre-chlorination, 441*f*
comparing ferrate preoxidation on settled and filtered samples of lake water, 441*f*
drinking water treatment, 435
effect of ferrate preoxidation on scanning of UVA after filtration, 443*f*
enhanced algae removal, 438–441
enhanced removal of algae by ferrate preoxidation by preoxidation time, 440*f*
ferrate preoxidation, 438, 439*f*
influences on surface architecture of algae cells, 442

jar tests, 437
materials and methods, 436–438
mechanism of coagulation, 438
redox potential of ferrate(VI), 435–436
scanning electron microscopy (SEM), 437
ultraviolet absorbance scanning, 437–438
waters, 436

Coagulation performance of ferrate
apparatus, 294–295
coagulation of kaolin and ferrate with phosphate/borate buffer, 303*f*
coagulation of kaolin with $FeCl_3$, 297*f*
coagulation of kaolin with ferrate, 297*f*
coagulation of kaolin with ferrate and $FeCl_3$, 301*f*
coagulation tests, 295
combination of sweep flocculation and charge interaction, 298, 299*f*
dependence of chemical dose and pH, 296, 298
dual role of ferrate (Fe(VI)) in water treatment, 293
experimental procedures, 295–296
flocculation index (FI), 294
floc strength tests, 296
lower pH values, 299–300
materials and methods, 293–296
oxidation performance complicating, 303
principle of photometric dispersion analysis (PDA), 294*f*
recovery factor, 296, 300
similarities and differences of ferrate and ferric chloride, 297*f*, 298
strength factor, 296, 300
variation of FI response during floc strength tests, 302*f*
variation of FI with dose and pH for $FeCl_3$, 299*f*
variation of FI with dose and pH for ferrate, 299*f*

Composite super-iron batteries
$BaFeO_4/MnO_2$, 210–212
$BaFeO_4$/Mn(VII), 210, 211*f*
K_2FeO_4/AgO or $BaFeO_4$/AgO composite cathodes, 214
K_2FeO_4/Mn(VII or VI), 207–210
See also Super-iron batteries

Conductivity measurements, ferryl(IV) ion decay, 169

Conductors
alkaline super-iron batteries, 224–227
discharge comparisons, 226*f*
effect of carbon based cathode conductive additive on discharge, 224*f*
effect of fluorinated graphite polymer on discharge, 225*f*
fluorinated polymer graphite, 225–227
See also Super-iron batteries

Cyanide
comparing rate of sulfide and, by ferrate(VI), 280*f*
contaminated water treatment, 279
inorganic contaminant analyses, 275–276
oxidation of aqueous solution, 192–193
oxidation products, 195
reactivity of ferrate(V) with, 161, 162*t*
remediation by on-line Fe(VI) treatment, 280*f*
simultaneous remediation of, and sulfide by on-line Fe(VI) treatment, 282*f*

Cyclic voltammetry (CV)
anode dissolution kinetics, 55–57
Fe(VI) analysis, 46

gold working electrode in NaOH–KOH system, 72*f*, 73*f*
iron in NaOH–KOH system, 78*f*
iron working electrode in NaOH–KOH system, 76*f*, 77*f*, 79*f*
platinum working electrode in NaOH–KOH system, 71*f*, 75*f*
See also Fe(VI)/Fe(III) system; Molten salt approach to ferrate(VI) synthesis
Cysteine
reactivity of ferrate(V) with, 161, 162*t*
second-order rate constants of ferrate with, 391*t*

D

Dairy waste lagoon effluent (DWLE)
treatment of estrogenic compounds in, 426–428
See also Estrogenic compounds
Detection of ferrate(VI). *See* Fluorescence technique
Dewatering wastewater biosolids
belt-pressing, 330
benefits of ferrate(VI) as sludge conditioner, 335–337
biosolid generation, 327
centrate color of dewatered sludge with ferrate treatment, 333*f*
centrifugation, 329
characteristics of solids after ferrate treatment, 331–337
common techniques, 327
cost, 327
dewaterability of centrifuge at different polymer doses, 334*f*
effect of ferrate on performance of vacuum filtering in, 336
effects of ferrate treatment on dewaterability by belt-pressing, 332, 335
effects of ferrate treatment on dewaterability by centrifuge, 332, 333*f*
effects of polymer addition on dewaterability of sludge with ferrate, 334
ferrate production, 328
ferrate(VI) as potential polymer alternative, 327–328
materials and methods, 328–330
microorganism removal rates of treated thickened sludge, 337*t*
pictures of dewatered sludge with ferrate treatments, 332*f*
procedures for dewatering tests, 329–330
sludge samples and sample treatments, 329
solids contents and drainage volume of belt-pressed sludge with ferrate, 335*f*
vacuum filtering, 331, 331*f*
Dibutyl phthalate (DBP)
analytical method, 366
DBP use in packaging materials, 365
degradation and reduction of ferrate with different weight ratios of K_2FeO_4:TiO_2, 371*f*
degradation at different ferrate concentrations, 369*f*
degradation at pH 9.0 by ferrate/TiO_2/UV, 367*f*
effect of ferrate concentration on, degradation by oxidation processes, 368, 370
effect of pH on ferrate-photocatalytic degradation of, 372–374
effect of TiO_2 dosage on DBP degradation in oxidation processes, 370, 372
experimental, 365–366
photocatalysis, 366

photocatalytic degradation of, by different electron acceptors, 374–375
photocatalytic degradation with TiO_2 at different ferrate concentrations, 369*f*
preliminary evaluation of ferrate/TiO_2/UV systems, 366–368
synergism between ferrate oxidation and photocatalysis, 367
Dimethylamine
ferrate oxidation, 196
rate constants of oxidation by pH, 193*f*
Dioxins, limit values of, in sludge, 311*t*
Direct electrochemical synthesis. *See* Electrochemical synthesis (in situ)
Disinfectant
ferrate(VI) potential in sludge, 313–314
See also Sludge treatment
Drinking water treatment
Coagulation-sedimentation, 435
ferrate and ferric chloride, 110
heavy metal removal, 457
See also Coagulation of algae-bearing water; Ferrate preoxidation and biological activated carbon filtration; Heavy metals; Urban runoff water

E

Electroanalytical techniques, Fe(VI) analysis, 44–47
Electrochemical behavior. *See* Fe(VI)/Fe(III) system
Electrochemical Fe(VI) water treatment
analyses of inorganic contaminants sulfide, cyanide and arsenite, 275–276
comparing rate of sulfide and cyanide oxidation, 280*f*
cyanide concentration measurement and calibration, 280*f*
Fe(VI) analyses, 274
Fe(VI) formation fundamentals, 273
Fe(VI) generation in on-line electrochemical flow cell, 274–275
hydrazine oxidation by Fe(VI), 287*f*
in-line purification configuration, 271
measurement of aniline, phenol and hydrazine, 282–283
on-line activated carbon water purification configuration, 272
on-line remediation configuration, 269–271
organic contaminated water purification, 282–287
oxidation of aniline and phenol by Fe(VI), 285*f*
oxidation of aniline or phenol by Fe(VI) and removal of oxidation products with activated carbon, 286*f*
oxidative treatment and removal of aniline, phenol and hydrazine, 284–287
quantitative analyses of aniline and phenol with UV/vis spectroscopy, 283*f*
remediation of arsenic by on-line, 281*f*
remediation of cyanide by on-line, 280*f*
remediation of sulfide by, at lower pH, 279*f*
remediation of sulfide by on-line, 278*f*

simultaneous oxidation of aniline, phenol and hydrazine by on-line, 288*f*
simultaneous remediation of cyanide and sulfide by on-line, 282*f*
step-wise addition of Fe(VI) to effluent and time arrangement, 277*f*
sulfide, cyanide and arsenite contamination with on-line, configuration, 276–281
sulfide concentration measurement and calibration, 278*f*
Electrochemical impedance spectroscopy (EIS)
anode dissolution, 57–62
electrical circuit to evaluate EIS data, 60*f*
pure iron anode, 60*f*, 61*f*
Electrochemical synthesis (in situ)
$BaFeO_4$, 4, 22–30
$BaFeO_4$ under various physical and chemical conditions, 24*t*
influence of internal cell temperature, 23–24, 25*f*
membrane effect on electrochemical synthesis, 27*f*
properties of $BaFeO_4$ by chemical and, 27, 28*t*
purity, yield and synthesis efficiency of $BaFeO_4$, 25*f*
refinements of cell size and electrolysis conditions, 25–26
scanning electron microscopy (SEM) of $BaFeO_4$ by chemical and, 30*f*
solid Fe(VI) salts, 3, 4
stability of $BaFeO_4$, 27–28, 29*f*
time dependent purity, 28
x-ray diffraction (XRD) and FTIR spectra of $BaFeO_4$ by chemical synthesis and, 25, 26*f*
See also Molten salt approach to ferrate(VI) synthesis
Electrolyses, batch, anode dissolution, 62–65
Electron acceptors, photocatalytic degradation of dibutyl phthalate by, 374–375
Electrospray ionization–mass spectrometry (ESI–MS), tetracycline before and after reaction with ferrate(VI), 413, 416*f*, 417*f*, 417
Encapsulated ferrate. *See* Trichloroethylene (TCE)
Endocrine disruptors
environmental impacts, 420–421
ferrate as sludge conditioner, 335–336
wastewater treatment, 310–312
See also Estrogenic compounds
Enzymes
iron in high oxidation states, 178
nonheme iron complexes, 177
Estradiol
destruction by ferrate(VI), 428, 430*f*
structure, 424*f*
See also Estrogenic compounds
Estrogenic compounds
chemical oxidation of steroidal estrogens (SE), 422
concentration of progesterone in dairy waste lagoon effluent (DWLE) vs. concentration of ferrate(VI), 429*f*
destruction of estradiol vs. log of percent concentration ferrate(VI), 430*f*
DWLE, 426–428
endocrine disruptor compounds (EDC), 420–421
environmental impacts of EDC, 421
estradiol reduction, 430, 430*f*
ferrate(VI) as alternative oxidant, 422

hormone analysis, 425–426
ion chromatogram showing quantitation of E2 and progesterone, 427*f*
mean levels in DWLE before and after ferrate(VI) or ferric chloride treatment, 428*t*
methods, 423–426
parent and daughter ions for quantitation and mass spectrometry parameters, 426*t*
percent starting hormone concentration vs. ferrate(VI), 429*f*
potencies of known and suspected EDCs, 421*t*
progesterone increase, 428, 431
quality assurance, 426
sample collection, 423
sample processing, 423, 425
steroidal estrogens (SE) removal from activated sludge, 421–422
structures of EDCs in study, 424*f*

Ethylenediaminetetraacetic acid (EDTA)
absorbance vs. time profile for Fe(VI), 342
complexing metal ions in groundwater, 340
ferrate(VI) as oxidizing agent in aqueous media, 341
half-life of oxidation reactions, 344
molecular structure, 340*f*
oxidation methods to remove, 340–341
pseudo-first-order rate constant vs. [EDTA], 342*f*
rate constants for Fe(VI) oxidation of amines, 343*t*
reaction of ferrate(VI) with, 259–261
reaction of ferrate(VI) with Fe^{III}EDTA, 262–265
reaction rates of Fe(VI) with, 341–344
reactivity order among amines, 343
recalcitrant compound, 339
uses, 340
See also Sulfamethoxazole (SMX); Transformation of Fe(VI) to Fe(III)

Eutectic NaOH–KOH system. *See* Molten salt approach to ferrate(VI) synthesis

Eutrophication
lake water, 435
See also Coagulation of algae-bearing water

Extended conductive matrixes, Fe(VI/III) reversible films on, 31–36

F

Facile charge transfer
Fe(III/VI) ferrate films, 229*f*
See also Rechargeable alkaline Fe(VI/III) thin film cathodes

Fe(III/VI) thin films
electrodeposition, 34
partial and full alleviation of Fe(III/VI) passivation, 34, 35
preparation on extended conductive matrixes, 31–36
preparation on smooth Pt, 31, 32*f*

Fe (iron). *See* Oxidation states of iron; Transformation of Fe(VI) to Fe(III)

Fenton reaction, chemical behavior of ferryl(IV) ion, 175

Ferrate (Fe(IV))
generation, 163–164
oxidizing agent in aqueous media, 341
reactivity, 164*t*
spectrum of Fe(IV)–pyrophosphate complex, 164*f*

Ferrate (Fe(V))
cysteine reaction, 161, 162*t*
decay, 159–161

first-order decay, 160
generation, 159
oxidation of compounds at pH=12.4, 162*t*
reaction with thiocyanate, 162*t*, 163
second-order decay, 160
visible spectra of, and Fe(VI) in aqueous basic solution, 160*f*
Ferrate (Fe(VI))
alkali and alkaline earth ferrates, 117–120
analyses, 274
anode dissolution synthesis, 53
benefits as sludge conditioner, 335–337
chemical synthesis of, salts, 125
comparison with other oxidants, 103
decomposition at different pH and hydrogen peroxide/ferrate ratios, 393, 395
determining concentration in dilute solutions, 146
development of high capacity battery, 69, 82, 146
dibutyl phthalate degradation, 368, 369*f*
dry methods, 104
dual role in water treatment, 293
effect of ferrate concentration on nonylphenol degradation, 396, 397*f*
effect on bacteria and viruses, 103
electrochemical method, 104–105
formation fundamentals, 273
generation in on-line electrochemical flow cell, 274–275
Mössbauer characteristics of, 119*t*
new dry process, 105, 106*f*
oxidation performance complicating coagulation effects, 303
oxidation potentials at several pH values, 103
oxidizing nature of, 391
physicochemical properties, 103
properties, 312–313
reduction potential, 352–353
second-order rate constants of ferrate with various compounds, 391*t*
strong oxidant, 95
synergism with photocatalytic oxidation, 366–368
synthetic processes in literature, 104–105
visible spectra of Fe(V) and, in aqueous basic solution, 160*f*
wastewater treatment, 82
wet methods, 104
See also Alkali ferrate synthesis; Anode dissolution kinetics; Coagulation performance of ferrate; Electrochemical Fe(VI) water treatment; Estrogenic compounds; Fe(VI)/Fe(III) system; Fe(VI) synthesis; Fluorescence technique; Heterogeneous photocatalytic reduction of iron(VI); Molten salt approach to ferrate(VI) synthesis; Nonylphenol (NP); Sludge treatment; Thermal stability of solid ferrates(VI); Transformation of Fe(VI) to Fe(III)
Ferrate preoxidation. *See* Coagulation of algae-bearing water; Heavy metals
Ferrate preoxidation and biological activated carbon filtration
batch study, 452, 453*f*
batch test, 448
characteristics of raw water, 448*t*
dynamic experiments, 448–449, 452
effect of ferrate and ozone on chemical oxygen demand

(COD) with time after BAC, 451*f*
effect of ozone and ferrate preoxidation on COD, UV-absorbance, and NH_4^+–N in biological activated carbon (BAC) reactor, 450*f*
effect of preoxidants on NH_4^+–N after BAC, 451*f*
enhancing drinking water treatment, 446–447
experiments, 447–448
parameters in different processes with ferrate preoxidation, 450*t*
Ferrate(VI) oxidation of nitrogen compounds
experimental, 179
formation of ferrate(VI)-aniline intermediate, 182*f*
loss of aniline-ferrate(VI) intermediate, 183*f*
proposed template reaction for *cis*-azobenzene formation, 183*f*
rate constants of Fe(VI) oxidation of hydrazines, 179*t*
rate constants of Fe(VI) oxidation of hydroxylamines, 180*t*
reactions with anilines, 181–184
reactions with hydrazines, 179–180
reactions with hydroxylamines, 180–181
reaction with *N,N*-dimethylaniline, 184*f*
Ferric chloride. *See* Coagulation performance of ferrate
Ferryl(IV) ion, aqueous
chemical behavior in Fenton reaction, 175
conductivity changes with decay, 170*f*
^{1}H NMR spectrum of products from cyclobutanol and, 171*f*
kinetics data for reactions in hydrogen atom transfer and hydride transfer pathways, 172*t*
Mössbauer and x-ray absorption spectroscopies, 169–171
oxidation of methyl *p*-tolyl sulfoxide with ozone, 172, 173*f*
oxidations by hydrogen atom transfer, 168
oxygen transfer to sulfoxides, 173–174
preparation and characterization, 168–169
reactions of, with organic substrates, 171–175
second-order rate constants for oxygen atom transfer from, 175*t*
second-order rate constants for reactions of, with organic substrates, 172*t*
stopped-flow conductivity measurements, 169
Fe(VI)/Fe(III) system
addition of Fe_2O_3 powder, 85
characteristic parameter of anodic peak, 89*t*
chemical composition and formal potential, 82
CVs (cyclic voltammograms) on SnO_2–Sb_2O_3/Ti electrodes in blank and working solutions, 84, 85*f*
CVs for Fe(III) compounds, 88*f*
CVs for $NaFeO_2$ and mixed powder microelectrode, 91*f*
disproportionation reaction of Fe(V) species, 90–91
electro-generation of ferrate from $Fe(OH)_3$ and $NaFeO_2$, 87, 88*f*
experimental, 83–84
molten basic electrolyte and activity of Fe_2O_3, 87
peak potential of cathodic peak varying by electrode materials, 91–92
powder microelectrode technique in NaOH, 87

reactions of Fe(III) compounds with saturated hypochlorite solution, 85, 86*f*
selective electro-generation of ferrate using Fe(III) compounds, 83
thermodynamic data of different valence Fe species, 82–83
two-electron transfer, 89–90
uses of open-hearth dust in iron/steel industry, 83
x-ray diffraction (XRD) of $NaFeO_2$ sample, 86*f*

Fe(VI) synthesis
alkaline hypochlorite synthesis of K_2FeO_4, 5–7
alternative syntheses of $BaFeO_4$, 3, 4
analysis, 38–40
aqueous phase reactants, 4
$BaFeO_4$ chromite analysis, 42
$BaFeO_4$ purity, yield and synthesis efficiency, 25*f*
chemistry of hexavalent iron, 3
direct electrochemical synthesis of $BaFeO_4$, 22–30
electrochemical analysis, 44–47
electrodeposition of super-iron films, 34
facile charge transfer in Fe(III/VI) ferrate films, 33
Fe(IV/III) reversible film preparation on extended conductive matrixes, 31–36
Fe(IV) potentiometric analysis, 44
Fe(IV) solution phase galvanostatic analysis, 44–45
Fe(VI) cyclic voltammetry analysis, 46
Fe(VI) Mössbauer analysis, 47–48
Fe(VI) solid cathode discharge analysis, 46–47
Fourier transform infrared (FTIR) spectra of solid alkali ferrates and alkali earth ferrates, 11*f*
FTIR analysis, 38–40
FTIR analysis of $BaFeO_4$, 39*f*
inductively coupled plasma (ICP) analysis, 40
influence of internal cell temperature on in-situ $BaFeO_4$ synthesis, 23–24, 25*f*
in situ electrochemical syntheses of $BaFeO_4$ at various physical and chemical conditions, 24*t*
in situ electrochemical synthesis, 3, 4
IR absorption of solid Ag_2FeO_4, Ag_2O, and K_2FeO_4, 20*f*
IR absorption of solid $Na_{1.1}K_{0.9}FeO_4$, K_2FeO_4, and Na_2FeO_4, 13*f*
IR spectra of zirconia coated and uncoated cathode materials, 37*f*
K_2FeO_4 chromite analysis, 41–42
long-term stability of K_2FeO_4, 8*f*
membrane effect on electrochemical synthesis of in situ chemically synthesized $BaFeO_4$, 27*f*
partial and full alleviation of Fe(III/VI) passivation in ferrate film, 34, 35
powder X-ray diffraction (XRD) pattern of K_2FeO_4 and Cs_2FeO_4, 10*f*
powder XRD pattern of SrFeO4 and $BaFeO_4$, 17*f*
preparation of Fe(III/VI) thin films, 30–36
preparation of Fe(IV/III) reversible film on smooth Pt substrate, 31, 32*f*
preparation of pure Li_2FeO_4, 13–14
preparation of Zr or Mn coated Fe(IV) salts, 36–38
properties of direct electrochemical and chemical synthesized $BaFeO_4$, 27, 28*t*

room temperature solubility of $BaFeO_4$, K_2FeO_4, and $SrFeO_4$, 17*f*
scanning electron microscopy (SEM) of chemical and electrochemical synthesized $BaFeO_4$, 30*f*
solid phase reactants, 4
solid state stability of Ag_2FeO_4, 21*f*
solid state stability of Cs_2FeO_4, K_2FeO_4, $Rb_{1.7}K_{0.3}FeO_4$ and $Na_{1.1}$K0.9FeO_4, 13, 14*f*
solid synthesis of $BaFeO_4$, 19, 21
solution precipitation synthesis of Ag_2FeO_4, 16, 18–19
solution precipitation synthesis of Cs_2FeO_4, Rb_2FeO_4, $K_xNa_{(2-x)}FeO_4$, and Li_2FeO_4, 7–14
solution precipitation synthesis of $SrFeO_4$ and $BaFeO_4$, 15–16
stability of $BaFeO_4$, 27–28, 29*f*
UV/vis analysis, 42–43
XRD analysis, 43–44
XRD and FTIR spectra of chemically and direct electrochemically synthesized $BaFeO_4$, 26*f*
zirconia alkaline cathode protection, 37
zirconia as pH sensor for high temperature aqueous systems, 36, 38
See also Molten salt approach to ferrate(VI) synthesis
Flocculation index (FI)
output of photometric dispersion analyzer (PDA), 294
recovery factor, 296, 300
strength factor, 296, 300
variation of response during floc strength test, 302*f*
variation with dose and pH for ferrate, 299*f*
variation with dose and pH for ferric chloride, 299*f*
See also Coagulation performance of ferrate
Fluid bed technology, alkali ferrate synthesis, 108–110
Fluorescence technique
absorption and emission spectra of 7-hydroxyl-6-methoxy coumarin (scopoletin), 150–151
absorption spectra of scopoletin by pH values, 150*f*
apparatus, 148
applicability of method, 153
2,2'-azinobis(3-ethylbenzothiazoline-6-sulfonate) (ABTS) reacting with Fe(VI), 147
calibration curves using fluorescence, 151, 153
emission intensity of scopoletin vs. [Fe(VI)], 152*f*
emission spectra of scopoletin by pH values, 151*f*
experimental, 148
kinetics of ferrate(VI) reactions, 146, 149
scopoletin as fluorescence agent, 147
second-order rate constant for reaction of ferrate(VI) with scopoletin, 149*t*
Fluorinated graphite conductor
cathode additive for super-iron batteries, 225–227
See also Conductors
Formamide, ferrate oxidation, 195
Formic acid
initial rate of Fe(VI) reduction vs. Fe(VI) concentrations, 359*f*
oxidation by Fe(VI), 358, 360–361
oxidation procedure, 353
photocatalytic oxidation of, under different conditions, 360*f*
photocatalytic reduction of Fe(VI) in suspensions with, 359*f*
release into environment, 352

second-order rate constants of ferrate with, 391*t*
setup for photocatalytic oxidation, 354*f*
uses, 352
See also Heterogeneous photocatalytic reduction of iron(VI)
Fourier transform infrared (FTIR), Fe(VI) analysis, 38–40
Frozen solution Mössbauer study. *See* Transformation of Fe(VI) to Fe(III)

G

Galvanostatic analysis, Fe(VI), 44–45
Glycine
reaction of ferrate(VI) with, 259, 260*f*
See also Transformation of Fe(VI) to Fe(III)
Gold electrode. *See* Molten salt approach to ferrate(VI) synthesis
Green preparation. *See* Potassium ferrate
Groundwater contamination. *See* Trichloroethylene (TCE)

H

Heavy metals
cadmium removal vs. pH with ferrate preoxidation, 463*f*
cadmium removal with ferrate preoxidation, 460*f*
coagulation, 457
drinking water treatments, 457
effect of ferrate dosage on removal, 460, 461*f*
effect of ferrate preoxidation on removal, 458–460
experimental methods, 457–458
ferrate(VI) technology for removal, 472, 473*t*
health concerns for lead and cadmium, 457
lead removal vs. pH with ferrate preoxidation, 462*f*
lead removal with ferrate preoxidation, 459*f*
limit values of, in sludge for land use, 310*t*
pH effect on removal, 460, 462
removal by ferrate preoxidation and humic acid, 463, 464*f*
removal efficiency, 318*t*
removal from storm water by potassium ferrate, 472, 473*t*
removal in river water by ferrate and ferric chloride, 319*f*
removal with ferrate(VI), 317–318
typical raw water quality, 458*t*
wastewater treatment, 310
See also Sludge treatment; Urban runoff water
Heterogeneous photocatalytic reduction of iron(VI)
ammonia, 351–352
ammonia and formic acid oxidation, 353–354
experimental, 353–354
formic acid, 352–353
initial rate of Fe(VI) reduction vs. Fe(VI) concentrations in formic acid, 359*f*
oxidation of ammonia, 356–358
oxidation of formic acid, 358, 360–361
photocatalytic oxidation of formic under different conditions, 360*f*
photocatalytic reduction of Fe(VI) in suspensions with formic acid, 359*f*
photocatalytic reduction of iron(VI), 354–355

rate of ammonia oxidation vs. molar ratios of Fe(VI) to ammonia, 357*f*
reduction of Fe species, 357, 360
setup for ammonia and formic acid oxidation and Fe(VI) reduction, 354*f*
using TiO_2 suspensions, 351

Hexavalent iron
chemistry, 3
See also Fe(VI) synthesis

High capacity battery
ferrate Fe(VI), 69, 82, 146, 198
See also Super-iron batteries

Humic acid. *See* Heavy metals

Hydrazine
measurement, 282–283
oxidation by Fe(VI), 285, 287*f*
oxidation of aqueous solution, 192–193
oxidation products, 195
oxidative treatment and removal from wastewater, 284–287
rate constants of oxidation by pH, 193*f*
reactions of ferrate(VI) with, 179–180
second-order rate constants of ferrate with, 391*t*
simultaneous oxidation of aniline, phenol and, 285, 287, 288*f*

Hydrogen atom transfer, oxidations of aqueous ferryl ion with organic substrates, 168

Hydrogen peroxide
effect on nonylphenol oxidation by ferrate, 399, 400*f*
ferrate decomposition by pH and hydrogen peroxide/ferrate ratios, 393, 395
second-order rate constants of ferrate with, 391*t*
See also Nonylphenol (NP)

7-Hydroxy-6-methoxy coumarin (scopoletin). *See* Fluorescence technique

Hydroxylamine
oxidation of aqueous solution, 192–193
oxidation products, 195
rate constants of oxidation by pH, 193*f*
second-order rate constants of ferrate with, 391*t*

Hydroxylamines, reactions of ferrate(VI) with, 180–181

Hygienization, sludge, 308–309

I

Inductively coupled plasma (ICP), Fe(VI) analysis, 40

Industrial waste water, ferrate and ferric chloride, 110–111

Inorganic contaminants
analyses of sulfide, cyanide and arsenite, 275–276
sulfide, cyanide and arsenite contaminated water treatment with on-line electrochemical Fe(VI), 276–281
See also Electrochemical Fe(VI) water treatment

In-situ electrochemical synthesis
$BaFeO_4$, 4, 22–30
solid Fe(VI) salts, 3
See also Electrochemical synthesis (in situ)

Iron electrode. *See* Molten salt approach to ferrate(VI) synthesis

Iron (Fe). *See* Heterogeneous photocatalytic reduction of iron(VI); Oxidation states of iron; Transformation of Fe(VI) to Fe(III)

Iron(III) complexes, dinuclear non-heme iron enzymes, 178

Iron(IV) complexes

attention of nonheme, 167–168
dinuclear non-heme iron enzymes, 178
intermediates in enzyme studies, 177
water oxidation by ferrate(VI), 189–190, 191*f*, 192*f*
See also Anode dissolution kinetics; Ferryl(IV) ion, aqueous; Fluorescence technique; Nonylphenol (NP)

K

Kinetics
aqueous ferryl(IV) ion reactions, 172*t*
decay of ferrate(V), 159–162
ferrate(V) oxidation of compounds, 162*t*
first-order rate constant for ferrate decomposition by pH, 411*t*
oxidation, of nitrogenous solutes, 192–194
oxidation of amines by ferrate(VI), 343
oxidation of sulfamethoxazole (SMX) by ferrate(VI), 344–345, 346*f*
rate constants of ferrate(VI) oxidation of hydrazines, 179*t*
rate constants of ferrate(VI) oxidation of hydroxylamines, 180*t*
reaction rates of Fe(VI) with ethylenediaminetetraacetic acid (EDTA), 341–344
reactivity of ferrate(IV) pyrophosphate, 163–164
second-order rate constants for aqueous ferryl(IV) ion reactions, 172*t*
second-order rate constants for oxygen atom transfer from aqueous ferryl(IV), 175*t*
second-order rate constants of ferrate with various compounds, 391*t*

L

Lead
ferrate dosage and, removal, 460, 461*f*
ferrate preoxidation and humic acid for removal, 463, 464*f*
health concerns, 457
pH effect on removal, 460, 462, 463*f*
removal by potassium ferrate, 472, 473*t*
See also Cadmium; Heavy metals
Lithium batteries. *See* Rechargeable super-iron lithium batteries
Lithium ferrate. *See* Alkali ferrates

M

Manganese
effect of ferrate dosage on removal, 472, 473*f*
removal by potassium ferrate, 472, 473*t*
See also Heavy metals; Urban runoff water
Manganese coated Fe(VI) salts, preparation, 36–38
Manganese mediation
Fe(VI) charge transfer, 205–212
See also Charge transfer
Metal-hydride anode
compatibility with rechargeable super-iron cathode films, 236, 238–239

See also Rechargeable alkaline Fe(VI/III) thin film cathodes
Metals, removal by potassium ferrate, 472, 473*t*
Methyl amine
oxidation of aqueous solution, 192–193
oxidation products, 195–196
rate constants of oxidation by pH, 193*f*
Microbial inactivation, ferrate as sludge conditioner, 336
Microorganism removal, ferrate as sludge conditioner, 337*t*
Micro pollutants
potential of degradation with ferrate(VI), 318, 320
wastewater treatment, 310–312
See also Sludge treatment
Molten salt approach to ferrate(VI) synthesis
cathodic charge for peak C_3 area vs. anodic vertex potential in NaOH–KOH, 77–78, 79*f*
CV (cyclic voltammetry) method, 69
CV for platinum electrode in eutectic NaOH–KOH melt, 70, 71*f*
CV of gold electrode in NaOH–KOH system, 71, 72*f*
CV of gold working electrode in NaOH–KOH with Fe_2O_3, 73*f*
CVs at various scan rates, 72–73
CVs of iron in NaOH–KOH, 78*f*
CVs of iron working electrode in NaOH–KOH, 76*f*, 77*f*
dependence of anodic peak current on square-root of potential scan rate, 73, 74*f*
dependence of C_3 peak current density on anodic vertex potential, 77, 79*f*
electrode surface with cathodically deposited iron, 73–74, 75*f*
experimental, 69–70
ferrate(VI) formation with iron working electrode, 75–76
identification of current peaks during cathodic potential scan, 75
See also Fe(VI) synthesis
Mössbauer studies
aqueous ferryl(IV) ion, 169–171
barium ferrate(VI) stability, 138, 140*f*, 140, 141*f*
characteristics of ferrate(VI), 119*t*
Fe(VI) analysis, 47–48, 107, 107*f*
frozen solution of ferrate(VI) reaction in NaOH solution with solid ethylenediaminetetraacetic acid (EDTA), 261*f*
frozen solution of ferrate(VI) reaction with glycine, 260*f*
frozen solution of ferrate(VI) reaction with solid EDTA in borate-phosphate buffer, 262*f*
identifying iron oxidation state, 113–114
intermediate iron species using rapid-freeze solution, 258
spectra of Na_4FeO_4, 116*f*
spectrum of frozen Fe(VI) solution, 118*f*
spectrum of iron sample with different oxidation states, 131*f*
spectrum of K_3MnO_4, 117*f*
See also Iron oxidation states; Transformation of Fe(VI) to Fe(III)
Mutagenic tests, urban runoff water, 475

N

NaOH–KOH eutectic system. *See* Molten salt approach to ferrate(VI) synthesis

NaOH solution. *See* Fe(VI)/Fe(III) system
Natural water, oxidation of nonylphenol by ferrate, 400–401
Nitrite
 oxidation of aqueous solution, 192–193
 rate constants of oxidation by pH, 193*f*
Nitrogen compounds
 second-order rate constants of ferrate with, 391*t*
 See also Ferrate(VI) oxidation of nitrogen compounds
Nitrogenous solutes
 oxidation kinetics of aqueous, 192–194
 pH dependence of rate constants of oxidation of, 193*f*
 product analysis of oxidations, 194–196
Nonheme iron(IV) complexes
 enzymes, 177
 interest, 167–168
Nonylphenol (NP)
 analytical methods for, 393, 394*t*
 chemical structure, 392*f*
 decomposition of ferrate, 393, 395
 decomposition of ferrate at different pHs and H_2O_2/FeO_4^{2-}, 395*f*
 effect of ferrate concentration on degradation of, 396
 effect of hydrogen peroxide on NP oxidation by ferrate, 399, 400*f*
 effect of pH on degradation, 396, 398
 environmental issues of NP and ethoxylates (NPEOs), 390
 experimental procedures, 392–393
 HPLC/MS conditions, 394*t*
 materials and methods, 392–393
 oxidation in natural water using ferrate, 400–401
 oxidizing nature of ferrate(VI), 391
 rate constant and removal efficiency at different ferrate doses, 397*t*
 rate constant and removal efficiency at different molar H_2O_2/FeO_4^{2-} ratios, 400*t*
 rate constant and removal efficiency at different pH, 399*t*
 second-order rate constants of ferrate with various compounds, 391*t*
 sewage treatment plant effluents, 390*t*
 solid phase extraction (SPE) methods, 394*f*
 time profile of NP concentration in natural water, 401*f*
 time profiles of ferrate and NP concentrations at different ferrate doses, 397*f*
 time profiles of ferrate and NP concentrations at different H_2O_2/FeO_4^{2-} ratios, 399*f*, 400*f*
 time profiles of ferrate and NP concentrations at different pH, 398*f*

O

Odors
 control with ferrate(VI), 315–316
 wastewater treatment, 309–310
 See also Sludge treatment
On-line delivery
 super-iron water purification, 269
 See also Electrochemical Fe(VI) water treatment
Organic contamination
 aniline, phenol and hydrazine measurements, 282–283
 limit values of, in sludge, 311*t*
 oxidative treatment and removal of aniline, phenol and hydrazine, 284–287

quantitative analyses of aniline and phenol by UV/vis spectroscopy, 283*f*
removal from storm water by ferrate(VI), 474
simultaneous oxidation of aniline, phenol and hydrazine, 285, 287, 288*f*
water purification, 282
See also Electrochemical Fe(VI) water treatment
Oxidants
comparing ferrate with other, 103
ferrate ion, 95
potentials of ferrate at pH values, 103
Oxidation. *See* Coagulation of algae-bearing water
Oxidations
ferrate(V), of compounds at pH=12.4, 162*t*
kinetics of water, by ferrate(VI), 189–190
See also Dibutyl phthalate (DBP); Ethylenediaminetetraacetic acid (EDTA); Ferrate(VI) oxidation of nitrogen compounds; Ferryl(IV) ion, aqueous; Nonylphenol (NP); Sulfamethoxazole (SMX); Transformation of Fe(VI) to Fe(III)
Oxidation states of iron
correlation between ^{57}Fe isomer shift and oxidation state of iron in oxo derivatives, 113*f*
interest in higher oxidation states, 112–114
iron(IV) compounds, 114–115
iron(V) compounds, 115, 117
iron(VI) compounds, 117–120
iron(VII) compounds, 120
iron(VIII) compounds, 121
Mössbauer characteristics of ferrate(VI), 119*t*
Mössbauer spectra of Na_4FeO_4 by temperature, 116*f*
Mössbauer spectrum of frozen Fe(VI) solution, 118*f*
Mössbauer spectrum of iron sample with different oxidation states, 121*f*
Mössbauer spectrum of K_3MnO_4 doped with iron, 117*f*
Oxygen atom transfer, aqueous ferryl(IV) ion, 173–174
Ozone treatment. *See* Ferrate preoxidation and biological activated carbon filtration

P

Passivation
Fe(III/VI) ferrate films, 229*f*
partial and full alleviation of Fe(III/VI), in ferrate film, 234, 235*f*
thick Fe(III/VI) charge transfer on smooth platinum, 228, 230
See also Rechargeable alkaline Fe(VI/III) thin film cathodes
pH
effect on nonylphenol degradation, 396, 398
ferrate decomposition by pH and hydrogen peroxide/ferrate ratios, 393, 395
ferrate-photocatalytic degradation of dibutyl phthalate, 372–374
ferrate(VI) as oxidant, 435–436
first-order rate constant for ferrate decomposition by, 411*t*
heavy metal removal by ferrate preoxidation, 460, 462, 463*f*
rate constant and removal efficiency of nonylphenol oxidation by ferrate, 399*t*

stability of ferrate(VI) by, 408, 410*f*, 411
See also Nonylphenol (NP)
Pharmaceuticals, ferrate(VI) oxidation, 320*f*
Phenol
measurement, 282–283
organics removal from storm water by ferrate(VI), 474
oxidation by Fe(VI) and removal of oxidation products by activated carbon, 286*f*
oxidation of aniline and, by Fe(VI), 285*f*
oxidative treatment and removal from wastewater, 284–287
quantitative analysis with UV/vis spectroscopy, 283*f*
second-order rate constants of ferrate with, 391*t*
simultaneous oxidation of aniline, phenol and hydrazine, 285, 287, 288*f*
See also Nonylphenol (NP)
Photocatalytic oxidation
evaluation alone and in combination with ferrate oxidation, 366–368
experimental, 366
synergism with ferrate oxidation, 367
See also Dibutyl phthalate (DBP)
Photocatalytic reduction
iron(VI), 354–355
See also Heterogeneous photocatalytic reduction of iron(VI)
Photometric dispersion analysis (PDA)
apparatus for coagulation tests, 294–295
comparing rate and magnitude of floc formation, 296
principle, 294*f*
See also Coagulation performance of ferrate
Phthalate esters
endocrine disruptor, 365
See also Dibutyl phthalate (DBP)
Platinum electrode. *See* Molten salt approach to ferrate(VI) synthesis
Platinum substrates. *See* Rechargeable alkaline Fe(VI/III) thin film cathodes
Platinum surface, Fe(VI/III) reversible films on, 31, 32*f*
Polycyclic aromatic hydrocarbons (PAH)
analysis of storm water samples, 470–471
ferrate(VI) technology for removal, 472, 474
runoff water, 468–469
See also Urban runoff water
Potassium ferrate
alkaline hypochlorite synthesis, 5–7
analytical methods, 97
concentrations of, in waste alkaline solution, 100*t*
concentrations of chloride and nitrate ions in waste alkaline solution, 98*t*
disadvantages of wet oxidation method, 95
discharge capacity of, 204*f*
encapsulation in paraffin wax, 379, 380–381, 382, 383*f*
energy capacity, 203*f*
experimental, 96–97
high temperature Mössbauer spectra showing $KFeO_2$ as decomposition product, 130*f*
long-term stability, 8*f*
materials and methods, 96–97
monitoring stability, 406, 408
Mössbauer spectra, 48*f*
purification of waste alkali, 97–98

removal of metals from storm water, 472, 473*t*
re-precipitation of, using purified waste alkali, 99–100
solubility of KCl and, in aqueous KOH, 99*f*
solubility of KCl in aqueous KOH, 98*f*
stability at room temperature, 132–134
thermal decomposition in air, 126–129
thermal decomposition in inert atmosphere, 132–133
thermogravimetric and differential scanning calorimetry (TG/DSC) curves of, in nitrogen, 133*f*
variable temperature in-situ XRD patterns of decomposition products, 131*f*
waste alkali purification, 96
water purification, 269
See also Alkali ferrates; Ethylenediaminetetraacetic acid (EDTA); Sulfamethoxazole (SMX); Super-iron batteries; Tetracycline; Thermal stability of solid ferrates(VI); Trichloroethylene (TCE)
Potassium permanganate, preoxidant, 435
Potentiometric analysis, Fe(VI), 44
Preoxidation
enhancing biodegradation, 446
See also Ferrate preoxidation and biological activated carbon filtration
Preoxidation by ferrate. *See* Coagulation of algae-bearing water
Progesterone
structure, 424*f*
treatment with ferrate(VI), 428, 431
See also Estrogenic compounds
Purification of water. *See* Electrochemical Fe(VI) water treatment

R

Radionuclides
decontamination by Fe(VI), 319*f*
removal with ferrate(VI), 317–318
See also Sludge treatment
Reactions of ferrate(VI). *See* Transformation of Fe(VI) to Fe(III)
Recalcitrant compounds
oxidation by Fe(VI), 339–341
See also Ethylenediaminetetraacetic acid (EDTA); Sulfamethoxazole (SMX)
Rechargeable alkaline Fe(VI/III) thin film cathodes
cathode charging potential of film on Pt substrate, 233*f*
cathode discharge potential during cycling, 233*f*
charge transfer of Fe(VI) cathode problems, 227
conductive matrix facilitated Fe(III/VI) charge transfer, 230–234
influence of normalized electroactivity and discharge current density on cycle life, 234*t*
irreversible alkaline charge transfer behavior of thick film, 231*f*
metal-hydride anode compatibility with, 236, 238–239
partial and full alleviation of Fe(III/VI) passivation in film, 235*f*
passivation of thick Fe(III/VI) charge transfer on smooth Pt, 228, 230

platinized titanium substrates for Fe(III/VI) charge transfer, 236
Pt substrates for Fe(III/VI) charge transfer, 232–234
quasi-reversible alkaline charge transfer behavior, 231*f*
quasi-reversible alkaline charge transfer behavior of film on gold-platinum co-deposited Ti substrate, 237*f*
quasi-reversible alkaline charge transfer behavior of film on platinized Ti, 237*f*
representation of facile charge transfer in Fe(III/VI) films, 229*f*
reversible, on smooth Pt, 228
reversible charge storage of Fe(VI) of Na_2FeO_4 nanofilm, 228*f*
two electrode rechargeable behavior, 238*f*
See also Super-iron batteries

Rechargeable super-iron lithium batteries
charge and discharge capacity vs. cycle number of composite $BaFeO_4$ electrodes, 249*f*
charge and discharge capacity vs. cycle number of composite K_2FeO_4 electrodes, 249*f*
chronopotentiogram of $BaFeO_4$ composite electrode, 248*f*
chronopotentiogram of K_2FeO_4composite electrode, 248*f*
correlation between charge and Li content of $BaFeO_4$ electrodes, 251*t*
cyclic voltammetry of K_2FeO_4 and $BaFeO_4$ cathodes with added carbon, 245, 247*f*
electrochemical behavior of Fe(VI) in nonaqueous electrolyte and reversible, 245, 247–252
electrochemical behavior of thin film K_2FeO_4 on Pt electrode, 252*f*
galvanostatic measurements of various Fe(VI) salts, 250*t*
Mössbauer spectroscopy analysis of $BaFeO_4$ electrodes, 252*t*
See also Super-iron batteries

Reduction. *See* Heterogeneous photocatalytic reduction of iron(VI)

Remediation configurations
in-line electrochemical Fe(VI) water, 271
on-line electrochemical Fe(VI), with activated carbon, 272
on-line electrochemical Fe(VI) water, 269–271
See also Electrochemical Fe(VI) water treatment

Rubidium ferrate. *See* Alkali ferrates

Runoff water. *See* Urban runoff water

S

Scanning electron microscopy (SEM)
algal cells before and after ferrate pretreatment, 442
ferrate-encapsulated paraffin before and after ferrate release, 384, 386

Scopoletin
absorption and emission spectra, 150–151
emission intensity vs. ferrate(VI) by pH, 152*f*
fluorescence agent, 147
molecular structure, 147*f*
See also Fluorescence technique

Sewage
nonylphenol in treatment plant effluents, 390
See also Nonylphenol (NP); Sludge treatment

Silicon rich steel (SRS). *See* Anode dissolution kinetics
Silver ferrate
  IR absorption, 20*f*
  solid state stability, 21*f*
  solution precipitation synthesis, 16, 18–19
Silver mediation, Fe(VI) charge transfer, 212–216
Sludge treatment
  benefits of ferrate(VI) as sludge conditioner, 335–337
  comparative performance of bacteria inactivation, 314*t*
  decontamination of radionuclides by Fe(VI), 319*f*
  dosage to remove 90% H_2S, 317*t*
  efforts in strategy development, 307
  ferrate(VI) ions, 308
  ferrate(VI) oxidation of pharmaceuticals, 320*f*
  ferrate(VI)'s potential as disinfectant, 313–314
  half-lives of reactions between Fe(VI) and pollutants, 315*f*
  heavy metals, 310
  issues when using ferrate(VI), 320
  limit values of heavy metals in sludge for use on land, 310*t*
  limit values of organic compounds and dioxins, 311*t*
  management, 308–312
  mass balance of sewage sludge, 321*t*
  micro pollutants and endocrine disruptors, 310–312
  odor control with ferrate(VI), 315–316
  odors, 309–310
  oxidation of sulfide by Fe(VI) in, 316*f*
  percentage inactivation of bacteria with ferrate(VI), 314*f*
  potential of degradation of organic micro-pollutants with ferrate(VI), 318, 320
  properties of ferrate(VI), 312–313
  removal efficiency of heavy metals, 318*t*
  removal of ammonia by Fe(VI) in, 317*f*
  removal of metal in river water by ferrate and ferric chloride, 319*f*
  revised European Union Directive, 307
  sludge hygienization, 308–309
  toxic heavy metal and radionuclides removal with ferrate(VI), 317–318
  See also Dewatering wastewater biosolids
Sodium hydroxide solution. *See* Fe(VI)/Fe(III) system
Sodium in ferrates. *See* Alkali ferrates
Solid cathode discharge, Fe(VI) analysis, 46–47
Solid ferrates(VI). *See* Thermal stability of solid ferrates(VI)
Solid phase reactants, $BaFeO_4$ synthesis, 4
Solid synthesis, $BaFeO_4$, 19, 21
Solution phase galvanostatic analysis, Fe(VI), 44–45
Solution precipitation synthesis
  Ag_2FeO_4, 16, 18–19
  Cs_2FeO_4, Rb_2FeO_4, $K_xNa_{(2-x)}FeO_4$, and Li_2FeO_4, 7–14
  IR absorption of solid Ag_2FeO_4 and K_2FeO_4, 20*f*
  powder XRD pattern of SrFeO4 and $BaFeO_4$, 17*f*
  room temperature solubility of $BaFeO_4$, K_2FeO_4, and $SrFeO_4$, 17*f*
  solid state stability of Ag_2FeO_4, 21*f*
  $SrFeO_4$ and $BaFeO_4$, 15–16
Stability

encapsulated and unencapsulated ferrate over time, 386, 387*f*
See also Thermal stability of solid ferrates(VI)
Steroidal estrogens (SE)
chemical oxidation, 422
compounds, 422, 424*f*
environmental concerns, 420–421
ferrate(VI) as oxidant, 422
hormone analysis, 425–426
removal from activated sludge, 422
Steroidal estrogens (SEs). *See* Estrogenic compounds
Sulfamethoxazole (SMX)
detection in secondary wastewater, 340
experimental, 341
ferrate(VI) as oxidizing agent in aqueous media, 341
mechanism with possible proton involvement, 345
molecular structure, 340*f*
oxidation methods to remove, 340–341
oxidation product analysis, 346–347
pH dependence of oxidation, 344–345
plot of SMX decrease and formation of oxygen in reaction with Fe(VI), 346*f*
rate constant vs. pH, 344*f*
rate law for oxidation by Fe(VI), 344
stoichiometry of oxidation by Fe(VI), 346
See also Ethylenediaminetetraacetic acid (EDTA)
Sulfide
comparing rate of, and cyanide by ferrate(VI), 280*f*
contaminated water treatment, 277–278
dosage to remove 90%, 317*t*
inorganic contaminant analyses, 275–276
oxidation by Fe(VI) in sludge treatment, 316
remediation by Fe(VI) treatment at lower pH, 279*f*
remediation by on-line Fe(VI) treatment, 278*f*
simultaneous remediation of cyanide and, by on-line Fe(VI) treatment, 282*f*
Sulfonamides
aquatic environment, 340
See also Sulfamethoxazole (SMX)
Sulfoxides, aqueous ferryl(IV) ion and oxygen transfer to, 173–174
Sulfur compounds
reactivity of ferrate(V) with, 161, 162*t*
second-order rate constants of ferrate with, 391*t*
Super-iron batteries
activation of $BaFeO_4$ and K_2FeO_4 by AgO, 214, 215*f*
alkaline Fe(VI) K_2FeO_4 and $BaFeO_4$ charge transfer, 202–220
alkaline Fe(VI) Na(K)FeO_4, Rb(K)FeO_4, Cs_2FeO_4, $SrFeO_4$ or Ag_2FeO_4 charge transfer, 220–224
$BaFeO_4$ cathodic capacity with electrolyte, 242*t*
$BaFeO_4$ cathodic capacity with temperature or particle size, 244*t*
$BaFeO_4$/MnO_2 composite, 210–212
$BaFeO_4$/Mn(VII) composite, 210
cell potential and energy capacity of alkaline, AAA cells, 211*f*
cell potential and energy capacity of alkaline cells with K_2FeO_4 composite cathodes, 208*f*
chemical mediation of Fe(VI) charge transfer, 205–212, 206–207

comparing discharge behavior of alkaline AAA cell of cathode composite, 209*t*
conductive matrix facilitated Fe(III/VI) charge transfer, 230–234
conductors for alkaline, 224–227
constant power comparison of MnO_2, $BaFeO_4$, and AgO/K_2FeO_4 cathodes, 214, 216
discharge capacity of K_2FeO_4, 204*f*
discharge efficiency of K_2FeO_4, $BaFeO_4$ alkaline, with $SrTiO_3$, $Ba(OH)_2$ additives, 216, 217*f*
discharge of $BaFeO_4$ vs. standard or high-power alkaline MnO_2 in AAA cylindrical cell configuration, 203*f*
electronic mediation of Fe(VI) charge transfer, 212–216
energy capacity of K_2FeO_4, $BaFeO_4$ and conventional MnO_2 cathode alkaline primary batteries with Zn anode, 203*f*
ferrate (Fe(VI)), 69, 82, 146, 198
fluorinated polymer graphite conductor, 225–227
fundamentals of primary alkaline, 199–202
influence of normalized electroactivity and discharge current density on cycle life, 234*t*
inhibition of Fe(VI) charge transfer, 205–206
insoluble cathodes, 198
irreversible alkaline charge transfer behavior, 231*f*
K_2FeO_4/Mn(VII or VI) composite, 207–210
metal-hydride anode compatibility with rechargeable, cathode films, 236, 238–239
Mn(IV), Co(III) modifiers and effects on, 216, 218*f*, 219–220
nonaqueous primary Fe(VI) charge transfer for super-iron lithium batteries, 239–245
nonaqueous reversible Fe(VI) charge transfer for rechargeable super-iron lithium batteries, 245, 247–252
passivation of thick Fe(III/VI) charge transfer on smooth Pt, 228–230
platinized titanium substrates for Fe(III/VI) charge transfer, 236
platinum substrates for Fe(III/VI) charge transfer, 232–234
primary nonaqueous, 239–245
quasi-reversible alkaline charge transfer, 231*f*
quasi-reversible alkaline charge transfer behavior, 237*f*
rechargeable alkaline Fe(VI/III) thin film cathodes, 227–239
reversibility, 198–199
reversible Fe(VI/III) ultra-thin film on smooth Pt, 228
theoretical 3-electron charge capacities of various Fe(VI) salts, 201*t*
titanate, Ba(II), Mn(IV) and Co(III) modifiers, 216–220
zirconia coating stabilization of alkaline K_2FeO_4 cathode, 202, 204
See also Rechargeable super-iron lithium batteries
Super-iron lithium batteries
constant load discharge of cells with Li_2FeO_4, K_2FeO_4, $SrFeO_4$ or $BaFeO_4$ Fe(VI) cathode and Li anode, 240*f*
constant load variation effect on $BaFeO_4$ cathode capacity in Li anode cell, 243*f*
discharge of Fe(VI)/MnO_2 co-cathode/Li anode cell with electrolyte, 246*f*

electrolyte effect on $BaFeO_4$ cathode capacity in Li anode cell, 243*f*
facilitated charge transfer, 241
Fe(VI) cathode/Li anode cells with electrolyte, 246*f*
primary nonaqueous, 239–245
specific capacity of $BaFeO_4$, 241
variation of $BaFeO_4$ cathodic capacity with electrolyte, 242*t*
variation of $BaFeO_4$ cathodic capacity with temperature or particle size, 244*t*
See also Super-iron batteries
Superoxide ion, second-order rate constants of ferrate with, 391*t*

T

Tetracycline
2,2'-azinobis(3-ethylbenzothiazoline-6-sulfonate) (ABTS) method for monitoring ferrate stability, 406, 408
calibration curves of Fe(VI) by ABTS method, 410*f*
chemical structure of, and ABTS, 407*f*
decomposition of Fe(VI) in aqueous solution, 410*f*
degradation efficiency as function of pH, 413, 414*f*, 415*f*
degradation of, 413, 417
degradation reactions, 408
effect of initial ferrate concentration on degradation, 413, 415*f*
electrospray ionization–mass spectrometry (ESI–MS) spectra of, and after reaction with ferrate(VI), 413, 416*f*, 417*f*, 417
first-order rate constant for ferrate decomposition by pH, 411*t*
materials and methods, 407–408
potassium ferrate as oxidant, 406
preparation of potassium ferrate, 407–408
sensitivity of ABTS method, 408–409, 411
stability of ferrate(VI) at various pH values, 408
stability of Fe(VI) at various pH values, 411
stoichiometry of ferrate to, 412–413
Thermal stability of solid ferrates(VI)
barium ferrate(VI) [$BaFeO_4$] synthesis, 134, 136
cesium ferrate(VI), 142
high temperature Mössbauer spectra exhibiting $KFeO_2$ as only Fe-product of K_2FeO_4 decomposition, 130*f*
Mössbauer spectroscopy of $BaFeO_4$, 138, 140*f*, 140
phase diagram of products by heating $BaFeO_4$ under various temperatures and oxygen pressures, 137*f*
room temperature Mössbauer spectra of fresh and aged K_2FeO_4, 135*f*
simultaneous TG/DTA curves from K_2FeO_4 in nitrogen, 134*f*
stability of $BaFeO_4$ at room temperature, 139–140, 140*f*, 141*f*
stability of K_2FeO_4 at room temperature, 132–134
TG and DSC curves of K_2FeO_4 in nitrogen atmosphere, 133*f*
TG and DSC curves of K_2FeO_4 in static air, 128*f*
TG/DTG/DTA curves of K_2FeO_4, 127*f*
thermal behavior of $BaFeO_4 \cdot 0.25H_2O$ by simultaneous TG/DTA, 138

thermal decomposition and electrochemical behavior of $BaFeO_4$, 137
thermal decomposition of ferrate(VI) salts, 126, 127*f*
thermal decomposition of K_2FeO_4 in air, 126–129
thermal decomposition of K_2FeO_4 in inert atmosphere, 132–133
variable temperature in-situ XRD patterns of decomposition products of K_2FeO_4 in static air, 131*f*

Thin films
preparation of Fe(VI/III) reversible on smooth Pt, 31, 32*f*
preparation of reversible Fe(VI/III) on extended conductive matrixes, 31–36
See also Rechargeable alkaline Fe(VI/III) thin film cathodes

Thiocyanate, reactivity of ferrate(V) with, 162*t*, 163

Titanium dioxide
dosage effect on dibutyl phthalate degradation, 370–372
photoinduced oxidation of water and compounds, 351
See also Heterogeneous photocatalytic reduction of iron(VI)

Titrimetric chromite analysis
$BaFeO_4$, 42
Fe(VI), 40–42
K_2FeO_4, 41–42

Transformation of Fe(VI) to Fe(III)
ethylenediaminetetraacetic acid (EDTA) as model complexing agent, 258
experimental, 258–259
frozen solution Mössbauer spectrum of reaction of ferrate(VI) and Fe^{III}EDTA, 262*f*
glycine as model for complex amino acids, 258
Mössbauer spectra of species from Fe^{III}EDTA in NaOH solution with and without ferrate(VI), 264*f*
Mössbauer spectroscopy with rapid freeze solution technique, 258
Mössbauer spectrum of frozen solution after ferrate(VI) reacted with EDTA, 261*f*
Mössbauer spectrum of frozen solution after ferrate(VI) reacted with EDTA in borate-phosphate buffer, 262*f*
Mössbauer spectrum of frozen solution after ferrate(VI) reacted with glycine, 260*f*
reaction between ferrate(VI) and EDTA, 259–261
reaction between ferrate(VI) and Fe^{III}EDTA, 262–265
reaction between ferrate(VI) and glycine, 259, 260*f*

Trichloroethylene (TCE)
encapsulation method, 380
encapsulation of potassium ferrate as oxidant, 379–380
experimental, 380–382
Fourier transform infrared spectroscopy, 382, 385*f*
instrumental analysis, 381–382
loading efficiency, 380–381
occurrence in groundwater, 379
percentage encapsulated ferrate released over time, 386*f*
preparation of ferrate encapsulated paraffin, 382, 383*f*
release profile of encapsulated ferrate, 384, 386
scanning electron micrographs of ferrate–encapsulated paraffin before and after ferrate release, 386*f*

stability study of encapsulated ferrate, 381
stability study of ferrate, 386, 387*f*
TCE oxidation by encapsulated ferrate, 381
TCE oxidation by ferrate, 387
x-ray diffraction (XRD) studies, 382, 384, 385*f*
Trimethylamine
ferrate oxidation, 196
rate constants of oxidation by pH, 193*f*

U

Ultraviolet absorbance (UVA)
ferrate preoxidation of cultured water samples, 442–443
scanning method, 437–438
Ultraviolet/visible spectroscopy, Fe(VI) analysis, 42–43
Urban runoff water
average concentration of trace metals in, 470*f*
chemical pollutants, 468–469
demand for water quality, 467
effect of ferrate dosage on Mn removal, 472, 473*f*
environmentally friendly treatment, 467
ferrate(VI) technology for enhancing storm water quality, 472–474
metal removal by potassium ferrate, 472, 473*t*
metals, 469–470
mutagenic tests, 475
origin, 467–468
polycyclic aromatic hydrocarbons (PAH), 468–469, 470–471
removal of organics by ferrate(VI), 474*f*
Urban waste water, ferrate and ferric chloride, 110

V

Vacuum filtering
dewatering technique, 331, 331*f*
effects of ferrate on sludge dewatering, 335
See also Dewatering wastewater biosolids

W

Waste alkali
concentrations of chloride and nitrate ions, 98*t*
concentrations of potassium ferrate in, 100*t*
disadvantages of wet oxidation method, 95
green preparation method, 95
purification, 96, 97–98
re-precipitation of potassium ferrate using, 99–100
solubility of KCl and K_2FeO_4 in aqueous KOH, 99*f*
solubility of KCl in aqueous KOH, 98*f*
See also Potassium ferrate
Wastewater treatment
development of strategies, 307
ferrate and ferric chloride, 110–111
ferrate (Fe(VI)), 82
See also Dewatering wastewater biosolids; Sludge treatment
Water
demand for potable, 467
See also Urban runoff water
Water, natural, oxidation of nonylphenol by ferrate, 400–401
Water oxidation

absorbance at time zero vs. pH, 191*f*
agreement of observed rate constants with predicted values, 191*f*
kinetics of, by ferrate(VI), 189–190
product analysis, 192
Water quality. *See* Ferrate preoxidation and biological activated carbon filtration; Urban runoff water
Water treatment
chlorination, 268
dewaterability of solids, 327–328
dual role of ferrate (Fe(VI)), 293
See also Coagulation of algae-bearing water; Dewatering wastewater biosolids; Electrochemical Fe(VI) water treatment; Trichloroethylene (TCE)
Wet oxidation method. *See* Waste alkali
White case iron (WCI). *See* Anode dissolution kinetics

X

X-ray absorption spectroscopy (XAS), aqueous ferryl(IV) ion, 170–171
X-ray diffraction (XRD)
chemically and direct electrochemically synthesized $BaFeO_4$, 26*f*
Fe(VI) analysis, 43–44
powder XRD pattern of K_2FeO_4 and Cs_2FeO_4, 10*f*
powder XRD pattern of $SrFeO_4$ and $BaFeO_4$, 17*f*
pure potassium ferrate and encapsulated in paraffin, 382, 384, 385*f*

Z

Zirconia
pH sensor for high temperature aqueous systems, 36, 38
stabilization of alkaline K_2FeO_4 cathode, 202, 204
See also Charge transfer
Zirconia coated Fe(VI) salts, preparation, 36–38